T0214978

D-Branes

D-branes represent a key theoretical tool in the understanding of strongly coupled superstring theory and M-theory. They have led to many striking discoveries, including the precise microphysics underlying the thermodynamic behaviour of certain black holes, and remarkable holographic dualities between large-N gauge theories and gravity.

This book provides a self-contained introduction to the technology of D-branes, presenting the recent developments and ideas in a pedagogical manner. It is suitable for use as a textbook in graduate courses on modern string theory and theoretical particle physics, and will also be an indispensable reference for seasoned practitioners. The introductory material is developed by first starting with the main features of string theory needed to get rapidly to grips with D-branes, uncovering further aspects while actually working with D-branes. Many advanced applications are covered, with discussions of open problems which could form the basis for new avenues of research. This title, first published in, has been reissued as 2003 an Open Access publication on Cambridge Core.

CLIFFORD V. JOHNSON obtained his BSc in physics at Imperial College, University of London, and his PhD in theoretical physics from the University of Southampton. He won a 1992 Lindeman Fellowship and a 1992 SERC NATO Fellowship, and became a member of the School of Natural Sciences at the Institute for Advanced Study, Princeton. He then spent a year teaching and doing research at the Physics Department of Princeton University, and went on to hold a postdoctoral position at the Institute for Theoretical Physics, University of California, Santa Barbara. From 1997 to 1999 he was Assistant Professor at the University of Kentucky. Dr Johnson is currently Reader in Theoretical Physics, Department of Mathematical Sciences, University of Durham, where he is a member of the Centre for Particle Theory.

CAMBRIDGE MONOGRAPHS ON
MATHEMATICAL PHYSICS

General editors: P. V. Landshoff, D. R. Nelson, S. Weinberg

J. Ambjørn, B. Durhuus and T. Jonsson *Quantum Geometry: A Statistical Field Theory Approach*

A. M. Anile *Relativistic Fluids and Magneto-Fluids*

J. A. de Azcárraga and J. M. Izquierdo *Lie Groups, Lie Algebras, Cohomology and Some Applications in Physics*[†]

V. Belinski and E. Verdaguer *Gravitational Solitons*

J. Bernstein *Kinetic Theory in the Early Universe*

G. F. Bertsch and R. A. Broglia *Oscillations in Finite Quantum Systems*

N. D. Birrell and P. C. W. Davies *Quantum Fields in Curved Space*[†]

M. Burgess *Classical Covariant Fields*

S. Carlip *Quantum Gravity in 2 + 1 Dimensions*

J. C. Collins *Renormalization*[†]

M. Creutz *Quarks, Gluons and Lattices*[†]

P. D. D'Eath *Supersymmetric Quantum Cosmology*

F. de Felice and C. J. S. Clarke *Relativity on Curved Manifolds*[†]

P. G. O. Freund *Introduction to Supersymmetry*[†]

J. Fuchs *Affine Lie Algebras and Quantum Groups*[†]

J. Fuchs and C. Schweigert *Symmetries, Lie Algebras and Representations: A Graduate Course for Physicists*

A. S. Galperin, E. A. Ivanov, V. I. Ogievetsky and E. S. Sokatchev *Harmonic Superspace*

R. Gambini and J. Pullin *Loops, Knots, Gauge Theories and Quantum Gravity*[†]

M. Göckeler and T. Schücker *Differential Geometry, Gauge Theories and Gravity*[†]

C. Gómez, M. Ruiz Altaba and G. Sierra *Quantum Groups in Two-dimensional Physics*

M. B. Green, J. H. Schwarz and E. Witten *Superstring Theory, volume 1: Introduction*[†]

M. B. Green, J. H. Schwarz and E. Witten *Superstring Theory, volume 2: Loop Amplitudes, Anomalies and Phenomenology*[†]

S. W. Hawking and G. F. R. Ellis *The Large-Scale Structure of Space-Time*[†]

F. Iachello and A. Aruna *The Interacting Boson Model*

F. Iachello and P. van Isacker *The Interacting Boson–Fermion Model*

C. Itzykson and J.-M. Drouffe *Statistical Field Theory, volume 1: From Brownian Motion to Renormalization and Lattice Gauge Theory*[†]

C. Itzykson and J.-M. Drouffe *Statistical Field Theory, volume 2: Strong Coupling, Monte Carlo Methods, Conformal Field Theory, and Random Systems*[†]

C. V. Johnson *D-Branes*

J. I. Kapusta *Finite-Temperature Field Theory*[†]

V. E. Korepin, A. G. Izergin and N. M. Boguliubov *The Quantum Inverse Scattering Method and Correlation Functions*[†]

M. Le Bellac *Thermal Field Theory*[†]

Y. Makeenko *Methods of Contemporary Gauge Theory*

N. H. March *Liquid Metals: Concepts and Theory*

I. M. Montvay and G. Münster *Quantum Fields on a Lattice*[†]

A. Ozorio de Almeida *Hamiltonian Systems: Chaos and Quantization*[†]

R. Penrose and W. Rindler *Spinors and Space-Time, volume 1: Two-Spinor Calculus and Relativistic Fields*[†]

R. Penrose and W. Rindler *Spinors and Space-Time, volume 2: Spinor and Twistor Methods in Space-Time Geometry*[†]

S. Pokorski *Gauge Field Theories*, 2nd edition

J. Polchinski *String Theory, volume 1: An Introduction to the Bosonic String*

J. Polchinski *String Theory, volume 2: Superstring Theory and Beyond*

V. N. Popov *Functional Integrals and Collective Excitations*[†]

R. G. Roberts *The Structure of the Proton*[†]

J. M. Stewart *Advanced General Relativity*[†]

A. Vilenkin and E. P. S. Shellard *Cosmic Strings and Other Topological Defects*[†]

R. S. Ward and R. O. Wells Jr *Twistor Geometry and Field Theories*[†]

[†] Issued as a paperback

D-Branes

CLIFFORD V. JOHNSON

University of Durham

CAMBRIDGE
UNIVERSITY PRESS

Shaftesbury Road, Cambridge CB2 8EA, United Kingdom

One Liberty Plaza, 20th Floor, New York, NY 10006, USA

477 Williamstown Road, Port Melbourne, VIC 3207, Australia

314–321, 3rd Floor, Plot 3, Splendor Forum, Jasola District Centre, New Delhi – 110025, India

103 Penang Road, #05-06/07, Visioncrest Commercial, Singapore 238467

Cambridge University Press is part of Cambridge University Press & Assessment, a department of the University of Cambridge.

We share the University's mission to contribute to society through the pursuit of education, learning and research at the highest international levels of excellence.

www.cambridge.org
Information on this title: www.cambridge.org/9781009401364

DOI: 10.1017/9781009401371

First published 2003
Reissued as OA 2023

A catalogue record for this publication is available from the British Library.

ISBN 978-1-009-40136-4 Hardback
ISBN 978-1-009-40139-5 Paperback

Cambridge University Press & Assessment has no responsibility for the persistence or accuracy of URLs for external or third-party internet websites referred to in this publication and does not guarantee that any content on such websites is, or will remain, accurate or appropriate.

*To my siblings Carol and Robert, my father Victor Reginald,
and especially my mother Delia, who have all shaped me.*

Contents

List of inserts   *xviii*

Preface   *xx*

1   Overview and overture   **1**
1.1   The classical dynamics of geometry   1
1.2   Gravitons and photons   7
1.3   Beyond classical gravity: perturbative strings   11
1.4   Beyond perturbative strings: branes   15
1.5   The quantum dynamics of geometry   19
1.6   Things to do in the meantime   20
1.7   On with the show   22

2   Relativistic strings   **24**
2.1   Motion of classical point particles   24
        2.1.1   Two actions   24
        2.1.2   Symmetries   26
2.2   Classical bosonic strings   27
        2.2.1   Two actions   27
        2.2.2   Symmetries   29
        2.2.3   String equations of motion   30
        2.2.4   Further aspects of the two dimensional
                perspective   31
        2.2.5   The stress tensor   35
        2.2.6   Gauge fixing   35
        2.2.7   The mode decomposition   37
        2.2.8   Conformal invariance as a residual symmetry   37
        2.2.9   Some Hamiltonian dynamics   38

2.3	Quantised bosonic strings	40
	2.3.1 The constraints and physical states	41
	2.3.2 The intercept and critical dimensions	42
	2.3.3 A glance at more sophisticated techniques	45
2.4	The sphere, the plane and the vertex operator	47
	2.4.1 States and operators	48
2.5	Chan–Paton factors	51
2.6	Unoriented strings	52
	2.6.1 Unoriented open strings	52
	2.6.2 Unoriented closed strings	54
	2.6.3 World-sheet diagrams	55
2.7	Strings in curved backgrounds	56
2.8	A quick look at geometry	61
	2.8.1 Working with the local tangent frames	61
	2.8.2 Differential forms	63
	2.8.3 Coordinate vs. orthonormal bases	65
	2.8.4 The Lorentz group as a gauge group	67
	2.8.5 Fermions in curved spacetime	68
	2.8.6 Comparison to differential geometry	68
3	**A closer look at the world-sheet**	**70**
3.1	Conformal invariance	70
	3.1.1 Diverse dimensions	70
	3.1.2 The special case of two dimensions	73
	3.1.3 States and operators	74
	3.1.4 The operator product expansion	75
	3.1.5 The stress tensor and the Virasoro algebra	76
3.2	Revisiting the relativistic string	80
3.3	Fixing the conformal gauge	85
	3.3.1 Conformal ghosts	85
	3.3.2 The critical dimension	86
3.4	The closed string partition function	87
4	**Strings on circles and T-duality**	**94**
4.1	Fields and strings on a circle	94
	4.1.1 The Kaluza–Klein reduction	95
	4.1.2 Closed strings on a circle	96
4.2	T-duality for closed strings	99
4.3	A special radius: enhanced gauge symmetry	100
4.4	The circle partition function	103
4.5	Toriodal compactifications	104

4.6	More on enhanced gauge symmetry	108
	4.6.1 Lie algebras and groups	108
	4.6.2 The classical Lie algebras	111
	4.6.3 Physical realisations with vertex operators	113
4.7	Another special radius: bosonisation	113
4.8	String theory on an orbifold	117
4.9	T-duality for open strings: D-branes	119
	4.9.1 Chan–Paton factors and Wilson lines	121
4.10	D-brane collective coordinates	123
4.11	T-duality for unoriented strings: orientifolds	125
5	**Background fields and world-volume actions**	**129**
5.1	T-duality in background fields	129
5.2	A first look at the D-brane world-volume action	131
	5.2.1 World-volume actions from tilted D-branes	133
5.3	The Dirac–Born–Infeld action	135
5.4	The action of T-duality	136
5.5	Non-Abelian extensions	136
5.6	D-branes and gauge theory	138
5.7	BPS lumps on the world-volume	138
6	**D-brane tension and boundary states**	**141**
6.1	The D-brane tension	142
	6.1.1 An open string partition function	142
	6.1.2 A background field computation	145
6.2	The orientifold tension	148
	6.2.1 Another open string partition function	148
6.3	The boundary state formalism	150
7	**Supersymmetric strings**	**155**
7.1	The three basic superstring theories	155
	7.1.1 Open superstrings: type I	155
	7.1.2 Closed superstrings: type II	160
	7.1.3 Type I from type IIB, the prototype orientifold	165
	7.1.4 The Green–Schwarz mechanism	166
7.2	The two basic heterotic string theories	169
	7.2.1 $SO(32)$ and $E_8 \times E_8$ from self-dual lattices	171
	7.2.2 The massless spectrum	172
7.3	The ten dimensional supergravities	174
7.4	Heterotic toroidal compactifications	176
7.5	Superstring toroidal compactification	178
7.6	A superstring orbifold: discovering the K3 manifold	179

7.6.1	The orbifold spectrum	180
7.6.2	Another miraculous anomaly cancellation	183
7.6.3	The K3 manifold	184
7.6.4	Blowing up the orbifold	185
7.6.5	Some other K3 orbifolds	189
7.6.6	Anticipating D-manifolds	191

8   **Supersymmetric strings and T-duality**   **192**

8.1	T-duality of supersymmetric strings	192
	8.1.1 T-duality of type II superstrings	192
	8.1.2 T-duality of type I superstrings	193
	8.1.3 T-duality for the heterotic strings	194
8.2	D-branes as BPS solitons	195
8.3	The D-brane charge and tension	197
8.4	The orientifold charge and tension	200
8.5	Type I from type IIB, revisited	201
8.6	Dirac charge quantisation	201
8.7	D-branes in type I	202

9   **World-volume curvature couplings**   **205**

9.1	Tilted D-branes and branes within branes	205
9.2	Anomalous gauge couplings	206
9.3	Characteristic classes and invariant polynomials	210
9.4	Anomalous curvature couplings	216
9.5	A relation to anomalies	218
9.6	D-branes and K-theory	220
9.7	Further non-Abelian extensions	221
9.8	Further curvature couplings	222

10   **The geometry of D-branes**   **224**

10.1	A look at black holes in four dimensions	224
	10.1.1 A brief study of the Einstein–Maxwell system	224
	10.1.2 Basic properties of Schwarzschild	225
	10.1.3 Basic properties of Reissner–Nordstrom	228
	10.1.4 Extremality, supersymmetry, and the BPS condition	228
	10.1.5 Multiple black holes and multicentre solutions	232
	10.1.6 Near horizon geometry and an infinite throat	233
	10.1.7 Cosmological constant; de Sitter and anti-de Sitter	233
	10.1.8 de-Sitter spacetime and the sphere	234
	10.1.9 Anti-de Sitter in various coordinate systems	235

	10.1.10 Anti-de Sitter as a hyperbolic slice	236
	10.1.11 Revisiting the extremal solution	237
10.2	The geometry of D-branes	238
	10.2.1 A family of 'p-brane' solutions	238
	10.2.2 The boost form of solution	239
	10.2.3 The extremal limit and coincident D-branes	240
10.3	Probing p-brane geometry with Dp-branes	243
	10.3.1 Thought experiment: building p with Dp	243
	10.3.2 Effective Lagrangian from the world-volume action	244
	10.3.3 A metric on moduli space	245
10.4	T-duality and supergravity solutions	246
	10.4.1 D$(p+1)$ from Dp	246
	10.4.2 D$(p-1)$ from Dp	248

11	**Multiple D-branes and bound states**	**249**
11.1	Dp and Dp' from boundary conditions	249
11.2	The BPS bound for the Dp–Dp' system	252
11.3	Bound states of fundamental strings and D-strings	254
11.4	The three-string junction	255
11.5	Aspects of D-brane bound states	258
	11.5.1 0–0 bound states	258
	11.5.2 0–2 bound states	258
	11.5.3 0–4 bound states	259
	11.5.4 0–6 bound states	260
	11.5.5 0–8 bound states	260

12	**Strong coupling and string duality**	**261**
12.1	Type IIB/type IIB duality	261
	12.1.1 D1-brane collective coordinates	261
	12.1.2 S-duality and $SL(2, \mathbb{Z})$	263
12.2	$SO(32)$ Type I/heterotic duality	264
	12.2.1 D1-brane collective coordinates	264
12.3	Dual branes from 10D string–string duality	265
	12.3.1 The heterotic NS-fivebrane	267
	12.3.2 The type IIA and type IIB NS5-brane	268
12.4	Type IIA/M-theory duality	271
	12.4.1 A closer look at D0-branes	271
	12.4.2 Eleven dimensional supergravity	271
12.5	$E_8 \times E_8$ heterotic string/M-theory duality	273
12.6	M2-branes and M5-branes	276
	12.6.1 Supergravity solutions	276

12.6.2 From D-branes and NS5-branes to M-branes
and back   277
12.7 U-duality   278
12.7.1 Type II strings on T^5 and $E_{6(6)}$ 278
12.7.2 U-duality and bound states   279

13   D-branes and geometry I   282
13.1 D-branes as probes of ALE spaces   282
13.1.1 Basic setup and a quiver gauge theory   282
13.1.2 The moduli space of vacua   285
13.1.3 ALE space as metric on moduli space   286
13.1.4 D-branes and the hyper-Kähler quotient   289
13.2 Fractional D-branes and wrapped D-branes   291
13.2.1 Fractional branes   291
13.2.2 Wrapped branes   292
13.3 Wrapped, fractional and stretched branes   294
13.3.1 NS5-branes from ALE spaces   295
13.3.2 Dual realisations of quivers   296
13.4 D-branes as instantons   300
13.4.1 Seeing the instanton with a probe   301
13.4.2 Small instantons   305
13.5 D-branes as monopoles   306
13.5.1 Adjoint Higgs and monopoles   309
13.5.2 BPS monopole solution from Nahm data   311
13.6 The D-brane dielectric effect   314
13.6.1 Non-Abelian world-volume interactions   314
13.6.2 Stable fuzzy spherical D-branes   316
13.6.3 Stable smooth spherical D-branes   318

14   K3 orientifolds and compactification   322
14.1 $\mathbb{Z}_N$ orientifolds and Chan–Paton factors   322
14.2 Loops and tadpoles for ALE $\mathbb{Z}_M$ singularities   324
14.2.1 One-loop diagrams and tadpoles   324
14.2.2 Computing the one-loop diagrams   325
14.2.3 Extracting the tadpoles   330
14.3 Solving the tadpole equations   333
14.3.1 T-duality relations   333
14.3.2 Explicit solutions   334
14.4 Closed string spectra   336
14.5 Open string spectra   339
14.6 Anomalies for $\mathcal{N} = 1$ in six dimensions   341

15   D-branes and geometry II   **345**
15.1   Probing p with $\mathrm{D}(p-4)$ 345
15.2   Probing six-branes: Kaluza–Klein monopoles
       and M-theory   346
15.3   The moduli space of 3D supersymmetric gauge theory   348
15.4   Wrapped branes and the enhançon mechanism   352
       15.4.1 Wrapping D6-branes   353
       15.4.2 The repulson geometry   354
       15.4.3 Probing with a wrapped D6-brane   356
15.5   The consistency of excision in supergravity   360
15.6   The moduli space of pure glue in 3D   362
       15.6.1 Multi-monopole moduli space   363

16   Towards M- and F-theory   **367**
16.1   The type IIB string and F-theory   367
       16.1.1 $SL(2, \mathbb{Z})$ duality   368
       16.1.2 The (p, q) strings   369
       16.1.3 String networks   371
       16.1.4 The self-duality of D3-branes   373
       16.1.5 (p, q) Fivebranes   375
       16.1.6 $SL(2, \mathbb{Z})$ and D7-branes   376
       16.1.7 Some algebraic geometry   379
       16.1.8 F-theory, and a dual heterotic description   383
       16.1.9 (p, q) Sevenbranes   384
       16.1.10 Enhanced gauge symmetry and singularities
               of K3   386
       16.1.11 F-theory at constant coupling   387
       16.1.12 The moduli space of $\mathcal{N} = 2\ SU(N)$ with $N_\mathrm{f} = 4$ 392
16.2   M-theory origins of F-theory   394
       16.2.1 M-branes and odd D-branes   396
       16.2.2 M-theory on K3 and heterotic on T^3 399
       16.2.3 Type IIA on K3 and heterotic on T^4 400
16.3   Matrix theory   400
       16.3.1 Another look at D0-branes   401
       16.3.2 The infinite momentum frame   402
       16.3.3 Matrix string theory   404

17   D-branes and black holes   **409**
17.1   Black hole thermodynamics   409
       17.1.1 The path integral and the Euclidean calculus   409
       17.1.2 The semiclassical approximation   411
       17.1.3 The temperature of black holes   412

17.2	The Euclidean action calculus	414
	17.2.1 The action for Schwarzschild	414
	17.2.2 The action for Reissner–Nordström	416
	17.2.3 The laws of thermodynamics	417
17.3	$D = 5$ Reissner–Nordström black holes	418
	17.3.1 Making the black hole	420
	17.3.2 Microscopic entropy and a 2D field theory	425
	17.3.3 Non-extremality and a 2D dilute gas limit	427
17.4	Near horizon geometry	429
17.5	Replacing T^4 with K3	432
	17.5.1 The geometry	432
	17.5.2 The microscopic entropy	433
	17.5.3 Probing the black hole with branes	434
	17.5.4 The enhançon and the second law	437
18	**D-branes, gravity and gauge theory**	**440**
18.1	The AdS/CFT correspondence	441
	18.1.1 Branes and the decoupling limit	441
	18.1.2 Sphere reduction and gauged supergravity	443
	18.1.3 Extracts from the dictionary	446
	18.1.4 The action, counterterms, and the stress tensor	449
18.2	The correspondence at finite temperature	452
	18.2.1 Limits of the non-extremal D3-brane	452
	18.2.2 The AdS–Schwarzschild black hole in global coordinates	453
18.3	The correspondence with a chemical potential	455
	18.3.1 Spinning D3-branes and charged AdS black holes	455
	18.3.2 The AdS–Reissner–Nordström black hole	459
	18.3.3 Thermodynamic phase structure	459
18.4	The holographic principle	464
19	**The holographic renormalisation group**	**467**
19.1	Renormalisation group flows from gravity	467
	19.1.1 A BPS domain wall and supersymmetry	469
19.2	Flowing on the Coulomb branch	472
	19.2.1 A five dimensional solution	472
	19.2.2 A ten dimensional solution	475
	19.2.3 Probing the geometry	475
	19.2.4 Brane distributions	478
19.3	An $\mathcal{N} = 1$ gauge dual RG flow	480
	19.3.1 The five dimensional solution	482

	19.3.2 The ten dimensional solution	486
	19.3.3 Probing with a D3-brane	487
	19.3.4 The Coulomb branch	488
	19.3.5 Kähler structure of the Coulomb branch	489
19.4	An $\mathcal{N} = 2$ gauge dual RG flow and the enhançon	494
	19.4.1 The five dimensional solution	494
	19.4.2 The ten dimensional solution	498
	19.4.3 Probing with a D3-brane	499
	19.4.4 The moduli space	500
19.5	Beyond gravity duals	502
20	**Taking stock**	**504**
	References	510
	Index	529

List of inserts

1.1 A reminder of Maxwell's field equations   4
1.2 Finding an inertial frame by freely falling   6
1.3 Maxwell written covariantly   8
1.4 Soliton properties and the kink solution   18
2.1 T is for tension   32
2.2 A rotating open string   33
2.3 Zero point energy from the exponential map   46
2.4 World-sheet perturbation theory diagrams   57
2.5 Yang–Mills theory with forms   66
3.1 Deformations, RG flows, and CFTs   84
3.2 Further aspects of conformal ghosts   87
3.3 Special points in the moduli space of tori   90
3.4 Partition functions   92
4.1 Affine Lie algebras   102
4.2 The Poisson resummation formula   107
4.3 The simply laced Lie algebras   112
4.4 Particles and Wilson lines   122
6.1 Vacuum energy   144
6.2 Translating closed to open   145
6.3 The boundary state as a coherent state   152
7.1 Gauge and gravitational anomalies   161
7.2 A list of anomaly polynomials   162
7.3 The Chern–Simons three-form   167
7.4 $SU(2)_\mathrm{L}$ versus $SU(2)_\mathrm{R}$ 180
7.5 Anticipating a string–string duality in $D = 6$ 186
7.6 A closer look at the Eguchi–Hanson space   188
8.1 A summary of forms and branes   198
9.1 The Dirac monopole as a gauge bundle   209

9.2   The first Chern class of the Dirac bundle   210
9.3   The Yang–Mills instanton as a gauge bundle   211
9.4   The BPST one-instanton connection   212
9.5   The Euler number of the sphere   216
10.1 Checking the Reissner–Nordström solution   226
10.2 A little hypersurface technology   229
12.1 Kaluza–Klein relations   274
12.2 Origins of $E_{6,(6)}$ and other U-duality Groups   280
13.1 The prototype BPS object   312
14.1 Jacobi's ϑ-functions   327
14.2 The abstruse indentity   328
14.3 Another string–string duality in $D = 6$ 343
15.1 The 'nut' of Taub–NUT   350
16.1 S^2 or $\mathbb{CP}^1$ from affine coordinates   380
17.1 pp-Waves as boosted Schwarzschild   423
17.2 The BTZ black hole   431
18.1 The large N limit and string theory   444
18.2 D3-brane distributions   457

Preface

In view of the exciting developments in our understanding of those particular aspects of fundamental physics that string theory seems to capture, it seems appropriate to collect together some of the key tools and ideas which helped move things forward. The developments included a true revolution, since the physical perspective changed so radically that it undermined the long-standing status of strings as the basic fundamental objects, and instead the idea has arisen that a string theory description is simply a special (albeit rather novel and beautiful) corner of a larger theory called 'M-theory'. This book is not an attempt at a history of the revolution, as we are (arguably) still in the midst of it, especially since we are in the awkward position of not knowing even one satisfactory intrinsic definition of M-theory, and have implicit knowledge of it only through interconnections of its various limits.

All revolutions are supposed to have a collection of characters who played a crucial role in it, 'heroes' if you will. Hence, one would be expected to proceed to list here the names of various individuals. While I was lucky to be in a position to observe a lot of the activity at first hand and collect many wonderful anecdotes about how some things came to be, I will decline to start listing names at this juncture. It is too easy to yield to the temptation to emphasise a few personalities in a short space (such as this preface), and the result can sometimes be at the expense of others, a practice which happens all too often elsewhere. This seems to me to be especially inappropriate in a field where the most striking characteristic of the contributions has been the *collective* effort of hundreds of thinkers all over the planet, often linked by e-mail and the web, often never having met each other in person.

There were marvellous weeks, back in 1995 and 1996 especially, where there was one key paper after another, from all over the world, driven by the fact that new ideas were pouring in from conversations everyone was excitedly having at blackboards, in the sand, over lunch, *via* e-mail, on the back of an envelope, *etc.* However, when one is speculating about aspects of fundamental physics which are not yet in the directly testable realm it should be noted that ideas – even radical ones – are cheap. Computational tools are needed to test them, and to provide access to the new regimes to which the ideas beckon. The collection of tools which filled this crucial role in this context was built around 'D-branes', and it was the change of perspective and computational power that they brought that unlocked that steady flow of marvellous papers. In my mind, they can indisputably be placed high on list of characters cast as heroes of the revolution. Indeed, many will speak of the feeling that often arose after working with them for a while in those exciting days, that the D-branes simply had a life and character of their own. They shaped the ideas and language of the field in a way that was directed by no single personality, and – most importantly – were a wonderful and sharp tool for investigating in detail the nature of the many bold conjectures which were made.

D-branes were discovered well before the revolution, of course, but in the Summer of 1995 it was shown by Joseph Polchinksi that they were relevant to strongly coupled string theory. I arrived as a postdoctoral researcher at the Institute for Theoretical Physics (Santa Barbara, California) in the following Autumn, and by then it was already clear that there were many people, both young and old, who could benefit from a refresher course on issues outside the realm of heterotic string theory (on which much of the focus had been up to then, with an eye on phenomenology) and an introduction to D-branes. Furthermore, there was some need for an agreement about language and conventions, since there had not been much in the way of texts or other notes which focused on the relevant aspects. (Polchinski's modern textbook[1] was still only partially written, and the manuscript had been seen only by a privileged few.)

Some of us begged Joe to give us some lectures at the ITP, and I (and probably others) quickly had the idea for a written set of notes that could be circulated to the world at large, as a basic toolbox. I suggested this to him, and he eventually agreed. During the lectures, I took such notes as I could and then together with Shyamoli Chaudhuri, we produced some notes with Joe, which we released[2] with his name listed as first author – breaking the strict alphabetical convention in this field – as it seemed to me highly inappropriate, given our roles as scribes, that his name might come last. Happily, the 'D-notes' (as I liked to call them) seemed to be well received by very many, and proved to be useful in forming a common point of departure for almost everyone working in the field.

I was fortunate enough to be asked tó give introductory lectures on D-branes over the following months and years, and this led me to write more notes to embellish the D-notes, finding new ways of explaining things, sometimes making illustrative links between different aspects, depending on the theme of the lecture series in which I was participating.

This book grew out of such lecture notes[3, 4], and contains my own bi-ased perspective on what aspects of D-branes ought to be included in an introductory text. Pressures of space mean that I have omitted a large number of remarkably interesting and useful material, and my choices will no doubt not suit everyone. I have made many efforts for it to be a stand alone handbook. It is intended that the person who knows little or no string theory (but with some background in quantum field the-ory and relativity) can open this book, and upon working through it, learn many things about string theory, and become adept at computing with D-branes, making no reference to another string theory text. Per-haps as a bonus, they will even learn various aspects of advanced topics in relativity, geometry and quantum gravity and quantum field theory since those are the meat and drink of D-brane physics. However, if they want a deeper knowledge of many aspects of string theory which are only sketched here due to lack of space, then they can consult the excellent text of Polchinski[1], and also that of Green, Schwarz and Witten[5], which is still an excellent text for many aspects of the subject. There are also many other sources, on the web (*e.g.,* www.arXiv.org) and elsewhere, of detailed reviews of various specialised topics, even other string theory books[6].

So, this is not intended to be a string theory textbook. It is instead a handbook or toolbox for concepts concerning branes in string theory, with emphasis on D-branes. However, since many of the applications are in what I like to call 'extreme string theory' – taking limits like strong coupling, low energy, large N, etc. – the reader will also learn important physics of those regimes and others, which are not covered in any other text at this time.

Over the years I have had the great benefit of lengthy conversations about string theory and D-branes with many people, out of which my intuition for these matters developed, and I would like to thank them all. Chief among these are Robert Myers, Joseph Polchinski, and Edward Witten, all of whose patience (and refreshing open-mindedness in the early days) is much appreciated. I also thank all of the people with whom I have collaborated in very many exciting research projects, and from whom I learned a great deal. Aspects of some of that work will appear in this text, and I would like it made clear that any inaccuracies in presenting the results are my own.

Parts of this book were written (or sometimes day-dreamt about) in many inspiring places, not all of which I can recall, but I should thank especially a number of institutions for providing facilities: The New York Public Library's Rose Reading Room, and Columbia University's Butler Library (New York, NY, USA), the Bodlean Library (Oxford, England), The Aspen Centre for Physics (Aspen, Co., USA), The Park City Mathematical Institute 2001 (Park City, Utah, USA), El Centro de Estudious Científicos (Valdivia, Chile), The Physics Department at Stellenbosch University, and the Stellenbosch Institute for Advanced Study (Stellenbosch, South Africa), The Perimeter Institute for Fundamental Research (Waterloo, Canada), The Village Vanguard (New York, USA), Broadway (New York, USA), and various United Airlines lounges world-wide.

Thanks to Ian Davies, James Gregory, Laur Järv, Ken Lovis, Rob Myers and David Page for reading and commenting on parts of the manuscript, and many people around the world for their useful remarks upon earlier notes which were absorbed into this book. Thanks also go to Jim Gates, Brian Greene, David Gross, Ted Jacobson and Lenny Susskind for their thoughts on a late title change, and on other important matters concerning the book, logistical and otherwise[†]. I'd like to thank all of my colleagues at the Department of Mathematical Sciences, University of Durham, for providing such a friendly and supportive working environment, and Carol, Delia and Robert Johnson for their encouragement. Thanks also to Elizabeth and Nich Butler for much appreciated culinary provision and other matters of hospitality over Christmas 2001.

I would especially like to thank Samantha Butler for her constant patience and support throughout this project, and beyond.

[†] A conversation with Brian led to a flirtation with the slightly irreverent idea of giving this book the simple title 'Volume III'. I abandoned this after a while, since it would produce confusion amongst those not aware of the affection held for (or existence of) the two-volume texts in references [1] and [5].

1
Overview and overture

Einstein's theory of the classical relativistic dynamics of gravity is remarkable, both in its simple elegance and in its profound statement about the nature of spacetime. Before we rush into the diverse matters which concern and motivate the search which leads to string theory and beyond, such as the nature of the quantum theory, the unification with other forces, etc., let us remind ourselves of some of the salient features of the classical theory. This will usefully foreshadow many of the concepts which we will encounter later.

1.1 The classical dynamics of geometry

Spacetime is of course a landscape of 'events', the points which make it up, and as such it is a classical (but of course relativistic) concept. Intuition from quantum mechanics points to a modification of this picture, and there are many concrete mechanisms in string theory which support this expectation and show that spacetime is at best a derived object or effective description. We shall see some of these mechanisms in the sequel. However, since string theory (as currently understood), seems to be devoid of a complete definition that does not require us to refer to spacetime, the language and concepts we will employ will have much in common with those used by professional practitioners of General Relativity, and of classical and quantum Field Theory. In fact, it will become clear to the newcomer that success in the physics of string theory is greatly aided by having technical facility in both of those fields. It is instructive to tour a little of the foundations of our modern approach to classical gravity and observe how the Relativist's and the Field Theorist's perspective are muddled together. String theory makes good and productive use of this sort of conflation.

It is useful to equip a description of spacetime with a set of coordinates x^μ, $\mu = 0, 1, \ldots, D - 1$, where $x^0 \equiv t$ (the time) and we shall remain open-minded and work in D dimensions for much of the discussion. The metric, with components $g_{\mu\nu}(x)$, is a function of the coordinates which allows for a local measure of the distance between points separated by an interval dx^μ:

$$ds^2 = g_{\mu\nu}(x)dx^\mu dx^\nu.$$

The metric is a tensor field since under an arbitrary change of variables $x^\mu \to x'^\mu(x)$ it transforms as

$$g_{\mu\nu} \longrightarrow g'_{\mu\nu} = g_{\alpha\beta}\frac{\partial x^\alpha}{\partial x'^\mu}\frac{\partial x^\beta}{\partial x'^\nu}. \tag{1.1}$$

Of course, 'distance' here means the more generalised Special Relativistic interval characterising how two events are separated, and it is negative, zero or positive, giving us timelike, null or spacelike separations, according to whether if it possible to connect the events by causal subluminal motion (appropriate to a massive particle), or by moving at the speed of light (massless particles), or not. This of course defines the signature of our metric as being 'mostly plus': $\{-++ + \cdots\}$ henceforth.

As a particle moves it sweeps out a path or 'world-line' $x^\mu(\tau)$ in spacetime (see figure 1.1), which is parametrised by τ. The wonderful thing is that what we would have said in pre-Einstein times was 'a particle moving under the influence of the gravitational force' is simply replaced by the statement 'a particle following a geodesic', a path which is determined by the metric in terms of the second order geodesic equation:

$$\frac{d^2 x^\lambda}{d\tau^2} = -\Gamma^\lambda_{\mu\nu}(g)\frac{dx^\mu}{d\tau}\frac{dx^\nu}{d\tau}, \tag{1.2}$$

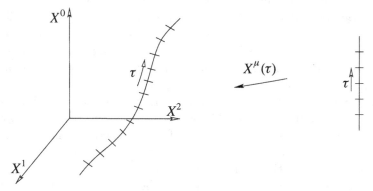

Fig. 1.1. A particle's world-line. The function $x^\mu(\tau)$ embeds the world-line, parametrised by τ, into spacetime, coordinatised by x^μ.

where the affine connection $\Gamma(g)$ is made out of first derivatives of the metric:

$$\Gamma^\lambda_{\mu\nu}(g) = \frac{1}{2}g^{\lambda\kappa}\left(\partial_\mu g_{\kappa\nu} + \partial_\nu g_{\kappa\mu} - \partial_\kappa g_{\mu\nu}\right).$$

Here and everywhere else, when working with curved spacetime we lower and raise indices with the metric and its inverse, (which has components $g^{\mu\nu}$ such that $g_{\mu\lambda}g^{\mu\alpha} = \delta^\alpha_\lambda$). Also note that $\partial_\mu \equiv \partial/\partial x^\mu$.

Switching language again we see that since the term on the left hand side of the equation (1.2) is what we think of as the 'acceleration', our Newtonian intuition determines the right hand side to be the 'applied force', attributed to gravity. In such language, $g_{\mu\nu}(x)$ is interpreted as a potential for the gravitational field.

In the purely geometrical language, there are no forces. There is only geometry, and the particle simply moves along geodesics. The above statement in equation (1.2) about how a particle moves in response to the metric is derivable from a simple action principle, which says that the motion minimises (more properly, extremises) the total path length that its motion sweeps out:

$$S = -m \int (-g_{\mu\nu}(x)dx^\mu dx^\nu)^{1/2} = -m \int_{\tau_i}^{\tau_f} (-g_{\mu\nu}(x)\dot{x}^\mu \dot{x}^\nu)^{1/2}d\tau\ , \quad (1.3)$$

where a dot denotes a derivative with respect to τ. (The reader might consider checking this by application of the Euler–Lagrange equations or by direct variation.)

The only question (which is of course one of the biggest) remaining is the nature of what determines the metric itself. This turns out to be governed by the distribution of stress-energy-momentum, and we must write field equations which determine how the one sources the other, just as we would in any field theory like Maxwell's electromagnetism (see insert 1.1).

The stress-energy-momentum contained in the matter is captured in the elegant package that is the tensor $T^{\mu\nu}(x)$, a second rank, symmetric, divergence-free tensor which for an observer with four-velocity **u**, encodes the energy density as $T_{\mu\nu}u^\mu u^\nu$, the momentum density as $-T_{\mu\nu}u^\mu x^\nu$, and shear pressures (stresses) as $T_{\mu\nu}x^\mu y^\nu$, where the unit vectors **x** and **y** are orthogonal to **u**.

Einstein's field equations are:

$$R_{\mu\nu} - \frac{1}{2}g_{\mu\nu}R = 8\pi G_{\mathrm{N}}T_{\mu\nu}\ , \quad (1.6)$$

where G_{N} is Newton's constant. As one would expect, the quantity on the left hand side is made up of the metric and its first and second derivatives,

---

Insert 1.1.   A reminder of Maxwell's field equations

'Maxwell's equations' are second order partial differential equations for the electromagnetic potentials $\vec{A}\,(\vec{x},t), \phi(\vec{x},t)$ from which the magnetic $(\vec{B}(\vec{x},t))$ and electric $(\vec{E}(\vec{x},t))$ fields can be derived:

$$\vec{E}(\vec{x},t) = -\,\vec{\nabla}\phi(\vec{x}\cdot t) - \frac{\partial\vec{A}(\vec{x},t)}{\partial t}$$
$$\vec{B}(\vec{x},t) = \vec{\nabla}\times\vec{A}(\vec{x},t). \tag{1.4}$$

In terms of the fields, Maxwell's equations are:

$$\vec{\nabla}\cdot\vec{E} = 4\pi\rho$$
$$\vec{\nabla}\cdot\vec{B} = 0$$
$$\vec{\nabla}\times\vec{E}+\frac{\partial\vec{B}}{\partial t} = 0$$
$$\vec{\nabla}\times\vec{B}-\frac{\partial\vec{E}}{\partial t} = 4\pi\,\vec{J}\,. \tag{1.5}$$

Here, the functions $\vec{J}(\vec{x},t)$ and $\rho(\vec{x},t)$, the *current density* and the *charge density* are the 'sources' in the field equations.

We have written the equations with the sources on the right hand side and the expression for the derivatives of the resulting fields (to which the sources give rise) on the left hand side. We will write these much more covariantly in insert 1.3.

---

where the Ricci scalar and tensor,

$$R \equiv g^{\mu\nu}R_{\mu\nu}, \qquad R_{\mu\nu} \equiv g^{\kappa\rho}g_{\lambda\rho}R^{\lambda}_{\mu\kappa\nu}, \tag{1.7}$$

are the only two contentful contractions of the Riemann tensor:

$$R^{\lambda}_{\mu\kappa\nu} \equiv \partial_{\mu}\Gamma^{\lambda}_{\kappa\nu} - \partial_{\nu}\Gamma^{\lambda}_{\kappa\mu} + \Gamma^{\rho}_{\kappa\mu}\Gamma^{\lambda}_{\rho\nu} - \Gamma^{\rho}_{\kappa\nu}\Gamma^{\lambda}_{\rho\mu}. \tag{1.8}$$

Except for the metric itself, the quantity on the left hand side of equation (1.6) is the unique rank two, divergenceless and symmetric tensor made from the metric (and its first and second derivatives), and hence can be allowed to be equated to the stress tensor.

When the stress tensor is zero, i.e. when there is no matter to act as a source, the vanishing of the left hand side is equivalent to the vanishing

$R_{\mu\nu} = 0$, and solutions of this equation are said to be 'Ricci-flat'. This includes highly non-trivial spacetimes such as Schwarzschild black holes, which follows from the non-linearity of the left hand side, representing the fact that the stress-energy in the gravitational field itself can act as its own source ('gravity gravitates').

The physical foundation behind the geometric approach is of course the Principle of Equivalence, which begins by observing that gravity is indistinguishable from acceleration, and tells one how to find a locally inertial frame: one must simply 'fall' under the influence of gravity (i.e. just follow a geodesic) and one does not feel one's own weight, and so one is in an inertial frame where the Laws of Special Relativity hold. See insert 1.2 for a reminder of this in equations. The sourceless field equations then follow from the recasting of the relative motion observed between frames on neighbouring geodesics in terms of an apparent 'tidal' force.

The full statement of the field equations to include sources is also guided by covariance, which means that it is a physical equation between tensors of the same type, and with the same divergenceless property (which is a physical statement of continuity). The equations are therefore true in all coordinate systems obtained by an arbitrary change of variables $x^\mu \to x'^\mu(x)$, since they transform as tensors in a way generalising the transformation of the metric in equation (1.1).

Note that the statement of divergencelessness is a covariant one too, i.e. $\nabla_\mu T^{\mu\nu} = 0$ uses the covariant derivative*, which is designed to yield a tensor after acting on one, say V:

$$\nabla_\kappa V^{\mu\cdots}_{\nu\cdots} \equiv \partial_\kappa V^{\mu\cdots}_{\nu\cdots} + \Gamma^\mu_{\lambda\kappa} V^{\lambda\cdots}_{\nu\cdots} + \cdots - \Gamma^\lambda_{\kappa\nu} V^{\mu\cdots}_{\lambda\cdots} - \cdots. \tag{1.9}$$

Finally, note that the field equations themselves may be derived from an action principle, the extremising of the Einstein–Hilbert action coupled to matter:

$$S = S_{\mathrm{M}} + S_{\mathrm{EH}}$$
$$S_{\mathrm{EH}} = \frac{1}{16\pi G_{\mathrm{N}}} \int d^D x \sqrt{-g}\, R$$
$$T^{\mu\nu} \equiv -\frac{2}{\sqrt{-g}} \frac{\delta S_{\mathrm{M}}}{\delta g_{\mu\nu}}, \tag{1.10}$$

where g is the determinant of the metric.

---

* In fact, this (not entirely unambiguous) procedure of replacing the ordinary derivative by the covariant derivative, together with the replacement of the Minkowski metric $\eta_{\mu\nu}$ by the curved spacetime metric $g_{\mu\nu}(x)$ is often called the principle of 'minimal coupling' as a procedure for how to generalise Special Relativistic quantities to curved spacetime.

Insert 1.2.   Finding an inertial frame by freely falling

In order to find an inertial frame, we must find coordinates so that at least locally, at a point x_o^ν, say, we can can do special relativity. This means that we perform a change of coordinates $x^\mu \to x'^\mu(x)$ so that when the metric changes, according to (1.1), the result is

$$g_{\mu\nu}(x_o^\nu) = \eta_{\mu\nu},$$

where $\eta_{\mu\nu}$ is the Minkowski metric, $\mathrm{diag}(-1, +1, \dots,)$. How accurately can we achieve this? In our coordinate transformation, we have in the neighbourhood of x_o^ν:

$$x^\mu(x^\nu) = x^\mu(x_o^\nu) + \frac{\partial x^\mu}{\partial x'^\nu}(x'^\nu - x_o'^\nu)$$

$$+ \frac{1}{2}\frac{\partial^2 x^\mu}{\partial x'^\nu \partial x'^\kappa}(x'^\nu - x_o'^\nu)(x'^\kappa - x_o'^\kappa)$$

$$+ \frac{1}{6}\frac{\partial^3 x^\mu}{\partial x'^\nu \partial x'^\kappa \partial x'^\lambda}(x'^\nu - x_o'^\nu)(x'^\kappa - x_o'^\kappa)(x'^\lambda - x_o'^\lambda) \dots$$

so we have, at first order, D^2 coefficients to adjust. Since $g'_{\mu\nu}$ has $D(D+1)/2$ components, we are left with

$$D^2 - \frac{D(D+1)}{2} = \frac{D(D-1)}{2}$$

transformations at our disposal. Happily, this is precisely the dimension of the Lorentz group, $SO(D-1,1)$ of rotations and boosts available in our inertial frame. At second order, we have $D^2(D+1)/2$ coefficients to adjust, which is precisely the same number of first derivatives $\partial g'_{\mu\nu}/\partial x'^\kappa$ of the metric that we need to adjust to zero, cancelling all of the 'forces' in the geodesic equation (1.2). At third order, we have $D^2(D+1)(D+2)/6$ coefficients to adjust, while there are $D^2(D+1)^2/4$ second derivatives of the metric, $\partial^2 g'_{\mu\nu}/\partial x'^\kappa \partial x'^\lambda$, to adjust, which is rather more. In fact, this failure to adjust

$$\frac{D^2(D+1)^2}{4} - \frac{D^2(D+1)(D+2)}{6} = \frac{D^2(D^2-1)}{12}$$

second derivatives is of course a statement of physics. This is precisely the number of independent components of the Riemann tensor $R^\lambda_{\kappa\mu\nu}$, which appears in the field equations determining the metric. So everything fits together rather nicely.

A favourite example of a stress tensor for a matter system is the Maxwell system of electromagnetism. Combining the electric potential ϕ and vector potential $\vec{A}$ into a four-vector $\mathbf{A}(\mathbf{x}) = (\phi, \vec{A})$, with components A_μ, the magnetic induction $\vec{B}$ and electric field $\vec{E}$ are captured in the rank two antisymmetric tensor field strength:

$$F_{\mu\nu} = \partial_\mu A_\nu - \partial_\nu A_\mu,$$

and an observer with four-velocity $\mathbf{u}$ reads the fields as:

$$E_\mu = F_{\mu\nu} u^\nu, \qquad B_\mu = \epsilon_{\mu\nu}{}^{\kappa\lambda} F_{\kappa\lambda} u^\nu, \tag{1.11}$$

where $\epsilon_{\mu\nu\kappa\lambda}$ is the totally antisymmetric tensor in four dimensions, with $\epsilon_{0123} = -1$. (See insert 1.3 for more on this covariant presentation of electromagnetism.) The action is:

$$S_{\mathrm{M}} = \int d^D x \mathcal{L} = -\frac{1}{16\pi} \int (-g)^{1/2} F_{\mu\nu} F^{\mu\nu} d^D x, \tag{1.12}$$

and so it is easily verified that the Euler–Lagrange equations

$$\frac{\partial \mathcal{L}}{\partial A_\mu} - \frac{\partial}{\partial x^\nu} \left(\frac{\partial \mathcal{L}}{\partial (\partial_\nu A_\mu)} \right) = 0,$$

give the field equations

$$\nabla_\nu F^{\mu\nu} = 0,$$

where we have used a very useful identity which is easily derived:

$$\delta(-g)^{1/2} = \tfrac{1}{2}(-g)^{1/2} g^{\mu\nu} \delta g_{\mu\nu}. \tag{1.13}$$

On the other hand, since

$$\frac{\partial \mathcal{L}}{\partial g_{\mu\nu}} = -\frac{(-g)^{1/2}}{8\pi} \left(g_{\lambda\beta} F^{\mu\lambda} F^{\nu\beta} - \tfrac{1}{4} g^{\mu\nu} F_{\sigma\rho} F^{\sigma\rho} \right) \tag{1.14}$$

the stress tensor is

$$T^{\mu\nu} = \frac{1}{4\pi} \left(g_{\lambda\beta} F^{\mu\lambda} F^{\nu\beta} - \tfrac{1}{4} g^{\mu\nu} F_{\sigma\rho} F^{\sigma\rho} \right). \tag{1.15}$$

1.2 Gravitons and photons

The quantum Field Theorist's most sacred tool is the idea of associating a particle to every sort of field, whether it be matter or force. So a force is

---

Insert 1.3.   Maxwell written covariantly

Probably most familiar is the flat space writing:

$$F^{\mu\nu} = \begin{pmatrix} 0 & E_1 & E_2 & E_3 \\ -E_1 & 0 & B_3 & -B_2 \\ -E_2 & -B_3 & 0 & B_1 \\ -E_3 & B_2 & -B_1 & 0 \end{pmatrix} \tag{1.16}$$

for the Maxwell tensor. In addition to the four-vector $\mathbf{A}(\mathbf{x}) = (\phi, \vec{A})$, one in general will have a four-current for the source, which combines the current and electric charge density: $\mathbf{J}(\mathbf{x}) = (\rho, \vec{J})$. With these definitions, Maxwell's equations take on a particularly simple covariant form:

$$\nabla_\nu F^{\mu\nu} = -4\pi J^\mu, \qquad \partial_\mu F_{\nu\kappa} + \partial_\nu F_{\kappa\mu} + \partial_\kappa F_{\mu\nu} = 0, \tag{1.17}$$

for the equations with sources, and the source-free equations (Bianchi identity). The energy-momentum tensor for electromagnetism is given in terms of $\mathbf{F}$ in equation (1.15), and is subject to the conservation equation (when the sources $J^\mu = 0$): $\nabla_\mu T^{\mu\nu} = 0$. This contains familiar physics. Specialising to flat space:

$$T_{00} = \frac{1}{8\pi}((\vec{E})^2 + (\vec{B})^2), \qquad T_{0i} = -\frac{1}{4\pi}(\vec{E} \times \vec{B}),$$

which is the familiar expression for the energy density and the momentum density (Poynting vector) of the electromagnetic field

---

mediated by a particle which propagates along in spacetime between objects carrying the charges of that interaction. There is great temptation to do this for gravity (by allowing all sources of stress-energy-momentum to emit and absorb appropriate quanta), but we immediately run into a conceptual log jam. On the one hand, we have just reminded ourselves of the beautiful picture that gravity is associated to the dynamics of spacetime itself, while on the other hand we would like to think of the gravitational force as mediated by gravitons which propagate on a spacetime background. A technical way of separating out this problem into manageable pieces (up to a point) is to study the linearised theory.

The idea is to treat the metric as split between the background which is say, flat spacetime given by the Minkowski metric $\eta_{\mu\nu}$, $\text{diag}(-1, +1, \ldots,)$, and some position dependent fluctuation $h_{\mu\nu}(x)$ which is to be small $h_{\mu\nu}(x) \ll 1$. Then the equations determining $h_{\mu\nu}(x)$ are derived from Einstein's equations (1.6) by substituting this ansatz:

$$g_{\mu\nu} = \eta_{\mu\nu} + h_{\mu\nu}(x),$$

and keeping only terms linear in $h_{\mu\nu}$.

Let us carry this out. We will raise and lower indices with $\eta_{\mu\nu}$, and note that $g^{\mu\nu}$ will continue to be the inverse metric, which is distinct from $\eta^{\mu\alpha}\eta^{\nu\beta}g_{\alpha\beta}$. Note also that $g^{\mu\nu} = \eta^{\mu\nu} - h^{\mu\nu}$, to the accuracy to which we are working. The affine connection becomes:

$$\Gamma^\rho_{\mu\nu} = \tfrac{1}{2}\eta^{\rho\alpha}\left(\partial_\mu h_{\nu\alpha} + \partial_\nu h_{\mu\alpha} - \partial_\alpha h_{\mu\nu}\right), \tag{1.18}$$

and to this order, the Ricci tensor and scalar are:

$$R_{\mu\nu} = \partial^\alpha \partial_{(\nu}h_{\mu)\alpha} - \tfrac{1}{2}\partial^\alpha\partial_\alpha h_{\mu\nu} - \tfrac{1}{2}\partial^\mu\partial_\nu h + O(h^2),$$
$$R = \partial^\alpha\partial^\beta h_{\alpha\beta} - \partial^\alpha\partial_\alpha h + O(h^2), \tag{1.19}$$

where $h = h^\mu_\mu$. Thus we learn that

$$R_{\mu\nu} - \tfrac{1}{2}\eta_{\mu\nu}R = \partial^\alpha\partial_{(\nu}h_{\mu)\alpha} - \tfrac{1}{2}\partial^\alpha\partial_\alpha h_{\mu\nu} - \tfrac{1}{2}\partial^\mu\partial_\nu h$$
$$-\tfrac{1}{2}\eta_{\mu\nu}\left(\partial^\alpha\partial^\beta h_{\alpha\beta} - \partial^\alpha\partial_\alpha h\right) + O(h^2).$$

Defining $\bar{\gamma}_{\mu\nu} = h_{\mu\nu} - \tfrac{1}{2}\eta_{\mu\nu}h$, we find our linearised field equations:

$$-\tfrac{1}{2}\partial^\alpha\partial_\alpha \bar{h}_{\mu\nu} + \partial^\alpha\partial_{(\mu}\bar{h}_{\mu)\alpha} - \tfrac{1}{2}\eta_{\mu\nu}\partial^\alpha\partial^\beta\bar{h}_{\beta\gamma} = 8\pi G_N T_{\mu\nu}. \tag{1.20}$$

There is an explicit gauge degree of freedom (recognisable from equation (1.1) as an infinitesimal coordinate transformation)

$$h_{\mu\nu} \to h_{\mu\nu} + \partial_\mu\xi_\nu + \partial_\nu\xi_\mu, \tag{1.21}$$

for arbitrary an arbitrary vector ξ_μ. Using this freedom, we *choose the gauge* $\partial^\nu\bar{h}_{\mu\nu} = 0$ (using a gauge transformation satisfying $\partial^\nu\partial_\nu\xi_\mu + \partial^\nu\bar{h}_{\mu\nu} = 0$), which implies

$$\partial^\alpha\partial_\alpha\bar{h}_{\mu\nu} = -16\pi G_N T_{\mu\nu}. \tag{1.22}$$

This is highly suggestive. Consider the system of electromagnetism, with equations of motion (1.17). The equations are invariant under the gauge transformation

$$A_\mu \to A_\mu + \partial_\mu\Lambda,$$

where Λ is an arbitrary scalar. We can use this freedom to choose a gauge $\partial_\mu A^\mu = 0$, (with a parameter satisfying $\partial_\mu \partial^\mu \Lambda + \partial^\nu A_\nu = 0$), which gives the simple equation

$$\partial_\mu \partial^\mu A_\nu = -4\pi J_\nu.$$

This is of a very similar form to what we achieved in equation (1.22) for the system of linearised gravity. The analogy is clear. The Maxwell system has yielded a field equation for a vector (spin one) particle (the photon $A_\mu(x)$) sourced by a vector current ($J_\mu(x)$), while the gravitational system yields the precisely analogous equation for a spin two particle (the graviton $h_{\mu\nu}(x)$) sourced by the stress tensor $T_{\mu\nu}(x)$.

This is the starting point for treating gravity on the same footing as field theory, and in many places later we will have cause to use the word or idea 'graviton', and it is in this sense (a spin two particle propagating on a reference background) that we will mean it. We have seen how to make the delicate journey from the Relativist's geometrical understanding of gravity to a perturbative Field Theorist's. To make the return journey, reconstructing a picture of, say the non-trivial spacetime metric due to a star by starting from the graviton picture is a bit harder, but roughly it is conceptually similar to the same problem in electromagnetism. How does one go from the picture of the photon moving along in spacetime to building up a picture of the strong magnetic fields around a pair of Helmholtz coils? Words and phrases which are offered include 'coherent state of photons', or 'condensation of photons', and these should invoke the idea that the coils' field cannot be constructed using only the perturbative photon picture. One can instead use the photon description to describe processes *in the background* of the Helmholtz field, and we can similarly do the same thing for gravity, describing the propagation of gravitons in the background fields produced by a star. In this way, we see that there is a possibility that there are situations where the conceptual separation between particle quanta and background *in principle* needs be no more dangerous in gravitation than it is in electromagnetism.

Eventually, however, we would like to compute beyond tree level, and the celebrated problems of the theory of gravity treated as a quantum theory will be encountered. Then, the linearised Einstein–Hilbert action

$$S = \frac{1}{16\pi G_{\rm N}} \int d^D x \left(\partial^\alpha \partial^\beta h_{\alpha\beta} - \partial^\alpha \partial_\alpha h \right), \qquad (1.23)$$

will eventually reveal itself to be non-renormalisable once we add interactions coming from the next order above linear. In particular, the process of recursively adding counterterms to the bare action in order to define physically measurable quantities does not terminate. As Field Theorists (and perhaps as Relativists) we would have cause to be discouraged, and it is a much celebrated statement that as String Theorists, we won't be.

1.3 Beyond classical gravity: perturbative strings

A reason for dwelling on some of the previous points is that it is customary to do a lot of moving back and forth between the picture of quanta moving on a flat background and other pictures, for example ones involving considerably curved background fields. This is not because string theorists have a clever collection of new technological tools for seeing how to move from one to the other (although as we shall see with the aid of supersymmetry, in some cases we can often keep track of many properties of objects in moving between pictures) but because as was said before, string theory is a developing subject which has borrowed and hybridised intuition from the Relativist's and the (perturbative and non-perturbative) quantum Field Theorist's worlds.

This borrowing is not to be taken as a sign of intellectual bankruptcy, but quite the opposite. The adoption of terminology and concepts from a wide range of other fields is as a result of the richness of genuinely novel physical phenomena, with (as a whole) no precise precedent or analogue, which the theory appears to be revealing. This is very similar to what happened almost precisely a century ago. The treatment of quanta in a context dependent manner either as a wave or as a particle, an understanding still called 'Wave–Particle Duality' by many, grew out of the attempt to grasp a new physical phenomenon – Quantum Mechanics – by reference to established physical concepts from the century before.

In the next chapter we will review how one proceeds to describe the relativistic string propagating in a flat background. There are two very broad categories, open strings which have end-points, and closed strings which do not. The basic input parameter is the mass per unit length of the string, its tension:

$$T = \frac{1}{2\pi\alpha'} \equiv \frac{1}{2\pi\ell_s^2}.$$

As is well known, the characteristic length scale of the string, ℓ_s, is traditionally very small compared to scales on which we do current-day physics. This means that string excitations will have a good description as point-particle-like states on scales much longer than ℓ_s. After quantisation, it rapidly becomes clear that the spectrum of string theory is rather rich and demands application. Since finite masses in the spectrum are set by the inverse of ℓ_s, the infinite tower of massive excitations of the string (see figure 1.2) will be very inaccessible at low energy (long distance, or infra-red (IR)). The tower is of course crucial to the properties of the

Fig. 1.2. The string spectrum has a massless sector separated by a gap (set by the tension) after which there is an infinite tower of massive states.

high energy (short distance, or ultra-violet (UV)) physics of the string. It is the massless part of the spectrum which is accessible at low energy and hence relevant to phenomenology.

For example, closed string theories describe a massless spin two particle which is identified with the graviton. The questions of non-renormalisability which arose in quantum field theory turn out to be circumvented by the remarkable ultra-violet properties of string theory, which give rise to an extremely well-behaved perturbative description of multi-loop processes involving gravitons[†]. The simple fact is that string theory is very unlike field theory at short distances, since it assembles together an infinity of increasingly massive excitations (in a particular way) which all play a role in the UV. The theory's supplying a satisfactory perturbative quantum theory of gravity is just the beginning of the many phenomena which arise from its properties as an extended object, as we shall see.

Other massless fields which arise in string spectra are Abelian and non-Abelian gauge fields, and various fermions and scalars, some of which one might expect give rise to the observed gauge interactions and matter fields. There is also a family of higher rank antisymmetric tensor fields generalising the photon on which we will focus in some detail. Remarkably, the value of one scalar excitation of interest, the dilaton Φ, determines the strength of the string self-interaction, $g_{\rm s} = e^{\Phi}$, and hence (since closed strings excitations can be gravitons) the value of $G_{\rm N}$. It is a striking fact that string theory dynamically determines its own coupling strength. (See figure 1.3.)

[†] Sadly, lack of space will prevent us from describing this here, and we refer the reader to a textbook on this[1, 5].

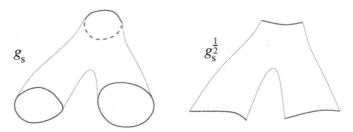

Fig. 1.3. The basic three-string interaction for closed strings, and its analogue for open strings. Its strength, g_s, along with the string tension, determines Newton's gravitational constant G_N.

Just as with the particle, it is straightforward to generalise the treatment of the string to motion in a curved background with metric $g_{\mu\nu}(x)$, and one can derive the analogue of classical geodesic equations of motion (if desired) for the string.

The string sweeps out a 'world-sheet' with coordinates $(\sigma^1, \sigma^2) \equiv (\tau, \sigma)$. The string's path in spacetime is described by $X^\mu(\tau, \sigma)$, giving the shape of the string's world-sheet in target spacetime (see figure 1.4). There is an 'induced metric' on the world-sheet given by $(\partial_a \equiv \partial/\partial\sigma^a)$:

$$h_{ab} = \partial_a X^\mu \partial_b X^\nu g_{\mu\nu}, \qquad (1.24)$$

with which we can perform meaningful measurements on the world-sheet as an object embedded in spacetime. Using this, we can define an action analogous to the one we thought of first for the particle, by asking that

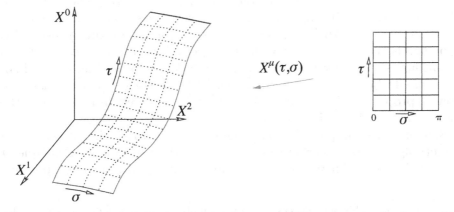

Fig. 1.4. A string's world-sheet. The function $X^\mu(\tau, \sigma)$ embeds the world-sheet, parametrised by (τ, σ), into spacetime, coordinatised by X^μ.

we extremise the area of the world-sheet:

$$S = -T \int dA = -T \int d\tau d\sigma \, (-\det h_{ab})^{1/2} \equiv \int d\tau d\sigma \, \mathcal{L}(\dot{X}, X'; \sigma, \tau).$$

$$(1.25)$$

Expanded, this is

$$S = -T \int d\tau d\sigma \left[\left(\frac{\partial X^\mu}{\partial \sigma} \frac{\partial X^\mu}{\partial \tau} \right)^2 - \left(\frac{\partial X^\mu}{\partial \sigma} \right)^2 \left(\frac{\partial X_\mu}{\partial \tau} \right)^2 \right]^{1/2}$$

$$= -T \int d\tau d\sigma \left[(X' \cdot \dot{X})^2 - X'^2 \dot{X}^2 \right]^{1/2},$$

$$(1.26)$$

where X' means $\partial X/\partial \sigma$.

This is very analogous to the case of the particle, and we will analyse it further in the next chapter. However, there is much more to the story than this. The thorny question arises concerning what dynamics govern the allowed metrics, and it is a riddle of considerable depth: the string has revealed itself as generating the basic quantum of gravity as one of its modes of oscillation. Our experience from before allows us to trust that there ought to be a manner in which one can treat the graviton (and hence the string that carries it) as a small disturbance on a fixed background, but there is an additional problem which we did not have last time. Since the string is also the source of gravity, and if it dynamically generates the strength of the coupling, it ought to also determine gravitational dynamics. How does it go about determining the gravitational background in which it is supposed to propagate? In the terms we used previously, where do the field equations governing the background come from?

The surprise turns out to be that internal quantum mechanical consistency of the string theory *does* make certain demands on the properties of spacetime, in ways that no previous theory has managed before. First of all, it requires that it only propagates in spacetimes of certain dimensionality (for example, 26 for bosonic strings, 10 for superstrings). Secondly, it demands that at low energy the background metric satisfies Einstein's equations (sourced by the stress tensor due to the other massless fields)! This should be contrasted with the case of a particle where the issue of how it propagates in a metric is completely divorced from whether the metric satisfies Einstein's equations.

Somehow, the simple generalisation of a particle to a string has captured something very new. Is there an analogue of the Equivalence Principle at work which gives Einstein's equations at low energy and then new physics[‡]

[‡] It is hoped that this new physics will cure a number of problems in strongly coupled gravity, like the loss of predictability of relativistic physics at spacetime singularities such as in black holes or at the Big Bang.

at high energy? Even though this remarkable fact is relatively old by now, there is no simple thought experiment which explains why a generalisation from a particle to a string quantum-mechanically demands the solution of field equations for which the underlying principle is covariance and equivalence.

1.4 Beyond perturbative strings: branes

The reader may have noticed that the word 'perturbative' was used a lot in the last section, even when describing the remarkable successes of string theory in the arena of quantum gravity. The Second Superstring Revolution gets its name from the remarkable change of perspective which occurred with breakthroughs in understanding of this very issue, and the resulting flow of ideas and results. A great deal of quite surprising insight was gained about the supersymmetric string theories (whose existence and consistency followed from discoveries in the First Superstring Revolution) in the limit of very strong coupling, much of which we will cover later.

The big question which arose time and again in string theory over the years before the revolution was the issue of its description beyond perturbation theory. Actually, there were possibly two problems and not just one, however they usually are discussed together, although they may be logically distinct. Motivated by analogy with field theory, string theorists sought for something like a field theory of strings, which would allow for the non-perturbative exploration of the landscape in which vacua lie, in a way which is familiar in field theory, allowing the study of important phenomena like tunnelling, instantons, solitons, etc. The idea was that there would be a 'string field' Σ whose role was to create and destroy a string in a particular configuration. This begins by being conceptually on a par with the successful ordinary field theory concept about the role of a field in creating and destroying particle quanta, but this view soon changes when one remembers that the string is like an infinite number of particles from the point of view of field theory. The ideally next simplest step would be to find a simple way of writing a kinetic energy and potential $V(\Sigma)$, which would allow a study of dynamics and hence 'second quantised' strings (to use another old misnomer). See figure 1.5. In principle, some type of field theory is not an altogether crazy thing to want to find. Given the success of the field theory framework, it would be an understatement to say that it would have been neglectful if the possibility had not been explored. There is another problem, however, into which experience with field theory seems to offer little insight. This is 'background independence'. In ordinary quantum field theory, a Lagrangian for the theory is defined with reference to a spacetime background. This

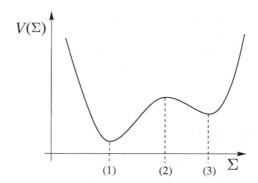

Fig. 1.5. A fanciful view of a slice through the infinite dimensional land-scape of non-perturbatively accessible string vacua. Σ represents the entire field content of a string theory, and $V(\Sigma)$ is a potential. Locations (1) and (3) represent perturbatively stable vacua, while (2) is unstable. Important physics may be found in the non-perturbative effects relating these vacua.

is of course fine, since the fields are supposed to propagate on this background. However, it is not clear that this luxury should be available to us in the string theory, since it is supposed to determine the background upon which it is propagating, given that it generates gravity and the value of G_{N}.

The search for string field theories were not entirely unsuccessful, but since they are very difficult to work with, at the time of writing, it is not clear what they have taught us. It is a remarkable achievement in itself that one *could* define a string field Σ, *and* find a sensible Lagrangian. Both the kinetic and potential are on the face of it, written in such a way that there is a chance of background independence since the 'derivative' and the means of multiplying together string fields do not seem to directly refer to spacetime. Sadly, the means of unpacking the Lagrangian to perform a computation require one to make reference to objects which originally were defined with perturbative intuition about backgrounds again, and so background independence is still not apparent.

This is not really a failure, if one reduces ones expectations about what a string field theory is supposed to do for us. It is possible to imagine that such a theory can tell us interesting physics involving various types of string vacua, and how they are inter-related, without ever addressing the background independence issue.

This possibility was regarded as unsatisfactory for a long time, since it made string theory seem logically incomplete, with no physical principle or mechanism to appeal to, given that it was supposed to be the theory of everything. Happily, the Second Revolution happened, and now we have a new possibility. String theory is not a theory of strings after all. There are two clear signs of this (which we will discuss later in detail). One is that there are extended objects in the theory ('D-branes') which carry[265] the basic charges of a special class of higher rank antisymmetric fields which the string theory necessarily describes, but cannot itself source. Coupled with this fact is that at arbitrarily strong coupling, these objects can become arbitrarily light (see insert 1.4), indeed lighter that the string itself, and so their behaviour dominates the low energy physics, undermining the fundamental role of the strings. A second sign is that some string theories are directly related at strong coupling (sometimes by a condensation of a tower of increasingly light D-particles) to a field theory – at low energy – which includes gravity. The short-distance completion of this gravitational theory does not seem to involve the dynamics of strings, and the new degrees of freedom are unknown. This unknown theory, whose existence is strongly suggested by the intricate web of strong/weak coupling dualities between the superstrings in diverse situations[151, 152, 153], is often called 'M-theory', and it seems that all of the superstring theories that we know of may be obtained as a limit of it. In this sense, we see that string theory is itself an *effective theory*, albeit a remarkably interesting one. All of the various string theories that we know are perturbative corners of a larger coupling space. See figure 1.6. From this new picture (in which in some cases the extended objects which become light at strong coupling are weakly coupled strings of an

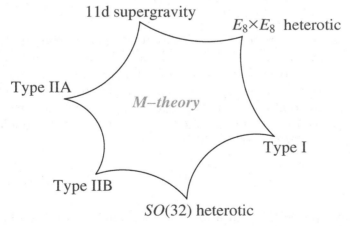

Fig. 1.6. A schematic diagram of the statement that all superstring theories, and eleven dimensional supergravity, are effective descriptions of certain dynamical corners of a larger theory, called 'M-theory'.

Insert 1.4. Soliton properties and the kink solution

Everybody's favourite soliton example is the kink solution of ϕ^4 theory in 1+1 dimensions. The mass m and the coupling λ combine into a dimensionless coupling $g = \lambda/m^2$, and we write:

$$\mathcal{L} = -\frac{1}{2}\partial^\mu\phi\partial_\mu\phi - U(\phi), \qquad U(\phi) = \frac{\lambda}{4}\left(\phi^2 - \frac{m^2}{\lambda}\right)^2.$$

The kink (or anti-kink) solution is

$$\phi_\pm(x) = \pm\frac{1}{\sqrt{g}}\tanh\left(\frac{m(x - x_0)}{\sqrt{2}}\right),$$

and so it is clearly an *interpolating solution* between the two vacua (located at $\pm\phi_0 = \pm 1/\sqrt{g}$) of the double well potential.

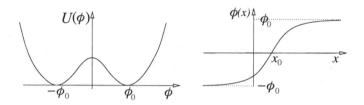

The parameter x_0 is a constant, corresponding to the ability to translate the solution. The configuration's mass-energy is:

$$E = \int_{-\infty}^{\infty}\left(\frac{1}{2}\left(\frac{\partial\phi_\pm}{\partial x}\right)^2 + U(\phi_\pm)\right)dx = \frac{2\sqrt{2}}{3}\frac{m}{g},$$

which is inversely proportional to the dimensionless coupling. So at weak coupling, this is a very heavy localised lump of energy. If we could trust this formula at strong coupling (and for various types of soliton in e.g. supersymmetric theories, we can), it is clear that for large g this solution becomes a light, sharply localised particle. In fact, it has a conserved charge, due to the existence of the topological current $j_\mu = (\sqrt{g}/2)\epsilon_{\mu\nu}\partial^\nu\phi$, which is:

$$Q = \int_{-\infty}^{\infty} j_0\,dx = \frac{\sqrt{g}}{2}\left(\phi(+\infty) - \phi(-\infty)\right) = \pm 1.$$

All of these properties will appear for solitons of theories which we shall study. The validity of the mass formula at strong coupling will allow various 'dualities' of supersymmetric theories to be uncovered.

entirely different type from the starting theory, giving a 'string–string' strong/weak coupling duality), it is clear that the string field theory approach would have had to produce a completely unlooked-for phenomenon, and convert the world-sheet expansion of one type of string (say a closed one) into the completely different type of world-sheet expansion of another type of string (say an open one). It would also have to point to new directions in which there is a perturbation theory not involving strings at all. Lastly, it would also have to be background independent.

Of course, this may yet happen (but we might not call it a field theory any more!), but another possibility is that string field theory (at least in the intuitive form in which it was conceived) will be useful as an effective theory (arising from M-theory) useful for the study of a restricted but important set of non-perturbative effects.

1.5 The quantum dynamics of geometry

The issue of background independence may be tied up with matters which the theory is only really still just touching on, and so it may have been premature to worry about it previously. This is the fact that there are dynamical signs that clearly show that string theory avoids a definite picture of some of the properties of spacetime which we would have thought were fixed, if we were field theorists.

Scattering of strings seems to show that attempts to confine the string to a small domain of spacetime are defeated by the strings' tendency to increasingly extend itself and spread out. From T-duality[14] (to be first encountered in chapter 4, but probably in every chapter beyond that), we learn that when a string theory is compactified on a circle, there is an ambiguity in the spectrum about whether the propagation is on a circle of radius R or radius ℓ_s^2/R. The standard 'momentum' states with energy in multiples of $1/R$ are joined by 'winding' states whose energy is in multiples of R/ℓ_s^2, coming from winding around the circle. The 'duality' exchanges these two types of mode. This is remarkable, especially if one considers the limit that $R \to 0$, since it says that an arbitrarily small circle compactification (reducing an effective spacetime dimension) is physically equivalent to having an arbitrarily large dimension (restoring an effective dimension). The outcome of this reasoning is that there appears to be an effective minimum distance arising in the dynamics of (perturbative) strings, of order the string scale ℓ_s. This is qualitatively just the sort of granularity of spacetime which one might have anticipated (and indeed it was) in thinking about expectations for a quantum theory of gravity. We can go even further, however.

As already mentioned, at strong coupling some string theories turn into something which at low energy is a field theory in one dimension *higher* than the target spacetime of the weakly coupled string. Since the string coupling is dynamically generated by the string itself, we arrive at the result that the dimension of spacetime itself is dynamical.

Also, the coordinates describing various objects like D-branes located in string theory's target space arise as not just numbers, but matrices[26]. For example, in superstring theory for N pointlike D-branes (known as D0-branes or D-particles), there are nine $N \times N$ matrices, $X^i(\tau)$, describing their world-lines parametrised by τ. When the D-branes are widely separated from each other, it is dynamically favourable for these matrices to be diagonal, and we have N copies of the usual coordinates x^μ describing the positions of N pointlike objects in nine spatial directions:

$$X^i(\tau) = \begin{pmatrix} x^i_1(\tau) & 0 & 0 & \cdots & \cdots \\ 0 & x^i_2(\tau) & 0 & \cdots & \cdots \\ 0 & 0 & x^i_3(\tau) & \cdots & \cdots \\ \vdots & \vdots & \ddots & \ddots & \cdots \\ \cdots & \cdots & \cdots & \cdots & x^i_N(\tau) \end{pmatrix}. \qquad (1.27)$$

When the branes are close together, there are dynamically favourable regimes when these matrices are non-commuting, and correspondingly, the spacetime coordinate interpretation is now in terms of a non-commutative picture. There is more here, actually. Since D0-branes turn out to be momentum modes, in a compact direction, of an eleven dimensional graviton, this picture turns out to be a sort of light cone formulation of the eleven dimensional theory. This is the beginning of the *Matrix Theory*[157] formulation of M-theory.

Spacetime is clearly a far more interesting place when the dynamics of string/M-theory are explored, and so it may be a while before we know even if we are asking the right sorts of questions about its nature. This includes the issue of background independence, and it may be that we have to wait for a complete formulation of M-theory (which may well have nothing to do with spacetime at all) before we get an answer.

1.6 Things to do in the meantime

While we wait for a complete formulation of M-theory to show up, there is a lot to do in the meantime. String theory's second revolution has provided us with a large number of tools to explore many regimes of fundamental physics, both old and new.

Gauge theories arise in string theories in many different (and often interrelated) ways, for example by dimensional reduction and the Kaluza–Klein mechanism (described in section 4.1.1), or as the collective dynamics on the world-volume of branes (described in section 4.10), or from gauge fields intrinsic to the structure of a closed string theory (described in section 7.2). So string theory is an arena for studying gauge theories. The very geometrical way in which string theories treat gauge fields allows for many gauge theory phenomena to be usefully recast in geometrical terms. This also means that known gauge theory phenomena, correctly interpreted in this context, can also teach us new things about the geometry of string theories. Many of the applications of D-branes which we will discuss later in this book are concerned with this powerful dialogue.

In this way, useful tools can be extracted for application to very concrete and pragmatic questions in the dynamics of strongly coupled gauge theory, of great concern to us of course in the physics being explored or shortly to be explored in experiments.

Since string theory is also a theory of gravity, it is exciting to learn that there are regimes where much progress may be made in the study of situations where hard questions about quantum gravity arise. The most celebrated example of this is the precise statistical interpretation of Bekenstein's thermodynamical black hole entropy[262], for a large class of black holes. This thermodynamical quantity can arise as the inevitable conclusion of semi-classical treatments of quantum gravity, where quantum fields are studied in a classical black hole background (a useful conceptual and technical compromise alluded to earlier). Such a treatment led Hawking[261] to realise that there is thermal radiation (at a specific temperature) from a black hole, after other suggestive properties[289, 292] led Bekenstein to the realisation that there is an entropy associated to the area of the horizon. The universal Bekenstein–Hawking entropy for a black hole is:

$$S = \frac{A}{4G_{\mathrm{N}}},\tag{1.28}$$

and is at the heart of the laws of black hole thermodynamics. This was a bit awkward, since there was no underling theory of quantum gravity to supply the 'statistical mechanics' which account for the precise relation between the entropy and the properties of the black hole. As we will describe in detail, for a large class of black holes, string theory provides the precise answer, in terms of D-brane constituents, and the gauge theories which describe them. In fact, (for a smaller class of black holes) the spacetime dynamics of individual D-branes conspires to provide a microscopic mechanism for the operation of the second law of thermodynamics as well[7].

One of the most profound insights of the revolution which might have the furthest-reaching consequences, is the identification of tractable regimes where a duality between gravitation and gauge theory can be found. This grew out of the above results concerning black holes, and even the ideas concerning the translation of gauge theory phenomena into geometry, but it is in some sense logically distinct from those. There is a very striking and intricate dynamical duality between the two, which again crosses dimensionality and is indicative of a very rich underlying picture. The 'AdS/CFT correspondence'[270, 271, 272], the title under which the simplest examples are known, is also the sharpest known example of what is known as the 'Holographic Principle'[286, 287], which states (roughly) that there should be a lower dimensional non-gravitational representation of the degrees of freedom of any quantum theory of gravity. Matrix theory is another example[158].

The idea of the principle arises from the realisation that any high energy density scattering used to probe the short distance degrees of freedom in a theory including gravity will ultimately create black holes. Black holes seem to exhibit all of their degrees of freedom on their horizon, an object which is of one dimension fewer than the parent theory. This suggests (but of course does not supply a definite constructive tip for how to find it) that there is a more economical description of theories of D-dimensional gravity in terms of a theory in $D - 1$ dimensions. The AdS/CFT correspondence manages this by relating a theory of gravity in an anti-de Sitter background (a highly symmetric spacetime with negative cosmological constant, reviewed in section 10.1.7) to a strongly coupled $SU(N)$ gauge theory (of large N) in one dimension fewer. This is remarkable, since theories of gravity and gauge theory are so very different in crucial dynamical respects, and we explore this in detail in chapter 18, showing how it arises from our study of D-branes, and exploring some of the consequences for new descriptions of strongly coupled gauge theory phenomena.

Exploring the correspondence in more complicated cases is of great interest, as it might give us insights and new tools which we can apply to more phenomenologically relevant gauge theories, and we spend some time discussing some examples of this.

1.7 On with the show

It is apparently an Irish saying that one will never plough a field by turning it over in one's mind, and so we should now begin the task of exploring things more carefully. In setting the scene, we have begun to unpack some of the more difficult concepts and some of the language which we will encounter many times as we go along. We will proceed by developing the basic language of string theory, uncovering many remarkable phenomena and vacua, using perturbation theory only. Certain perturbative hints of non-perturbative physics will appear from time to time, and with the help of D-branes and supersymmetry, we later uncover such physics using many 'duality' relations. Much later, we combine these techniques and ideas to probe and map out aspects of M-theory, and also to study certain aspects of duality in field theory. It will be an exciting journey.

2

Relativistic strings

This chapter is devoted to an introduction to bosonic strings and their quantisation. There is no attempt made at performing a rigourous or exhaustive derivation of some of the various formulae we will encounter, since that would take us well away from the main goal. That goal is to understand some of how string theory incorporates some of the familiar spacetime physics that we know from low energy field theory, and then rapidly proceed to the point where many of the remarkable properties which make strings so different from field theory are manifest. That will be a good foundation for appreciating just what D-branes really are. The careful reader who needs to know more of the details behind some of what we will introduce is invited to consult texts devoted to the study of string theory.

2.1 Motion of classical point particles

Let us start by reminding ourselves about a description of a point particle. We already touched on it in section 1.1, but we want to take it a bit further now, in preparation for doing the same thing for the analogous formulation for strings. The particle moves in the 'target spacetime' (with coordinates $(t \equiv X^0, X^1, \ldots, X^{D-1})$) sweeping out a 'world-line' (see figure 1.1, page 2) parametrised by τ. We want to write an action principle which yields equations of motion for the allowed paths, $X^\mu(\tau)$.

2.1.1 Two actions

The most obvious action is the total path length swept out in spacetime. The infinitesimal path length traversed is:

$$d\ell = (-ds^2)^{1/2} = (-dX^\mu dX^\nu \eta_{\mu\nu})^{1/2} = (-dX^\mu dX_\mu)^{1/2}, \qquad (2.1)$$

and we have assumed that the particle is massive and hence that $ds^2 < 0$. The massless case will be discussed below. So the action is

$$S_{\text{o}} = -m \int d\ell = -m \int d\tau (-\dot{X}^\mu \dot{X}_\mu)^{1/2}, \tag{2.2}$$

where a dot denotes differentiation with respect to τ. Let us vary the action:

$$\delta S_{\text{o}} = m \int d\tau (-\dot{X}^\mu \dot{X}_\mu)^{-1/2} \dot{X}^\nu \delta \dot{X}_\nu = m \int d\tau u^\nu \delta \dot{X}_\nu$$

$$= -m \int d\tau \dot{u}^\nu \delta X_\nu, \tag{2.3}$$

where the last step used integration by parts, and

$$u^\nu \equiv (-\dot{X}^\mu \dot{X}_\mu)^{-1/2} \dot{X}^\nu. \tag{2.4}$$

So for δX arbitrary, we get $\dot{u}^\nu = 0$, which is Newton's Law of motion:

$$\frac{d^2 X^\mu}{d\tau^2} = 0, \tag{2.5}$$

where we have used $d\ell/d\tau = (-\dot{X}^\mu \dot{X}_\mu)^{1/2}$. There is another action from which we can derive the same physics. Consider the action

$$S = \frac{1}{2} \int d\tau \left(\eta^{-1} \dot{X}^\mu \dot{X}_\mu - \eta m^2 \right), \tag{2.6}$$

for some independent function $\eta(\tau)$ defined on the world-line.

N.B. In preparation for the coming treatment of strings, think of the function η as related to the particle's 'world-line metric', $\gamma_{\tau\tau}$, as $\eta(\tau) = [-\gamma_{\tau\tau}(\tau)]^{1/2}$. The function $\gamma(\tau)$ ensures world-line reparametrisation invariance:

$$ds^2 = \gamma_{\tau\tau} d\tau d\tau = \gamma_{\tau'\tau'} d\tau' d\tau'.$$

This is all a bit redundant in $0 + 1$ dimensions, but the structure will make more sense when we consider the $1+1$ dimensions of the string's world-sheet.

If we vary S with respect to η:

$$\delta S = \frac{1}{2} \int d\tau \left[-\eta^{-2} \dot{X}^\mu \dot{X}_\mu - m^2 \right] \delta\eta. \tag{2.7}$$

So for $\delta\eta$ arbitrary, we get an equation of motion

$$\eta^2 m^2 + \dot{X}^\mu \dot{X}_\mu = 0, \tag{2.8}$$

which we can solve with $\eta = m^{-1}(-\dot{X}^\mu \dot{X}_\mu)^{1/2}$. Upon substituting this into our expression (2.6) defining S, we get:

$$S = -\frac{1}{2} \int d\tau \left\{ m(-\dot{X}^\mu \dot{X}_\mu)^{1/2} + (-\dot{X}^\mu \dot{X}_\mu)^{1/2} m^{-1} m^2 \right\} = S_\mathrm{o}, \tag{2.9}$$

showing that the two actions are equivalent.

Notice, however, that the action S allows for a treatment of the massless, $m = 0$, case, in contrast to S_o. Another attractive feature of S is that it does not use the awkward square root that S_o does in order to compute the path length. The use of the 'auxiliary' parameter η allows us to get away from that.

2.1.2 Symmetries

There are two notable symmetries of the action.

- Spacetime Lorentz/Poincaré:

$$X^\mu \to X'^\mu = \Lambda^\mu{}_\nu X^\nu + A^\mu,$$

where Λ is an $SO(1,3)$ Lorentz matrix and A^μ is an arbitrary constant four-vector. This is a trivial global symmetry of S (and also S_o), following from the fact that we wrote them in covariant form.

- world-line reparametrisations:

$$\delta X = \zeta(\tau) \frac{dX(\tau)}{d\tau}$$

$$\delta\eta = \frac{d}{d\tau} \left[\zeta(\tau)\eta(\tau) \right],$$

for some parameter $\zeta(\tau)$. This is a non-trivial local or 'gauge' symmetry of S. This large extra symmetry on the world-line (and its analogue when we come to study strings) is very useful. We can, for example, use it to pick a nice gauge where we set $\eta = m^{-1}$. This gives a nice simple action, resulting in a simple expression for the conjugate momentum to X^μ:

$$\Pi^\mu = \frac{\partial \mathcal{L}}{\partial \dot{X}_\mu} = m\dot{X}^\mu. \tag{2.10}$$

We will use this much later.

2.2 Classical bosonic strings

Turning to strings, we parametrise the 'world-sheet' which the string sweeps out with coordinates $(\sigma^1, \sigma^2) = (\tau, \sigma)$. The latter is a spatial coordinate, and for now, we take the string to be an open one, with $0 \leq \sigma \leq \pi$ running from one end to the other. The string's evolution in spacetime is described by the functions $X^\mu(\tau, \sigma)$, $\mu = 0, \ldots, D-1$, giving the shape of the string's world-sheet in target spacetime (see figure 1.4, p. 13).

2.2.1 Two actions

As we already discussed in section 1.3, using the induced metric on the world-sheet which we recall here:

$$h_{ab} = \partial_a X^\mu \partial_b X^\nu \eta_{\mu\nu}, \tag{2.11}$$

we can measure distances on the world-sheet as an object embedded in spacetime, and hence define an action analogous to the one for the particle: the total area swept out by the world-sheet (equation (1.25)), which we repeat here:

$$S_{\rm o} = -T \int dA = -T \int d\tau d\sigma \, (-\det h_{ab})^{1/2} \equiv \int d\tau d\sigma \, \mathcal{L}(\dot{X}, X'; \sigma, \tau). \tag{2.12}$$

$$S_{\rm o} = -T \int d\tau d\sigma \left[\left(\frac{\partial X^\mu}{\partial \sigma} \frac{\partial X^\mu}{\partial \tau} \right)^2 - \left(\frac{\partial X^\mu}{\partial \sigma} \right)^2 \left(\frac{\partial X_\mu}{\partial \tau} \right)^2 \right]^{1/2}$$

$$= -T \int d\tau d\sigma \left[(X' \cdot \dot{X})^2 - X'^2 \dot{X}^2 \right]^{1/2}, \tag{2.13}$$

where X' means $\partial X/\partial \sigma$ and a dot means differentiation with respect to τ. This is the Nambu–Goto action.

Varying the action, we have generally:

$$\delta S_{\rm o} = \int d\tau d\sigma \left\{ \frac{\partial \mathcal{L}}{\partial \dot{X}^\mu} \delta \dot{X}^\mu + \frac{\partial \mathcal{L}}{\partial X'^\mu} \delta X'^\mu \right\}$$

$$= \int d\tau d\sigma \left\{ -\frac{\partial}{\partial \tau} \frac{\partial \mathcal{L}}{\partial \dot{X}^\mu} - \frac{\partial}{\partial \sigma} \frac{\partial \mathcal{L}}{\partial X'^\mu} \right\} \delta X^\mu$$

$$+ \int d\tau \left\{ \frac{\partial \mathcal{L}}{\partial X'^\mu} \delta X'^\mu \right\} \Big|_{\sigma=0}^{\sigma=\pi}. \tag{2.14}$$

Requiring this to be zero, we get:

$$\frac{\partial}{\partial \tau} \frac{\partial \mathcal{L}}{\partial \dot{X}^\mu} + \frac{\partial}{\partial \sigma} \frac{\partial \mathcal{L}}{\partial X'^\mu} = 0 \quad \text{and} \quad \frac{\partial \mathcal{L}}{\partial X'^\mu} = 0 \quad \text{at} \quad \sigma = 0, \pi, \tag{2.15}$$

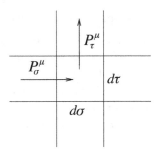

Fig. 2.1. The infinitesimal momenta on the world sheet.

which are statements about the conjugate momenta:

$$\frac{\partial}{\partial \tau} P_\tau^\mu + \frac{\partial}{\partial \sigma} P_\sigma^\mu = 0 \quad \text{and} \quad P_\sigma^\mu = 0 \quad \text{at} \quad \sigma = 0, \pi. \qquad (2.16)$$

Here, P_σ^μ is the momentum running along the string (i.e. in the σ direction) while P_τ^μ is the momentum running transverse to it. The total spacetime momentum is given by integrating up the infinitesimal (see figure 2.1):

$$dP^\mu = P_\tau^\mu d\sigma + P_\sigma^\mu d\tau. \qquad (2.17)$$

Actually, we can choose any slice of the world-sheet in order to compute this momentum. A most convenient one is a slice $\tau = $ constant, revealing the string in its original paramaterisation: $P^\mu = \int P_\tau^\mu d\sigma$, but any other slice will do.

Similarly, one can define the angular momentum:

$$M^{\mu\nu} = \int (P_\tau^\mu X^\nu - P_\tau^\nu X^\mu) d\sigma. \qquad (2.18)$$

It is a simple exercise to work out the momenta for our particular Lagrangian:

$$P_\tau^\mu = T \frac{\dot{X}^\mu X'^2 - X'^\mu (\dot{X} \cdot X')}{\sqrt{(\dot{X} \cdot X')^2 - \dot{X}^2 X'^2}}$$

$$P_\sigma^\mu = T \frac{X'^\mu \dot{X}^2 - \dot{X}^\mu (\dot{X} \cdot X')}{\sqrt{(\dot{X} \cdot X')^2 - \dot{X}^2 X'^2}}. \qquad (2.19)$$

It is interesting to compute the square of P_σ^μ from this expression, and one finds that

$$P_\sigma^2 \equiv P_\sigma^\mu P_{\mu\sigma} = -2T^2 \dot{X}^2. \qquad (2.20)$$

This is our first (perhaps) non-intuitive classical result. We noticed that P_σ vanishes at the endpoints, in order to prevent momentum from flowing off the ends of the string. The equation we just derived implies that $\dot{X}^2 = 0$ at the endpoints, which is to say that they move at the speed of light.

Just like we did in the point particle case, we can introduce an equivalent action which does not have the square root form that the current one has. Once again, we do it by introducing a independent metric, $\gamma_{ab}(\sigma, \tau)$, on the world-sheet, and write the 'Polyakov' action:

$$S = -\frac{1}{4\pi\alpha'} \int d\tau d\sigma (-\gamma)^{1/2} \gamma^{ab} \partial_a X^\mu \partial_b X^\nu \eta_{\mu\nu}$$
$$= -\frac{1}{4\pi\alpha'} \int d^2\sigma \, (-\gamma)^{1/2} \gamma^{ab} h_{ab}. \tag{2.21}$$

If we vary γ, we get

$$\delta S = -\frac{1}{4\pi\alpha'} \int d^2\sigma \left\{ -\frac{1}{2}(-\gamma)^{1/2} \delta\gamma \gamma^{ab} h_{ab} + (-\gamma)^{1/2} \delta\gamma^{ab} h_{ab} \right\}. \tag{2.22}$$

Using the fact that $\delta\gamma = \gamma\gamma^{ab}\delta\gamma_{ab} = -\gamma\gamma_{ab}\delta\gamma^{ab}$, (which we already used in higher dimensions, see equation (1.13)) we get

$$\delta S = -\frac{1}{4\pi\alpha'} \int d^2\sigma \, (-\gamma)^{1/2} \delta\gamma^{ab} \left\{ h_{ab} - \frac{1}{2}\gamma_{ab}\gamma^{cd} h_{cd} \right\}. \tag{2.23}$$

Therefore we have

$$h_{ab} - \frac{1}{2}\gamma_{ab}\gamma^{cd} h_{cd} = 0, \tag{2.24}$$

from which we can derive

$$\gamma^{ab} h_{ab} = 2(-h)^{1/2}(-\gamma)^{-1/2}, \tag{2.25}$$

and so substituting into S, we recover (just as in the point-particle case) that it reduces to the Nambu–Goto action, S_o.

2.2.2 Symmetries

Let us again study the symmetries of the action.
- Spacetime Lorentz/Poincaré:

$$X^\mu \to X'^\mu = \Lambda^\mu{}_\nu X^\nu + A^\mu,$$

where Λ is an $SO(1,3)$ Lorentz matrix and A^μ is an arbitrary constant four-vector. Just as before this is a trivial global symmetry of S (and also S_o), following from the fact that we wrote them in covariant form.

- world-sheet reparametrisations:

$$\delta X^\mu = \zeta^a \partial_a X^\mu$$
$$\delta \gamma^{ab} = \zeta^c \partial_c \gamma^{ab} - \partial_c \zeta^a \gamma^{cb} - \partial_c \zeta^b \gamma^{ac}, \tag{2.26}$$

for two parameters $\zeta^a(\tau, \sigma)$. This is a non-trivial local or 'gauge' symmetry of S. This is a large extra symmetry on the world-sheet of which we will make great use.

- Weyl invariance:

$$\gamma_{ab} \rightarrow \gamma'_{ab} = e^{2\omega} \gamma_{ab}, \tag{2.27}$$

specified by a function $\omega(\tau, \sigma)$. This ability to do local rescalings of the metric results from the fact that we did not have to choose an overall scale when we chose γ^{ab} to rewrite S_o in terms of S. This can be seen especially if we rewrite the relation (2.25) as $(-h)^{-1/2} h_{ab} = (-\gamma)^{-1/2} \gamma_{ab}$.

N.B. We note here for future use that there are just as many parameters needed to specify the local symmetries (three) as there are independent components of the world-sheet metric. This is very useful, as we shall see.

2.2.3 String equations of motion

We can get equations of motion for the string by varying our action (2.21) with respect to the X^μ:

$$\delta S = \frac{1}{2\pi\alpha'} \int d^2\sigma \, \partial_a \left\{ (-\gamma)^{1/2} \gamma^{ab} \partial_b X_\mu \right\} \delta X^\mu$$
$$- \frac{1}{2\pi\alpha'} \int d\tau \, (-\gamma)^{1/2} \partial_\sigma X_\mu \delta X^\mu \Big|_{\sigma=0}^{\sigma=\pi}, \tag{2.28}$$

which results in the equations of motion:

$$\partial_a \left((-\gamma)^{1/2} \gamma^{ab} \partial_b X^\mu \right) \equiv (-\gamma)^{1/2} \nabla^2 X^\mu = 0, \tag{2.29}$$

with *either*:
$$\left. \begin{array}{l} X'^\mu(\tau, 0) = 0 \\ X'^\mu(\tau, \pi) = 0 \end{array} \right\} \qquad \begin{array}{l} \text{Open string} \\ \text{(Neumann b.c.s)} \end{array} \tag{2.30}$$

or:

$$\left. \begin{aligned} X'^{\mu}(\tau,0) &= X'^{\mu}(\tau,\pi) \\ X^{\mu}(\tau,0) &= X^{\mu}(\tau,\pi) \\ \gamma_{ab}(\tau,0) &= \gamma_{ab}(\tau,\pi) \end{aligned} \right\} \quad \begin{aligned} &\text{Closed string} \\ &\text{(periodic b.c.s)} \end{aligned} \tag{2.31}$$

We shall study the equation of motion (2.29) and the accompanying boundary conditions a lot later. We are going to look at the standard Neumann boundary conditions mostly, and then consider the case of Dirichlet conditions later, when we uncover D-branes, using T-duality. Notice that we have taken the liberty of introducing closed strings by imposing periodicity (see also insert 2.1 (p. 32)).

2.2.4 Further aspects of the two dimensional perspective

The action (2.21) may be thought of as a two dimensional model of D bosonic fields $X^{\mu}(\tau,\sigma)$. This two dimensional theory has reparameterisation invariance, as it is constructed using the metric $\gamma_{ab}(\tau,\sigma)$ in a covariant way. It is natural to ask whether there are other terms which we might want to add to the theory which have similar properties.

With some experience from General Relativity two other terms spring effortlessly to mind. One is the Einstein–Hilbert action (supplemented with a boundary term):

$$\chi = \frac{1}{4\pi} \int_{\mathcal{M}} d^2\sigma \, (-\gamma)^{1/2} R + \frac{1}{2\pi} \int_{\partial\mathcal{M}} ds K, \tag{2.32}$$

where R is the two dimensional Ricci scalar on the world-sheet $\mathcal{M}$ and K is the trace of the extrinsic curvature tensor on the boundary $\partial\mathcal{M}$. This latter quantity may be less familiar to some, and we will use it a lot in diverse dimensions much later in this book. (There is a discussion of it in insert 10.2 (p. 229), and we will not worry about it in detail here lest we get sidetracked.)

The other term is:

$$\Theta = \frac{1}{4\pi\alpha'} \int_{\mathcal{M}} d^2\sigma \, (-\gamma)^{1/2}, \tag{2.33}$$

which is the cosmological term. What is the role of these terms here? Well, under a Weyl transformation (2.27), it can be seen that $(-\gamma)^{1/2} \to e^{2\omega}(-\gamma)^{1/2}$ and $R \to e^{-2\omega}(R - 2\nabla^2\omega)$, and so χ is invariant, (because R changes by a total derivative which is cancelled by the variation of K) but Θ is not.

So we will include χ, but not Θ in what follows. Let us anticipate something that we will do later, which is to work with Euclidean signature to help make sense of the topological statements to follow: γ_{ab} with signature $(-+)$ has been replaced by g_{ab} with signature $(++)$. Now, since as

Insert 2.1. T is for tension

As a first non-trivial example (and to learn that T, a mass per unit length, really is the string's tension) let us consider a closed string lying in the (X^1, X^2) plane.

$$X^0 = 2R\tau;$$
$$X^1 = R\sin 2\sigma$$
$$X^2 = R\cos 2\sigma.$$

We have made it by arranging that the $\sigma = 0$, π ends meet, that momentum flows across that join. An examination of the equations of motion shows that this configuration is not a solution, and there are terms which do not vanish corresponding to the fact that the string does not want to stay at rest: since the string has tension, it is likely to want to shrink its length away if put into this shape. So let us think of this as a snapshot of such a situation, ignoring the non-vanishing terms which involve time derivative. It is worth taking the time to use this to show that one gets

$$P^\mu_\tau = T\,(2R, 0, 0), \quad P^\mu_\sigma = T\,(0, -2R\cos 2\sigma, 2R\sin 2\sigma),$$

which is interesting, as a sketch shows.

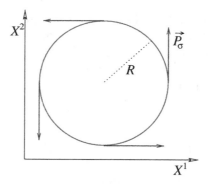

There is momentum flowing around the string (which is lying in a circle of radius R). The total momentum is

$$P^\mu = \int_0^\pi d\sigma\, P^\mu_\tau.$$

The only non-zero component is the mass-energy: $M = 2\pi RT = $ length$\times T$.

Insert 2.2. A rotating open string

As a second non-trivial example consider the following open string rotating at a constant angular velocity in the (X^1, X^2) plane. Such a configuration is:

$$X^0 = \tau; \quad X^1 = A\left(\sigma - \frac{\pi}{2}\right)\cos\omega\tau, \quad X^2 = A\left(\sigma - \frac{\pi}{2}\right)\sin\omega\tau,$$

where it should be checked that the equations of motion fix $A = \frac{2}{\pi\omega}$. This is what it looks like (the spinning string is shown in frozen snapshots).

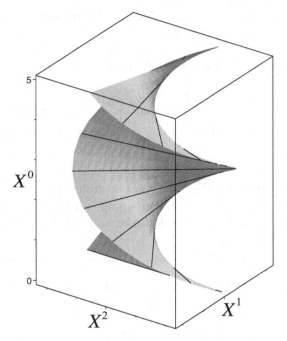

It is again a worthwhile exercise to compute P^μ, and also $M^{\mu\nu}$. With $J \equiv M^{12}$ and $M \equiv P^0$, some algebra shows that

$$\frac{|J|}{M^2} = \frac{1}{2\pi T} = \alpha'.$$

This parameter, α', is the slope of the celebrated 'Regge' trajectories: the straight line plots of J vs. M^2 seen in nuclear physics in the 1960s. There remains the determination of the intercept of this straight line graph with the J-axis. It turns out to be one for the bosonic string as we shall see.

we said earlier, the full string action resembles two dimensional gravity coupled to D bosonic 'matter' fields X^μ, and the equations of motion are, of course,

$$R_{ab} - \frac{1}{2}\gamma_{ab}R = T_{ab}. \tag{2.34}$$

The left hand side vanishes identically in two dimensions, and so there are no dynamics associated to (2.32). The quantity χ depends only on the topology of the world-sheet (it is the Euler number) and so will only matter when comparing world sheets of different topology. This will arise when we compare results from different orders of string perturbation theory and when we consider interactions.

We can see this in the following. Let us add our new term to the action, and consider the string action to be:

$$S = \frac{1}{4\pi\alpha'} \int_{\mathcal{M}} d^2\sigma\, g^{1/2} g^{ab} \partial_a X^\mu \partial_b X_\mu$$
$$+ \lambda \left\{ \frac{1}{4\pi} \int_{\mathcal{M}} d^2\sigma\, g^{1/2} R + \frac{1}{2\pi} \int_{\partial\mathcal{M}} ds K \right\}, \tag{2.35}$$

where λ is – for now – and arbitrary parameter that we have not fixed to any particular value.

N.B. It will turn out that λ is not a free parameter. In the full string theory, it has dynamical meaning, and will be equivalent to the expectation value of one of the massless fields – the 'dilaton' – described by the string.

So what will λ do? Recall that it couples to Euler number, so in the full path integral defining the string theory:

$$\mathcal{Z} = \int \mathcal{D}X \mathcal{D}g\, e^{-S}, \tag{2.36}$$

resulting amplitudes will be weighted by a factor $e^{-\lambda\chi}$, where $\chi = 2 - 2h - b - c$. Here, h, b, c are the numbers of handles, boundaries and crosscaps, respectively, on the world sheet. Consider figure 2.2. An emission and reabsorption of an open string results in a change $\delta\chi = -1$, while for a closed string it is $\delta\chi = -2$. Therefore, relative to the tree level open string diagram (disc topology), the amplitudes are weighted by e^λ and $e^{2\lambda}$, respectively. The quantity $g_s \equiv e^\lambda$ therefore will be called the closed string coupling. Note that it is the square of the open string coupling, which justifies the labelling we gave of the two three-string diagrams in figure 1.3.

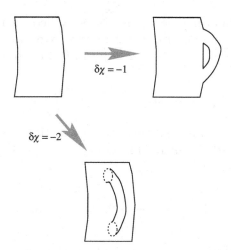

Fig. 2.2. World-sheet topology change due to emission and reabsorption of open and closed strings.

2.2.5 The stress tensor

Let us also note that we can define a two dimensional energy-momentum tensor:

$$T^{ab}(\tau, \sigma) \equiv -\frac{2\pi}{\sqrt{-\gamma}} \frac{\delta S}{\delta \gamma_{ab}} = -\frac{1}{\alpha'} \left\{ \partial^a X_\mu \partial^b X^\mu - \frac{1}{2} \gamma^{ab} \gamma_{cd} \partial^c X_\mu \partial^d X^\mu \right\}. \tag{2.37}$$

Notice that

$$T_a^a \equiv \gamma_{ab} T^{ab} = 0. \tag{2.38}$$

This is a consequence of Weyl symmetry. Reparametrisation invariance, $\delta_\gamma S' = 0$, translates here into (see discussion after equation (2.34))

$$T^{ab} = 0. \tag{2.39}$$

These are the classical properties of the theory we have uncovered so far. Later on, we shall attempt to ensure that they are true in the quantum theory also, with interesting results.

2.2.6 Gauge fixing

Now recall that we have three local or 'gauge' symmetries of the action:

$$\text{2D reparametrisations}: \ \sigma, \tau \to \tilde\sigma(\sigma, \tau), \tilde\tau(\sigma, \tau),$$
$$\text{Weyl}: \ \gamma_{ab} \to \exp(2\omega(\sigma, \tau))\gamma_{ab}. \tag{2.40}$$

The two dimensional metric γ_{ab} is also specified by three independent functions, as it is a symmetric 2×2 matrix. We may therefore use the gauge symmetries (see equations (2.26) and (2.27)) to choose γ_{ab} to be a particular form:

$$\gamma_{ab} = \eta_{ab}e^{\phi} = \begin{pmatrix} -1 & 0 \\ 0 & 1 \end{pmatrix} e^{\phi}, \tag{2.41}$$

i.e. the metric of two dimensional Minkowski, times a positive function known as a *conformal factor*. In this 'conformal' gauge, our X^{μ} equations of motion (2.29) become:

$$\left(\frac{\partial^2}{\partial\sigma^2} - \frac{\partial^2}{\partial\tau^2} \right) X^{\mu}(\tau, \sigma) = 0, \tag{2.42}$$

the two dimensional wave equation. (In fact, the reader should check that the conformal factor cancels out entirely of the action in equation (2.21).) As the wave equation is $\partial_{\sigma+}\partial_{\sigma-} X^{\mu} = 0$, we see that the full solution to the equation of motion can be written in the form:

$$X^{\mu}(\sigma, \tau) = X_L^{\mu}(\sigma^+) + X_R^{\mu}(\sigma^-), \tag{2.43}$$

where $\sigma^{\pm} \equiv \tau \pm \sigma$.

N.B. Write $\sigma^{\pm} = \tau \pm \sigma$. This gives metric $ds^2 = -d\tau^2 + d\sigma^2 \rightarrow -d\sigma^+ d\sigma^-$. So we have $\eta_{-+} = \eta_{+-} = -1/2$, $\eta^{-+} = \eta^{+-} = -2$ and $\eta_{++} = \eta_{--} = \eta^{++} = \eta^{--} = 0$. Also, $\partial_{\tau} = \partial_+ + \partial_-$ and $\partial_{\sigma} = \partial_+ - \partial_-$.

Our constraints on the stress tensor become:

$$T_{\tau\sigma} = T_{\sigma\tau} \equiv \frac{1}{\alpha'}\dot{X}^{\mu}X'_{\mu} = 0$$

$$T_{\sigma\sigma} = T_{\tau\tau} = \frac{1}{2\alpha'}\left(\dot{X}^{\mu}\dot{X}_{\mu} + X'^{\mu}X'_{\mu} \right) = 0, \tag{2.44}$$

or

$$T_{++} = \frac{1}{2}(T_{\tau\tau} + T_{\tau\sigma}) = \frac{1}{\alpha'}\partial_+X^{\mu}\partial_+X_{\mu} \equiv \frac{1}{\alpha'}\dot{X}_L^2 = 0$$

$$T_{--} = \frac{1}{2}(T_{\tau\tau} - T_{\tau\sigma}) = \frac{1}{\alpha'}\partial_-X^{\mu}\partial_-X_{\mu} \equiv \frac{1}{\alpha'}\dot{X}_R^2 = 0, \tag{2.45}$$

and T_{-+} and T_{+-} are identically zero.

2.2.7 The mode decomposition

Our equations of motion (2.43), with our boundary conditions (2.30) and (2.31) have the simple solutions:

$$X^\mu(\tau,\sigma) = x^\mu + 2\alpha' p^\mu \tau + i(2\alpha')^{1/2} \sum_{n\neq 0} \frac{1}{n}\alpha_n^\mu e^{-in\tau}\cos n\sigma, \qquad (2.46)$$

for the open string and

$$X^\mu(\tau,\sigma) = X_R^\mu(\sigma^-) + X_L^\mu(\sigma^+)$$

$$X_R^\mu(\sigma^-) = \frac{1}{2}x^\mu + \alpha' p^\mu \sigma^- + i\left(\frac{\alpha'}{2}\right)^{1/2} \sum_{n\neq 0}\frac{1}{n}\alpha_n^\mu e^{-2in\sigma^-}$$

$$X_L^\mu(\sigma^+) = \frac{1}{2}x^\mu + \alpha' p^\mu \sigma^+ + i\left(\frac{\alpha'}{2}\right)^{1/2} \sum_{n\neq 0}\frac{1}{n}\tilde{\alpha}_n^\mu e^{-2in\sigma^+}, \qquad (2.47)$$

for the closed string, where, to ensure a real solution we impose $\alpha_{-n}^\mu = (\alpha_n^\mu)^*$ and $\tilde{\alpha}_{-n}^\mu = (\tilde{\alpha}_n^\mu)^*$. Note that x^μ and p^μ are the centre of mass position and momentum, respectively. In each case, we can identify p^μ with the zero mode of the expansion:

$$\text{open string:} \qquad \alpha_0^\mu = (2\alpha')^{1/2}p^\mu;$$

$$\text{closed string:} \qquad \alpha_0^\mu = \left(\frac{\alpha'}{2}\right)^{1/2}p^\mu. \qquad (2.48)$$

N.B. Notice that the mode expansion for the closed string (2.47) is simply that of a pair of independent left and right moving travelling waves going around the string in opposite directions. The open string expansion (2.46) on the other hand, has a standing wave for its solution, representing the left and right moving sector reflected into one another by the Neumann boundary condition (2.30).

2.2.8 Conformal invariance as a residual symmetry

Actually, we have not gauged away all of the local symmetry by choosing the gauge (2.41). We can do a left–right decoupled change of variables:

$$\sigma^+ \to f(\sigma^+) = \sigma'^+; \quad \sigma^- \to g(\sigma^-) = \sigma'^-. \qquad (2.49)$$

Then, as

$$\gamma'_{ab} = \frac{\partial \sigma^c}{\partial \sigma'^a}\frac{\partial \sigma^d}{\partial \sigma'^b}\gamma_{cd}, \qquad (2.50)$$

we have

$$\gamma'_{+-} = \left(\frac{\partial f(\sigma^+)}{\partial \sigma^+} \frac{\partial g(\sigma^-)}{\partial \sigma^-} \right)^{-1} \gamma_{+-}. \tag{2.51}$$

However, we can undo this with a Weyl transformation of the form

$$\gamma'_{+-} = \exp(2\omega_{\mathrm{L}}(\sigma^+) + 2\omega_{\mathrm{R}}(\sigma^-))\gamma_{+-}, \tag{2.52}$$

if $\exp(-2\omega_{\mathrm{L}}(\sigma^+)) = \partial_+ f(\sigma^+)$ and $\exp(-2\omega_{\mathrm{R}}(\sigma^-)) = \partial_- g(\sigma^-)$. So we still have a residual 'conformal' symmetry. As f and g are independent arbitrary functions on the left and right, we have an infinite number of conserved quantities on the left and right. This is because the conservation equation $\nabla_a T^{ab} = 0$, together with the result $T_{+-} = T_{-+} = 0$, turns into:

$$\partial_- T_{++} = 0 \quad \text{and} \quad \partial_+ T_{--} = 0, \tag{2.53}$$

but since $\partial_- f = 0 = \partial_+ g$, we have

$$\partial_-(f(\sigma^+)T_{++}) = 0 \quad \text{and} \quad \partial_+(g(\sigma^-)T_{--}) = 0, \tag{2.54}$$

resulting in an infinite number of conserved quantities. The fact that we have this infinite dimensional conformal symmetry is the basis of some of the most powerful tools in the subject, for computing in perturbative string theory. We will return to it not too far ahead.

2.2.9 Some Hamiltonian dynamics

Our Lagrangian density is

$$\mathcal{L} = -\frac{1}{4\pi\alpha'} \left(\partial_\sigma X^\mu \partial_\sigma X_\mu - \partial_\tau X^\mu \partial_\tau X_\mu \right), \tag{2.55}$$

from which we can derive that the conjugate momentum to X^μ is

$$\Pi^\mu = \frac{\delta\mathcal{L}}{\delta(\partial_\tau X^\mu)} = \frac{1}{2\pi\alpha'}\dot{X}^\mu. \tag{2.56}$$

So we have the equal time Poisson brackets:

$$[X^\mu(\sigma), \Pi^\nu(\sigma')]_{\mathrm{P.B.}} = \eta^{\mu\nu}\delta(\sigma - \sigma'), \tag{2.57}$$

$$[\Pi^\mu(\sigma), \Pi^\nu(\sigma')]_{\mathrm{P.B.}} = 0, \tag{2.58}$$

with the following results on the oscillator modes:

$$[\alpha_m^\mu, \alpha_n^\nu]_{\mathrm{P.B.}} = [\tilde{\alpha}_m^\mu, \tilde{\alpha}_n^\nu]_{\mathrm{P.B.}} = im\delta_{m+n}\eta^{\mu\nu}$$

$$[p^\mu, x^\nu]_{\mathrm{P.B.}} = \eta^{\mu\nu}; \quad [\alpha_m^\mu, \tilde{\alpha}_n^\nu]_{\mathrm{P.B.}} = 0. \tag{2.59}$$

We can form the Hamiltonian density

$$\mathcal{H} = \dot{X}^\mu \Pi_\mu - \mathcal{L} = \frac{1}{4\pi\alpha'} \left(\partial_\sigma X^\mu \partial_\sigma X_\mu + \partial_\tau X^\mu \partial_\tau X_\mu \right), \qquad (2.60)$$

from which we can construct the Hamiltonian H by integrating along the length of the string. This results in:

$$H = \int_0^\pi d\sigma \, \mathcal{H}(\sigma) = \frac{1}{2} \sum_{-\infty}^\infty \alpha_{-n} \cdot \alpha_n \qquad \text{(open)} \qquad (2.61)$$

$$H = \int_0^{2\pi} d\sigma \, \mathcal{H}(\sigma) = \frac{1}{2} \sum_{-\infty}^\infty \left(\alpha_{-n} \cdot \alpha_n + \tilde{\alpha}_{-n} \cdot \tilde{\alpha}_n \right) \qquad \text{(closed).}$$

(We have used the notation $\alpha_n \cdot \alpha_n \equiv \alpha_n^\mu \alpha_{n\mu}$.) The constraints $T_{++} = 0 = T_{--}$ on our energy-momentum tensor can be expressed usefully in this language. We impose them mode by mode in a Fourier expansion, defining:

$$L_m = \frac{T}{2} \int_0^\pi e^{-2im\sigma} T_{--} d\sigma = \frac{1}{2} \sum_{-\infty}^\infty \alpha_{m-n} \cdot \alpha_n, \qquad (2.62)$$

and similarly for $\bar{L}_m$, using T_{++}. Using the Poisson brackets (2.59), these can be shown to satisfy the 'Virasoro' algebra:

$$[L_m, L_n]_{\text{P.B.}} = i(m-n)L_{m+n}; \quad [\bar{L}_m, \bar{L}_n]_{\text{P.B.}} = i(m-n)\bar{L}_{m+n};$$
$$[\bar{L}_m, L_n]_{\text{P.B.}} = 0. \qquad (2.63)$$

Notice that there is a nice relation between the zero modes of our expansion and the Hamiltonian:

$$H = L_0 \quad \text{(open)}; \quad H = L_0 + \bar{L}_0 \quad \text{(closed).} \qquad (2.64)$$

So to impose our constraints, we can do it mode by mode and ask that $L_m = 0$ and $\bar{L}_m = 0$, for all m. Looking at the zeroth constraint results in something interesting. Note that

$$L_0 = \frac{1}{2}\alpha_0^2 + 2 \times \frac{1}{2} \sum_{n=1}^\infty \alpha_{-n} \cdot \alpha_n$$

$$= \alpha' p^\mu p_\mu + \sum_{n=1}^\infty \alpha_{-n} \cdot \alpha_n$$

$$= -\alpha' M^2 + \sum_{n=1}^\infty \alpha_{-n} \cdot \alpha_n. \qquad (2.65)$$

Requiring L_0 to be zero – diffeomorphism invariance – results in a (space-time) mass relation:

$$M^2 = \frac{1}{\alpha'} \sum_{n=1}^{\infty} \alpha_{-n} \cdot \alpha_n \qquad \text{(open)}, \qquad (2.66)$$

where we have used the zero mode relation (2.48) for the open string. A similar exercise produces the mass relation for the closed string:

$$M^2 = \frac{2}{\alpha'} \sum_{n=1}^{\infty} (\alpha_{-n} \cdot \alpha_n + \tilde{\alpha}_{-n} \cdot \tilde{\alpha}_n) \qquad \text{(closed)}. \qquad (2.67)$$

These formulae (2.66) and (2.67) give us the result for the mass of a state in terms of how many oscillators are excited on the string. The masses are set by the string tension $T = (2\pi\alpha')^{-1}$, as they should be. Let us not dwell for too long on these formulae however, as they are significantly modified when we quantise the theory, since we have to understand the infinite constant which we ignored.

2.3 Quantised bosonic strings

For our purposes, the simplest route to quantisation will be to promote everything we met previously to operator statements, replacing Poisson Brackets by commutators in the usual fashion: $[\ ,\ ]_{\text{P.B.}} \to -i[\ ,\ ]$. This gives:

$$[X^\mu(\tau,\sigma), \Pi^\nu(\tau,\sigma')] = i\eta^{\mu\nu}\delta(\sigma-\sigma'); \quad [\Pi^\mu(\tau,\sigma), \Pi^\nu(\tau,\sigma')] = 0$$
$$[\alpha_m^\mu, \alpha_n^\nu] = [\tilde{\alpha}_m^\mu, \tilde{\alpha}_n^\nu] = m\delta_{m+n}\eta^{\mu\nu}$$
$$[x^\nu, p^\mu] = i\eta^{\mu\nu}; \quad [\alpha_m^\mu, \tilde{\alpha}_n^\nu] = 0. \qquad (2.68)$$

> **N.B.** One of the first things that we ought to notice here is that $\sqrt{m}\alpha_{\pm m}^\mu$ are like creation and annihilation operators for the harmonic oscillator. There are actually D independent families of them – one for each spacetime dimension – labelled by μ.

In the usual fashion, we will define our Fock space such that $|0;k\rangle$ is an eigenstate of p^μ with centre of mass momentum k^μ. This state is annihilated by α_m^ν.

What about our operators, the L_m? Well, with the usual 'normal ordering' prescription that all annihilators are to the right, the L_m are all fine when promoted to operators, except the Hamiltonian, L_0. It needs more careful definition, since α_n^μ and α_{-n}^μ *do not commute.* Indeed, as an operator, we have that

$$L_0 = \frac{1}{2}\alpha_0^2 + \sum_{n=1}^\infty \alpha_{-n} \cdot \alpha_n + \text{constant}, \tag{2.69}$$

where the apparently infinite constant is composed of the infinite sum $(1/2) \sum_{n=1}^\infty n$ for each of the D families of oscillators. As is of course to be anticipated, this infinite constant can be regulated to give a finite answer, corresponding to the total zero point energy of all of the harmonic oscillators in the system.

2.3.1 The constraints and physical states

For now, let us not worry about the value of the constant, and simply impose our constraints on a state $|\phi\rangle$ as*:

$$
\begin{aligned}
(L_0 - a)|\phi\rangle &= 0; & L_m|\phi\rangle &= 0 \quad \text{for } m > 0, \\
(\bar{L}_0 - a)|\phi\rangle &= 0; & \bar{L}_m|\phi\rangle &= 0 \quad \text{for } m > 0,
\end{aligned}
\tag{2.70}
$$

where our regulated constant is set by a, which is to be computed. There is a reason why we have not also imposed this constraint for the L_{-m}s. This is because the Virasoro algebra (2.63) in the quantum case is:

$$[L_m, L_n] = (m-n)L_{m+n} + \frac{D}{12}(m^3 - m)\delta_{m+n}; \quad [\bar{L}_m, L_n] = 0;$$

$$[\bar{L}_m, \bar{L}_n] = (m-n)\bar{L}_{m+n} + \frac{D}{12}(m^3 - m)\delta_{m+n}. \tag{2.71}$$

There is a central term in the algebra, which produces a non-zero constant when $m = n$. Therefore, imposing both L_m and L_{-m} would produce an inconsistency. Note now that the first of our constraints (2.70) produces a modification to the mass formulae:

$$M^2 = \frac{1}{\alpha'}\left(\sum_{n=1}^\infty \alpha_{-n} \cdot \alpha_n - a\right) \qquad \text{(open)} \tag{2.72}$$

$$M^2 = \frac{2}{\alpha'}\left(\sum_{n=1}^\infty (\alpha_{-n} \cdot \alpha_n + \tilde{\alpha}_{-n} \cdot \tilde{\alpha}_n) - 2a\right) \qquad \text{(closed)}.$$

---

* This assumes that the constant a on each side are equal. At this stage, we have no other choice. We have isomorphic copies of the same string modes on the left and the right, for which the values of a are by definition the same. When we have more than one consistent conformal field theory to choose from, then we have the freedom to consider having non-isomorphic sectors on the left and right. This is how the heterotic string is made, for example, as we shall see later.

Notice that we can denote the (weighted) number of oscillators excited as $N = \sum \alpha_{-n} \cdot \alpha_n \ (= \sum n N_n)$ on the left and $\bar{N} = \sum \tilde{\alpha}_{-n} \cdot \tilde{\alpha}_n \ (= \sum n \bar{N}_n)$ on the right. N_n and $\bar{N}_n$ are the true count, on the left and right, of the number of copies of the oscillator labelled by n is present.

There is an extra condition in the closed string case. While $L_0 + \bar{L}_0$ generates time translations on the world sheet (being the Hamiltonian), the combination $L_0 - \bar{L}_0$ generates translations in σ. As there is no physical significance to where on the string we are, the physics should be invariant under translations in σ, and we should impose this as an operator condition on our physical states:

$$(L_0 - \bar{L}_0)|\phi\rangle = 0, \tag{2.73}$$

which results in the 'level-matching' condition $N = \bar{N}$, equating the number of oscillators excited on the left and the right. This is indeed the difference between the two equations in (2.70).

In summary then, we have two copies of the open string on the left and the right, in order to construct the closed string. The only extra subtlety is that we should use the correct zero mode relation (2.48) and match the number of oscillators on each side according to the level matching condition (2.73).

2.3.2 The intercept and critical dimensions

Let us consider the spectrum of states level by level, and uncover some of the features, focusing on the open string sector. Our first and simplest state is at level 0, i.e. no oscillators excited at all. There is just some centre of mass momentum that it can have, which we shall denote as k^μ. Let us write this state as $|0; k\rangle$. The first of our constraints (2.70) leads to an expression for the mass:

$$(L_0 - a)|0; k\rangle = 0 \qquad \Rightarrow \alpha' k^2 = a, \qquad \text{so} \qquad M^2 = -\frac{a}{\alpha'}. \tag{2.74}$$

This state is a tachyonic state, having negative mass-squared (assuming $a > 0$.

The next simplest state is that with momentum k^μ, and one oscillator excited. We are also free to specify a polarisation vector ζ^μ. We denote this state as $|\zeta, k\rangle \equiv (\zeta \cdot \alpha_{-1})|0; k\rangle$; it starts out the discussion with D independent states. The first thing to observe is the norm of this state:

$$\begin{aligned}
\langle \zeta; k||\zeta; k'\rangle &= \langle 0; k|\zeta^* \cdot \alpha_1 \zeta \cdot \alpha_{-1}|0; k'\rangle \\
&= \zeta_\mu^* \zeta_\nu \langle 0; k|\alpha_1^\mu \alpha_{-1}^\nu|0; k'\rangle \\
&= \zeta \cdot \zeta \langle 0; k|0; k'\rangle = \zeta \cdot \zeta (2\pi)^D \delta^D(k - k'), \tag{2.75}
\end{aligned}$$

where we have used the commutator (2.68) for the oscillators. From this we see that the timelike ζs will produce a state with *negative norm*. Such states cannot be made sense of in a unitary theory, and are often called[†] 'ghosts'.

Let us study the first constraint:

$$(L_0 - a)|\zeta; k\rangle = 0 \quad \Rightarrow \quad \alpha' k^2 + 1 = a, \qquad M^2 = \frac{1-a}{\alpha'}. \quad (2.76)$$

The next constraint gives:

$$(L_1)|\zeta; k\rangle = \sqrt{\frac{\alpha'}{2}} k \cdot \alpha_1 \zeta \cdot \alpha_{-1}|0; k\rangle = 0 \quad \Rightarrow, \qquad k \cdot \zeta = 0. \quad (2.77)$$

Actually, at level one, we can also make a special state of interest: $|\psi\rangle \equiv L_{-1}|0; k\rangle$. This state has the special property that it is orthogonal to any physical state, since $\langle\phi|\psi\rangle = \langle\psi|\phi\rangle^* = \langle 0; k|L_1|\phi\rangle = 0$. It also has $L_1|\psi\rangle = 2L_0|0; k\rangle = \alpha' k^2|0; k\rangle$. This state is called a 'spurious' state.

So we note that there are three interesting cases for the level one physical state we have been considering.

1. $a < 1 \Rightarrow M^2 > 0$:

   - momentum k is timelike,
   - we can choose a frame where it is $(k, 0, 0, \ldots)$,
   - spurious state is not physical, since $k^2 \neq 0$,
   - $k \cdot \zeta = 0$ removes the timelike polarisation; $D - 1$ states left.

2. $a > 1 \Rightarrow M^2 < 0$:

   - momentum k is spacelike,
   - we can choose a frame where it is $(0, k_1, k_2, \ldots)$,
   - spurious state is not physical, since $k^2 \neq 0$,
   - $k \cdot \zeta = 0$ removes a spacelike polarisation; $D - 1$ tachyonic states left, one which is including ghosts.

3. $a = 1 \Rightarrow M^2 = 0$:

   - momentum k is null,
   - we can choose a frame where it is $(k, k, 0, \ldots)$,
   - spurious state is physical *and* null, since $k^2 = 0$,

[†] These are not to be confused with the ghosts of the friendly variety – Faddeev–Popov ghosts. These negative norm states are problematic and need to be removed.

- $k \cdot \zeta = 0$ and $k^2 = 0$ remove two polarisations; $D - 2$ states left.

So if we choose case (3), we end up with the special situation that we have a massless vector in the D dimensional target spacetime. It even has an associated gauge invariance: since the spurious state is physical and null, and therefore we can add it to our physical state with no physical consequences, defining an equivalence relation:

$$|\phi\rangle \sim |\phi\rangle + \lambda|\psi\rangle \qquad \Rightarrow \qquad \zeta^\mu \sim \zeta^\mu + \lambda k^\mu. \qquad (2.78)$$

Case (1), while interesting, corresponds to a massive vector, where the extra state plays the role of a longitudinal component. Case (2) seems bad. We shall choose case (3), where $a = 1$.

It is interesting to proceed to level two to construct physical and spurious states, although we shall not do it here. The physical states are massive string states. If we insert our level one choice $a = 1$ and see what the condition is for the spurious states to be both physical and null, we find that there is a condition on the spacetime dimension[‡]: $D = 26$.

In summary, we see that $a = 1$, $D = 26$ for the open bosonic string gives a family of extra null states, giving something analogous to a point of 'enhanced gauge symmetry' in the space of possible string theories. This is called a 'critical' string theory, for many reasons. We have the 24 states of a massless vector we shall loosely called the photon, A_μ, since it has a $U(1)$ gauge invariance (2.78). There is a tachyon of $M^2 = -1/\alpha'$ in the spectrum, which will not trouble us unduly. We will actually remove it in going to the superstring case. Tachyons will reappear from time to time, representing situations where we have an unstable configuration (as happens in field theory frequently). Generally, it seems that we should think of tachyons in the spectrum as pointing us towards an instability, and in many cases, the source of the instability is manifest.

Our analysis here extends to the closed string, since we can take two copies of our result, use the appropriate zero mode relation (2.48), and level matching. At level zero we get the closed string tachyon which has $M^2 = -4/\alpha'$. At level zero we get a tachyon with mass given by $M^2 = -4/\alpha'$, and at level 1 we get 24^2 massless states from $\alpha^\mu_{-1}\tilde{\alpha}^\nu_{-1}|0; k\rangle$. The traceless symmetric part is the graviton, $G_{\mu\nu}$ and the antisymmetric part, $B_{\mu\nu}$, is sometimes called the Kalb–Ramond field, and the trace is the dilaton, Φ.

[‡] We get a condition on the spacetime dimension here because level two is the first time it can enter our formulae for the norms of states, via the central term in the Virasoro algebra (2.71).

2.3.3 A glance at more sophisticated techniques

Later we shall do a more careful treatment of our gauge fixing procedure (2.41) by introducing Faddeev–Popov ghosts (b, c) to ensure that we stay on our chosen gauge slice in the full theory. Our resulting two dimensional conformal field theory will have an extra sector coming from the (b, c) ghosts.

The central term in the Virasoro algebra (2.71) represents an anomaly in the transformation properties of the stress tensor, spoiling its properties as a tensor under general coordinate transformations. Generally:

$$\left(\frac{\partial \sigma'^+}{\partial \sigma^+}\right)^2 T'_{++}(\sigma'^+) = T_{++}(\sigma^+) - \frac{c}{12}\left\{\frac{2\partial_\sigma^3 \sigma' \partial_\sigma \sigma' - 3\partial_\sigma^2 \sigma' \partial_\sigma^2 \sigma'}{2\partial_\sigma \sigma' \partial_\sigma \sigma'}\right\}, \quad (2.79)$$

where here c is a number, the *central charge* which depends upon the content of the theory. In our case, we have D bosons, which each contribute 1 to c, for a total anomaly of D.

The ghosts do two crucial things: They contribute to the anomaly the amount -26, and therefore we can retain all our favourite symmetries for the dimension $D = 26$. They also cancel the contributions to the vacuum energy coming from the oscillators in the $\mu = 0, 1$ sector, leaving $D - 2$ transverse oscillators' contribution.

The regulated value of $-a$ is the vacuum or 'zero point' energy (z.p.e.) of the transverse modes of the theory. This zero point energy is simply the Casimir energy arising from the fact that the two dimensional field theory is in a box. The box is the infinite strip, for the case of an open string, or the infinite cylinder, for the case of the closed string (see figure 2.3).

A periodic (integer moded) boson such as the types we have here, X^μ, each contribute $-1/24$ to the vacuum energy (see insert 2.3 (p. 46) on a quick way to compute this). So we see that in 26 dimensions, with only

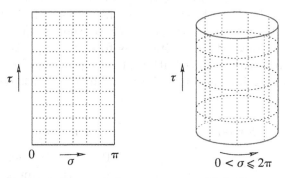

Fig. 2.3. String world-sheets as boxes upon which lives two dimensional conformal field theory.

Insert 2.3.   Zero point energy from the exponential map

After doing the transformation to the z-plane, it is interesting to note that the Fourier expansions we have been working with to define the modes of the stress tensor become Laurent expansions on the complex plane, e.g.

$$T_{zz}(z) = \sum_{m=-\infty}^{\infty} \frac{L_m}{z^{m+2}}.$$

One of the most straightforward exercises is to compute the zero point energy of the cylinder or strip (for a field of central charge c) by starting with the fact that the plane has no Casimir energy. One simply plugs the exponential change of coordinates $z = e^w$ into the anomalous transformation for the energy momentum tensor and compute the contribution to T_{ww} starting with T_{zz}:

$$T_{ww} = -z^2 T_{zz} - \frac{c}{24},$$

which results in the Fourier expansion on the cylinder, in terms of the modes:

$$T_{ww}(w) = -\sum_{m=-\infty}^{\infty} \left(L_m - \frac{c}{24}\delta_{m,0} \right) e^{i\sigma - \tau}.$$

---

24 contributions to count (see previous paragraph), we get that $-a = 24 \times (-1/24) = -1$. (Notice that from equation (2.69), this implies that $\sum_{n=1}^{\infty} n = -1/12$, which is in fact true (!) in ζ-function regularisation.)

Later, we shall have world-sheet fermions ψ^μ as well, in the supersymmetric theory. They each contribute $1/2$ to the anomaly. World sheet superghosts will cancel the contributions from ψ^0, ψ^1. Each anti-periodic fermion will give a z.p.e. contribution of $-1/48$.

Generally, taking into account the possibility of both periodicities for either bosons or fermions:

$$\text{z.p.e.} = \frac{1}{2}\omega \quad \text{for boson;} \qquad -\frac{1}{2}\omega \quad \text{for fermion} \qquad (2.80)$$

$$\omega = \frac{1}{24} - \frac{1}{8}(2\theta - 1)^2 \qquad \begin{cases} \theta = 0 & \text{(integer modes)} \\ \theta = \frac{1}{2} & \text{(half–integer modes).} \end{cases}$$

This is a formula that we shall use many times in what is to come.

2.4 The sphere, the plane and the vertex operator

The ability to choose the conformal gauge, as first discussed in section 2.2.6, gives us a remarkable amount of freedom, which we can put to good use. The diagrams in figure 2.3 represent free strings coming in from $\tau = -\infty$ and going out to $\tau = +\infty$. Let us first focus on the closed string, the cylinder diagram. Working with Euclidean signature by taking $\tau \to -i\tau$, the metric on it is

$$ds^2 = d\tau^2 + d\sigma^2, \qquad -\infty < \tau < +\infty \qquad 0 < \sigma \leq 2\pi.$$

We can do the change of variables

$$z = e^{\tau - i\sigma}, \tag{2.81}$$

with the result that the metric changes to

$$ds^2 = d\tau^2 + d\sigma^2 \longrightarrow |z|^{-2}dzd\bar{z}.$$

This is conformal to the metric of the complex plane: $d\hat{s}^2 = dzd\bar{z}$, and so we can use this as our metric on the world-sheet, since a conformal factor $e^{\phi} = |z|^{-2}$ drops out of the action, as we already noticed.

The string from the infinite past $\tau = -\infty$ is mapped to the origin while the string in the infinite future $\tau = +\infty$ is mapped to the 'point' at infinity. Intermediate strings are circles of constant radius $|z|$. See figure 2.4. The more forward-thinking reader who prefers to have the $\tau = +\infty$ string at the origin can use the complex coordinate $\tilde{z} = 1/z$ instead.

One can even ask that *both* strings be placed at finite distance in z. Then we need a conformal factor which goes like $|z|^{-2}$ at $z = 0$ as before, but like $|z|^2$ at $z = \infty$. There is an infinite set of functions which do that, but one particularly nice choice leaves the metric:

$$ds^2 = \frac{4R^2 dzd\bar{z}}{(R^2 + |z|^2)^2}, \tag{2.82}$$

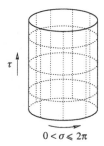
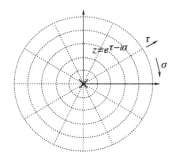
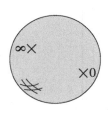

Fig. 2.4. The cylinder diagram is conformal to the complex plane and the sphere.

which is the familiar expression for the metric on a round S^2 with radius R, resulting from adding the point at infinity to the plane. See figure 2.4. The reader should check that the precise analogue of this process will relate the strip of the open string to the upper half plane, or to the disc. The open strings are mapped to points on the real axis, which is equivalent to the boundary of the disc. See figure 2.5.

We can go even further and consider the interaction with three or more strings. Again, a clever choice of function in the conformal factor can be made to map any tubes or strips corresponding to incoming strings to a point on the interior of the plane, or on the surface of a sphere (for the closed string) or the real axis of the upper half-plane of the boundary of the disc (for the open string). See figure 2.6.

2.4.1 States and operators

There is one thing which we might worry about. Have we lost any information about the state that the string was in by performing this reduction of an entire string to a point? Should we not have some sort of marker with which we label each point with the properties of the string it came from? The answer is in the affirmative, and the object which should be inserted at these points is called a 'vertex operator'. Let us see where it comes from.

As we learned in the previous subsection, we can work on the complex plane with coordinate z. In these coordinates, our mode expansions (2.46) and (2.47) become:

$$X^\mu(z, \bar{z}) = x^\mu - i \left(\frac{\alpha'}{2}\right)^{1/2} \alpha_0^\mu \ln z\bar{z} + i \left(\frac{\alpha'}{2}\right)^{1/2} \sum_{n \neq 0} \frac{1}{n} \alpha_n^\mu \left(z^{-n} + \bar{z}^{-n}\right),$$

$$(2.83)$$

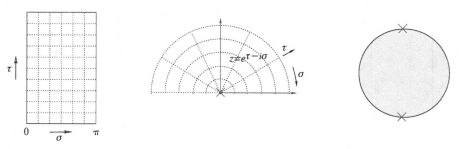

Fig. 2.5. The strip diagram is conformal to the upper half of the complex plane and the disc.

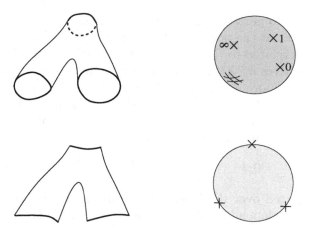

Fig. 2.6. Mapping any number of external string states to the sphere or disc using conformal transformations.

for the open string, and for the closed:

$$X^\mu(z, \bar z) = X_L^\mu(z) + X_R^\mu(\bar z)$$

$$X_L^\mu(z) = \frac{1}{2} x^\mu - i \left(\frac{\alpha'}{2}\right)^{1/2} \alpha_0^\mu \ln z + i \left(\frac{\alpha'}{2}\right)^{1/2} \sum_{n \neq 0} \frac{1}{n} \alpha_n^\mu z^{-n}$$

$$X_R^\mu(\bar z) = \frac{1}{2} x^\mu - i \left(\frac{\alpha'}{2}\right)^{1/2} \tilde\alpha_0^\mu \ln \bar z + i \left(\frac{\alpha'}{2}\right)^{1/2} \sum_{n \neq 0} \frac{1}{n} \tilde\alpha_n^\mu \bar z^{-n}, \quad (2.84)$$

where we have used the zero mode relations (2.48). In fact, notice that:

$$\partial_z X^\mu(z) = -i \left(\frac{\alpha'}{2}\right)^{1/2} \sum_n \alpha_n^\mu z^{-n-1}$$

$$\partial_{\bar z} X^\mu(\bar z) = -i \left(\frac{\alpha'}{2}\right)^{1/2} \sum_n \tilde\alpha_n^\mu \bar z^{-n-1}, \quad (2.85)$$

and that we can invert these to get (for the closed string)

$$\alpha_{-n}^\mu = \left(\frac{2}{\alpha'}\right)^{1/2} \oint \frac{dz}{2\pi} z^{-n} \partial_z X^\mu(z) \qquad \tilde\alpha_{-n}^\mu = \left(\frac{2}{\alpha'}\right)^{1/2} \oint \frac{dz}{2\pi} \bar z^{-n} \partial_{\bar z} X^\mu(z), \quad (2.86)$$

which are non-zero for $n \geq 0$. This is suggestive: equations (2.85) define left–moving (holomorphic) and right-moving (anti-holomorphic) fields. We previously employed the objects on the left in (2.86) in making states by acting, e.g. $\alpha_{-1}^\mu |0; k\rangle$. The form of the right hand side suggests that this is equivalent to performing a contour integral around an insertion of a pointlike operator at the point z in the complex plane (see figure 2.7). For example, α^μ_{-1} is related to the residue $\partial_z X^\mu(0)$, while the α^μ_{-m} correspond to higher derivatives $\partial_z^m X^\mu(0)$. This is course makes sense, as higher levels correspond to more oscillators excited on the string, and hence higher frequency components, as measured by the higher derivatives. The state with no oscillators excited (the tachyon), but with some momentum k, simply corresponds in this dictionary to the insertion of

$$|0; k\rangle \qquad \Longleftrightarrow \qquad \int d^2 z \; : e^{ik\cdot X} : \qquad (2.87)$$

We have integrated over the insertions' position on the sphere since the result should not depend upon our parameterisation. This is reasonable, as it is the simplest form that allows the right behaviour under translations: A translation by a constant vector, $X^\mu \to X^\mu + A^\mu$, results in a multiplication of the operator (and hence the state) by a phase $e^{i\mathbf{k}\cdot\mathbf{A}}$. The normal ordering signs :: are there to remind us that the expression means to expand and keep all creation operators to the left, when expanding in terms of the $\alpha_{\pm m}$s.

The closed string level one vertex operator corresponds to the emission or absorption of $G_{\mu\nu}$, $B_{\mu\nu}$ and Φ:

$$\zeta_{\mu\nu}\alpha^\mu_{-1}\tilde{\alpha}^\nu_{-1}|0; k\rangle \qquad \Longleftrightarrow \qquad \int d^2 z \; : \zeta_{\mu\nu}\partial_z X^\mu \partial_{\bar{z}} X^\nu e^{ik\cdot X} : \qquad (2.88)$$

where the symmetric part of $\zeta_{\mu\nu}$ is the graviton and the antisymmetric part is the antisymmetric tensor.

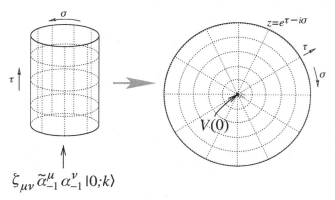

$$\zeta_{\mu\nu}\,\tilde{\alpha}^\mu_{-1}\,\alpha^\nu_{-1}\,|0;k\rangle$$

Fig. 2.7. The correspondence between states and operator insertions. A closed string (graviton) state $\zeta_{\mu\nu}\alpha^\mu_{-1}\tilde{\alpha}^\nu_{-1}|0; k\rangle$ is set up on the closed string at $\tau = -\infty$ and it propagates in. This is equivalent to inserting a graviton vertex operator $V^{\mu\nu}(z) =: \zeta_{\mu\nu}\partial_z X^\mu \partial_{\bar{z}} X^\nu e^{ik\cdot X} :$ at $z = 0$.

For the open string, the story is similar, but we get two copies of the relations (2.86) for the single set of modes α^μ_{-n} (recall that there are no $\tilde{\alpha}$s). This results in, for example the relation for the photon:

$$\zeta_\mu \alpha^\mu_{-1}|0;k\rangle \qquad \Longleftrightarrow \qquad \int dl \ : \zeta_\mu \partial_t X^\mu e^{ik\cdot X} :, \qquad (2.89)$$

where the integration is over the position of the insertion along the real axis. Also, ∂_t means the derivative tangential to the boundary. The tachyon is simply the boundary insertion of the momentum : $e^{ik\cdot X}$: alone.

2.5 Chan–Paton factors

Let us endow the string endpoints with a slightly more interesting property. We can add non-dynamical degrees of freedom to the ends of the string without spoiling spacetime Poincaré invariance or world-sheet conformal invariance. These are called 'Chan–Paton' degrees of freedom[22] and by declaring that their Hamiltonian is zero, we guarantee that they stay in the state that we put them into. In addition to the usual Fock space labels we have been using for the state of the string, we ask that each end be in a state i or j for i, j from 1 to N (see figure 2.8). We use a family of $N \times N$ matrices, λ^a_{ij}, as a basis into which to decompose a string wavefunction

$$|k;a\rangle = \sum_{i,j=1}^{N} |k,ij\rangle \lambda^a_{ij}. \qquad (2.90)$$

These wavefunctions are called 'Chan–Paton factors'. Similarly, all open string vertex operators carry such factors. For example, consider the tree-level (disc) diagram for the interaction of four oriented open strings in figure 2.9. As the Chan–Paton degrees of freedom are non-dynamical, the right end of string number 1 must be in the same state as the left end of string number 2, etc., as we go around the edge of the disc. After summing over all the possible states involved in tying up the ends, we are left with a trace of the product of Chan–Paton factors,

$$\lambda^1_{ij}\lambda^2_{jk}\lambda^3_{kl}\lambda^4_{li} = \text{Tr}(\lambda^1\lambda^2\lambda^3\lambda^4). \qquad (2.91)$$

i $\qquad\qquad\qquad\qquad\qquad$ j

Fig. 2.8. An open string with Chan–Paton degrees of freedom.

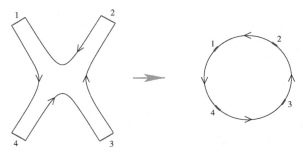

Fig. 2.9. A four-point scattering of open strings, and its conformally related disc amplitude.

All open string amplitudes will have a trace like this and are invariant under a global (on the world-sheet) $U(N)$:

$$\lambda^i \ \rightarrow \ U\lambda^i U^{-1}, \tag{2.92}$$

under which the endpoints transform as $\mathbf{N}$ and $\bar{\mathbf{N}}$.

Notice that the massless vector vertex operator $V^{a\mu} = \lambda^a_{ij}\partial_t X^\mu \exp \times (ik \cdot X)$ transforms as the adjoint under the $U(N)$ symmetry. *This means that the global symmetry of the world-sheet theory is promoted to a gauge symmetry in spacetime.* It is a gauge symmetry because we can make a different $U(N)$ rotation at separate points $X^\mu(\sigma, \tau)$ in spacetime.

2.6 Unoriented strings

2.6.1 Unoriented open strings

There is an operation of world-sheet parity Ω which takes $\sigma \rightarrow \pi - \sigma$, on the open string, and acts on $z = e^{\tau - i\sigma}$ as $z \leftrightarrow -\bar{z}$. In terms of the mode expansion (2.83), $X^\mu(z, \bar{z}) \rightarrow X^\mu(-\bar{z}, -z)$ yields

$$
\begin{aligned}
x^\mu &\rightarrow x^\mu \\
p^\mu &\rightarrow p^\mu \\
\alpha^\mu_m &\rightarrow (-1)^m \alpha^\mu_m.
\end{aligned}
\tag{2.93}
$$

This is a global symmetry of the open string theory and so we can, if we wish, also consider the theory that results when it is gauged, by which we mean that only Ω-invariant states are left in the spectrum. We must also consider the case of taking a string around a closed loop. It is allowed to come back to itself only up to an over all action of Ω, which is to swap the ends. This means that we must include unoriented world-sheets in our analysis. For open strings, the case of the Möbius strip is a useful example to keep in mind. It is on the same footing as the cylinder when we consider gauging Ω. The string theories which result from gauging Ω are understandably called 'unoriented string theories'.

Let us see what becomes of the string spectrum when we perform this projection. The open string tachyon is even under Ω and so survives the projection. However, the photon, which has only one oscillator acting, does not:

$$\Omega|k\rangle = +|k\rangle$$
$$\Omega\alpha^{\mu}_{-1}|k\rangle = -\alpha^{\mu}_{-1}|k\rangle. \tag{2.94}$$

We have implicitly made a choice about the sign of Ω as it acts on the vacuum. The choice we have made in writing equation (2.94) corresponds to the symmetry of the vertex operators (2.89): the resulting minus sign comes from the orientation reversal on the tangent derivative ∂_t (see figure 2.10).

Fortunately, we have endowed the string's ends with Chan–Paton factors, and so there is some additional structure which can save the photon. While Ω reverses the Chan–Paton factors on the two ends of the string, it can have some additional action:

$$\Omega\lambda_{ij}|k, ij\rangle \;\rightarrow\; \lambda'_{ij}|k, ji\rangle, \quad \lambda' = M\lambda^{T}M^{-1}. \tag{2.95}$$

This form of the action on the Chan–Paton factor follows from the requirement that it be a symmetry of the amplitudes which have factors like those in equation (2.91).

If we act twice with Ω, this should square to the identity on the fields, and leave only the action on the Chan–Paton degrees of freedom. States should therefore be invariant under:

$$\lambda \to M M^{-T} \lambda M^{T} M^{-1}. \tag{2.96}$$

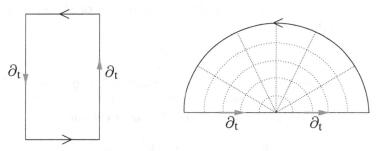

Fig. 2.10. The action of Ω on the photon vertex operator can be deduced from seeing how exchanging the ends of the string changes the sign of the tangent derivative, ∂_t.

Now it should be clear that the λ must span a complete set of $N \times N$ matrices: If strings with ends labelled ik and jl are in the spectrum for *any* values of k and l, then so is the state ij. This is because jl implies lj by CPT, and a splitting–joining interaction in the middle gives $ik + lj \rightarrow ij + lk$.

Now equation (2.96) and Schur's lemma require MM^{-T} to be proportional to the identity, so M is either symmetric or antisymmetric. This gives two distinct cases, modulo a choice of basis[24]. Denoting the $n \times n$ unit matrix as I_n, we have the symmetric case:

$$M = M^T = I_N. \tag{2.97}$$

In order for the photon $\lambda_{ij}\alpha^{\mu}_{-1}|k, ij\rangle$ to be even under Ω and thus survive the projection, λ must be antisymmetric to cancel the minus sign from the transformation of the oscillator state. So $\lambda = -\lambda^T$, giving the gauge group $SO(N)$. For the antisymmetric case, we have:

$$M = -M^T = i \begin{bmatrix} 0 & I_{N/2} \\ -I_{N/2} & 0 \end{bmatrix}. \tag{2.98}$$

For the photon to survive, $\lambda = -M\lambda^T M$, which is the definition of the gauge group $USp(N)$. Here, we use the notation that $USp(2) \equiv SU(2)$. Elsewhere in the literature this group is often denoted $Sp(N/2)$.

2.6.2 Unoriented closed strings

Turning to the closed string sector. For closed strings, we see that the mode expansion (2.84) for $X^{\mu}(z, \bar{z}) = X^{\mu}_L(z) + X^{\mu}_R(\bar{z})$ is invariant under a world-sheet parity symmetry $\sigma \rightarrow -\sigma$, which is $z \rightarrow -\bar{z}$. (We should note that this is a little different from the choice of Ω we took for the open strings, but more natural for this case. The two choices are related to each other by a shift of π.) This natural action of Ω simply reverses the left- and right-moving oscillators:

$$\Omega: \qquad \alpha^{\mu}_n \leftrightarrow \tilde{\alpha}^{\mu}_n. \tag{2.99}$$

Let us again gauge this symmetry, projecting out the states which are odd under it. Once again, since the tachyon contains no oscillators, it is even and is in the projected spectrum. For the level one excitations:

$$\Omega\alpha^{\mu}_{-1}\tilde{\alpha}^{\nu}_{-1}|k\rangle = \tilde{\alpha}^{\mu}_{-1}\alpha^{\nu}_{-1}|k\rangle, \tag{2.100}$$

and therefore it is only those states which are symmetric under $\mu \leftrightarrow \nu$ – the graviton and dilaton – which survive the projection. The antisymmetric tensor is projected out of the theory.

2.6.3 World-sheet diagrams

As stated before, once we have gauged Ω, we must allow for unoriented world-sheets, and this gives us rather more types of string world-sheet than we have studied so far. Figure 2.11 depicts the two types of one-loop diagram we must consider when computing amplitudes for the open string. The annulus (or cylinder) is on the left, and can be taken to represent an open string going around in a loop. The Möbius strip on the right is an open string going around a loop, but returning with the ends reversed. The two surfaces are constructed by identifying a pair of opposite edges on a rectangle, one with and the other without a twist.

Figure 2.12 shows an example of two types of closed string one-loop diagram we must consider. On the left is a torus, while on the right is a Klein bottle, which is constructed in a similar way to a torus save for a twist introduced when identifying a pair of edges.

In both the open and closed string cases, the two diagrams can be thought of as descending from the oriented case after the insertion of the normalised projection operator $\frac{1}{2}\text{Tr}(1 + \Omega)$ into one-loop amplitudes.

Similarly, the unoriented one-loop open string amplitude comes from the annulus and Möbius strip. We will discuss these amplitudes in more detail later.

The lowest order unoriented amplitude is the projective plane $\mathbb{RP}^2$, which is a disk with opposite points identified (see figure 2.13). Shrinking

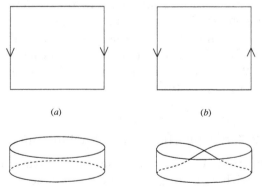

Fig. 2.11. (*a*) Constructing a cylinder or annulus by identifying a pair of opposite edges of a rectangle. (*b*) Constructing a Möbius strip by identifying after a twist.

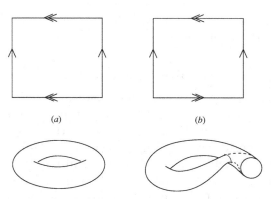

Fig. 2.12. (*a*) Constructing a torus by identifying opposite edges of a rectangle. (*b*) Constructing a Klein bottle by identifying after a twist.

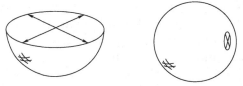

Fig. 2.13. Constructing the projective plane $\mathbb{RP}^2$ by identifying opposite points on the disk. This is equivalent to a sphere with a crosscap insertion.

the identified hole down, we recover the fact that $\mathbb{RP}^2$ may be thought of as a sphere with a crosscap inserted, where the crosscap is the result of shrinking the identified hole. Actually, a Möbius strip can be thought of as a disc with a crosscap inserted, and a Klein bottle is a sphere with two crosscaps. Since a sphere with a hole (one boundary) is the same as a disc, and a sphere with one handle is a torus, we can classify all world-sheet diagrams in terms of the number of handles, boundaries and crosscaps that they have. Insert 2.4 (p.57) summaries all the world-sheet perturbation theory diagrams up to one loop.

2.7 Strings in curved backgrounds

So far, we have studied strings propagating in the (uncompactified) target spacetime with metric $\eta_{\mu\nu}$. While this alone is interesting, it is curved backgrounds of one sort or another which will occupy much of this book, and so we ought to see how they fit into the framework so far.

Insert 2.4.   World-sheet perturbation theory diagrams

It is worthwhile summarising all of the string theory diagrams up to one-loop in a table. Recall that each diagram is weighted by a factor $g_s^\chi = g_s^{2h-2+b+c}$ where h, b, c are the numbers of handles, boundaries and crosscaps, respectively.

	g_s^{-2}	g_s^{-1}	g_s^0
closed oriented	sphere S^2 (plane)		torus T^2
open oriented		disc D_2 (half-plane)	cylinder C_2 (annulus)
closed unoriented		projective plane $\mathbb{RP}^2$	Klein bottle KB
open unoriented			Möbius strip MS

A natural generalisation of our action is simply to study the 'σ-model' action:

$$S_\sigma = -\frac{1}{4\pi\alpha'} \int d^2\sigma \, (-\gamma)^{1/2} \gamma^{ab} G_{\mu\nu}(X) \partial_a X^\mu \partial_b X^\nu. \qquad (2.101)$$

Comparing this to what we had before (2.21), we see that from the two dimensional point of view this still looks like a model of D bosonic fields X^μ, but with *field dependent* couplings given by the non-trivial spacetime metric $G_{\mu\nu}(X)$. This is an interesting action to study.

A first objection to this is that we seem to have cheated somewhat: strings are supposed to generate the graviton (and ultimately any curved backgrounds) dynamically. Have we cheated by putting in such a background by hand? Or a more careful, less confrontational question might be: is it consistent with the way strings generate the graviton to introduce curved backgrounds in this way?

Well, let us see. Imagine, to start off, that the background metric is only locally a small deviation from flat space: $G_{\mu\nu}(X) = \eta_{\mu\nu} + h_{\mu\nu}(X)$, where h is small.

Then, in conformal gauge, we can write in the Euclidean path integral (2.36):

$$e^{-S_\sigma} = e^{-S} \left(1 + \frac{1}{4\pi\alpha'} \int d^2z \, h_{\mu\nu}(X) \partial_z X^\mu \partial_{\bar{z}} X^\nu + \cdots \right), \qquad (2.102)$$

and we see that if $h_{\mu\nu}(X) \propto g_s \zeta_{\mu\nu} \exp(ik \cdot X)$, where ζ is a symmetric polarisation matrix, we are simply inserting a graviton emission vertex operator. So we are indeed consistent with that which we have already learned about how the graviton arises in string theory. Furthermore, the insertion of the full $G_{\mu\nu}(X)$ is equivalent in this language to inserting an exponential of the graviton vertex operator, which is another way of saying that a curved background is a 'coherent state' of gravitons.

It is clear that we should generalise our success, by including σ-model couplings which correspond to introducing background fields for the antisymmetric tensor and the dilaton:

$$S_\sigma = \frac{1}{4\pi\alpha'} \int d^2\sigma \, g^{1/2} \left\{ (g^{ab} G_{\mu\nu}(X) + i\epsilon^{ab} B_{\mu\nu}(X)) \partial_a X^\mu \partial_b X^\nu + \alpha' \Phi R \right\}, \qquad (2.103)$$

where $B_{\mu\nu}$ is the background antisymmetric tensor field and Φ is the background value of the dilaton. The coupling for $B_{\mu\nu}$ is a rather straightforward generalisation of the case for the metric. The power of α' is there to counter the scaling of the dimension one fields X^μ, and the antisymmetric tensor accommodates the antisymmetry of B. For the dilaton, a coupling to the two dimensional Ricci scalar is the simplest way of writing a reparametrisation invariant coupling when there is no index structure. Correspondingly, there is no power of α' in this coupling, as it is already dimensionless.

N.B. It is worth noting that α' is rather like $\hbar$ for this two dimensional theory, since the action is very large if $\alpha' \to 0$, and so this is a good limit to expand around. In this sense, the dilaton coupling is a one-loop term. Another thing to notice is that the $\alpha' \to 0$ limit is also like a 'large spacetime radius' limit. This can be seen by scaling lengths by $G_{\mu\nu} \to r^2 G_{\mu\nu}$, which results in an expansion in α'/r^2. Large radius is equivalent to small α'.

The next step is to do a full analysis of this new action and ensure that in the quantum theory, one has Weyl invariance, which amounts to the tracelessness of the two dimensional stress tensor. Calculations (which we will not discuss here) reveal that:

$$T^a{}_a = -\frac{1}{2\alpha'}\beta^G_{\mu\nu}g^{ab}\partial_a X^\mu \partial_b X^\nu - \frac{i}{2\alpha'}\beta^B_{\mu\nu}\epsilon^{ab}\partial_a X^\mu \partial_b X^\nu - \frac{1}{2}\beta^\Phi R, \quad (2.104)$$

$$\beta^G_{\mu\nu} = \alpha'\left(R_{\mu\nu} + 2\nabla_\mu\nabla_\nu\Phi - \frac{1}{4}H_{\mu\kappa\sigma}H_\nu{}^{\kappa\sigma}\right) + O(\alpha'^2),$$

$$\beta^B_{\mu\nu} = \alpha'\left(-\frac{1}{2}\nabla^\kappa H_{\kappa\mu\nu} + \nabla^\kappa\Phi H_{\kappa\mu\nu}\right) + O(\alpha'^2), \quad (2.105)$$

$$\beta^\Phi = \alpha'\left(\frac{D-26}{6\alpha'} - \frac{1}{2}\nabla^2\Phi + \nabla_\kappa\Phi\nabla^\kappa\Phi - \frac{1}{24}H_{\kappa\mu\nu}H^{\kappa\mu\nu}\right) + O(\alpha'^2),$$

with $H_{\mu\nu\kappa} \equiv \partial_\mu B_{\nu\kappa} + \partial_\nu B_{\kappa\mu} + \partial_\kappa B_{\mu\nu}$. For Weyl invariance, we ask that each of these β-functions for the σ-model couplings actually vanish. (See insert 3.1 for further explanation of this.) The remarkable thing is that these resemble *spacetime field equations for the background fields*. These field equations can be derived from the following spacetime action:

$$S = \frac{1}{2\kappa_0^2}\int d^D X (-G)^{1/2}e^{-2\Phi}\left[R + 4\nabla_\mu\Phi\nabla^\mu\Phi - \frac{1}{12}H_{\mu\nu\lambda}H^{\mu\nu\lambda}\right.$$
$$\left. -\frac{2(D-26)}{3\alpha'} + O(\alpha')\right]. \quad (2.106)$$

N.B. Now we note something marvellous: Φ is a background field which appears in the closed string theory σ-model multiplied by the Euler density. So comparing to equation (2.35) (and discussion following), we recover the remarkable fact that the string coupling g_s is not fixed, but is in fact given by the value of one of the background fields in the theory: $g_s = e^{\langle\Phi\rangle}$. So the only free parameter in the theory is the string tension.

Turning to the open string sector, we may also write the effective action which summarises the leading order (in α') open string physics at tree level:

$$S = -\frac{C}{4} \int d^D X\, e^{-\Phi} \mathrm{Tr} F_{\mu\nu} F^{\mu\nu} + O(\alpha'), \qquad (2.107)$$

with C a dimensionful constant which we will fix later. It is of course of the form of the Yang–Mills action, where $F_{\mu\nu} = \partial_\mu A_\nu - \partial_\nu A_\mu$. The field A_μ is coupled in σ-model fashion to the boundary of the world sheet by the boundary action:

$$\int_{\partial\mathcal{M}} d\tau\, A_\mu \partial_t X^\mu, \qquad (2.108)$$

mimicking the form of the vertex operator (2.89).

One should note the powers of e^Φ in the above actions. Recall that the expectation value of e^Φ sets the value of g_s. We see that the appearance of Φ in the actions are consistent with this, as we have $e^{-2\Phi}$ in front of all of the closed string parts, representing the sphere (g_s^{-2}) and $e^{-\Phi}$ for the open string, representing the disc (g_s^{-1}).

Notice that if we make the following redefinition of the background fields:

$$\tilde{G}_{\mu\nu}(X) = e^{2\Omega(X)} G_{\mu\nu} = e^{4(\Phi_0 - \Phi)/(D-2)} G_{\mu\nu}, \qquad (2.109)$$

and use the fact that the new Ricci scalar can be derived using:

$$\tilde{R} = e^{-2\Omega} \left[R - 2(D-1)\nabla^2\Omega - (D-2)(D-1)\partial_\mu\Omega\partial^\mu\Omega \right], \qquad (2.110)$$

the action (2.106) becomes:

$$S = \frac{1}{2\kappa^2} \int d^D X (-\tilde{G})^{1/2} \left[\tilde{R} - \frac{4}{D-2}\nabla_\mu\tilde{\Phi}\nabla^\mu\tilde{\Phi} \right. \qquad (2.111)$$

$$\left. - \frac{1}{12} e^{-8\tilde{\Phi}/(D-2)} H_{\mu\nu\lambda} H^{\mu\nu\lambda} - \frac{2(D-26)}{3\alpha'} e^{4\tilde{\Phi}/(D-2)} + O(\alpha') \right],$$

with $\tilde{\Phi} = \Phi - \Phi_0$. Looking at the part involving the Ricci scalar, we see that we have the form of the standard Einstein–Hilbert action (i.e. we have removed the factor involving the dilaton Φ), with Newton's constant set by

$$\kappa \equiv \kappa_0 e^{\Phi_0} = (8\pi G_N)^{1/2}. \tag{2.112}$$

The standard terminology to note here is that the action (2.106) written in terms of the original fields is called the 'string frame action', while the action (2.111) is referred to as the 'Einstein frame action'. It is in the latter frame that one gives meaning to measuring quantities like gravitational mass-energy. It is important to note the means, equation (2.109), to transform from the fields of one to another, depending upon dimension.

2.8 A quick look at geometry

Now that we are firmly in curved spacetime, it is probably a good idea to gather some concepts, language and tools which will be useful to us in many places later on. We have already reminded ourselves in chapter 1 of aspects of the classical differential geometry that is used to formulate the dynamics of gravity, introducing the metric, affine connection, Riemann tensors, etc. We will have reason to use another very pleasant way of writing of the various geometrical objects which appear in dynamical gravity, so we will quickly review it now, visiting a few other useful objects like differential forms along the way.

2.8.1 Working with the local tangent frames

We can introduce 'vielbeins' which locally diagonalise the metric[§]:

$$g_{\mu\nu}(x) = \eta_{ab} e_\mu^a(x) e_\nu^b(x).$$

The vielbeins form a basis for the tangent space at the point x, and orthonormality gives

$$e_\mu^a(x) e^{\mu b}(x) = \eta^{ab}.$$

These are interesting objects, connecting curved and tangent space, and transforming appropriately under the natural groups of each (see figure 2.14). It is a covariant vector under general coordinate transformations $x \to x'$:

$$e_\mu^a \to e_\mu'^a = \frac{\partial x^\nu}{\partial x'^\mu} e_\nu^a,$$

[§] 'Vielbein' means 'many legs', adapted from the German. In $D = 4$ it is called a 'vierbein'. We shall offend the purists henceforth and not capitalise nouns taken from the German language into physics, such as 'ansatz', 'bremsstrahlung' and 'gedankenexperiment'.

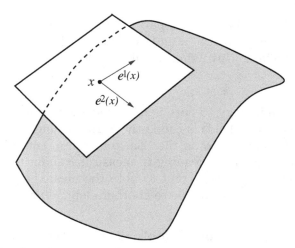

Fig. 2.14. The local tangent frame to curved spacetime is a copy of Minkowski space, upon which the Lorentz group acts naturally.

and a contravariant vector under local Lorentz:

$$e_\mu^a(x) \rightarrow e_\mu'^a(x) = \Lambda^a{}_b(x)e_\mu^b(x),$$

where $\Lambda^a{}_b(x)\Lambda^c{}_d(x)\eta_{ac} = \eta_{bd}$ defines Λ as being in the Lorentz group $SO(1, D{-}1)$.

So we have the expected freedom to define our vielbein up to a local Lorentz transformation in the tangent frame. In fact the condition Λ is a Lorentz transformation guarantees that the metric is invariant under local Lorentz:

$$g_{\mu\nu} = \eta_{ab}e'^a{}_\mu e'^b{}_\nu. \tag{2.113}$$

Notice that we can naturally define a family of inverse vielbiens as well, by raising and lowering indices in the obvious way, $e_a^\mu = \eta_{ab}g^{\mu\nu}e_\nu^b$. (We use the same symbol for the vielbien, but the index structure will make it clear what we mean.) Clearly,

$$g^{\mu\nu} = \eta^{ab}e_a^\mu e_b^\nu, \quad e_b^\mu e_\mu^a = \delta_b^a. \tag{2.114}$$

In fact, the vielbien may be thought of as simply the matrix of coefficients of the transformation (discussed in insert 1.2) which finds a locally inertial frame $\xi^a(x)$ from the general coordinates x^μ at the point $x = x_\mathrm{o}$:

$$e_\mu^a(x) = \left.\frac{\partial \xi^a(x)}{\partial x^\mu}\right|_{x=x_\mathrm{o}},$$

which, by construction, has the transformation properties ascribed to it above.

As a not-unrelated aside, note that the prototype contravariant vector in curved spacetime is in fact the object whose components are the infinitessimal coordinate displacements, dx^μ, since by the elementary chain rule, under $x \to x'$:

$$dx^\mu \to dx'^\mu = \frac{\partial x'^\mu}{\partial x^\nu} dx^\nu. \tag{2.115}$$

They are often thought of as the coordinate basis elements, $\{dx^\mu\}$, for the 'cotangent' space at the point x, and are a natural dual coordinate basis to that of the tangent space, the objects $\{\partial/\partial x^\mu\}$, via the perhaps obvious relation:

$$\frac{\partial}{\partial x^\mu} \cdot dx^\nu = \delta_\mu^\nu. \tag{2.116}$$

Of course, the $\{\partial/\partial x^\mu\}$ are the prototype covariant vectors:

$$\frac{\partial}{\partial x^\mu} \to \frac{\partial}{\partial x'^\mu} = \frac{\partial x^\nu}{\partial x'^\mu} \frac{\partial}{\partial x^\nu}. \tag{2.117}$$

The things we usually think of as vectors in curved spacetime have a natural expansion in terms of these bases:

$$V = V^\mu \frac{\partial}{\partial x^\mu}, \qquad \text{or} \quad V = V_\mu dx^\mu,$$

where the latter is sometimes called a 'covector', and is also in fact a one-form.

2.8.2 Differential forms

Since we've seen some one-forms appearing, let's pause to introduce them properly, if briefly. As might be apparent, it is the dx^μ which are useful for constructing p-forms, objects whose components are rank p tensors which are totally antisymmetric[¶].

As already stated, the dx^μ are themselves the basis for one-forms. Any one-form A has components A_μ and is expanded $A = A_\mu dx^\mu$. To make higher rank forms, we need the idea of the *wedge* product $\wedge$. The basis for two-forms for example, is made by the antisymmetric tensor product

$$dx^\mu \wedge dx^\nu \equiv dx^\mu \otimes dx^\nu - dx^\nu \otimes dx^\mu = -dx^\nu \wedge dx^\mu,$$

and we may then define a two-form F to have totally antisymmetric components $F_{\mu\nu}$, so that $F = (F_{\mu\nu}/2)dx^\mu \wedge dx^\nu$. After noting paranthetically

---

[¶] We will not give an exhaustive account of these objects here, but enough detail to get an intuitive feel for what we need. We shall uncover more features as we need them.

and for completeness that ordinary functions are zero-forms, the generalisation to higher rank forms is obvious: we make a basis for a p-form by making a totally antisymmetric combination of tensor multiplications of the one-forms, by adding together the results of taking products in all possible permutations, including a result with a minus sign if the permutation is odd, and a plus sign if it is even, giving us for example:

$$dx^{\mu_1} \wedge dx^{\mu_2} \wedge dx^{\mu_3}$$
$$\equiv dx^{\mu_1} \otimes dx^{\mu_2} \otimes dx^{\mu_3} + dx^{\mu_2} \otimes dx^{\mu_3} \otimes dx^{\mu_1} + dx^{\mu_3} \otimes dx^{\mu_1} \otimes dx^{\mu_2}$$
$$-dx^{\mu_1} \otimes dx^{\mu_3} \otimes dx^{\mu_2} - dx^{\mu_3} \otimes dx^{\mu_2} \otimes dx^{\mu_1} - dx^{\mu_2} \otimes dx^{\mu_1} \otimes dx^{\mu_3}.$$

So in general we have, for rank p:

$$dx^{\mu_1} \wedge dx^{\mu_2} \wedge \cdots \wedge dx^{\mu_p},$$

with which we can define a p-form $G_{(p)}$ with totally antisymmetric components $G_{\mu_1\mu_2\cdots\mu_p}$. We have:

$$G_{(p)} = \frac{1}{p!} G_{\mu_1\mu_2\cdots\mu_p} dx^{\mu_1} \wedge dx^{\mu_2} \wedge \cdots \wedge dx^{\mu_p}.$$

It is natural to define the 'exterior derivative' which makes a $(p+1)$-form from a p-form:

$$dG_{(p)} = \frac{1}{p!} \frac{\partial}{\partial x^\nu} \left(G_{\mu_1\mu_2\cdots\mu_p} \right) dx^\nu \wedge dx^{\mu_1} \wedge dx^{\mu_2} \wedge \cdots \wedge dx^{\mu_p}.$$

Notice that d^2 always gives zero, since (as the reader should check) this would give a symmetric combination of partial derivatives, which is being summed with the antisymmetric basis, which can't help but give zero.

A form G which can be written everywhere as the result of having acted with d on a form of lower rank is said to be 'exact'. A form H for which $dH = 0$ is 'closed'. Exact forms are trivially closed, since $d^2 = 0$, and so the interesting exercise is to find the closed forms on a space which are not exact. This is a problem of *cohomology*, and we shall have some more to say about this matter in chapter 9.

Forms are extremely natural objects to integrate over some manifold, M. In fact, a manifold of dimension p has a natural form defined on it, of rank p, which is simply the volume form, $\omega = dx^1 \wedge \cdots \wedge dx^p$. All p-forms on M are made by taking this object and multiplying it by some function. So the meaning of integrating a p-form on a manifold of dimension p is simply the standard multiple integration of the function:

$$\int_M F_{(p)} \equiv \int_M \frac{1}{p!} F_{\mu_1\cdots\mu_p} dx^{\mu_1} \wedge \cdots \wedge dx^{\mu_p}$$
$$= \int_M F_{1\cdots p} dx^1 \wedge \cdots \wedge dx^p = \int_M F_{1\cdots p} d^p x,$$

where the reader should notice that this required no metric on the manifold to be defined at all. Putting this observation together with the statements about cohomology, it should be apparent that forms give tools for computing topological properties of manifolds, since they can be integrated on various submanifolds to give numbers, and we never have to specify a metric.

The wedge or exterior product between a p-form and a q-form, which gives a $(p + q)$ form, is straightforward to define. On components, the result is:

$$(A_{(p)} \wedge B_{(q)})_{\mu_1 \cdots \mu_{p+q}} = \frac{(p+q)!}{p!q!} A_{[\mu_1 \cdots \mu_p} B_{\mu_{p+1} \cdots \mu_{p+q}]}.$$

It is worth noting that

$$A_{(p)} \wedge B_{(q)} = (-1)^{pq} B_{(q)} \wedge A_{(p)}.$$

More subtle is the observation that the space of independent p-forms on a D-dimensional spacetime is in fact of the same dimension as that of the $D - p$-forms. There is a map which takes one into the other, called 'Hodge duality', which takes any p-form and gives back a $(D - p)$-form. On the basis it is:

$$^*(dx^{\mu_1} \wedge dx^{\mu_2} \wedge \cdots \wedge dx^{\mu_p}) =$$
$$\frac{(-g)^{1/2}}{(D - p)!} \epsilon^{\mu_1 \mu_2 \cdots \mu_p}{}_{\mu_{p+1} \mu_{p+2} \cdots \mu_D} dx^{\mu_{p+1}} \wedge dx^{\mu_{p+2}} \wedge \cdots \wedge dx^{\mu_D},$$

from which its action on components of any form gives:

$$^*G_{\mu_1 \cdots \mu_{D-p}} = \frac{(-g)^{1/2}}{p!} \epsilon_{\mu_1 \cdots \mu_{D-p}}{}^{\nu_1 \cdots \nu_p} G_{\nu_1 \cdots \nu_p}.$$

Notice that it is the totally antisymmetric tensor (normalised to unity for its non-zero components) which appears in this definition, and indices are raised and lowered with the metric.

A most useful object is the *'inner product'* between two p-forms, $A_{(p)}$ and $B_{(p)}$, which yields a number. It is defined as:

$$(A_{(p)}, B_{(p)}) \equiv \int_{\mathcal{M}} A_{(p)} \wedge {}^* B_{(p)} = p! \int_{\mathcal{M}} (-g)^{1/2} A_{\mu_1 \mu_2 \cdots} B^{\mu_1 \mu_2 \cdots} dx^1 \wedge \cdots dx^D.$$

2.8.3 Coordinate vs. orthonormal bases

Yet another way of thinking of the vielbiens is as a means of converting that coordinate basis into a basis for the tangent space which is orthonormal, via $\{e^a = e^a_\mu(x) dx^\mu\}$. We see that we have defined a natural family of

---

Insert 2.5.   Yang–Mills theory with forms

Just in case differential forms which we are briefly introducing have not been encountered before, let us familiarise ourselves with how they work using Yang–Mills theory as an example. The gauge potential, which is valued in the Lie algebra of some gauge group G can be written as a matrix-valued one-form: $A = t^a A^a_\mu dx^\mu$, where the t^a are generators of the Lie algebra. (The index a here is a label of generators in the adjoint representation of the Yang–Mills gauge group G.) Recall also that the generators of the Lie algebra satisfy

$$[t^a, t^b] = i f^{ab}{}_c t^c,$$

where the $f^{ab}{}_c$ are the '*structure constants*'. We shall discuss some Lie algebra and group theory more carefully in section 4.6.1.

We write the Yang–Mills field strength as a matrix-valued 2-form:

$$F = dA + A \wedge A = F^a t^a = \frac{1}{2} t^a F^a_{\mu\nu} dx^\mu \wedge dx^\nu,$$

$$\text{where} \quad F^a_{\mu\nu} = \partial_\mu A^a_\nu - \partial_\nu A^a_\mu + i f^a{}_{bc} A^b_\mu A^c_\nu.$$

Note that we'll sometimes suppress the $\wedge$ and write $F = dA + A^2$ for short.

A gauge transformation is

$$A \to \Sigma A \Sigma^{-1} - d\Sigma\Sigma^{-1}, \quad \Sigma \in G,$$

or infinitesimally, writing $\Sigma = e^{-\Lambda}$, it is:

$$\delta A = d\Lambda + [A, \Lambda].$$

The field strength transforms under this as

$$F \to \Sigma F \Sigma^{-1}; \quad \text{or} \quad \delta F = [F, \Lambda].$$

The action for the theory is

$$S_{\text{YM}} = \int d^D x \sqrt{-g} \left(-\frac{1}{4g^2_{\text{YM}}} \text{Tr}(F^2) \right),$$

where by $\text{Tr}(F^2)$ we mean $F^a_{\mu\nu} F^{b\mu\nu} \text{Tr}(t^a t^b)$ and the trace is on the gauge indices. Here g^2_{YM} is the Yang–Mills coupling.

one-forms. Similarly, using the inverse vielbiens, we can make an orthonormal basis for the dual tangent space via $e_a = e_a^\mu \partial/\partial x^\mu$.

As an example, for the two-sphere, S^2, of radius R, the metric in standard polar coordinates (θ, ϕ) is $ds^2 = R^2(d\theta^2 + \sin^2\theta d\phi^2)$ and so we have:

$$e_\theta^1 = R, \quad e_\phi^2 = R\sin\theta, \quad \text{i.e.} \quad e^1 = Rd\theta, \quad e^2 = R\sin\theta \, d\phi. \quad (2.118)$$

The things we think of as vectors, familiar from flat space, now have two natural settings. In the local frame, there is the usual vector property, under which the vector has Lorentz contravariant components $V^a(x)$. But we can now relate this component to another object which has an index which is contravariant under general coordinate transformations, V^μ. These objects are related by our handy vielbiens: $V^a(x) = e_\mu^a(x)V^\mu$.

2.8.4 The Lorentz group as a gauge group

The standard covariant derivative which we defined earlier in equation (1.9), e.g. on a contravariant vector V^μ, has a counterpart for $V^a = e_\mu^a V^\mu$:

$$D_\nu V^\mu = \partial_\nu V^\mu + \Gamma_{\nu\kappa}^\mu V^\kappa \quad \Rightarrow \quad D_\nu V^a = \partial_\nu V^a + \omega^a{}_{b\nu} V^b,$$

where $\omega^a{}_{b\nu}$ is the *spin connection*, which we can write as a 1-form in either basis:

$$\omega^a{}_b = \omega^a{}_{b\mu} dx^\mu = \omega^a{}_{b\mu} e_c^\mu e_\nu^c dx^\nu = \omega^a{}_{bc} e^c.$$

We can think of the two Minkowski indices (a, b) from the space tangent structure as labelling components of ω as an $SO(D-1, 1)$ matrix in the fundamental representation. So in the analogy with Yang–Mills theory, (see insert 2.5), ω_μ is rather like a gauge potential and the gauge group is the Lorentz group.

Actually, the most natural appearance of the spin connection is in the *structure equations* of Cartan. One defines the torsion T^a, and the curvature $R^a{}_b$, both two-forms, as follows:

$$T^a \equiv \frac{1}{2} T^a{}_{bc} e^a \wedge e^b = de^a + \omega^a{}_b \wedge e^b$$

$$R^a{}_b \equiv \frac{1}{2} R^a{}_{bcd} e^c \wedge e^d = d\omega^a{}_b + \omega^a{}_c \wedge \omega^c{}_b. \quad (2.119)$$

Now consider a Lorentz transformation $e^a \to e'^a = \Lambda^a{}_b e^b$. It is amusing to work out how the torsion changes. Writing the result as $T'^a = \Lambda^a{}_b T^b$, the reader might like to check that this implies that the spin connection must transform as (treating everything as $SO(1, D-1)$ matrices):

$$\omega \to \Lambda\omega\Lambda^{-1} - d\Lambda \cdot \Lambda^{-1}, \quad \text{i.e.} \quad \omega_\mu \to \Lambda\omega_\mu\Lambda^{-1} - \partial_\mu\Lambda \cdot \Lambda^{-1}, \quad (2.120)$$

or infinitessimally we can write $\Lambda = e^{-\Theta}$, and it is:

$$\delta\omega = d\Theta + [\omega, \Theta]. \tag{2.121}$$

A further check shows that the curvature two-form does

$$R \to R' = \Lambda R \Lambda^{-1}, \quad \text{or} \quad \delta R = [R, \Theta], \tag{2.122}$$

which is awfully nice. This shows that the curvature two-form is the analogue of the Yang–Mills field strength two-form in insert 2.5. The following rewriting makes it even more suggestive:

$$R^a{}_b = \frac{1}{2}R^a{}_{b\mu\nu}dx^\mu \wedge dx^\nu, \quad R^a{}_{b\mu\nu} = \partial_\mu \omega^a{}_{b\nu} - \partial_\nu \omega^a{}_{b\mu} + [\omega_\mu, \omega_\nu]^a{}_b.$$

2.8.5 Fermions in curved spacetime

Another great thing about this formalism is that it allows us to discuss fermions in curved spacetime. Recall first of all that we can represent the Lorentz group with the Γ-matrices as follows. The group's algebra is:

$$[J_{ab}, J_{cd}] = -i(\eta_{ad}J_{bc} + \eta_{bc}J_{ad} - \eta_{ac}J_{bd} - \eta_{db}J_{ac}), \tag{2.123}$$

with $J_{ab} = -J_{ba}$, and we can define via the Clifford algebra:

$$\{\Gamma^a, \Gamma^b\} = 2\eta^{ab}, \quad J^{ab} = -\frac{i}{4}\left[\Gamma^a, \Gamma^b\right], \tag{2.124}$$

where the curved space Γ-matrices are related to the familiar flat (tangent) spacetime ones as $\Gamma^a = e^a_\mu(x)\Gamma^\mu(x)$, giving $\{\Gamma^\mu, \Gamma^\nu\} = 2g^{\mu\nu}$. With the Lorentz generators defined in this way, it is now natural to couple a fermion ψ to spacetime. We write a covariant derivative as

$$D_\mu\psi(x) = \partial_\mu\psi(x) + \frac{i}{2}J_{ab}\omega^{ab}{}_\mu(x)\psi(x), \tag{2.125}$$

and since the curved space Γ-matrices are now covariantly constant, we can write a sensible Dirac equation using this: $\Gamma^\mu D_\mu\psi = 0$.

2.8.6 Comparison to differential geometry

Let us make the connection to the usual curved spacetime formalism now, and fix what ω is in terms of the vielbiens (and hence the metric). Asking that the torsion vanishes is equivalent to saying that the vielbeins are covariantly constant, so that $D_\mu e^a_\nu = 0$. This gives $D_\mu V^a = e^{a\nu}D_\mu V_\nu$, allowing the two definitions of covariant derivatives to be simply related by using the vielbeins to convert the indices.

The fact that the metric is covariantly constant in terms of curved spacetime indices relates the affine connection to the metric connection, and in this language makes ω^{ab} antisymmetric in its indices. Finally, we get that

$$\omega^a{}_{b\mu} = e^a_\nu \nabla_\mu e^\nu_b = e^a_\nu (\partial_\mu e^\nu_b + \Gamma^\nu_{\mu\kappa} e^\kappa_b).$$

We can now write covariant derivatives for objects with mixed indices (appropriately generalising the rule for terms to add depending upon the index structure), for example, on a vielbien:

$$D_\mu e^a_\nu = \partial_\mu e^a_\nu - \Gamma^\kappa_{\mu\nu} e^a_\kappa + \omega_\mu{}^a{}_b e^b_\nu. \tag{2.126}$$

Revisiting our two-sphere example, with bases given in equation (2.118), we can see that

$$0 = de^1 + \omega^1{}_2 \wedge e^2 = 0 + \omega^1{}_2 \wedge e^2,$$
$$0 = de^2 + \omega^2{}_1 \wedge e^1 = R\cos\theta d\theta \wedge d\phi + \omega^2{}_1 \wedge e^1, \tag{2.127}$$

from which we see that $\omega^1{}_2 = -\cos\theta\, d\phi$. The curvature is:

$$R^1{}_2 = d\omega^1{}_2 = \sin\theta d\theta \wedge d\phi = \frac{1}{R^2} e^1 \wedge e^2 = R^1{}_{212} e^1 \wedge e^2. \tag{2.128}$$

Notice that we can recover our friend the usual Riemann tensor if we pulled back the tangent space indices (a, b) on $R^a{}_{b\mu\nu}$ to curved space indices using the vielbiens e^μ_a.

One last thing to note is the usefulness of forms for writing volume elements for integration:

$$dV \equiv e = e^1 \wedge e^2 \wedge \cdots \wedge e^D = (-g)^{1/2} dx^1 \wedge dx^2 \wedge \cdots \wedge dx^D = (-g)^{1/2} d^D x.$$

Commonly, we will take the totally antisymmetric symbol ϵ and make a tensor out of it by multiplying by $(-g)^{1/2}$, defining:

$$\varepsilon_{\mu_1 \cdots \mu_D} = (-g)^{1/2} \epsilon_{\mu_1 \cdots \mu_D},$$

and the reader should check that this is a tensor, noting that the factor of the tensor density $(-g)^{1/2}$ will produce just the right non-tensorial parts to cancel those of the permutation symbol.

We can write the Einstein–Hilbert Lagrangian as:

$$\mathcal{L} \sim eR, \tag{2.129}$$

where R is the Ricci scalar, with $de^a + \omega e^a = 0$ as an additional condition.

3

A closer look at the world-sheet

The careful reader has patiently suspended disbelief for a while now, allowing us to race through a somewhat rough presentation of some of the highlights of the construction of consistent relativistic strings. This enabled us, by essentially stringing lots of oscillators together, to go quite far in developing our intuition for how things work, and for key aspects of the language.

Without promising to suddenly become rigourous, it seems a good idea to revisit some of the things we went over quickly, in order to unpack some more details of the operation of the theory. This will allow us to develop more tools and language for later use, and to see a bit further into the structure of the theory.

3.1 Conformal invariance

We saw in section 2.2.8 that the use of the symmetries of the action to fix a gauge left over an infinite dimensional group of transformations which we could still perform and remain in that gauge. These are conformal transformations, and the world-sheet theory is in fact conformally invariant. It is worth digressing a little and discussing conformal invariance in arbitrary dimensions first, before specialising to the case of two dimensions. We will find a surprising reason to come back to conformal invariance in higher dimensions much later, so there is a point to this.

3.1.1 Diverse dimensions

Imagine[275] that we do a change of variables $x \to x'$. Such a change, if invertible, is a 'conformal transformation' if the metric is invariant up to

70

an overall scale $\Omega(x)$, which can depend on position:

$$g'_{\mu\nu}(x') = \Omega(x)g_{\mu\nu}(x). \qquad (3.1)$$

The name comes from the fact that angles between vectors are unchanged.
If we consider the infinitessimal change

$$x^\mu \to x'^\mu = x^\mu + \epsilon^\mu(x), \qquad (3.2)$$

then from equation (1.1), we get:

$$g'_{\mu\nu} = g_{\mu\nu} - (\partial_\mu \epsilon_\nu + \partial_\nu \epsilon_\mu), \qquad (3.3)$$

and so we see that in order for this to be a conformal transformation,

$$\partial_\mu \epsilon_\nu + \partial_\nu \epsilon_\mu = F(x)g_{\mu\nu}, \qquad (3.4)$$

where, by taking the trace of both sides, it is clear that:

$$F(x) = \frac{2}{D}g^{\mu\nu}\partial_\mu \epsilon_\nu.$$

It is enough to consider our metric to be Minkowski space, in Cartesian
coordinates, i.e. $g_{\mu\nu} = \eta_{\mu\nu}$. We can take one more derivative ∂_κ of the
expression (3.4), and then do the permutation of indices $\kappa \to \mu, \mu \to
\nu, \nu \to \kappa$ twice, generating two more expressions. Adding together any
two of those and subtracting the third gives:

$$2\partial_\mu \partial_\nu \epsilon_\kappa = \partial_\mu F \eta_{\nu\kappa} + \partial_\nu F \eta_{\kappa\mu} - \partial_\kappa F \eta_{\mu\nu}, \qquad (3.5)$$

which yields

$$2\Box\epsilon_\kappa = (2 - D)\partial_\kappa F. \qquad (3.6)$$

We can take another derivative this expression to get $2\partial_\mu \Box\epsilon_\kappa = (2 -
D)\partial_\mu \partial_\kappa F$, which should be compared to the result of acting with $\Box$ on
equation (3.4) to eliminate ϵ leaving:

$$\eta_{\mu\nu}\Box F = (2 - D)\partial_\mu \partial_\nu F \implies (D - 1)\Box F = 0, \qquad (3.7)$$

where we have obtained the last result by contraction.

For general D we see that the last equations above ask that $\partial_\mu \partial_\nu F = 0$,
and so F is linear in x. This means that ϵ is quadratic in the coordinates,
and of the form:

$$\epsilon_\mu = A_\mu + B_{\mu\nu}x^\nu + C_{\mu\nu\kappa}x^\nu x^\kappa, \qquad (3.8)$$

where C is symmetric in its last two indices.

Table 3.1.   *The finite form of the conformal transformations and their infinitesimal generators*

Operation	Action	Generator
translations	$x'^\mu = x^\mu + A^\mu$	$P_\mu = -i\partial_\mu$
rotations	$x'^\mu = M^\mu{}_\nu x^\nu$	$L_{\mu\nu} = i(x_\mu \partial_\nu - x_\nu \partial_\mu)$
dilations	$x'^\mu = \lambda x^\mu$	$D = -ix^\mu \partial_\mu$
special conformal transformations	$x'^\mu = \dfrac{x^\mu - b^\mu x^2}{1 - 2(\mathbf{x} \cdot \mathbf{b}) - b^\mu x^2}$	$K_\mu = -i(2x_\mu x^\nu \partial_\nu - x^2 \partial_\mu)$

The parameter A_μ is obviously a translation. Placing the B term in equation (3.8) back into equation (3.4) yields that $B_{\mu\nu}$ is the sum of an antisymmetric part $\omega_{\mu\nu} = -\omega_{\nu\mu}$ and a trace part λ:

$$B_{\mu\nu} = \omega_{\mu\nu} + \lambda\eta_{\mu\nu}. \tag{3.9}$$

This represents a scale transformation by $1 + \lambda$ and an infinitessimal rotation. Finally, direct substitution shows that

$$C_{\mu\nu\kappa} = \eta_{\mu\kappa}b_\nu + \eta_{\mu\nu}b_\kappa - \eta_{\nu\kappa}b_\mu, \tag{3.10}$$

and so the infinitesimal transformation which results is of the form

$$x'^\mu = x^\mu + 2(\mathbf{x} \cdot \mathbf{b})x^\mu - b^\mu x^2, \tag{3.11}$$

which is called a 'special conformal transformation'. Its finite form can be written as:

$$\frac{x'^\mu}{x'^2} = \frac{x^\mu}{x^2} - b^\mu, \tag{3.12}$$

and so it looks like an inversion, then a translation, and then an inversion. We gather together all the transformations, in their finite form, in table 3.1.

Poincaré and dilatations together form a subgroup of the full conformal group, and it is indeed a special theory that has the full conformal invariance given by enlargement by the special conformal transformations.

It is interesting to examine the commutation relations of the generators, and to do so, we rewrite them as

$$J_{-1,\mu} = \tfrac{1}{2}(P_\mu - K_\mu), \qquad J_{0,\mu} = \tfrac{1}{2}(P_\mu + K_\mu),$$
$$J_{-1,0} = D, \quad J_{\mu\nu} = L_{\mu\nu}, \tag{3.13}$$

with $J_{ab} = -J_{ba}$, $a, b = -1, 0, \ldots, D$, and the commutators are:

$$[J_{ab}, J_{cd}] = -i(\eta_{ad}J_{bc} + \eta_{bc}J_{ad} - \eta_{ac}J_{bd} - \eta_{db}J_{ac}). \qquad (3.14)$$

Note that we have defined an extra value for our indices, and η is now diag$(-1, -1, +1, \ldots)$. This is the algebra of the group $SO(D, 2)$ with $\frac{1}{2}(D+2)(D+1)$ parameters.

3.1.2 The special case of two dimensions

As we have already seen in section 2.2.8, the conformal transformations are equivalent to conformal mappings of the plane to itself, which is an infinite dimensional group. This might seem puzzling, since from what we saw just above, one might have expected $SO(2, 2)$, or in the case where we have Euclideanised the world-sheet, $SO(3, 1)$, a group with six parameters. Actually, this group is a very special subgroup of the infinite family, which is distinguished by the fact that the mappings are invertible. These are the *global* conformal transformations. Imagine that $w(z)$ takes the plane into itself. It can at worst have zeros and poles, (the map is not unique at a branch point, and is not invertible if there is an essential singularity) and so can be written as a ratio of polynomials in z. However, for the map to be invertible, it can only have a single zero, otherwise there would be an ambiguity determining the pre-image of zero in the inverse map. By working with the coordinate $\tilde{z} = 1/z$, in order to study the neighbourhood of infinity, we can conclude that it can only have a single simple pole also. Therefore, up to a trivial overall scaling, we have

$$z \to w(z) = \frac{az + b}{cz + d}, \qquad (3.15)$$

where a, b, c, d are complex numbers, with for invertability, the determinant of the matrix

$$\begin{pmatrix} a & b \\ c & d \end{pmatrix}$$

should be non-zero, and after a scaling we can choose $ad - bc = 1$. This is the group $SL(2, \mathbb{C})$ which is indeed isomorphic to $SO(3, 1)$. In fact, since a, b, c, d is indistinguishable from $-a, -b, -c, -d$, the correct statement is that we have invariance under $SL(2, \mathbb{C})/\mathbb{Z}_2$.

For the open string we have the upper half-plane, and so we are restricted to considering maps which preserve (say) the real axis of the complex plane. The result is that a, b, c, d must be real numbers, and the resulting group of invertible transformations is $SL(2, \mathbb{R})/\mathbb{Z}_2$. Correspondingly, the infinite part of the algebra is also reduced in size by half, as the holomorphic and antiholomorphic parts are no longer independent.

N.B. Notice that the dimension of the group $SL(2,\mathbb{C})$ is six, equivalent to three complex parameters. Often, in computations involving a number of operators located at points, z_i, a conventional gauge fixing of this invariance is to set three of the points to three values: $z_1 = 0, z_2 = 1, z_3 = \infty$. Similarly, the dimension of $SL(2,\mathbb{R})$ is three, and the convention used there is to set three (real) points on the boundary to $z_1 = 0, z_2 = 1, z_3 = \infty$.

3.1.3 States and operators

A very important class of fields in the theory are those which transform under the $SO(2,D)$ conformal group as follows:

$$\phi(x^\mu) \longrightarrow \phi(x'^\mu) = \left|\frac{\partial x}{\partial x'}\right|^{\frac{\Delta}{D}} \phi(x^\mu) = \Omega^{\frac{\Delta}{2}} \phi(x^\mu). \tag{3.16}$$

Here, $\left|\frac{\partial x}{\partial x'}\right|$ is the Jacobian of the change of variables. (Δ is the dimension of the field, as mentioned earlier.) Such fields are called 'quasi-primary', and the correlation functions of some number of the fields will inherit such transformation properties:

$$\langle\phi_1(x_1)\dots\phi_n(x_n)\rangle = \left|\frac{\partial x}{\partial x'}\right|^{\frac{\Delta_1}{D}}_{x=x_1} \dots \left|\frac{\partial x}{\partial x'}\right|^{\frac{\Delta_n}{D}}_{x=x_n} \langle\phi_1(x_1')\dots\phi_n(x_n')\rangle. \tag{3.17}$$

In two dimensions, the relation is

$$\phi(z,\bar{z}) \longrightarrow \phi(z',\bar{z}') = \left(\frac{\partial z}{\partial z'}\right)^h \left(\frac{\partial \bar{z}}{\partial \bar{z}'}\right)^{\bar{h}} \phi(z,\bar{z}), \tag{3.18}$$

where $\Delta = h + \bar{h}$, and we see the familiar holomorphic factorisation. This mimics the transformation properties of the metric under $z \to z'(z)$:

$$g'_{z\bar{z}} = \left(\frac{\partial z}{\partial z'}\right)\left(\frac{\partial \bar{z}}{\partial \bar{z}'}\right) g_{z\bar{z}},$$

the conformal mappings of the plane. This is an infinite dimensional family, extending the expected six of $SO(2,2)$, which is the subset which is globally well-defined. The transformations (3.18) define what is called a 'primary field', and the quasi-primaries defined earlier are those restricted to $SO(2,2)$. So a primary is automatically a quasi-primary, but not vice versa.

In any dimension, we can use the definition (3.16) to construct a definition of a conformal field theory (CFT). First, we have a notion of a vacuum $|0\rangle$ that is $SO(2, D)$ invariant, in which all the fields act. In such a theory, all of the fields can be divided into two categories: a field is either quasi-primary, or it is a linear combination of quasi-primaries and their derivatives. Conformal invariance imposes remarkably strong constraints on how the two- and three-point functions of the quasi-primary fields must behave. Obviously, for fields placed at positions x_i, translation invariance means that they can only depend on the differences $x_i - x_j$.

3.1.4 The operator product expansion

In principle, we ought to be imagining the possibility of constructing a new field at the point x^μ by colliding together two fields at the same point. Let us label the fields as ϕ_k, then we might expect something of the form:

$$\lim_{x \to y} \phi_i(x)\phi_j(y) = \sum_k c_{ij}{}^k(x - y)\phi_k(y), \qquad (3.19)$$

where the coefficients $c_{ij}{}^k(x-y)$ depend only on which operators (labelled by i, j) enter on the left. Given the scaling dimensions Δ_i for ϕ_i, we see that the coordinate behaviour of the coefficient should be:

$$c_{ij}{}^k(x - y) \sim \frac{1}{(x - y)^{\Delta_i + \Delta_j - \Delta_k}}.$$

This 'operator product expansion' (OPE) in conformal field theory is actually a convergent series, as opposed to the case of the OPE in ordinary field theory where it is merely an asymptotic series. An asymptotic series has a family of exponential contributions of the form $\exp(-L/|x - y|)$, where L is a length scale appropriate to the problem. Here, conformal invariance means that there is no length scale in the theory to play the role of L in an asymptotic expansion, and so the convergence properties of the OPE are stronger. In fact, the radius of convergence of the OPE is essentially the distance to the next operator insertion.

The OPE only really has sensible meaning if we define the operators as acting with a specific time ordering, and so we should specify that $x^0 > y^0$ in the above. In two dimensions, after we have continued to Euclidean time and work on the plane, the equivalent of time ordering is radial ordering (see figure 2.4). All OPE expressions written later will be taken to be appropriately time ordered.

Actually, the OPE is a useful way of giving us a definition of a normal ordering prescription in this operator language*. It follows from Wick's theorem, which says that the time ordered expression of a product of operators is equal to the normal ordered expression plus the sum of all contractions of pairs of operators in the expressions. The contraction is a number, which is computed by the correlator of the contracted operators.

$$\phi_i(x)\phi_j(y) = \; : \phi_i(x)\phi_j(y) : + \langle\!\langle \phi_i(x)\phi_j(y) \rangle\!\rangle. \qquad (3.20)$$

Actually, we can compute the OPE between objects made out of products of operators with this sort of way of thinking about it. We'll compute some examples later (for example in equations (3.37) and (3.39)) so that it will be clear that it is quite straightforward.

3.1.5 The stress tensor and the Virasoro algebra

The stress-energy-momentum tensor's properties can be seen directly from conformal invariance in many ways, because of its definition as a conjugate to the metric via equation (1.10) which we reproduce here:

$$T^{\mu\nu} \equiv -\frac{2}{\sqrt{-g}}\frac{\delta S}{\delta g_{\mu\nu}}. \qquad (3.21)$$

A change of variables of the form (3.2) gives, using equation (3.3):

$$S \longrightarrow S - \frac{1}{2}\int d^D x \sqrt{-g}\, T^{\mu\nu}\delta g_{\mu\nu} = S + \frac{1}{2}\int d^D x \sqrt{-g}\, T^{\mu\nu}\left(\partial_\mu\epsilon_\nu + \partial_\nu\epsilon_\mu\right).$$

In view of equation (3.4), this is:

$$S \longrightarrow S + \frac{1}{D}\int d^D x \sqrt{-g}\, T^\mu{}_\mu \partial_\nu \epsilon^\nu$$

for a conformal transformation. So if the action is conformally invariant, then the stress tensor must be traceless, $T^\mu{}_\mu = 0$.

We can formulate this more carefully using Noether's theorem, and also extract some useful information. Since the change in the action is

$$\delta S = \int d^D x \, \sqrt{-g}\, \partial_\mu\epsilon_\nu T^{\mu\nu},$$

given that the stress tensor is conserved, we can integrate by parts to write this as

$$\delta S = \int_\partial \epsilon_\nu T^{\mu\nu} dS_\mu.$$

---

* For free fields, this definition of normal ordering is equivalent to the definition in terms of modes, where the annihilators are placed to the right.

We see that the current $j^\mu = T^{\mu\nu}\epsilon_\mu$, with ϵ_ν given by equation (3.4) is associated to the conformal transformations. The charge constructed by integrating over an equal time slice

$$Q = \int d^{D-1}x J^0,$$

is conserved, and it is responsible for infinitessimal conformal transformations of the fields in the theory, defined in the standard way:

$$\delta_\epsilon \phi(x) = \epsilon[Q, \phi]. \tag{3.22}$$

In two dimensions, infinitesimally, a coordinate transformation can be written as

$$z \to z' = z + \epsilon(z), \qquad \bar{z} \to \bar{z}' = \bar{z} + \bar{\epsilon}(\bar{z}).$$

As we saw in the previous chapter, or can be verified using the above discussion, the tracelessness condition yields $T_{z\bar{z}} = T_{\bar{z}z} = 0$ and the conservation of the stress tensor is

$$\partial_z T_{zz}(z) = 0 = \partial_{\bar{z}} T_{\bar{z}\bar{z}}(\bar{z}).$$

For simplicity, we shall often use the shorthand: $T(z) \equiv T_{zz}(z)$ and $\bar{T}(\bar{z}) \equiv T_{\bar{z}\bar{z}}(\bar{z})$. On the plane, an equal time slice is over a circle of constant radius, and so we can define

$$Q = \frac{1}{2\pi i} \oint (T(y)\epsilon(y)dy + \bar{T}(\bar{y})\bar{\epsilon}(\bar{y})d\bar{y}).$$

Infinitesimal transformations can then be constructed by an appropriate definition of the commutator $[Q, \phi(z)]$ of Q with a field ϕ.

Notice that this commutator requires a definition of two operators at a point, and so our previous discussion of the OPE comes into play here. We also have the added complication that we are performing a y-contour integration around one of the operators, inserted at z or $\bar{z}$. Under the integral sign, the OPE requires that $|z| < |y|$, when we have $Q\phi(y)$, or that $|z| > |y|$ if we have $\phi(y)Q$. The commutator requires the difference between these two, and after consulting figure 3.1, can be seen in the limit $y \to z$ to simply result in the y contour integral around the point z of the OPE $T(z)\phi(y)$ (with a similar discussion for the antiholomorphic case):

$$\delta_{\epsilon,\bar{\epsilon}}\phi(z,\bar{z}) = \frac{1}{2\pi i} \oint (\{T(y)\phi(z,\bar{z})\}\epsilon(y)dy + \{\bar{T}(\bar{y})\phi(z,\bar{z})\}\bar{\epsilon}(\bar{y})d\bar{y}). \tag{3.23}$$

The result should simply be the infinitesimal version of the defining equation (3.18), which the reader should check is:

$$\delta_{\epsilon,\bar{\epsilon}}\phi(z,\bar{z}) = \left(h\frac{\partial\epsilon}{\partial z}\phi + \epsilon\frac{\partial\phi}{\partial z} \right) + \left(\bar{h}\frac{\partial\bar{\epsilon}}{\partial\bar{z}}\phi + \bar{\epsilon}\frac{\partial\phi}{\partial\bar{z}} \right). \tag{3.24}$$

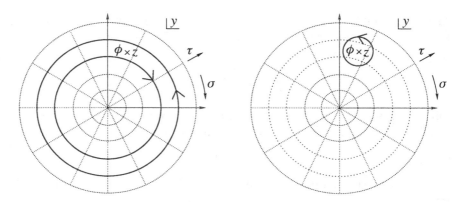

Fig. 3.1. Computing the commutator between the generator Q, defined as a contour in the y-plane, and the operator ϕ, inserted at z. The result in the limit $y \to z$ is on the right.

This defines the operator product expansions $T(z)\phi(z,\bar{z})$ and $\bar{T}(\bar{z})\phi(z,\bar{z})$ for us as:

$$T(y)\phi(z,\bar{z}) = \frac{h}{(y-z)^2}\phi(z,\bar{z}) + \frac{1}{(y-z)}\partial_z\phi(z,\bar{z}) + \cdots$$

$$\bar{T}(\bar{y})\phi(z,\bar{z}) = \frac{\bar{h}}{(\bar{y}-\bar{z})^2}\phi(z,\bar{z}) + \frac{1}{(\bar{y}-\bar{z})}\partial_{\bar z}\phi(z,\bar{z}) + \cdots, \qquad (3.25)$$

where the ellipsis indicates that we have ignored parts which are regular (analytic). These OPEs constitute an alternative definition of a primary field with holomorphic and antiholomorphic weights $h,\bar{h}$, often referred to simply as an $(h,\bar{h})$ primary.

We are at liberty to Laurent expand the infinitesimal transformation around $(z,\bar{z}) = 0$:

$$\epsilon(z) = -\sum_{n=-\infty}^{\infty} a_n z^{n+1}, \qquad \bar{\epsilon}(\bar{z}) = -\sum_{n=-\infty}^{\infty} \bar{a}_n \bar{z}^{n+1},$$

where the $a_n, \bar{a}_n$ are coefficients. The quantities which appear as generators, $\ell_n = z^{n+1}\partial_z, \bar{\ell}_n = \bar{z}^{n+1}\partial_{\bar{z}}$, satisfy the commutation relations

$$[\ell_n, \ell_m] = (n-m)\ell_{n+m},$$
$$[\ell_n, \bar{\ell}_m] = 0,$$
$$[\bar{\ell}_n, \bar{\ell}_m] = (n-m)\bar{\ell}_{n+m}, \qquad (3.26)$$

which is the classical version of the Virasoro algebra we saw previously in equation (2.63), or the quantum case in equation (2.71) with the central extension, $c = \bar{c} = 0$.

Now we can compare with what we learned here. It should be clear after some thought that $\ell_{-1}, \ell_0, \ell_1$ and their antiholomorphic counterparts form the six generators of the global conformal transformations generating $SL(2, \mathbb{C}) = SL(2, \mathbb{R}) \times SL(2, \mathbb{R})$. In fact, $\ell_{-1} = \partial_z$ and $\bar{\ell}_{-1} = \partial_{\bar{z}}$ generate translations, $\ell_0 + \bar{\ell}_0$ generates dilations, $i(\ell_0 - \bar{\ell}_0)$ generates rotations, while $\ell_1 = z^2 \partial_z$ and $\bar{\ell}_1 = \bar{z}^2 \partial_{\bar{z}}$ generate the special conformal transformations.

Let us note some useful pieces of terminology and physics here. Recall that we had defined physical states to be those annihilated by the $\ell_n, \bar{\ell}_n$ with $n > 0$. Then ℓ_0 and $\bar{\ell}_0$ will measure properties of these physical states. Considering them as operators, we can find a basis of ℓ_0 and $\bar{\ell}_0$ eigenstates, with eigenvalues h and $\bar{h}$ (two independent numbers), which are the 'conformal weights' of the state: $\ell_0|h\rangle = h|h\rangle$, $\bar{\ell}_0|\bar{h}\rangle = \bar{h}|\bar{h}\rangle$. Since the sum and difference of these operators are the dilations and the rotations, we can characterise the scaling dimension and the spin of a state or field as $\Delta = h + \bar{h}$, $s = h - \bar{h}$.

It is worth noting here that the stress-tensor itself is *not* in general a primary field of weight $(2, 2)$, despite the suggestive fact that it has two indices. There can be an anomalous term, allowed by the symmetries of the theory:

$$T(z)T(y) = \frac{c}{2}\frac{1}{(z-y)^4} + \frac{2}{(z-y)^2}T(y) + \frac{1}{z-y}\partial_y T(y),$$

$$\bar{T}(\bar{z})\bar{T}(\bar{y}) = \frac{\bar{c}}{2}\frac{1}{(\bar{z}-\bar{y})^4} + \frac{2}{(\bar{z}-\bar{y})^2}\bar{T}(\bar{y}) + \frac{1}{\bar{z}-\bar{y}}\partial_{\bar{y}}\bar{T}(\bar{y}). \quad (3.27)$$

The holomorphic conformal anomaly c and its antiholomorphic counterpart $\bar{c}$, can in general be non-zero. We shall see this occur below.

It is worthwhile turning some of the above facts into statements about commutation relation between the modes of $T(z), \bar{T}(\bar{z})$, which we remind the reader are defined as:

$$T(z) = \sum_{n=-\infty}^{\infty} L_n z^{-n-2}, \qquad L_n = \frac{1}{2\pi i}\oint dz\, z^{n+1}T(z),$$

$$\bar{T}(\bar{z}) = \sum_{n=-\infty}^{\infty} \bar{L}_n \bar{z}^{-n-2}, \qquad \bar{L}_n = \frac{1}{2\pi i}\oint d\bar{z}\, \bar{z}^{n+1}\bar{T}(\bar{z}). \quad (3.28)$$

In these terms, the resulting commutator between the modes is that displayed in equation (2.71), with D replaced by $\bar{c}$ and c on the right and left.

The definition (3.24) of the primary fields ϕ translates into

$$[L_n, \phi(y)] = \frac{1}{2\pi i}\oint dz\, z^{n+1}T(z)\phi(y) = h(n+1)y^n\phi(y) + y^{n+1}\partial_y\phi(y). \quad (3.29)$$

It is useful to decompose the primary into its modes:

$$\phi(z) = \sum_{n=-\infty}^{\infty} \phi_n z^{-n-h}, \qquad \phi_n = \frac{1}{2\pi i} \oint dz \; z^{h+n-1} \phi(z). \qquad (3.30)$$

In terms of these, the commutator between a mode of a primary and of the stress tensor is:

$$[L_n, \phi_m] = [n(h-1) - m]\phi_{n+m}, \qquad (3.31)$$

with a similar antiholomorphic expression. In particular this means that our correspondence between states and operators can be made precise with these expressions. $L_0|h\rangle = h|h\rangle$ matches with the fact that $\phi_{-h}|0\rangle = |h\rangle$ would be used to make a state, or more generally $|h, \bar{h}\rangle$, if we include both holomorphic and antiholomorphic parts. The result $[L_0, \phi_{-h}] = h\phi_{-h}$ guarantees this.

In terms of the finite transformation of the stress tensor under $z \to z'$, the result (3.27) is

$$T(z) = \left(\frac{\partial z'}{\partial z}\right)^2 T(z') + \frac{c}{12} \left(\frac{\partial z'}{\partial z}\right)^{-2} \left[\frac{\partial z'}{\partial z}\frac{\partial^3 z'}{\partial z^3} - \frac{3}{2}\left(\frac{\partial^2 z'}{\partial z^2}\right)^2\right], \qquad (3.32)$$

where the quantity multiplying $c/12$ is called the 'Schwarzian derivative', $S(z, z')$. It is interesting to note (and the reader should check) that for the $SL(2, \mathbb{C})$ subgroup, the proper global transformations, $S(z, z') = 0$. This means that the stress tensor is in fact a quasi-primary field, but not a primary field.

3.2 Revisiting the relativistic string

Now we see the full role of the energy-momentum tensor which we first encountered in the previous chapter. Its Laurent coefficients there, L_n and $\bar{L}_n$, realised there in terms of oscillators, satisfied the Virasoro algebra, and so its role is to generate the conformal transformations. We can use it to study the properties of various operators in the theory of interest to us.

First, we translate our result of equation (2.44) into the appropriate coordinates here:

$$T(z) = -\frac{1}{\alpha'} : \partial_z X^\mu(z) \partial_z X_\mu(z) :,$$

$$\bar{T}(\bar{z}) = -\frac{1}{\alpha'} : \partial_{\bar{z}} X^\mu(\bar{z}) \partial_{\bar{z}} X_\mu(\bar{z}) :. \qquad (3.33)$$

We can use here our definition (3.20) of the normal ordering at the operator level here, which we construct with the OPE. To do this, we need to know the result for the OPE of ∂X^μ with itself. This we can get by observing that the propagator of the field $X^\mu(z, \bar{z}) = X(z) + \bar{X}(\bar{z})$ is

$$\langle X(z)^\mu X^\nu(y) \rangle = -\frac{\alpha'}{2} \eta^{\mu\nu} \log(z - y),$$

$$\langle \bar{X}(\bar{z})^\mu \bar{X}^\nu(\bar{y}) \rangle = -\frac{\alpha'}{2} \eta^{\mu\nu} \log(\bar{z} - \bar{y}). \tag{3.34}$$

By taking a couple of derivatives, we can deduce the OPE of $\partial_z X^\mu(z)$ or $\partial_{\bar{z}} \bar{X}^\mu(\bar{z})$:

$$\partial_z X^\mu(z) \partial_y X^\nu(y) = -\frac{\alpha'}{2} \frac{\eta^{\mu\nu}}{(z - y)^2} + \cdots$$

$$\partial_{\bar{z}} \bar{X}^\nu(\bar{z}) \partial_{\bar{y}} \bar{X}^\mu(\bar{y}) = -\frac{\alpha'}{2} \frac{\eta^{\mu\nu}}{(\bar{z} - \bar{y})^2} + \cdots. \tag{3.35}$$

So in the above, we have, using our definition of the normal ordered expression using the OPE (see discussion below equation (3.20)):

$$T(z) = -\frac{1}{\alpha'} : \partial_z X^\mu(z) \partial_z X_\mu(z) := -\frac{1}{\alpha'} \lim_{y \to z} \left[\partial_z X^\mu(z) \partial_z X_\mu(y) - \frac{D}{(\bar{z} - \bar{y})^2} \right], \tag{3.36}$$

with a similar expression for the antiholomorphic part. It is now straightforward to evaluate the OPE of $T(z)$ and $\partial_z X^\nu(y)$. We simply extract the singular part of the following:

$$T(z) \partial_y X^\nu(y) = \frac{1}{\alpha'} : \partial_z X^\mu(z) \partial_z X_\mu(z) : \partial_y X^\nu(y)$$

$$= 2 \cdot \frac{1}{\alpha'} \partial_z X^\mu(z) \langle \partial_z X_\mu(z) \partial_z X^\nu(y) \rangle + \cdots$$

$$= \partial_z X^\nu(z) \frac{1}{(z - y)^2} + \cdots. \tag{3.37}$$

In the above, we were instructed by Wick to perform the two possible contractions to make the correlator. The next step is to Taylor expand for small $(z - y)$: $X^\nu(z) = X^\nu(y) + (z - y)\partial_y X^\nu(y) + \cdots$, substitute into our result, to give:

$$T(z) \partial_y X^\nu(y) = \frac{\partial_y X^\nu(y)}{(z - y)^2} + \frac{\partial_y^2 X^\nu(y)}{z - y} + \cdots, \tag{3.38}$$

and so we see from our definition in equation (3.25) that that the field $\partial_z X^\nu(z)$ is a primary field of weight $h = 1$, or a $(1, 0)$ primary field, since from the OPEs (3.35), its OPE with $\bar{T}$ obviously vanishes. Similarly, the antiholomorphic part is a $(0,1)$ primary. Notice that we should have suspected this to be true given the OPE we deduced in (3.35).

Another operator we used last chapter was the normal ordered exponentiation $V(z) =: \exp(i\mathbf{k} \cdot \mathbf{X}(z))$:, which allowed us to represent the momentum of a string state. Here, the normal ordering means that we should not contract the various Xs which appear in the expansion of the exponential with each other. We can extract the singular part to define the OPE with $T(z)$ by following our noses and applying the Wick procedure as before:

$$
\begin{aligned}
T(z)V(y) &= \frac{1}{\alpha'} : \partial_z X^\mu(z)\partial_z X_\mu(z) :: e^{i\mathbf{k}\cdot\mathbf{X}(y)} : \\
&= \frac{1}{\alpha'}(\langle\partial_z X^\mu(z) i\mathbf{k} \cdot \mathbf{X}(y)\rangle)^2 : e^{i\mathbf{k}\cdot\mathbf{X}(y)} : \\
&\quad + 2 \cdot \frac{1}{\alpha'}\partial_z X^\mu(z)\langle\partial_z X_\mu(z) i\mathbf{k} \cdot \mathbf{X}(y)\rangle : e^{i\mathbf{k}\cdot\mathbf{X}(y)} : \\
&= \frac{\alpha' k^2}{4}\frac{1}{(z-y)^2} : e^{i\mathbf{k}\cdot\mathbf{X}(y)} : + \frac{i\mathbf{k} \cdot \partial_z\mathbf{X}(z)}{(z-y)} : e^{i\mathbf{k}\cdot\mathbf{X}(y)} : \\
&= \frac{\alpha' k^2}{4}\frac{V(y)}{(z-y)^2} + \frac{\partial_y V(y)}{(z-y)}.
\end{aligned}
\tag{3.39}
$$

We have Taylor expanded in the last line, and throughout we only displayed explicitly the singular parts. The expressions tidy up themselves quite nicely if one realises that the worst singularity comes from when there are two contractions with products of fields using up both pieces of $T(z)$. Everything else is either non-singular, or sums to reassemble the exponential after combinatorial factors have been taken into account. This gives the first term of the second line. The second term of that line comes from single contractions. The factor of two comes from making two choices to contract with one or other of the two identical pieces of $T(z)$, while there are other factors coming from the n ways of choosing a field from the term of order n from the expansion of the exponential. After dropping the non-singular term, the remaining terms (with the n) reassemble the exponential again. (The reader is advised to check this explicitly to see how it works.) The final result (when combined with the antiholomorphic counterpart) shows that $V(y)$ is a primary field of weight $(\alpha' k^2/4, \alpha' k^2/4)$.

Now we can pause to see what this all means. Recall from section 2.4.1 that the insertion of states is equivalent to the insertion of operators into the theory, so that:

$$S \to S' = S + \lambda \int d^2z \mathcal{O}(z, \bar{z}). \qquad (3.40)$$

In general, we may consider such an operator insertion for a general theory. For the theory to remain conformally invariant, the operator must be a *marginal* operator, which is to say that $\mathcal{O}(z, \bar{z})$ must at least have dimension $(1, 1)$ do that the integrated operator is dimensionless. In principle, the dimension of the operator after the deformation (i.e. in the new theory defined by S') can change, and so the full condition for the operator is that it must remain $(1, 1)$ after the insertion (see insert 3.1). It in fact defines a direction in the space of couplings, and λ can be thought of as an infinitessimal motion in that direction. The statement of the existence of a marginal operator is then referred to the existence of a 'flat direction'.

In the first instance, we recall that the use of the tachyon vertex operator $V(z, \bar{z})$ corresponds to the addition of $\int d^2z\, V(z, \bar{z})$ to the action. We wish the theory to remain conformal (preserving the relativistic string's symmetries, as stressed in chapter 1), and so $V(z, \bar{z})$ must be $(1,1)$. In fact, since our conformal field theory is actually free, we need do no more to check that the tachyon vertex is marginal. So we require that $(\alpha' k^2/4, \alpha' k^2/4) = (1, 1)$. Therefore we get the result that $M^2 \equiv -k^2 = -4/\alpha'$, the result that we obtained previously for the tachyon.

Another example is the level one closed string vertex operator:

$$: \partial_z X^\mu \partial_{\bar{z}} X^\nu \exp(i\mathbf{k} \cdot \mathbf{X}) : .$$

It turns out that there are no further singularities in contracting this with the stress tensor, and so the weight of this operator is $(1 + \alpha' k^2/4, 1 + \alpha' k^2/4)$. So, marginality requires that $M^2 \equiv -k^2 = 0$, which is the massless result that we encountered earlier.

Another computation that the reader should consider doing is to work out explicitly the $T(z)T(y)$ OPE, and show that it is of the form (3.27) with $c = D$, as each of the D bosons produces a conformal anomaly of unity. This same is true from the antiholomorphic sector, giving $\bar{c} = D$. Also, for open strings, we get the same amount for the anomaly. This result was alluded to in chapter 2. This is problematic, since this conformal anomaly prevents the full operation of the string theory. In particular, the anomaly means that the stress tensor's trace does not in fact vanish quantum mechanically.

This is all repaired in the next section, since there is another sector which we have not yet considered.

Insert 3.1.   Deformations, RG flows, and CFTs

A useful picture to have in mind for later use is of a conformal field theory as a 'fixed point' in the space of theories coordinatised by the coefficients of possible operators such as in equation (3.40). (There is an infinite set of such perturbations and so the space is infinite dimensional.) In the usual reasoning using the renomalisation group (RG), once the operator is added with some value of the coupling, the theory (i.e. the value of the coupling) flows along an RG trajectory as the energy scale μ is changed. The 'β-function', $\beta(\lambda) \equiv \mu \partial \lambda / \partial \mu$ characterises the behaviour of the coupling. One can imagine the existence of 'fixed points' of such flows, where $\beta(\lambda) = 0$ and the coupling tends to a specific value, as shown in the diagram.

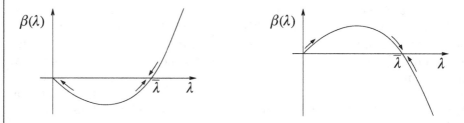

On the left, $\bar{\lambda}$ is an 'infra-red (IR) fixed point', since the coupling is driven to it for decreasing μ, while on the right, $\bar{\lambda}$ is an 'ultra-violet (UV) fixed point', since the coupling is driven to it for increasing μ. The origins of each diagram of course define a fixed point of the opposite type to that at $\bar{\lambda}$. A conformal field theory is then clearly such a fixed point theory, where the scale dependence of all couplings exactly vanishes. A *'marginal operator'* is an operator which when added to the theory, does not take it away from the fixed point. A *'relevant operator'* deforms a theory increasingly as μ goes to the IR, while an *'irrelevant operator'* is increasingly less important in the IR. This behaviour is reversed on going to the UV. When applied to a fixed point, such non-marginal operators can be used to deform fixed point theories away from the conformal point, often allowing us to find other interesting theories, as we will do in later chapters. $D = 4$ Yang–Mills theories, for sufficiently few flavours of quark (like QCD), have negative β-function, and so behave roughly as the neighbourhood of the origin in the left diagram. *'Asymptotic freedom'* is the process of being driven to the origin (zero coupling) in the UV. Later, we will see examples of both type of fixed point theory.

3.3 Fixing the conformal gauge

It must not be forgotten where all of the riches of the previous section – the conformal field theory – came from. We made a gauge choice in equation (2.41) from which many excellent results followed. However, despite everything, we saw that there is in fact a conformal anomaly equal to D (or a copy each on both the left and the right hand side, for the closed string). The problem is that we have not made sure that the gauge fixing was performed properly. This is because we are fixing a local symmetry, and it needs to be done dynamically in the path integral, just as in gauge theory. This is done with Faddeev–Popov ghosts in a very similar way to the methods used in field theory. Let us not go into the details of it here, but assume that the interested reader can look into the many presentations of the procedure in the literature. The key difference with field theory approach is that it introduces two ghosts, c^a and b_{ab} which are rank one and rank two tensors on the world sheet. The action for them is:

$$S^{\text{gh}} = -\frac{1}{4\pi} \int d^2\sigma \sqrt{g} g^{ab} c^c \nabla_a b_{bc}, \tag{3.41}$$

and so b_{ab} and c^a, which are anticommuting, are conjugates of each other.

3.3.1 Conformal ghosts

Once the conformal gauge has been chosen, (see equation (2.41)) picking the diagonal metric, we have

$$S^{\text{gh}} = -\frac{1}{2\pi} \int d^2z \left(c(z)\partial_{\bar{z}} b(z) + \bar{c}(\bar{z})\partial_z \bar{b}(\bar{z}) \right). \tag{3.42}$$

From equation (3.41), the stress tensor for the ghost sector is:

$$T^{\text{gh}}(z) =: c(z)\partial_z b(z) : + : 2(\partial_z c(z))b(z) :, \tag{3.43}$$

with a similar expression for $\bar{T}_{\text{ghost}}(\bar{z})$. Just as before, as the ghosts are free fields, with equations of motion $\partial_z c = 0 = \partial_z b$, we can Laurent expand them as follows:

$$b(z) = \sum_{n=-\infty}^{\infty} b_n z^{-n-2}, \qquad c(z) = \sum_{n=-\infty}^{\infty} c_n z^{-n+1}, \tag{3.44}$$

which follows from the property that b is of weight 2 and c is of weight -1, a fact which might be guessed from the structure of the action (3.41). The quantisation yields

$$\{b_m, c_n\} = \delta_{m+n}. \tag{3.45}$$

and the stress tensor is

$$L_n^{\text{gh}} = \sum_{m=-\infty}^{\infty} (2n - m) : b_m c_{n-m} : -\delta_{n,0}, \tag{3.46}$$

where we have a normal ordering constant -1, as in the previous sector,

$$[L_m^{\text{gh}}, b_n] = (m - n)b_{m+n}, \qquad [L_m^{\text{gh}}, c_n] = -(2m + n)c_{m+n}. \tag{3.47}$$

The OPE for the ghosts is given by

$$b(z)c(y) = \frac{1}{(z - y)} + \cdots, \qquad c(z)b(y) = \frac{1}{(z - y)} + \cdots,$$
$$b(z)b(y) = O(z - y), \qquad c(z)c(y) = O(z - y), \tag{3.48}$$

where the second expression is obtained from the first by the anticommuting property of the ghosts. The second line also follows from the anticommuting property. There can be no non-zero result for the singular parts there.

As with everything for the closed string, we must supplement the above expressions with very similar ones referring to $\bar{z}$, $\bar{c}(\bar{z})$ and $\bar{b}(\bar{z})$. For the open string, we carry out the same procedures as before, defining everything on the upper half-plane, reflecting the holomorphic into the antiholomorpic parts, defining a single set of ghosts (see also insert 3.2).

3.3.2 *The critical dimension*

Now comes the fun part. We can evaluate the conformal anomaly of the ghost system, by using the techniques for computation of the OPE that we refined in the previous section. We can do it for the ghosts in as simple a way as for the ordinary fields, using the expression (3.43) above. In the following, we will focus on the most singular part, to isolate the conformal anomaly term. This will come from when there are two contractions in each term. The next level of singularity comes from one contraction, and so on:

$$T^{\text{gh}}(z)T^{\text{gh}}(y)$$
$$= (: \partial_z b(z)c(z) : + : 2b(z)\partial_z c(z) :)(: \partial_y b(y)c(y) : + : 2b(y)\partial_y c(y) :)$$
$$= : \partial_z b(z)c(z) :: \partial_y b(y)c(y) : +2 : b(z)\partial_z c(z) :: \partial_y b(y)c(y) :$$
$$+ 2 : \partial_z b(z)c(z) :: b(y)\partial_y c(y) : +4 : b(z)\partial_z c(z) :: b(y)\partial_y c(y) :$$
$$= \langle\partial_z b(z)c(y)\rangle\langle c(z)\partial_y b(y)\rangle + 2\langle b(z)c(y)\rangle\langle\partial_z c(z)\partial_y b(y)\rangle$$
$$+ 2\langle\partial_z b(z)\partial_y c(y)\rangle\langle c(z)b(y)\rangle + 4\langle b(z)\partial_y c(y)\rangle\langle\partial_z c(z)b(y)\rangle$$
$$= -\frac{13}{(z - y)^4}, \tag{3.49}$$

<hr>

Insert 3.2.   Further aspects of conformal ghosts

Notice that the flat space expression (3.42) is also consistent with the stress tensor

$$T(z) =: \partial_z b(z)c(z) : -\kappa : \partial_z[b(z)c(z)] :, \qquad (3.50)$$

for arbitrary κ, with a similar expression for the antiholomorphic sector. It is a useful exercise to use the OPEs of the ghosts given in equation (3.48) to verify that this gives b and c conformal weights $h = \kappa$ and $h = 1 - \kappa$, respectively. The case we studied above was $\kappa = 2$. Further computation (recommended) reveals that the conformal anomaly of this system is $c = 1 - 3(2\kappa - 1)^2$, with a similar expression for the antiholomorphic version of the above.

The case of fermionic ghosts will be of interest to us later. In that case, the action and stress tensor are just like before, but with $b \rightarrow \beta$ and $c \rightarrow \gamma$, where β and γ, are *fermionic*. Since they are fermionic, they have singular OPEs

$$\beta(z)\gamma(y) = -\frac{1}{(z - y)} + \cdots, \qquad \gamma(z)\beta(y) = \frac{1}{(z - y)} + \cdots . \ (3.51)$$

A computation gives conformal anomaly $3(2\kappa - 1)^2 - 1$, which in the case $\kappa = 3/2$, gives an anomaly of 11. In this case, they are the 'superghosts', required by supersymmetry in the construction of superstrings later on.

<hr>

and so comparing with equation (3.27), we see that the ghost sector has conformal anomaly $c = -26$. A similar computation gives $\bar{c} = -26$.

So recalling that the 'matter' sector, consisting of the D bosons, has $c = \bar{c} = D$, we have achieved the result that the conformal anomaly vanishes in the case $D = 26$. This also applies to the open string in the obvious way.

3.4 The closed string partition function

We have all of the ingredients we need to compute our first one-loop diagram[†]. It will be useful to do this as a warm up for more complicated

<hr>

[†] Actually, we've had them for some time now, essentially since chapter 2.

examples later, and in fact we will see structures in this simple case which will persist throughout.

Consider the closed string diagram of figure 3.2(a). This is a vacuum diagram, since there are no external strings. This torus is clearly a one loop diagram and in fact it is easily computed. It is distinguished topologically by having two completely independent one-cycles. To compute the path integral for this we are instructed, as we have seen, to sum over all possible metrics representing all possible surfaces, and hence all possible tori.

Well, the torus is completely specified by giving it a flat metric, and a complex structure, τ, with $\mathrm{Im}\tau \geq 0$. It can be described by the lattice given by quotienting the complex w-plane by the equivalence relations

$$w \sim w + 2\pi n; \quad w \sim w + 2\pi m\tau, \tag{3.52}$$

for any integers m and n, as shown in figure 3.2(b). The two one-cycles can be chosen to be horizontal and vertical. The complex number τ specifies the *shape* of a torus, which cannot be changed by infinitesimal diffeomorphisms of the metric, and so we must sum over all all of them. Actually, this naive reasoning will make us overcount by a lot, since in fact there are a lot of τs which define the same torus. For example, clearly for a torus with given value of τ, the torus with $\tau + 1$ is the same torus, by the equivalence relation (3.52). The full family of equivalent tori can be reached from any τ by the 'modular transformations':

$$\begin{aligned} T: & \quad \tau \to \tau + 1 \\ S: & \quad \tau \to -\frac{1}{\tau}, \end{aligned} \tag{3.53}$$

which generate the group $SL(2, \mathbb{Z})$, which is represented here as the group

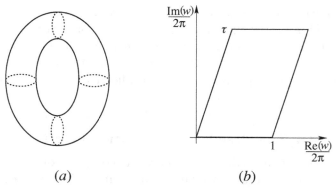

(a) (b)

Fig. 3.2. (a) A closed string vacuum diagram. (b) The flat torus and its complex structure.

of 2×2 unit determinant matrices with integer elements:

$$SL(2, \mathbb{Z}): \quad \tau \rightarrow \frac{a\tau + b}{c\tau + d}; \quad \text{with} \quad \begin{pmatrix} a & b \\ c & d \end{pmatrix}, \quad ad - bc = 1. \quad (3.54)$$

(It is worth noting that the map between tori defined by S exchanges the two one-cycles, therefore exchanging space and (Euclidean) time.) The full family of inequivalent tori is given not by the upper half-plane $H_\perp$ (i.e. τ such that $\text{Im}\tau \geq 0$) but the quotient of it by the equivalence relation generated by the group of modular transformations. This is $\mathcal{F} = H_\perp / PSL(2, \mathbb{Z})$, where the P reminds us that we divide by the extra $\mathbb{Z}_2$ which swaps the sign on the defining $SL(2, \mathbb{Z})$ matrix, which clearly does not give a new torus. The commonly used fundamental domain in the upper half-plane corresponding to the inequivalent tori is drawn in figure 3.3. Any point outside that can be mapped into it by a modular transformation.

The fundamental region $\mathcal{F}$ is properly defined as follows: Start with the region of the upper half-plane which is in the interval $(-\frac{1}{2}, +\frac{1}{2})$ and above the circle of unit radius. we must then identify the two vertical edges, and also the two halves of the remaining segment of the circle. This produces a space which is smooth everywhere except for two points about which there are conical singularities, described in insert 3.3.

The string propagation on our torus can be described as follows. Imagine that the string is of length 1, and lies horizontally. Mark a point on the string. Running time upwards, we see that the string propagates for a time $t = 2\pi\text{Im}\tau \equiv 2\pi\tau_2$. Once it has got to the top of the diagram, we see that

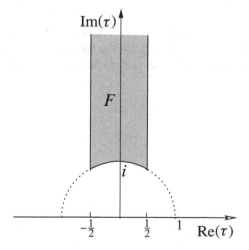

Fig. 3.3. The space of inequivalent tori.

Insert 3.3.   Special points in the moduli space of tori

Actually, there are two very special points of interest on $\mathcal{F}$, depicted in figure 3.3. They can be clearly seen in the figure. The point $\tau = i$ and the point $\tau = e^{\frac{2\pi i}{3}}$, which is one sharp corner (its mirror image is also visible). The significance of these points is that they are fixed points of certain elements of $SL(2, \mathbb{Z})$. The point $\tau = i$ is fixed by the element S, while the other point is fixed by the element ST.

These points are 'orbifold' singularities, a term that will become more widely used here after chapter 4. For our purposes here, this means that they have a conical deficit angle. For example, the point $\tau = i$, because it is at the tip of a region formed by folding the plane in half (remember we identified the two halves of the circle segment), has a deficit angle of π. In other words, because of the folding, one only needs to go half way around a circle in order to return to where one started. Similalry, the other orbifold point has a deficit angle of $4\pi/3$: one only needs to go a third of the way around a circle in order to return to where one started.

One may visualise the significance of these points, recalling that we make the tori from lattices in the plane. The lattices for these two points have special, and familiar, symmetry. The $\tau = i$ point is simply a *square* lattice, and S is in fact just a $\pi/2$ rotation. Notice that $S^4 = 1$, which fits with this fact nicely. The $\tau = e^{\frac{2\pi i}{3}}$ point is an *hexagonal* lattice, and ST is a rotation by $\pi/3$, which dovetails nicely with the relation $(ST)^6 = 1$. We draw the lattice below, with appropriate basis vectors. It might be worth studying the action of S and ST, and considering the tori to which they correspond.

our marked point has shifted rightwards by an amount $x = 2\pi \text{Re}\tau \equiv 2\pi \tau_1$. We actually already have studied the operators that perform these two operations. The operator for time translations is the Hamiltonian (2.64), $H = L_0 + \bar{L}_0 - (c + \bar{c})/24$ while the operator for translations along the string is the momentum $P = L_0 - \bar{L}_0$ discussed above equation (2.73). Recall that $c = \bar{c} = D - 2 = 24$. So our vacuum path integral is

$$Z = \text{Tr}\left\{ e^{-2\pi\tau_2 H} e^{2\pi i \tau_1 P} \right\} = \text{Tr} q^{L_0 - \frac{c}{24}} \bar{q}^{\bar{L}_0 - \frac{\bar{c}}{24}}. \tag{3.55}$$

Here, $q \equiv e^{2\pi i \tau}$, and the trace means a sum over everything which is discrete and an integral over everything which is continuous, which in this case, is simply τ. This is easily evaluated, as the expressions for L_0 and $\bar{L}_0$ give a family of simple geometric sums (see insert 3.4 (p. 92)), and the result can be written as:

$$Z = \int_{\mathcal{F}} \frac{d^2\tau}{\tau_2^2} Z(q), \quad \text{where} \tag{3.56}$$

$$Z(q) = |\tau_2|^{-12} (q\bar{q})^{-1} \left| \prod_{n=1}^{\infty} (1 - q^n)^{-24} \right|^2 = (\sqrt{\tau_2} \eta \bar{\eta})^{-24}, \tag{3.57}$$

is the 'partition function', with Dedekind's function

$$\eta(q) \equiv q^{\frac{1}{24}} \prod_{n=1}^{\infty} (1 - q^n); \quad \eta\left(-\frac{1}{\tau}\right) = \sqrt{-i\tau}\, \eta(\tau). \tag{3.58}$$

This is a pleasingly simple result. One very interesting property it has is that it is actually 'modular invariant'. It is invariant under the T transformation in equation (3.52), since under $\tau \to \tau + 1$, we get that $Z(q)$ picks up a factor $\exp(2\pi i(L_0 - \bar{L}_0))$. This factor is precisely unity, as follows from the level matching formula (2.73). Invariance of $Z(q)$ under the S transformation $\tau \to -1/\tau$ follows from the property mentioned in equation (3.58), after a few steps of algebra, and using the result $S: \tau_2 \to \tau_2/|\tau|^2$.

Modular invariance of the partition function is a crucial property. It means that we are correctly integrating over all inequivalent tori, which is required of us by diffeomorphism invariance of the original construction. Furthermore, we are counting each torus only once, which is of course important.

Note that $Z(q)$ really deserves the name 'partition function' since if it is expanded in powers of q and $\bar{q}$, the powers in the expansion – after multiplication by $4/\alpha'$ – refer to the (mass)^2 level of excitations on the left

---

Insert 3.4.   Partition functions

It is not hard to do the sums. Let us look at one dimension, and so one family of oscillators α_n. We need to consider

$$\operatorname{Tr} q^{L_0} = \operatorname{Tr} q^{\sum_{n=0}^{\infty} \alpha_{-n} \alpha_n}.$$

We can see what the operator $q^{\sum_{n=0}^{\infty} \alpha_{-n} \alpha_n}$ means if we write it explicitly in a basis of all possible multiparticle states of the form $\alpha_{-n}|0\rangle$, $(\alpha_{-n})^2|0\rangle$, etc.:

$$q^{\alpha_{-n} \alpha_n} = \begin{pmatrix} 1 & & & & \\ & q^n & & & \\ & & q^{2n} & & \\ & & & q^{3n} & \\ & & & & \ddots \end{pmatrix},$$

and so clearly $\operatorname{Tr} q^{\alpha_{-n} \alpha_n} = \sum_{i=1}^{\infty} (q^n)^i = (1 - q^n)^{-1}$, which is remarkably simple! The final sum over all modes is trivial, since

$$\operatorname{Tr} q^{\sum_{n=0}^{\infty} \alpha_{-n} \alpha_n} = \prod_{n=0}^{\infty} \operatorname{Tr} q^{\alpha_{-n} \alpha_n} = \prod_{n=0}^{\infty} (1 - q^n)^{-1}.$$

We get a factor like this for all 24 dimensions, and we also get contributions from both the left and right to give the result.

Notice that if our modes were fermions, ψ_n, things would be even simpler. We would not be able to make multiparticle states $(\psi_{-n})^2|0\rangle$, (Pauli), and so we only have a 2×2 matrix of states to trace in this case, and so we simply get

$$\operatorname{Tr} q^{\psi_{-n} \psi_n} = (1 + q^n).$$

Therefore the partition function is

$$\operatorname{Tr} q^{\sum_{n=0}^{\infty} \psi_{-n} \psi_n} = \prod_{n=0}^{\infty} \operatorname{Tr} q^{\psi_{-n} \psi_n} = \prod_{n=0}^{\infty} (1 + q^n).$$

We will encounter such fermionic cases later.

and right, while the coefficient in the expansion gives the degeneracy at that level. The degeneracy is the number of partitions of the level number into positive integers. For example, at level three this is three, since we have $\alpha_{-3}, \alpha_{-1}\alpha_{-2}$, and $\alpha_{-1}\alpha_{-1}\alpha_{-1}$.

The overall factor of $(q\bar{q})^{-1}$ sets the bottom of the tower of masses. Note for example that at level zero we have the tachyon, which appears only once, as it should, with $M^2 = -4/\alpha'$. At level one, we have the massless states, with multiplicity 24^2, which is appropriate, since there are 24^2 physical states in the graviton multiplet $(G_{\mu\nu}, B_{\mu\nu}, \Phi)$. Introducing a common piece of terminology, a term $q^{w_1}\bar{q}^{w_2}$, represents the appearance of a 'weight' (w_1, w_2) field in the 1+1 dimensional conformal field theory, denoting its left-moving and right-moving weights or 'conformal dimensions'.

4

Strings on circles and T-duality

In this chapter we shall study the spectrum of strings propagating in a spacetime that has a compact direction. The theory has all of the properties we might expect from the knowledge that at low energy we are placing gravity and field theory on a compact space. Indeed, as the compact direction becomes small, the parts of the spectrum resulting from momentum in that direction become heavy, and hence less important, but there is much more. The spectrum has additional sectors coming from the fact that closed strings can wind around the compact direction, contributing states whose mass is proportional to the radius. Thus, they become light as the circle shrinks. This will lead us to T-duality, relating a string propagating on a large circle to a string propagating on a small circle[14]. This is just the first of the remarkable symmetries relating two string theories in different situations that we shall encounter here. It is a crucial consequence of the fact that strings are extended objects. Studying its consequences for open strings will lead us to D-branes, since T-duality will relate the Neumann boundary conditions we have already encountered to Dirichlet ones[9, 11], corresponding to open strings ending on special hypersurfaces in spacetime.

4.1 Fields and strings on a circle

Let us remind ourselves of what happens in field theory, for the case of placing gravity on a spacetime with a compact direction. This will help us appreciate the extra features encountered in the case of strings, and will also prepare for remarks to be made in a variety of cases much later. We start with the idea of Kaluza, later refined by Klein.

94

4.1.1 The Kaluza–Klein reduction

Imagine that we are in five dimensions, with metric components G_{MN}, $M, N = 0, \ldots, 4$, and that the spacetime is actually of topology $\mathbb{R}^4 \times S^1$, and so has one compact direction. So we will have the usual four dimensional coordinates on $\mathbb{R}^4$, $(x^\mu, \mu = 0, \ldots, 3)$ and a periodic coordinate, $x^4 = x^4 + 2\pi R$, where R is the radius of the circle.

Now as we have seen before, the five dimensional coordinate transformation $x^M \to x'^M = x^M + \epsilon^M(x)$ is an invariance of our five dimensional theory, under which

$$G_{MN} \to G'_{MN} = G_{MN} - \partial_M \epsilon_N - \partial_N \epsilon_M. \tag{4.1}$$

The metric has the natural decomposition into $G^{(5)}_{\mu\nu}$, $G^{(5)}_{44}$, and $G^{(5)}_{\mu 4}$, where the superscript is necessary to distinguish similar-looking quantities in four dimensions, as we shall see.

Let us consider the class of transformations $\epsilon_4(x^\mu)$, $\epsilon_\mu = 0$, which corresponds to an x^μ-dependent isometry (rotation) of the circle. Then $G^{(5)}_{\mu\nu}$ and $G^{(5)}_{44}$ are invariant, and

$$G^{(5)}_{\mu 4} \to G'^{(5)}_{\mu 4} = G^{(5)}_{\mu 4} - \partial_\mu \epsilon_4(x). \tag{4.2}$$

However, from the four dimensional point of view, $G^{(5)}_{44}$ is a scalar, $G^{(5)}_{\mu\nu}$ is proportional to the metric, and $G^{(5)}_{\mu 4}$ is a vector, proportional to what we will call A_μ, and so equation (4.2) is simply a $U(1)$ gauge transformation: $A_\mu \to A_\mu - \partial_\mu \Lambda(x)$. So the $U(1)$ of electromagnetism can be thought of as resulting from compactifying gravity, the gauge field being an internal component of the metric. The idea of using this, as a first attempt at unifying gravity with electromagnetism, was that R is small enough that the world would be effectively four dimensional on larger scales, so an observer would have to work hard to see it. On distance scales much longer than that set by R, physical quantities in the theory would be effectively x^4-independent.

Let us be a bit more precise. Explicitly, we can write the most general metric consistent with the translation invariance in x^4 as

$$ds^2 = G^{(5)}_{MN} dx^M dx^N = G^{(4)}_{\mu\nu} dx^\mu dx^\nu + G_{44} \left(dx^4 + A_\mu dx^\mu \right)^2, \tag{4.3}$$

and we write $G_{44} = e^{2\phi}$. The five dimensional Ricci scalar decomposes as

$$R^{(5)} = R^{(4)} - 2e^{-\phi} \nabla^2 e^\phi - \tfrac{1}{4} e^{2\phi} F_{\mu\nu} F^{\mu\nu}, \tag{4.4}$$

where $F_{\mu\nu} = \partial_\mu A_\nu - \partial_\nu A_\mu$. Notice for future reference that the lower dimensional metric components in the $0, 1, 2, 3$ directions are a modification of the higher dimensional metric components:

$$G^{(4)}_{\mu\nu} = G^{(5)}_{\mu\nu} - e^{2\phi} A_\mu A_\nu,$$

which is an important observation for later. So, suppressing the x^4 dependence of the fields, we get

$$S = \frac{1}{16\pi G^N_{(5)}} \int (-G_{(5)})^{1/2} R^{(5)} d^5 x$$

$$= \frac{1}{16\pi G^N_{(4)}} \int (-G_{(4)})^{1/2} \left(R^{(4)} - \tfrac{3}{2}\partial_\mu \phi \partial^\mu \phi - \tfrac{1}{4}e^{3\phi} F_{\mu\nu} F^{\mu\nu} \right) d^4 x,$$

where we have defined $\tilde{G}^{(4)}_{\mu\nu} = e^{\phi} G^{(4)}_{\mu\nu}$ and used equation (2.110). Now we have a relation between the five dimensional and four dimensional Newton constants:

$$\frac{2\pi R}{G^N_{(5)}} = \frac{1}{G^N_{(4)}}, \tag{4.5}$$

and the gauge coupling is set by ϕ and Newton's constant.

Let us be more careful about following how the x^4-independence of the theory arises. Since momentum in x^4 is quantised as $p_4 = n/R$, any scalar (or component of a field) in $D = 5$ (which obeys $\partial^M \partial_M \phi = 0$) can be expanded:

$$\phi(x^\mu) = \sum_{n \in \mathbb{Z}} \phi_n(x^\mu) e^{inx^4/R}, \tag{4.6}$$

giving

$$\partial^\mu \partial_\mu \phi_n - \frac{n^2}{R^2} \phi = 0, \tag{4.7}$$

and so we see that the ϕ_n appear in four dimensions as a family of scalars of mass $m = n/R$, and $U(1)$ charge n. We get a tower of states which becomes extremely heavy for very small R, and are therefore hard to excite. We shall see this sort of spectrum arise in the closed string theory as well (since it contains gravity at low energy), but accompanied by new features.

4.1.2 Closed strings on a circle

The mode expansion (2.84) for the closed string theory can be written as:

$$X^\mu(z, \bar{z}) = \frac{x^\mu}{2} + \frac{\tilde{x}^\mu}{2} - i\sqrt{\frac{\alpha'}{2}}(\alpha_0^\mu + \tilde{\alpha}_0^\mu)\tau + \sqrt{\frac{\alpha'}{2}}(\alpha_0^\mu - \tilde{\alpha}_0^\mu)\sigma + \text{oscillators.} \tag{4.8}$$

We have already identified the spacetime momentum of the string:

$$p^\mu = \frac{1}{\sqrt{2\alpha'}}(\alpha_0^\mu + \tilde{\alpha}_0^\mu). \qquad (4.9)$$

If we run around the string, i.e. take $\sigma \to \sigma + 2\pi$, the oscillator terms are periodic and we have

$$X^\mu(z, \bar{z}) \to X^\mu(z, \bar{z}) + 2\pi\sqrt{\frac{\alpha'}{2}}(\alpha_0^\mu - \tilde{\alpha}_0^\mu). \qquad (4.10)$$

So far, we have studied the situation of non-compact spatial directions for which the embedding function $X^\mu(z, \bar{z})$ is single-valued, and therefore the above change must be zero, giving

$$\alpha_0^\mu = \tilde{\alpha}_0^\mu = \sqrt{\frac{\alpha'}{2}}p^\mu. \qquad (4.11)$$

Indeed, momentum p^μ takes a continuum of values reflecting the fact that the direction X^μ is non-compact.

Let us consider the case that we have a compact direction, say X^{25}, of radius R. Our direction X^{25} therefore has period $2\pi R$. The momentum p^{25} now takes the discrete values n/R, for $n \in \mathbb{Z}$. Now, under $\sigma \sim \sigma + 2\pi$, $X^{25}(z, \bar{z})$ is not single valued, and can change by $2\pi w R$, for $w \in \mathbb{Z}$. Solving the two resulting equations gives:

$$\alpha_0^{25} + \tilde{\alpha}_0^{25} = \frac{2n}{R}\sqrt{\frac{\alpha'}{2}}$$

$$\alpha_0^{25} - \tilde{\alpha}_0^{25} = \sqrt{\frac{2}{\alpha'}}wR \qquad (4.12)$$

and so we have:

$$\alpha_0^{25} = \left(\frac{n}{R} + \frac{wR}{\alpha'}\right)\sqrt{\frac{\alpha'}{2}} \equiv P_{\mathrm{L}}\sqrt{\frac{\alpha'}{2}}$$

$$\tilde{\alpha}_0^{25} = \left(\frac{n}{R} - \frac{wR}{\alpha'}\right)\sqrt{\frac{\alpha'}{2}} \equiv P_{\mathrm{R}}\sqrt{\frac{\alpha'}{2}}. \qquad (4.13)$$

We can use this to compute the formula for the mass spectrum in the remaining uncompactified 24+1 dimensions, using the fact that $M^2 = -p_\mu p^\mu$, where now $\mu = 0, \ldots, 24$.

$$M^2 = -p^\mu p_\mu = \frac{2}{\alpha'}(\alpha_0^{25})^2 + \frac{4}{\alpha'}(N - 1)$$

$$= \frac{2}{\alpha'}(\tilde{\alpha}_0^{25})^2 + \frac{4}{\alpha'}(\bar{N} - 1), \qquad (4.14)$$

where $N, \bar{N}$ denote the total levels on the left- and right-moving sides, as before. These equations follow from the left and right $L_0, \bar{L}_0$ constraints. Recall that the sum and difference of these give the Hamiltonian and the level-matching formulae. Here, they are modified, and a quick computation gives:

$$M^2 = \frac{n^2}{R^2} + \frac{w^2 R^2}{\alpha'^2} + \frac{2}{\alpha'}\left(N + \tilde{N} - 2\right)$$
$$nw + N - \tilde{N} = 0. \tag{4.15}$$

The key features here are that there are terms in addition to the usual oscillator contributions. In the mass formula, there is a term giving the familiar contribution of the Kaluza–Klein tower of momentum states for the string (see section 4.1.1), and a new term from the tower of winding states. This latter term is a very stringy phenomenon. Notice that the level matching term now also allows a mismatch between the number of left and right oscillators excited, in the presence of discrete winding and momenta.

In fact, notice that we can get our usual massless Kaluza–Klein states* by taking

$$n = w = 0; \qquad N = \bar{N} = 1, \tag{4.16}$$

exciting an oscillator in the compact direction. There are two ways of doing this, either on the left or the right, and so there are two $U(1)$s following from the fact that there is an internal component of the metric and also of the antisymmetric tensor field. We can choose to identify the two gauge fields of this $U(1) \times U(1)$ as follows:

$$A_{\mu(\mathrm{R})} \equiv \frac{1}{2}(G - B)_{\mu,25}; \quad A_{\mu(\mathrm{L})} \equiv \frac{1}{2}(G + B)_{\mu,25}.$$

We have written these states out explicitly, together with the corresponding spacetime fields, and the vertex operators (at zero momentum), below.

field	state	operator
$G_{\mu\nu}$	$(\alpha^\mu_{-1}\tilde{\alpha}^\nu_{-1} + \alpha^\nu_{-1}\tilde{\alpha}^\mu_{-1})\vert 0; k\rangle$	$\partial X^\mu \bar{\partial} X^\nu + \partial X^\nu \bar{\partial} X^\mu$
$B_{\mu\nu}$	$(\alpha^\mu_{-1}\tilde{\alpha}^\nu_{-1} - \alpha^\nu_{-1}\tilde{\alpha}^\mu_{-1})\vert 0; k\rangle$	$\partial X^\mu \bar{\partial} X^\nu - \partial X^\nu \bar{\partial} X^\mu$
$A_{\mu(R)}$	$\alpha^\mu_{-1}\tilde{\alpha}^{25}_{-1}\vert 0; k\rangle$	$\partial X^\mu \bar{\partial} X^{25}$
$A_{\mu(L)}$	$\tilde{\alpha}^\mu_{-1}\alpha^{25}_{-1}\vert 0; k\rangle$	$\partial X^{25} \bar{\partial} X^\mu$
$\phi \equiv \frac{1}{2}\log G_{25,25}$	$\alpha^{25}_{-1}\tilde{\alpha}^{25}_{-1}\vert 0; k\rangle$	$\partial X^{25} \bar{\partial} X^{25}$

---

* We shall sometimes refer to Kaluza–Klein states as 'momentum' states, to distinguish them from 'winding' states, in what follows.

So we have these 25-dimensional massless states which are basically the components of the graviton and antisymmetric tensor fields in 26 dimensions, now relabelled. (There is also of course the dilaton Φ, which we have not listed.) There is a pair of gauge fields giving a $U(1)_\mathrm{L} \times U(1)_\mathrm{R}$ gauge symmetry, and in addition a massless scalar field ϕ. Actually, ϕ is a massless scalar which can have any background vacuum expectation value (vev), which in fact sets the radius of the circle. This is because the square root of the metric component $G_{25,25}$ is indeed the measure of the radius of the X^{25} direction.

4.2 T-duality for closed strings

Let us now study the generic behaviour of the spectrum (4.15) for different values of R. For larger and larger R, momentum states become lighter, and therefore it is less costly to excite them in the spectrum. At the same time, winding states become heavier, and are more costly. For smaller and smaller R, the reverse is true, and it is gets cheaper to excite winding states while it is momentum states which become more costly.

We can take this further: as $R \to \infty$, all of the winding states, i.e. states with $w \neq 0$, become infinitely massive, while the $w = 0$ states with all values of n go over to a continuum. This fits with what we expect intuitively, and we recover the fully uncompactified result.

Consider instead the case $R \to 0$, where all of the momentum states, i.e. states with $n \neq 0$, become infinitely massive. If we were studying field theory we would stop here, as this would be all that would happen – the surviving fields would simply be independent of the compact coordinate, and so we have performed a dimension reduction. In closed string theory things are quite different: the pure winding states (i.e. $n = 0$, $w \neq 0$, states) form a continuum as $R \to 0$, following from our observation that it is very cheap to wind around the small circle. *Therefore, in the $R \to 0$ limit, an effective uncompactified dimension actually reappears!*

Notice that the formula (4.15) for the spectrum is invariant under the exchange

$$n \leftrightarrow w \qquad \text{and} \qquad R \leftrightarrow R' \equiv \alpha'/R. \qquad (4.17)$$

The string theory compactified on a circle of radius R' (with momenta and windings exchanged) is the 'T-dual' theory[14], and the process of going from one theory to the other will be referred to as 'T-dualising'.

The exchange takes (see (equation 4.13))

$$\alpha_0^{25} \to \alpha_0^{25}, \quad \tilde{\alpha}_0^{25} \to -\tilde{\alpha}_0^{25}. \qquad (4.18)$$

The dual theories are identical in the fully interacting case as well (after a shift of the coupling to be discussed shortly)[15]. Simply rewrite the radius

R theory by performing the exchange

$$X^{25}(z, \bar{z}) = X^{25}(z) + X^{25}(\bar{z}) \longrightarrow X'^{25}(z, \bar{z}) = X^{25}(z) - X^{25}(\bar{z}). \quad (4.19)$$

The energy-momentum tensor and other basic properties of the conformal field theory are invariant under this rewriting, and so are therefore all of the correlation functions representing scattering amplitudes, etc. The only change, as follows from equation (4.18), is that the zero mode spectrum in the new variable is that of the α'/R theory.

So these theories are physically identical. T-duality, relating the *R* and α'/R theories, is an exact symmetry of perturbative closed string theory. Shortly, we shall see that it is non-perturbatively exact as well.

N.B. The transformation (4.19) can be regarded as a spacetime parity transformation acting only on the right-moving (in the world sheet sense) degrees of freedom. We shall put this picture to good use in what is to come.

4.3 A special radius: enhanced gauge symmetry

Given the relation we deduced between the spectra of strings on radii *R* and α'/R, it is clear that there ought to be something interesting about the theory at the radius $R = \sqrt{\alpha'}$. The theory should be self-dual, and this radius is the 'self-dual radius'. There is something else special about this theory besides just self-duality.

At this radius we have, using (4.13),

$$\alpha_0^{25} = \frac{(n + w)}{\sqrt{2}}; \qquad \tilde{\alpha}_0^{25} = \frac{(n - w)}{\sqrt{2}}, \quad (4.20)$$

and so from the left and right we have:

$$\begin{aligned} M^2 = -p^\mu p_\mu &= \frac{2}{\alpha'}(n + w)^2 + \frac{4}{\alpha'}(N - 1) \\ &= \frac{2}{\alpha'}(n - w)^2 + \frac{4}{\alpha'}(\bar{N} - 1). \end{aligned} \quad (4.21)$$

So if we look at the massless spectrum, we have the conditions:

$$(n + w)^2 + 4N = 4; \qquad (n - w)^2 + 4\bar{N} = 4. \quad (4.22)$$

As solutions, we have the cases $n = w = 0$ with $N = 1$ and $\bar{N} = 1$ from before. These are include the vectors of the $U(1) \times U(1)$ gauge symmetry of the compactified theory.

Now, however, we see that we have more solutions. In particular:

$$n = -w = \pm 1, \quad N = 1, \ \bar{N} = 0; \qquad n = w = \pm 1, \quad N = 0, \ \bar{N} = 1.$$
$$(4.23)$$

The cases where the excited oscillators are in the non-compact direction yield two pairs of massless vector fields. In fact, the first pair go with the left $U(1)$ to make an $SU(2)$, while the second pair go with the right $U(1)$ to make another $SU(2)$. Indeed, they have the correct ± 1 charges under the Kaluza–Klein $U(1)$s in order to be the components of the W-bosons for the $SU(2)_L \times SU(2)_R$ 'enhanced gauge symmetries'. The term is appropriate since there is an extra gauge symmetry at this special radius, given that new massless vectors appear there.

When the oscillators are in the compact direction, we get two pairs of massless bosons. These go with the massless scalar ϕ to fill out the massless adjoint Higgs field for each $SU(2)$. These are the scalars whose vevs give the W-bosons their masses when we are away from the special radius.

In fact, this special property of the string theory is succinctly visible at all mass levels, by looking at the partition function (4.30). At the self-dual radius, it can be rewritten as a sum of squares of 'characters' of the $su(2)$ affine Lie algrebra:

$$Z(q, R = \sqrt{\alpha'}) = |\chi_1(q)|^2 + |\chi_2(q)|^2, \qquad (4.24)$$

where

$$\chi_1(q) \equiv \eta^{-1} \sum_n q^{n^2}, \quad \chi_2(q) \equiv \eta^{-1} \sum_n q^{(n+1/2)^2}. \qquad (4.25)$$

It is amusing to expand these out (after putting in the other factors of $(\eta\bar{\eta})^{-1}$ from the uncompactified directions) and find the massless states we discussed explicitly above.

It does not matter if an affine Lie algebra has not been encountered before by the reader. We can take this as an illustrative example, arising in a natural and instructive way. See insert 4.1 for further discussion[12]. In the language of two dimensional conformal field theory, there are additional left- and right-moving currents (i.e. fields with weights $(1,0)$ and $(0,1)$) present. We can construct them as vertex operators by exponentiating some of the existing fields. The full set of vertex operators of the $SU(2)_L \times SU(2)_R$ spacetime gauge symmetry:

$$SU(2)_L: \ \bar{\partial}X^\mu \partial X^{25}(z), \ \ \bar{\partial}X^\mu \exp(\pm 2iX^{25}(z)/\sqrt{\alpha'})$$
$$SU(2)_R: \ \partial X^\mu \bar{\partial}X^{25}(z), \ \ \partial X^\mu \exp(\pm 2iX^{25}(\bar{z})/\sqrt{\alpha'}), \qquad (4.26)$$

corresponding to the massless vectors we constructed by hand above.

Insert 4.1.   Affine Lie algebras

The key structure of an affine Lie algebra is just what we have seen
arise naturally in this self-duality example. In addition to all of the
nice structures that the conformal field theory has – most pertinently,
the Virasoro algebra – there is a family of unit weight operators,
often constructed as vertex operators as we saw in equation (4.26),
which form the Lie algebra of some group G. They are unit weight as
measured either from the left or the right, and so we can have such
structures on either side. Let us focus on the left. Then, as $(1,0)$
operators, $J^a(z)$, (a is a label) we have:

$$[L_n, J^a_m] = m J^a_{n+m}, \qquad (4.27)$$

where

$$J^a_n = \frac{1}{2\pi i} \oint dz \; z^{-n-1} J^a(z), \qquad (4.28)$$

and

$$[J^a_n, J^b_m] = i f^{ab}{}_c J^c_{n+m} + m k d^{ab} \delta_{n+m}, \qquad (4.29)$$

where it should be noticed that the zero modes of these currents
form a Lie algebra, with structure constants $f^{ab}{}_c$. The constants d^{ab}
define the inner product between the generators $(t^a, t^b) = d^{ab}$. Since
in bosonic string theory a mode with index -1 creates a state that
is massless in spacetime, J^a_{-1} can be placed either on the left with
$\tilde{\alpha}^\mu_{-1}$ on the right (or vice versa) to give a state $J^a_{-1} \tilde{\alpha}^\mu_{-1} |0\rangle$ which is
a massless vector $A^{\mu a}$ in the adjoint of G, for which the low energy
physics must be Yang–Mills theory.

The full algebra is called an 'affine Lie algebra', or a 'current
algebra', and sometimes a 'Kac–Moody' algebra[275]. In a standard
normalisation, k is an integer and is called the 'level' of the affinisa-
tion. In the case that we first see this sort of structure, the string at
a self-dual radius, the level is 1. The currents in this case are:

$$J^3(z) = i \alpha'^{-1/2} \partial_z X^{25}(z),$$
$$J^1(z) = \; : \cos(2\alpha'^{-1/2} X^{25}(z)) :, \quad J^2(z) =: \sin(2\alpha'^{-1/2} X^{25}(z)) :$$

which satisfy the algebra in (4.29) with $f^{abc} = \epsilon^{abc}$, $k = 1$, and
$d^{ab} = \frac{1}{2}\delta^{ab}$, as appropriate to the fundamental representation.

The vertex operator for the change of radius, $\partial X^{25} \bar{\partial} X^{25}$, corresponding to the field ϕ, transforms as a $(\mathbf{3},\mathbf{3})$ under $SU(2)_{\mathrm{L}} \times SU(2)_{\mathrm{R}}$, and therefore a rotation by π in one of the $SU(2)$s transforms it into minus itself. The transformation $R \to \alpha'/R$ is therefore the $\mathbb{Z}_2$ Weyl subgroup of the $SU(2) \times SU(2)$. Since T-duality is part of the spacetime gauge theory, this is a clue that it is an exact symmetry of the closed string theory, if we assume that non-perturbative effects preserve the spacetime gauge symmetry. We shall see that this assumption seems to fit with non-perturbative discoveries to be described later.

4.4 The circle partition function

It is useful to consider the partition function of the theory on the circle. This is a computation as simple as the one we did for the uncompactified theory earlier, since we have done the hard work in working out L_0 and $\bar{L}_0$ for the circle compactification. Each non-compact direction will contribute a factor of $(\eta \bar{\eta})^{-1}$, as before, and the non-trivial part of the final τ-integrand, coming from the compact X^{25} direction is:

$$Z(q, R) = (\eta \bar{\eta})^{-1} \sum_{n,w} q^{\frac{\alpha'}{4} P_{\mathrm{L}}^2} \bar{q}^{\frac{\alpha'}{4} P_{\mathrm{R}}^2}, \qquad (4.30)$$

where $P_{\mathrm{L,R}}$ are given in (4.13). Our partition function is manifestly T-dual, and is in fact also modular invariant. Under T, it picks us a phase $\exp(\pi i (P_{\mathrm{L}}^2 - P_{\mathrm{R}}^2))$, which is again unity, as follows from the second line in (4.15): $P_{\mathrm{L}}^2 - P_{\mathrm{R}}^2 = 2nw$. Under S, the role of the time and space translations as we move on the torus are exchanged, and this in fact exchanges the sums over momentum and winding. T-duality ensures that the S-transformation properties of the exponential parts involving $P_{\mathrm{L,R}}$ are correct, while the rest is S invariant as we have already discussed.

It is a useful exercise to expand this partition function out, after combining it with the factors from the other non-compact dimensions first, to see that at each level the mass (and level matching) formulae (4.15) which we derived explicitly is recovered.

In fact, the modular invariance of this circle partition function is part of a very important larger story. The left and right momenta $P_{\mathrm{L,R}}$ are components of a special two dimensional lattice, $\Gamma_{1,1}$. There are two basis vectors $k = (1/R, 1/R)$ and $\hat{k} = (R, -R)$. We make the lattice with arbitrary integer combinations of these, $nk + w\hat{k}$, whose components are $(P_{\mathrm{L}}, P_{\mathrm{R}})$. (cf. equation (4.13)). If we define the dot products between our basis vectors to be $k \cdot \hat{k} = 2$ and $k \cdot k = 0 = \hat{k} \cdot \hat{k}$, our lattice then has a Lorentzian signature, and since $P_{\mathrm{L}}^2 - P_{\mathrm{R}}^2 = 2nw \in 2\mathbb{Z}$, it is called 'even'. The 'dual' lattice $\Gamma_{1,1}^*$ is the set of all vectors whose dot product with (P_L, P_R) gives an integer. In fact, our lattice is self-dual, which is to say that $\Gamma_{1,1} = \Gamma_{1,1}^*$. It is the 'even' quality which guarantees invariance under T as we have seen, while it is the 'self-dual' feature which ensures invariance under S. In fact, S is just a change of basis in the lattice, and the self-duality feature translates into the fact that the Jacobian for this is unity.

4.5 Toriodal compactifications

It will be very useful later on for us to outline how things work more generally. The case of compactification on the circle encountered above can be easily generalised to compactification on the torus $T^d \simeq (S^1)^d$. Let us denote the compact dimensions by X^m, where $m, n = 1, \ldots, d$. Their periodicity is specified by

$$X^m \sim X^m + 2\pi R^{(m)} \mathsf{n}^m,$$

where the n^m are integers and $R^{(m)}$ is the radius of the mth circle. The metric on the torus, G_{mn}, can be diagonalised into standard unit Euclidean form by the veilbeins e_m^a where $a, b = 1, \ldots, d$:

$$G_{mn} = \delta_{ab} e_m^a e_n^b,$$

and it is convenient to use tangent space coordinates $X^a = X^m e_m^a$ so that the equivalence can be written:

$$X^a \sim X^a + 2\pi e_m^a \mathsf{n}^m.$$

We have defined for ourselves a lattice $\Lambda = \{e_m^a \mathsf{n}^m, \mathsf{n}^m \in \mathbb{Z}\}$. We now write our torus in terms of this as

$$T^d \equiv \frac{\mathbb{R}^d}{2\pi\Lambda}.$$

There are of course conjugate momenta to the X^a, which we denote as p^a. They are quantised, since moving from one lattice point to another, producing a change in the vector X by $\delta X \in 2\pi\Lambda$ are physically equivalent, and so single-valuedness of the wavefunction imposes $\exp(ip \cdot X) = \exp(ip \cdot [X + \delta X])$, i.e.

$$p \cdot \delta X \in 2\pi\mathbb{Z},$$

from which we see that clearly

$$p^n = G^{mn} \mathsf{n}_m,$$

where n_m are integers. In other words, the momenta live in the dual lattice, Λ^*, of Λ, defined by

$$\Lambda^* \equiv \{e^{*am}\mathsf{n}_m, \mathsf{n}_m \in \mathbb{Z}\},$$

where the inverse veilbiens $e^{*am}\mathsf{n}_m$ are defined in the usual way using the inverse metric:

$$e^{*am} \equiv e_m^a G^{mn}, \quad \text{or} \quad e^{*am}e_m^b = \delta^{ab}.$$

Of course we can have winding sectors as well, since as we go around the string via $\sigma \to \sigma + 2\pi$, we can change to a new point on the lattice characterised by a set of integers w^m, the winding number. Let us write out the string mode expansions. We have

$$X^a(\tau, \sigma) = X_L^a(\tau - \sigma) + X_R^a(\tau + \sigma), \quad \text{where}$$

$$X_L^a = x_L^a - i\sqrt{\frac{\alpha'}{2}}p_L^a(\tau - \sigma) + \text{oscillators} \qquad x_L^a = \frac{x^a}{2} - \theta^a$$

$$p_L^a = p^a + \frac{w^a R^{(a)}}{\alpha'} \equiv e^{*am}\mathsf{n}_m + \frac{1}{\alpha'}e_m^a w^m, \tag{4.31}$$

for the left, while on the right we have

$$X_R^a = x_R^a - i\sqrt{\frac{\alpha'}{2}}p_R^a(\tau + \sigma) + \text{oscillators} \qquad x_R^a = \frac{x^a}{2} + \theta^a$$

$$p_R^a = p^a - \frac{w^a R^{(a)}}{\alpha'} \equiv e^{*am}\mathsf{n}_m - \frac{1}{\alpha'}e_m^a w^m. \tag{4.32}$$

The action of the manifest T-duality symmetry is simply to act with a right-handed parity, as before, swopping $p_L \leftrightarrow p_L$ and $p_R \leftrightarrow -p_R$, and hence momenta and winding and $X_L \leftrightarrow X_L$ and $X_R \leftrightarrow -X_R$.

To see more, let us enlarge our bases for the two separate lattices Λ, Λ^* into a singe one, via:

$$\hat{e}_m = \frac{1}{\alpha'}\begin{pmatrix} e_m^a \\ -e_m^a \end{pmatrix}, \quad \hat{e}^{*m} = \begin{pmatrix} e^{*am} \\ e^{*am} \end{pmatrix},$$

and now we can write

$$\hat{p} = \begin{pmatrix} p_L^a \\ p_R^a \end{pmatrix} = \hat{e}_m w^m + \hat{e}^{*m}\mathsf{n}_m,$$

which lives in a $(d+d)$-dimensional lattice which we will call $\Gamma_{d,d}$. We can choose the metric on this space to be of Lorentzian signature (d, d), which is achieved by

$$G = \begin{pmatrix} \delta_{ab} & 0 \\ 0 & -\delta_{ab} \end{pmatrix},$$

and using this we see that

$$\hat{e}_m \cdot \hat{e}_n = 0 = \hat{e}^{*m} \cdot \hat{e}^{*n}$$
$$\hat{e}_m \cdot \hat{e}^{*n} = \frac{2}{\alpha'} \delta_n^m, \tag{4.33}$$

which shows that the lattice is *self-dual*, since (up to a trivial overall scaling), the structure of the basis vectors of the dual is identical to that of the original: $\Gamma_{d,d}^* = \Gamma_{d,d}$. Furthermore, we see that the inner product between any two momenta is given by

$$(\hat{e}_m w^m + \hat{e}^{*m} \mathsf{n}_m) \cdot (\hat{e}_n w'^n + \hat{e}^{*n} \mathsf{n}_m) = \frac{2}{\alpha'}(w^m \mathsf{n}'_m + \mathsf{n}_m w'^m). \tag{4.34}$$

In other words, the lattice is *even*, because the inner product gives even integer multiples of $2/\alpha'$.

It is these properties that guarantee that the string theory is modular invariant[173]. The partition function for this compactification is the obvious generalisation of the expression given in (4.30):

$$Z_{T^d} = (\eta\bar{\eta})^{-d} \sum_{\Gamma_{d,d}} q^{\frac{\alpha'}{4} p_{\mathrm{L}}^2} \bar{q}^{\frac{\alpha'}{4} p_{\mathrm{R}}^2}, \tag{4.35}$$

where the $p_{\mathrm{L,R}}$ are given in (4.32). Recall that the modular group is generated by $T : \tau \to \tau + 1$, and $S : \tau \to -1/\tau$. So T-invariance follows from the fact that its action produces a factor $\exp(i\pi\alpha'(p_{\mathrm{L}}^2 - p_{\mathrm{R}}^2)/2) = \exp(i\pi\alpha'(\hat{p}^2)/2)$ which is unity because the lattice is even, as shown in equation (4.34).

Invariance under S follows by rewriting the partition function $Z(-1/\tau)$ using the Poisson resummation formula given in insert 4.2, to get the result that

$$Z_\Gamma\left(-\frac{1}{\tau}\right) = \mathrm{vol}(\Gamma^*) Z_{\Gamma^*}(\tau).$$

The volume of the lattice's unit cell is unity, for a self-dual lattice, since $\mathrm{vol}(\Lambda)\mathrm{vol}(\Lambda^*) = 1$ for any lattice and its dual, and therefore S-invariance is demonstrated, and we can define a consistent string compactification.

Insert 4.2. The Poisson resummation formula

A very useful trick is the following. Assume that we have a function $f(x)$ defined on $\mathbb{R}^n$. Then its Fourier transform is given as

$$f(x) = \int \frac{d^n k}{(2\pi)^n} e^{ik\cdot x} \hat{f}(k).$$

The formula we need is written in terms of this. If we sum over a lattice $\Lambda \subset \mathbb{R}^n$, then:

$$\sum_{n\in\Lambda} f(n) = \int \sum_{n\in\Lambda} \frac{d^n k}{(2\pi)^n} e^{ik\cdot m} \hat{f}(k) = \text{vol}(\Lambda^*) \sum_{m\in\Lambda^*} \hat{f}(2\pi m).$$

We shall meet two very important examples of large even and self-dual lattices later in subsection 7.2. They are associated to the construction of the modular invariant partition functions of the ten dimensional $E_8 \times E_8$ and $SO(32)$ heterotic strings[20].

There is a large space of inequivalent lattices of the type under discussion, given by the shape of the torus (specified by background parameters in the metric G) and the fluxes of the B-field through it. We can work out this 'moduli space' of compactifications. It would naively seem to be simply $O(d,d)$, since this is the space of rotations naturally acting, taking such lattices into each other, i.e. starting with some reference lattice Γ_0, $\Gamma' = G\Gamma_0$ should be a different lattice. We must remember that the physics cares only about the values of p_L^2 and p_R^2, and so therefore we must count as equivalent any choices related by the $O(d) \times O(d)$ which acts independently on the left and right momenta: $G \sim G'G$, for $G' \in O(d) \times O(d)$. So at least *locally*, the space of lattices is isomorphic to

$$\mathcal{M} = \frac{O(d,d)}{O(d) \times O(d)}. \tag{4.36}$$

A quick count of the dimension of this space gives $2d(2d-1)/2 - 2 \times d(d-1)/2 = d^2$, which fits nicely, since this is the number of independent components contained in the metric G_{mn}, $(d(d+1)/2)$ and the antisymmetric tensor field B_{mn}, $(d(d-1)/2)$, for which we can switch on constant values (sourced by winding).

There are still a large number of discrete equivalences between the lattices, which follows from the fact that there is a discrete subgroup of

$O(d, d)$, called $O(d, d, \mathbb{Z})$, which maps our reference lattice Γ_0 into itself: $\Gamma_0 \sim G''\Gamma_0$. This is the set of discrete linear transformations generated by the subgroups of $SL(2d, \mathbb{Z})$ which preserves the inner product given in equations (4.33). This group includes the T-dualities on all of the d circles, linear redefinitions of the axes, and discrete shifts of the B-field. The full space of torus compactifications is often denoted:

$$\mathcal{M} = O(d, d, \mathbb{Z}) \backslash O(d, d) / [O(d) \times O(d)], \tag{4.37}$$

where we divide by one action under left multiplication, and the other under right.

Now we see that there is a possibility of much more than just the $SU(2)_\mathrm{L} \times SU(2)_\mathrm{R}$ enhanced gauge symmetry which we got in the case of a single circle. We can have this large symmetry from any of the d circles, of course but there is more, since there are extra massless states that can be made by choices of momenta from more than one circle, corresponding to weight one vertex operators. This will allow us to make very large enhanced gauge groups, up to rank d, as we shall see later in section 7.2.

4.6 More on enhanced gauge symmetry

The reader is probably keen to see more of where some of the structures of sections 4.3, 4.4, and 4.5 come from, and so we will pause here to study a little about Lie groups and algebras.

4.6.1 Lie algebras and groups

Lie algebras are usually described in terms of a basis of *generators*, t^a, which have a specific antisymmetric product:

$$[t^a, t^b] = i f^{ab}{}_c t^c, \tag{4.38}$$

where the $f^{ab}{}_c$ are often called the *structure constants*. This product must satisfy the Jacobi identity, which states that:

$$[t^a, [t^b, t^c]] + [t^b, [t^c, t^a]] + [t^c, [t^a, t^b]] = 0.$$

Once we have the algebra, we can form the group G by exponentiating the generators, to make a group element

$$g = e^{i\lambda_a t^a}.$$

N.B. One of the reasons why Lie groups are interesting is that *the group elements form a manifold*, and so there is a lot of familiar geometry to be found in their description. For example, one can think of the Lie algebra as the vector space that is simply the tangent space to the group manifold, G, and keep in mind a picture like that in figure 2.14. The natural way to make the Lie algebra from the group elements g is *via* the Maurer–Cartan forms, $g^{-1}dg$ which give a family of one-forms which are valued in the Lie algebra. We won't use this much, but the curious reader can look ahead to insert 7.4, where we make this explicit for $SU(2)$, which is the manifold S^3.

There is also an inner product between the generators, which is defined as $(t^a, t^b) = d^{ab}$, which is positive if the group is compact. We can lower and raise indices with this fellow, and having done this on the structure constants to get f^{abc}, there is an additional condition that they are totally antisymmetric in all of their indices. We shall restrict our attention mostly to the *simple* Lie algebras, for which a choice can be made to make d^{ab} proportional to δ^{ab}.

Most familiar is of course the representation of the algebra in (4.38) by matrices, for which we can use the notation t_R^a, where R stands for a representation, and the matrix elements are denoted $t_{R,ij}^a$. The antisymmetric product is then the familiar matrix commutator, and the inner product is matrix multiplication with the trace. Then we have $\mathrm{Tr}(t_R^a t_R^b) = T_R \delta^{ab}$, where T_R is a number which depends on the representation. Note that we can define the *Casimir invariant* of the representation R as $t_R^a t_R^a = Q_R \mathbf{1}$.

The Jacobi identity above translates into

$$f^{abd} f^{cde} + f^{bcd} f^{ade} + f^{cad} f^{bde} = 0.$$

A most convenient matrix representation of the algebra is given by

$$(t_A^a)_{bc} = -if^a{}_{bc},$$

and for this we see that we get

$$[t_A^a, t_A^b] = if^a{}_{bc} t_A^c,$$

and so we see that the structure constants themselves form a representation of the Lie algebra. This is the *adjoint representation*. Notice that the dimension of the representation is the number of generators of the group.

It is useful to divide the generators t^a into two families. There is the maximal set of commuting generators, which are denoted H^i, where $i = 1, \ldots, r$ with r being the *rank* of the group, and there are the rest, denoted E^α of reasons to be given very shortly.

The set H^i, for which

$$[H^i, H^j] = 0,$$

is the *Cartan subalgebra*, and the H^i are often said to form the *maximal torus*, which we shall discuss more later. These elements are the generalisation of J_3 from the familiar case of $SU(2)$. For a representation of dimension d, we can think of the H^i as $d \times d$ matrices. We will pick a specific basis for these and keep in that basis to describe everything else. Being all mutually commutative, they may be simultaneously diagonalised, and there are d distinct eigenvalues for each H. Consider the nth entry along a diagonal. Each of the H^i supplies a component, w^i, of a vector w in a space $\mathbb{R}^r$. There are d such *weight vectors*.

Everything else can be given an assignment of 'charges' corresponding to the H-eigenvalues, via

$$[H^i, E^\alpha] = \alpha^i E^\alpha.$$

We can think of the α^i as components of an r-dimensional vector known as a root. It is a vector in the space $\mathbb{R}^r$ mentioned above. Every root is uniquely associated to a generator E^α. The remaining parts of the Lie algebra are:

$$[E^\alpha, E^\beta] = \begin{cases} \epsilon(\alpha, \beta) E^{\alpha, \beta} & \text{if } \alpha + \beta \text{ is a root,} \\ 2\alpha \cdot H / \alpha \cdot \alpha & \text{if } \alpha + \beta = 0, \\ 0 & \text{otherwise,} \end{cases}$$

where the dot product is defined with the relevant part of the inner product form, d_{ij}, and $\epsilon(\alpha, \beta)$ is ± 1. It is worth noting that the roots are the weights of the adjoint representation.

The E^α are the generalisations of the $J^\pm$ familiar from $SU(2)$, the raising and lowering operators. One can decompose weights into three classes, whether they are positive, negative, or zero. This is given by whether or not the first non-zero entry is positive, negative or zero (i.e. all components zero). There is a unique highest weight in any representation. Specialising to the weights of the adjoint representation, the roots, divides the E^α into raising operators, if α is positive, and lowering operators if α is negative. One can build the whole representation of the groups starting with the highest weight and acting with the lowering operators, while acting on a highest weight with a raising operator gives zero.

The *simple roots* are the positive roots that cannot be written as the sum of two positive roots, and they form a linearly independent set. The number of them is equal to the rank of the group, r. Using these, it can be shown that the entire structure of the group may be reconstructed. A useful way of specifying the simple roots is to give their relative lengths and the angles between them, which turn out to be restricted to between 90° and 180°. The *Dynkin diagram* is a very useful way of giving that information in an easy to read form. Each simple root is a node in the diagram. There are links between nodes if the angle between them is not 90°. There is a single line if the angle is 120°, a double line if the angle is 135° and a triple line if it is 150°. To denote the odd root which is shorter than the rest, it is often a practice to make the note a different shade of colour in the diagram.

4.6.2 The classical Lie algebras

Let us list the *classical Lie algebras* of Cartan's classification.

- $SU(n)$ Denoted A_{n-1} in Cartan's classification. The generators are traceless $n \times n$ Hermitian matrices, and the group elements of $SU(n)$ are unit determinant unitary matrices.

- $SO(n)$ If $n = 2k + 1$ this is denoted B_k, while if $n = 2k$ it is D_k. The generators are $n \times n$ antisymmetric Hermitian matrices, and the group elements of $SO(n)$ are real orthogonal matrices.

- $Sp(k) = USp(2k)$ This is denoted C_k in the classification. The generators are Hermitian $2k \times 2k$ matrices t satisfying

$$MtM^{-1} = -t^T,$$

where T denotes the transpose and

$$M = i \begin{pmatrix} 0 & I_k \\ -I_k & 0 \end{pmatrix},$$

where I_k is the $k \times k$ identity matrix. The groups is the set of unitary matrices u satisfying

$$MuM^{-1} = u^{-T},$$

where $-T$ denotes the inverse of the transpose.

We will often have cause to encounter some non-compact groups closely related to these. We obtain them by multiplying some generators by an i. In this way we will get the set of traceless imaginary matrices to make the group of real matrices of unit determinant, $SL(n)$ by continuing $SU(n)$. We have already encountered $O(n, m)$, which is a continuation of $O(n + m)$ made by such a continuation.

Insert 4.3.   The simply laced Lie algebras

It turns out that for the Lie algebras A_n, D_k, E_6, E_7 and E_8, all of the roots are the same length. These are called the *simply laced* algebras. It is very useful to know a bit about their structure, as manifest in the Dynkin diagrams given below.

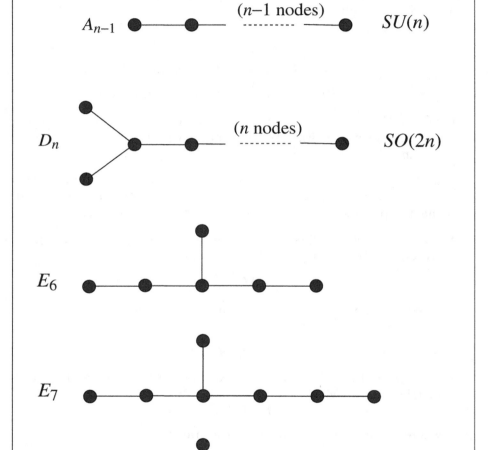

4.6.3 Physical realisations with vertex operators

Now we can return to some of the physical objects that we saw arising in the string theory and make contact with some of the structures we saw above. Recall that we represented the weights as vectors in $\mathbb{R}^r$, where r was the rank of the Lie algebra, arising as charges under the commuting generators or maximal torus given by the H^i. These vectors came with a specific set of entries, and we could build all representations out of them, by adding vectors. The set of points in $\mathbb{R}^r$ made in this way is the *Lie algebra lattice*, and it can be placed on a very physical footing in the context of toroidal compactification in the following way.

If we placed r directions X^i on a torus T^r, the weight (0,1) objects $H^i(z) = i\alpha'^{-1/2}\partial_z X^i$ parameterise the very object we have been working with: the maximal torus. The weight vectors that we had, with the additive structure allowing us to reach other points in the lattice, building up other representations, are simply the momenta, which are the zero modes of the $H^i(z)$, which are also additive.

In general, we can make states corresponding to the weight vector w^i with the vertex operator $\exp(2i\alpha'^{-1/2}w \cdot \phi)$. So now we see how to get a gauge symmetry, following the discussion in insert 4.1, we need to have vertex operators of weight (0,1) to go with the $H^i(z)$. These can be made with the vertex operators if the $w^2 = 2$. So we see that we need the simply laced algebras to do this. They are listed in insert 4.3, together with their Dynkin diagrams.

4.7 Another special radius: bosonisation

Before proceeding with the T-duality discussion, let us pause for a moment to remark upon something which will be useful later. In the case that $R = \sqrt{(\alpha'/2)}$, something remarkable happens. The partition function is:

$$Z\left(q, R = \sqrt{\frac{\alpha'}{2}}\right) = (\eta\bar{\eta})^{-1} \sum_{n,w} q^{\frac{1}{2}\left(n+\frac{w}{2}\right)^2} \bar{q}^{\frac{1}{2}\left(n-\frac{w}{2}\right)^2}. \tag{4.39}$$

Note that the allowed momenta at this radius are (cf. equation (4.13)):

$$\alpha_0^{25} = P_{\mathrm{L}}\sqrt{\frac{\alpha'}{2}} = \left(n + \frac{w}{2}\right)$$

$$\tilde{\alpha}_0^{25} = P_{\mathrm{R}}\sqrt{\frac{\alpha'}{2}} = \left(n - \frac{w}{2}\right), \tag{4.40}$$

and so they span both integer and half-integer values. Now when P_{L} is an integer, then so is P_{R} and vice versa, and so we have two distinct sectors,

integer and half-integer. In fact, we can rewrite our partition function as
a set of sums over these separate sectors:

$$Z_{R=\sqrt{\alpha'/2}}$$

$$= \frac{1}{2} \left\{ \left| \frac{1}{\eta} \sum_n q^{\frac{1}{2}n^2} \right|^2 + \left| \frac{1}{\eta} \sum_n (-1)^n q^{\frac{1}{2}n^2} \right|^2 + \left| \frac{1}{\eta} \sum_n q^{\frac{1}{2}\left(n+\frac{1}{2}\right)^2} \right|^2 \right\}. \qquad (4.41)$$

The middle sum is rather like the first, except that there is a -1 whenever
n is odd. Taking the two sums together, it is just like we have performed
the sum (trace) over all the integer momenta, but placed a projection
onto even momenta, using the projector

$$P = \frac{1}{2}(1 + (-1)^n). \qquad (4.42)$$

In fact, an investigation will reveal that the third term can be written
with a partner just like it save for an insertion of $(-1)^n$ also, but that
latter sum vanishes identically. This all has a specific meaning which we
will uncover shortly.

Notice that the partition function can be written in yet another nice
way, this time as

$$Z_{R=\sqrt{\alpha'/2}} = \frac{1}{2} \left(|f_4^2(q)|^2 + |f_3^2(q)|^2 + |f_2^2(q)|^2 \right), \qquad (4.43)$$

where, for here and for future use, let us define

$$f_1(q) \equiv\; = q^{\frac{1}{24}} \prod_{n=1}^{\infty} (1 - q^n) \equiv \eta(\tau)$$

$$f_2(q) \equiv\; = \sqrt{2} q^{\frac{1}{24}} \prod_{n=1}^{\infty} (1 + q^n)$$

$$f_3(q) \equiv\; = q^{-\frac{1}{48}} \prod_{n=1}^{\infty} (1 + q^{n-\frac{1}{2}})$$

$$f_4(q) \equiv\; = q^{-\frac{1}{48}} \prod_{n=1}^{\infty} (1 - q^{n-\frac{1}{2}}), \qquad (4.44)$$

and note that

$$f_2\left(-\frac{1}{\tau}\right) = f_4(\tau); \quad f_3\left(-\frac{1}{\tau}\right) = f_3(\tau); \qquad (4.45)$$

$$f_3(\tau + 1) = f_4(\tau); \quad f_2(\tau + 1) = f_2(\tau). \qquad (4.46)$$

While the rewriting as (4.43) might not look like much at first glance, this is in fact the partition function of a single Dirac fermion in two dimensions: $Z(R = \sqrt{\alpha'/2}) = Z_{\text{Dirac}}$. We have arrived at the result that a boson (at a special radius) is in fact equivalent to a fermion. This is called 'bosonisation' or 'fermionisation', depending upon one's perspective. How can this possibly be true?

The action for a Dirac fermion, $\Psi = (\Psi_L, \Psi_R)^T$ (which has two components in two dimensions) is, in conformal gauge:

$$S_{\text{Dirac}} = \frac{i}{2\pi} \int d^2\sigma \ \bar{\Psi}\gamma^a \partial_a \Psi = \frac{i}{\pi} \int d^2\sigma \ \bar{\Psi}_L \bar{\partial}\Psi_L - \frac{i}{\pi} \int d^2\sigma \ \bar{\Psi}_R \partial\Psi_R,$$
(4.47)

where we have used

$$\gamma^0 = i \begin{pmatrix} 0 & 1 \\ 1 & 0 \end{pmatrix}, \quad \gamma^1 = i \begin{pmatrix} 0 & -1 \\ 1 & 0 \end{pmatrix}.$$

Now, as a fermion goes around the cylinder $\sigma \to \sigma + 2\pi$, there are two types of boundary condition it can have. It can be periodic, and hence have integer moding, in which case it is said to be in the 'Ramond' (R) sector. It can instead be antiperiodic, have half-integer moding, and is said to be in the 'Neveu–Schwarz' (NS) sector.

In fact, these two sectors in this theory map to the two sectors of allowed momenta in the bosonic theory: integer momenta to NS and half-integer to R. The various parts of the partition function can be picked out and identified in fermionic language. For example, the contribution:

$$\left| f_3^2(q) \right|^2 \equiv \left| q^{-\frac{1}{24}} \right|^2 \left| \prod_{n=1}^{\infty} (1 + q^{n-\frac{1}{2}})^2 \right|^2,$$

looks very fermionic, (recall insert 3.4 (p. 92)) and is in fact the trace over the contributions from the NS sector fermions as they go around the torus. It is squared because there are two components to the fermion, Ψ and $\bar{\Psi}$. We have the squared modulus beyond that since we have the contribution from the left and the right.

The $f_4(q)$ contribution on the other hand, arises from the NS sector with a $(-)^F$ inserted, where F counts the number of fermions at each level. The $f_2(q)$ contribution comes from the R sector, and there is a vanishing contribution from the R sector with $(-1)^F$ inserted. We see that that the projector

$$P = \frac{1}{2}(1 + (-1)^F)$$
(4.48)

is the fermionic version of the projector (4.42) we identified previously. Notice that there is an extra factor of two in front of the R sector contribution due to the definition of f_2. This is because the R ground state is in fact degenerate. The modes Ψ_0 and $\bar{\Psi}_0$ define two ground states which map into one another. Denote the vacuum by $|s\rangle$, where s can take the values $\pm\frac{1}{2}$. Then

$$\Psi_0|-\tfrac{1}{2}\rangle = 0; \qquad \bar{\Psi}_0|+\tfrac{1}{2}\rangle = 0;$$

$$\bar{\Psi}_0|-\tfrac{1}{2}\rangle = |+\tfrac{1}{2}\rangle; \qquad \Psi_0|+\tfrac{1}{2}\rangle = |-\tfrac{1}{2}\rangle, \tag{4.49}$$

and Ψ_0 and $\bar{\Psi}_0$ therefore form a representation of the two dimensional Clifford algebra. We will see this in more generality later on. In D dimensions there are $D/2$ components, and the degeneracy is $2^{D/2}$.

As a final check, we can see that the zero point energies work out nicely too. The mnemonic (2.80) gives us the zero point energy for a fermion in the NS sector as $-1/48$, we multiply this by two since there are two components and we see that that we recover the weight of the ground state in the partition function. For the Ramond sector, the zero point energy of a single fermion is $1/24$. After multiplying by two, we see that this is again correctly obtained in our partition function, since $-1/24 + 1/8 = 1/12$. It is awfully nice that the function $f_2^2(q)$ has the extra factor of $2q^{1/8}$, just for this purpose.

This partition function is again modular invariant, as can be checked using elementary properties of the f-functions (4.46): f_2 transforms into f_4 under the S transformation, while under T, f_4 transforms into f_3.

At the level of vertex operators, the correspondence between the bosons and the fermions is given by:

$$\begin{aligned}\Psi_{\mathrm{L}}(z) &= e^{i\beta X_{\mathrm{L}}^{25}(z)}; \quad \bar{\Psi}_{\mathrm{L}}(z) = e^{-i\beta X_{\mathrm{L}}^{25}(z)}; \\ \Psi_{\mathrm{R}}(\bar{z}) &= e^{i\beta X_{\mathrm{R}}^{25}(\bar{z})}; \quad \bar{\Psi}_{\mathrm{R}}(\bar{z}) = e^{-i\beta X_{\mathrm{R}}^{25}(\bar{z})},\end{aligned} \tag{4.50}$$

where $\beta = \sqrt{2/\alpha'}$. This makes sense, for the exponential factors define fields single-valued under $X^{25} \to X^{25} + 2\pi R$, at our special radius $R = \sqrt{\alpha'/2}$. We also have

$$\Psi_{\mathrm{L}}(z)\bar{\Psi}_{\mathrm{L}}(z) = \partial_z X^{25}; \quad \Psi_{\mathrm{R}}(\bar{z})\bar{\Psi}_{\mathrm{R}}(\bar{z}) = \partial_{\bar{z}} X^{25}, \tag{4.51}$$

which shows how to combine two $(0, 1/2)$ fields to make a $(0, 1)$ field, with a similar structure on the left. Notice also that the symmetry $X^{25} \to -X^{25}$ swaps $\Psi_{\mathrm{L(R)}}$ and $\bar{\Psi}_{\mathrm{L(R)}}$, a symmetry of interest in the next subsection. We will return to this bosonisation/fermionisation relation in later sections, where it will be useful to write vertex operators in various ways in the supersymmetric theories.

4.8 String theory on an orbifold

There is a rather large class of string vacua, called 'orbifolds'[23], with many applications in string theory. We ought to study them, as many of the basic structures which will occur in their definition appear in more complicated examples later on.

The circle S^1, parametrised by X^{25}, has the obvious $\mathbb{Z}_2$ symmetry R_{25} : $X^{25} \to -X^{25}$. This symmetry extends to the full spectrum of states and operators in the complete theory of the string propagating on the circle. Some states are even under R_{25}, while others are odd. Just as we saw before in the case of Ω, it makes sense to ask whether we can define another theory from this one by truncating the theory to the sector which is even. This would define string theory propagating on the 'orbifold' space $S^1/\mathbb{Z}_2$.

In defining this geometry, note that it is actually a line segment, where the endpoints of the line are actually 'fixed points' of the $\mathbb{Z}_2$ action. The point $X^{25} = 0$ is clearly such a point and the other is $X^{25} = \pi R \sim -\pi R$, where R is the radius of the original S^1. A picture of the orbifold space is given in figure 4.1. In order to check whether string theory on this space is sensible, we ought to compute the partition function for it. We can work this out by simply inserting the projector

$$P = \frac{1}{2}(1 + R_{25}),$$ (4.52)

which will have the desired effect of projecting out the R_{25}-odd parts of the circle spectrum. So we expect to see two pieces to the partition function: a part that is $\frac{1}{2}$ times Z_{circle}, and another part which is Z_{circle} with R_{25} inserted. Noting that the action of R_{25} is

$$R_{25} : \begin{cases} \alpha_n^{25} \to -\alpha_n^{25} \\ \tilde{\alpha}_n^{25} \to -\tilde{\alpha}_n^{25} \end{cases},$$ (4.53)

the partition function is:

$$Z_{\text{orbifold}} = \frac{1}{2}\left[Z(R, \tau) + 2\left(|f_2(q)|^{-2} + |f_3(q)|^{-2} + |f_4(q)|^{-2}\right)\right].$$ (4.54)

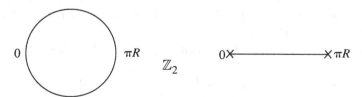

Fig. 4.1. A $\mathbb{Z}_2$ orbifold of a circle, giving a line segment with two fixed points.

The f_2 part is what one gets if one works out the projected piece, but there are two extra terms. From where do they come? One way to see that those extra pieces must be there is to realise that the first two parts on their own cannot be modular invariant. The first part is of course already modular invariant on its own, while the second part transforms (4.46) into f_4 under the S transformation, so it has to be there too. Meanwhile, f_4 transforms into f_3 under the T-transformation, and so that must be there also, and so on.

While modular invariance is a requirement, as we saw, what is the physical meaning of these two extra partition functions? What sectors of the theory do they correspond to and how did we forget them?

The sectors we forgot are very stringy in origin, and arise in a similar fashion to the way we saw windings appear in earlier sections. There, the circle may be considered as a quotient of the real line $\mathbb{R}$ by a translation $X^{25} \rightarrow X^{25} + 2\pi R$. There, we saw that as we go around the string, $\sigma \rightarrow \sigma + 2\pi$, the embedding map $X^{25}(\sigma)$ is allowed to change by any amount of the lattice, $2\pi R w$. Here, the orbifold further imposes the equivalence $X^{25} \sim -X^{25}$, and therefore, as we go around the string, we ought to be allowed:

$$X^{25}(\sigma + 2\pi, \tau) = -X^{25}(\sigma, \tau) + 2\pi w R,$$

for which the solution to the Laplace equation is:

$$X^{25}(z, \bar{z}) = x^{25} + i\sqrt{\frac{\alpha'}{2}} \sum_{n=-\infty}^{\infty} \frac{1}{\left(n + \frac{1}{2}\right)} \left(\alpha^{25}_{n+\frac{1}{2}} z^{n+\frac{1}{2}} + \tilde{\alpha}^{25}_{n+\frac{1}{2}} \bar{z}^{n+\frac{1}{2}} \right),$$

$$(4.55)$$

with $x^{25} = 0$ or πR, no zero mode α^{25}_0 (hence no momentum), and no winding: $w = 0$.

This is a configuration of the string allowed by our equations of motion and boundary conditions and therefore has to be included in the spectrum. We have two identical copies of these 'twisted sectors' corresponding to strings trapped at 0 and πR in spacetime. They are trapped, since x^{25} is fixed and there is no momentum.

Notice that in this sector, where the boson $X^{25}(w, \bar{w})$ is antiperiodic as one goes around the cylinder, there is a zero point energy of $1/16$ from the twisted sector: it is a weight $(1/16, 1/16)$ field, in terms of where it appears in the partition function.

Schematically therefore, the complete partition function ought to be

$$Z_{\text{orbifold}} = \text{Tr}_{\text{untwisted}} \left(\frac{(1 + R_{25})}{2} q^{L_0 - \frac{1}{24}} \bar{q}^{\bar{L}_0 - \frac{1}{24}} \right)$$

$$+ \text{Tr}_{\text{twisted}} \left(\frac{(1 + R_{25})}{2} q^{L_0 - \frac{1}{24}} \bar{q}^{\bar{L}_0 - \frac{1}{24}} \right) \qquad (4.56)$$

to ensure modular invariance, and indeed, this is precisely what we have in (4.54). The factor of two in front of the twisted sector contribution is because there are two identical twisted sectors, and we must sum over all sectors.

In fact, substituting in the expressions for the f-functions, one can discover the weight $(1/16, 1/16)$ twisted sector fields contributing to the vacuum of the twisted sector. This simply comes from the $q^{-1/48}$ factor in the definition of the $f_{3,4}$-functions. They appear inversely, and for example on the left, we have $1/48 = -c/24 + 1/16$, where $c = 1$.

Finally, notice that the contribution from the twisted sectors do not depend upon the radius R. This fits with the fact that the twisted sectors are trapped at the fixed points, and have no knowledge of the extent of the circle.

4.9 T-duality for open strings: D-branes

Let us now consider the $R \to 0$ limit of the open string spectrum. Open strings do not have a conserved winding around the periodic dimension and so they have no quantum number comparable to w, so something different must happen, as compared to the closed string case. In fact, it is more like field theory: when $R \to 0$ the states with non-zero internal momentum go to infinite mass, but there is no new continuum of states coming from winding. So we are left with a theory in one dimension fewer. A puzzle arises when one remembers that theories with open strings have closed strings as well, so that in the $R \to 0$ limit the closed strings live in D spacetime dimensions but the open strings only in $D - 1$.

This is perfectly fine, though, since the interior of the open string is indistinguishable from the closed string and so should still be vibrating in D dimensions. The distinguished part of the open string are the endpoints, and these are restricted to a $D - 1$ dimensional hyperplane.

This is worth seeing in more detail. Write the open string mode expansion as

$$X^\mu(z, \bar{z}) = X^\mu(z) + X^\mu(\bar{z}),$$

$$X^\mu(z) = \frac{x^\mu}{2} + \frac{x'^\mu}{2} - i\alpha' p^\mu \ln z + i \left(\frac{\alpha'}{2}\right)^{1/2} \sum_{n \neq 0} \frac{1}{n} \alpha_n^\mu z^{-n},$$

$$X^\mu(\bar{z}) = \frac{x^\mu}{2} - \frac{x'^\mu}{2} - i\alpha' p^\mu \ln \bar{z} + i \left(\frac{\alpha'}{2}\right)^{1/2} \sum_{n \neq 0} \frac{1}{n} \alpha_n^\mu \bar{z}^{-n}, \qquad (4.57)$$

where x'^μ is an arbitrary number which cancels out when we make the usual open string coordinate. Imagine that we place X^{25} on a circle of radius R. The T-dual coordinate is

$$X'^{25}(z, \bar{z}) = X^{25}(z) - X^{25}(\bar{z})$$

$$= x'^{25} - i\alpha' p^{25} \ln\left(\frac{z}{\bar{z}}\right) + i(2\alpha')^{1/2} \sum_{n \neq 0} \frac{1}{n} \alpha_n^{25} e^{-in\tau} \sin n\sigma$$

$$= x'^{25} + 2\alpha' p^{25}\sigma + i(2\alpha')^{1/2} \sum_{n \neq 0} \frac{1}{n} \alpha_n^{25} e^{-in\tau} \sin n\sigma$$

$$= x'^{25} + 2\alpha' \frac{n}{R}\sigma + i(2\alpha')^{1/2} \sum_{n \neq 0} \frac{1}{n} \alpha_n^{25} e^{-in\tau} \sin n\sigma. \qquad (4.58)$$

Notice that there is no dependence on τ in the zero mode sector. This is where momentum usually comes from in the mode expansion, and so we have no momentum. In fact, since the oscillator terms vanish at the endpoints $\sigma = 0, \pi$, we see that *the endpoints do not move in the X'^{25} direction!* Instead of the usual Neumann boundary condition $\partial_n X \equiv \partial_\sigma X = 0$, we have $\partial_t X \equiv i\partial_\tau X = 0$. More precisely, we have the Dirichlet condition that the ends are at a fixed place:

$$X'^{25}(\pi) - X'^{25}(0) = \frac{2\pi\alpha' n}{R} = 2\pi n R'. \qquad (4.59)$$

In other words, the values of the coordinate X'^{25} at the two ends are equal up to an integral multiple of the periodicity of the dual dimension, corresponding to a string that winds as in figure 4.2.

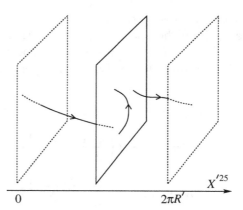

Fig. 4.2. Open strings with endpoints attached to a hyperplane. The dashed planes are periodically identified. The strings shown have winding numbers zero and one.

This picture is consistent with the fact that under T-duality, the definition of the normal and tangential derivatives get exchanged:

$$\partial_n X^{25}(z, \bar{z}) = \frac{\partial X^{25}(z)}{\partial z} + \frac{\partial X^{25}(\bar{z})}{\partial \bar{z}} = \partial_t X'^{25}(z, \bar{z})$$

$$\partial_t X^{25}(z, \bar{z}) = \frac{\partial X^{25}(z)}{\partial z} - \frac{\partial X^{25}(\bar{z})}{\partial \bar{z}} = \partial_n X'^{25}(z, \bar{z}). \tag{4.60}$$

Notice that this all pertains to just the direction which we T-dualised, X^{25}. So the ends are still free to move in the other 24 spatial dimensions, which constitutes a hyperplane called a 'D-brane'. There are 24 spatial directions, so we shall denote it a D24-brane.

4.9.1 Chan–Paton factors and Wilson lines

This picture becomes even more rich when we include Chan–Paton factors[25]. Consider the case of $U(N)$, the oriented open string. When we compactify the X^{25} direction, we can include a Wilson line

$$A_{25} = \text{diag}\{\theta_1, \theta_2, \ldots, \theta_N\}/2\pi R,$$

which generically breaks $U(N) \to U(1)^N$. (See insert 4.4 (p. 122) for a short discussion.) Locally this is pure gauge,

$$A_{25} = -i\Lambda^{-1}\partial_{25}\Lambda, \quad \Lambda = \text{diag}\{e^{iX^{25}\theta_1/2\pi R}, e^{iX^{25}\theta_2/2\pi R}, \ldots, e^{iX^{25}\theta_1/2\pi R}\}. \tag{4.61}$$

We can gauge A_{25} away, but since the gauge transformation is not periodic, the fields pick up a phase

$$\text{diag}\left\{e^{-i\theta_1}, e^{-i\theta_2}, \ldots, e^{-i\theta_N}\right\} \tag{4.62}$$

under $X^{25} \to X^{25} + 2\pi R$.

What is the effect in the dual theory? From the phase (4.62) the open string momenta are now fractional. As the momentum is dual to winding number, we conclude that the fields in the dual description have fractional winding number, i.e. their endpoints are no longer on the same hyperplane. Indeed, a string whose endpoints are in the state $|ij\rangle$ picks up a phase $e^{i(\theta_j - \theta_i)}$, so their momentum is $(2\pi n + \theta_j - \theta_i)/2\pi R$. Modifying the endpoint calculation (4.59) then gives

$$X'^{25}(\pi) - X'^{25}(0) = (2\pi n + \theta_j - \theta_i)R'. \tag{4.67}$$

In other words, up to an arbitrary additive constant, the endpoint in state i is at position

$$X'^{25} = \theta_i R' = 2\pi\alpha' A_{25,ii}. \tag{4.68}$$

Insert 4.4.   Particles and Wilson lines

The following illustrates an interesting gauge configuration which arises when spacetime has the non-trivial topology of a circle (with coordinate X^{25}) of radius R. Consider the case of $U(1)$. Let us make the following choice of constant background gauge potential:

$$A_{25}(X^\mu) = -\frac{\theta}{2\pi R} = -i\Lambda^{-1}\frac{\partial\Lambda}{\partial X^{25}}, \qquad (4.63)$$

where $\Lambda(X^{25}) = e^{-\frac{i\theta X^{25}}{2\pi R}}$. This is clearly pure gauge, but only locally. There still exists non-trivial physics. Form the gauge invariant quantity ('Wilson line'):

$$W_q = \exp\left(iq \oint dX^{25} A_{25}\right) = e^{-iq\theta}. \qquad (4.64)$$

Where does this observable show up? Imagine a point particle of charge q under the $U(1)$. Its action can be written (see section 4.2) as:

$$S = \int d\tau \left\{\frac{1}{2}\dot{X}^\mu \dot{X}_\mu - iqA_\mu \dot{X}^\mu\right\} = \int d\tau \mathcal{L}. \qquad (4.65)$$

The last term is just $-iq \int A = -iq \int A_\mu dx^\mu$, in the language of forms. This is the natural coupling of a world volume to an antisymmetric tensor, as we shall see.) Recall that in the path integral we are computing e^{-S}. So if the particle does a loop around X^{25} circle, it will pick up a phase factor of W_q. Notice: the conjugate momentum to X^μ is

$$\Pi^\mu = i\frac{\partial\mathcal{L}}{\partial\dot{X}^\mu} = i\dot{X}^\mu, \qquad \text{except for} \qquad \Pi^{25} = i\dot{X}^{25} - \frac{q\theta}{2\pi R} = \frac{n}{R},$$

where the last equality results from the fact that we are on a circle. Now we can of course gauge away A with the choice Λ^{-1}, but it will be the case that as we move around the circle, i.e. $X^{25} \to X^{25} + 2\pi R$, the particle (and all fields) of charge q will pick up a phase $e^{iq\theta}$. So the canonical momentum is shifted to:

$$p^{25} = \frac{n}{R} + \frac{q\theta}{2\pi R}. \qquad (4.66)$$

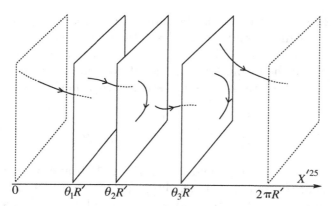

Fig. 4.3. Three D-branes at different positions, with various strings attached.

We have in general N hyperplanes at different positions as depicted in figure 4.3.

4.10 D-brane collective coordinates

Clearly, the whole picture goes through if several coordinates

$$X^m = \{X^{25}, X^{24}, \ldots, X^{p+1}\} \qquad (4.69)$$

are periodic, and we rewrite the periodic dimensions in terms of the dual coordinates. The open string endpoints are then confined to N $(p+1)$-dimensional hyperplanes, the D$(p+1)$-branes. The Neumann conditions on the world-sheet, $\partial_n X^m(\sigma^1, \sigma^2) = 0$, have become Dirichlet conditions $\partial_t X'^m(\sigma^1, \sigma^2) = 0$ for the dual coordinates. In this terminology, the original 26 dimensional open string theory theory contains N D25-branes. A 25-brane fills space, so the string endpoint can be anywhere: it just corresponds to an ordinary Chan–Paton factor.

It is natural to expect that the hyperplane is dynamical rather than rigid[8]. For one thing, this theory still has gravity, and it is difficult to see how a perfectly rigid object could exist. Rather, we would expect that the hyperplanes can fluctuate in shape and position as dynamical objects. We can see this by looking at the massless spectrum of the theory, interpreted in the dual coordinates.

Taking for illustration the case where a single coordinate is dualised, consider the mass spectrum. The $D-1$ dimensional mass is

$$M^2 = (p^{25})^2 + \frac{1}{\alpha'}(N-1)$$
$$= \left(\frac{[2\pi n + (\theta_i - \theta_j)]R'}{2\pi\alpha'}\right)^2 + \frac{1}{\alpha'}(N-1). \qquad (4.70)$$

Note that $[2\pi n + (\theta_i - \theta_j)]R'$ is the minimum length of a string winding between hyperplanes i and j. Massless states arise generically only for non-winding (i.e. $n = 0$) open strings whose end points are on the same hyperplane, since the string tension contributes an energy to a stretched string. We have therefore the massless states (with their vertex operators):

$$\alpha^{\mu}_{-1}|k; ii\rangle, \quad V = \partial_t X^{\mu},$$
$$\alpha^{m}_{-1}|k; ii\rangle, \quad V = \partial_t X^{25} = \partial_n X'^{25}. \tag{4.71}$$

The first of these is a gauge field living on the D-brane, with $p + 1$ components tangent to the hyperplane, $A^{\mu}(\xi^a)$, $\mu, a = 0, \ldots, p$. Here, $\xi^{\mu} = x^{\mu}$ are coordinates on the D-branes' world-volume. The second was the gauge field in the compact direction in the original theory. In the dual theory it becomes the transverse position of the D-brane (see equation (4.68)). From the point of view of the world-volume, it is a family of scalar fields, $\Phi^m(\xi^a)$, $(m = p + 1, \ldots, D - 1)$ living there.

We saw this in equation (4.68) for a Wilson line, which was a constant gauge potential. Now imagine that, as genuine scalar fields, the Φ^m vary as we move around on the world-volume of the D-brane. This therefore embeds the brane into a variable place in the transverse coordinates. This is simply describing a specific *shape* to the brane as it is embedded in spacetime. The $\Phi^m(\xi^a)$ are exactly analogous to the embedding coordinate map $X^{\mu}(\sigma, \tau)$ with which we described strings in the earlier sections.

The values of the gauge field backgrounds describe the shape of the branes as a soliton background, then. Meanwhile their quanta describe fluctuations of that background. This is the same phenomenon which we found for our description of spacetime in string theory. We started with strings in a flat background and discover that a massless closed string state corresponds to fluctuations of the geometry. Here we found first a flat hyperplane, and then discovered that a certain open string state corresponds to fluctuations of its shape. Remarkably, these open string states are simply gauge fields, and this is one of the reasons for the great success of D-branes. There are other branes in string theory (as we shall see) and they have other types of field theory describing their collective dynamics. D-branes are special, in that they have a beautiful description using gauge theory. Ultimately, we can use the long experience of working with gauge theories to teach us much about D-branes, and later, the geometry of D-branes and the string theories in which they live can teach us a lot about gauge theories. This is the basis of the dialogue between gauge theory and geometry which dominates the field at present.

It is interesting to look at the $U(N)$ symmetry breaking in the dual picture where the brane can move transverse to their world-volumes. When no D-branes coincide, there is just one massless vector each, or $U(1)^N$ in all, the generic unbroken group. If k D-branes coincide, there are new massless states because strings which are stretched between these branes can achieve vanishing length. Thus, there are k^2 vectors, forming the adjoint of a $U(k)$ gauge group[25, 26]. This coincident position corresponds to $\theta_1 = \theta_2 = \cdots = \theta_k$ for some subset of the original $\{\theta\}$, so in the original theory the Wilson line left a $U(k)$ subgroup unbroken. At the same time, there appears a set of k^2 massless scalars: the k positions are promoted to a matrix. This is not intuitive at first, but plays an important role in the dynamics of D-branes[26]. We will examine many consequences of this later in this book. Note that if all N branes are coincident, we recover the $U(N)$ gauge symmetry.

Although this picture seems quite odd, and will become more so in the unoriented theory, note that all we have done is to rewrite the original open string theory in terms of variables which are more natural in the limit $R \ll \sqrt{\alpha'}$. Various obscure features of the small-radius limit become clear in the T-dual picture.

Observe that, since T-duality interchanges Neumann and Dirichlet boundary conditions, a further T-duality in a direction tangent to a Dp-brane reduces it to a D$(p-1)$-brane, while a T-duality in a direction orthogonal turns it into a D$(p+1)$-brane.

4.11 T-duality for unoriented strings: orientifolds

The $R \to 0$ limit of an unoriented theory also leads to a new extended object. Recall that the effect of T-duality can also be understood as a one-sided parity transformation. For closed strings, the original coordinate is $X^m(z, \bar{z}) = X^m(z) + X^m(\bar{z})$. We have already discussed how to project string theory with these coordinates by Ω. The dual coordinate is $X'^m(z, \bar{z}) = X^m(z) - X^m(\bar{z})$. The action of world sheet parity reversal is to exchange $X^\mu(z)$ and $X^\mu(\bar{z})$. This gives for the dual coordinate:

$$X'^m(z, \bar{z}) \leftrightarrow -X'^m(\bar{z}, z). \tag{4.72}$$

This is the product of a world-sheet and a spacetime parity operation.
In the unoriented theory, strings are invariant under the action of Ω, while in the dual coordinate the theory is invariant under the product of world-sheet parity and a spacetime parity. This generalisation of the usual unoriented theory is known as an 'orientifold', a term that mixes the term 'orbifold' with orientation reversal.

Imagine that we have separated the string wavefunction into its internal part and its dependence on the centre of mass, x^m. Furthermore, take the internal wavefunction to be an eigenstate of Ω. The projection then determines the string wavefunction at $-x^m$ to be the same as at x^m, up to a sign. The various components of the metric and antisymmetric tensor satisfy, for example,

$$G_{\mu\nu}(x^\mu, -x^m) = G_{\mu\nu}(x^\mu, x^m), \quad B_{\mu\nu}(x^\mu, -x^m) = -B_{\mu\nu}(x^\mu, x^m),$$
$$G_{\mu n}(x^\mu, -x^m) = -G_{\mu n}(x^\mu, x^m), \quad B_{\mu n}(x^\mu, -x^m) = B_{\mu n}(x^\mu, x^m),$$
$$G_{mn}(x^\mu, -x^m) = G_{mn}(x^\mu, x^m), \quad B_{mn}(x^\mu, -x^m) = -B_{mn}(x^\mu, x^m). \quad (4.73)$$

In other words, when we have k compact directions, the T-dual spacetime is the torus T^{25-k} moded by a $\mathbb{Z}_2$ reflection in the compact directions. So we are instructed to perform an orbifold construction, modified by the extra sign. In the case of a single periodic dimension, for example, the dual spacetime is the line segment $0 \leq x^{25} \leq \pi R'$. The reader should remind themselves of the orbifold construction in section 4.8. At the ends of the interval, there are fixed 'points', which are in fact spatially 24-dimensional planes. Looking at the projections (4.73) in this case, we see that on these fixed planes, the projection is just like we did for the Ω-projection of the 25+1 dimensional theory in section 2.6: the theory is unoriented there, and half the states are removed. These orientifold fixed planes are called 'O-planes' for short. For this case, we have two O24-planes. (For k directions we have 2^k O$(25 - k)$-planes arranged on the vertices of a hypercube.) In particular, we can usefully think of the original case of $k = 0$ as being on an O25-plane.

While the theory is unoriented on the O-plane, away from the orientifold fixed planes, the local physics is that of the *oriented* string theory. The projection relates the physics of a string at some point x^m to the string at the image point $-x^m$.

In string perturbation theory, orientifold planes are not dynamical. Unlike the case of D-branes, there are no string modes tied to the orientifold plane to represent fluctuations in its shape. Our heuristic argument in the previous subsection that gravitational fluctuations force a D-brane to move dynamically does not apply to the orientifold fixed plane. This is because the identifications (4.73) become *boundary conditions* at the fixed plane, such that the incident and reflected gravitational waves cancel. For the D-brane, the reflected wave is higher order in the string coupling.

The orientifold construction was discovered via T-duality[8] and independently from other approaches[27, 10]. One can of course consider more general orientifolds which are not simply T-duals of toroidal compactifications. The idea is simply to combine a group of discrete symmetries with Ω

such that the resulting group of operations (the 'orientifold group', G_Ω) is itself a symmetry of some string theory. One then has the right to ask what the nature of the projected theory obtained by dividing by G_Ω might be. This is a fruitful way of construction interesting and useful string vacua[28]. We shall have more to say about this later, since in superstring theory we shall find that O-planes, like D-branes , are sources of various closed string sector fields. Therefore there will be additional consistency conditions to be satisfied in constructing an orientifold, amounting to making sure that the field equations are satisfied.

So far our discussion of orientifolds was just for the closed string sector. Let us see how things are changed in the presence of open strings. In fact, the situation is similar. Again, let us focus for simplicity on a single compact dimension. Again there is one orientifold fixed plane at 0 and another at $\pi R'$. Introducing $SO(N)$ Chan–Paton factors, a Wilson line can be brought to the form

$$\mathrm{diag}\{\theta_1, -\theta_1, \theta_2, -\theta_2, \ldots, \theta_{N/2}, -\theta_{N/2}\}. \tag{4.74}$$

Thus in the dual picture there are $\frac{1}{2}N$ D-branes on the line segment $0 \le X'^{25} < \pi R'$, and $\frac{1}{2}N$ at their image points under the orientifold identification.

Strings can stretch between D-branes and their images, as shown in figure 4.4. The generic gauge group is $U(1)^{N/2}$, where all branes are separated. As in the oriented case, if m D-branes are coincident there is a $U(m)$ gauge group. However, now if the m D-branes in addition lie at one

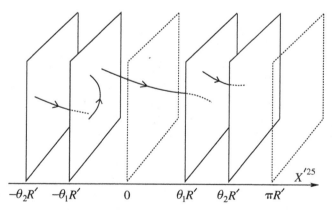

Fig. 4.4. Orientifold planes at 0 and $\pi R'$. There are D-branes at $\theta_1 R'$ and $\theta_2 R'$, and their images at $-\theta_1 R'$ and $-\theta_2 R'$. Ω acts on any string by a combination of a spacetime reflection through the planes and reversing the orientation arrow.

of the fixed planes, then strings stretching between one of these branes and one of the image branes also become massless and we have the right spectrum of additional states to fill out $SO(2m)$. The maximal $SO(N)$ is restored if all of the branes are coincident at a single orientifold plane. Note that this maximally symmetric case is asymmetric between the two fixed planes. Similar considerations apply to $USp(N)$. As we saw before, the difference between the appearance of the two groups is in a sign on the matrix M as it acts on the string wavefunction. Later, we shall see that this sign is correlated with the sign of the charge and tension of the orientifold plane.

We should emphasise that there are $\frac{1}{2}N$ dynamical D-branes but an N-valued Chan–Paton index. An interesting case is when $k + \frac{1}{2}$ D-branes lie on a fixed plane, which makes sense because the number $2k + 1$ of indices is integer. A brane plus image can move away from the fixed plane, but the number of branes remaining is always half-integer. This anticipates a discussion which we shall have about fractional branes much later, in section 13.2, even outside the context of orientifolds.

5

Background fields and world-volume actions

T-duality is clearly a remarkable phenomenon that is highly indicative of the different view string theory has of spacetime from that of field theories. This heralds a rather rich landscape of possibilities for new physics, and indeed T-duality will govern much of what we will study in the rest of this book, either directly or indirectly. So far, we have uncovered it at the level of the string spectrum, and have used it to discover D-branes and orientifolds. However, we have so far restricted ourselves to flat spacetime backgrounds, with none of the other fields in the string spectrum switched on. In this chapter, we shall study the action of T-duality when the massless fields of the string theory take on non-trivial values, giving us curved backgrounds and/or gauge fields on the world-volume of the D-branes. It is also important to uncover further aspects of the dynamics of D-branes in non-trivial backgrounds, and we shall also uncover an action to describe this here.

5.1 T-duality in background fields

The first thing to notice is that T-duality acts non-trivial on the dilaton, and therefore modifies the string coupling[16, 17]. After dimensional reduction on a circle of radius R, the effective 25-dimensional string coupling read off from the reduced string frame supergravity action is now $g_s = e^{\Phi}(2\pi R)^{-1/2}$. Since the resulting 25-dimensional theory is supposed to have the same physics, by T-duality, as a theory with a dilaton $\tilde{\Phi}$, compactified on a circle of radius R', it is required that this coupling is equal to $\tilde{g}_s = e^{\tilde{\Phi}}(2\pi R')^{-1/2}$, the string coupling of the dual 25-dimensional theory:

$$e^{\tilde{\Phi}} = e^{\Phi}\frac{\alpha'^{1/2}}{R}.$$

(5.1)

This is just part of a larger statement about the T-duality transformation properties of background fields in general. Starting with background fields $G_{\mu\nu}$, $B_{\mu\nu}$ and Φ, let us first T-dualise in one direction, which we shall label X^{25}, as before. In other words, X^{25} is a direction which is a circle of radius R, and the dual circle X'^{25} is a circle of radius $R' = \alpha'/R$.

We may start with the two dimensional sigma model (2.103) with background fields $G_{\mu\nu}$, $B_{\mu\nu}$, Φ, and assume that locally, all of the fields are independent of the direction X^{25}. In this case, we may write an equivalent action by introducing a Lagrange multiplier, which we shall call X'^{25}:

$$S_\sigma = \frac{1}{4\pi\alpha'} \int d^2\sigma g^{1/2} \left\{ g^{ab} \left[G_{25,25}v_a v_b + 2G_{25,\mu}v_a \partial_b X^\mu + G_{\mu\nu} \partial_a X^\mu \partial_b X^\nu \right] \right.$$

$$\left. + i\epsilon^{ab} \left[2B_{25,\mu}v_a \partial_b X^\mu + B_{\mu\nu} \partial_a X^\mu \partial_b X^\nu + 2X'^{25} \partial_a v_b \right] + \alpha' R\Phi \right\}. \quad (5.2)$$

Since the equation of motion for the Lagrange multiplier is

$$\frac{\partial \mathcal{L}}{\partial X'^{25}} = i\epsilon^{ab} \partial_a v_b = 0,$$

we can write a solution as $v_b = \partial_b \phi$ for any scalar ϕ, which we might as well call X^{25}, since upon substitution of this solution back into the action, we get our original action in (2.103).

Instead, we can find the equation of motion for the quantity v_a:

$$\frac{\partial \mathcal{L}}{\partial v_a} - \frac{\partial}{\partial \sigma_b} \left(\frac{\partial \mathcal{L}}{\partial(\partial_b v_a)} \right) = 0 \quad (5.3)$$

$$= g^{ab} \left[G_{25,25} v_b + G_{25,\mu} \partial_b X^\mu \right] + i\epsilon^{ab} \left[B_{25,\mu} \partial_b X^\mu + \partial_b X'^{25} \right],$$

which, upon solving it for v_a and substituting back into the equations gives an action of the form (2.103), but with fields $\tilde{G}_{\mu\nu}$ and $\tilde{B}_{\mu\nu}$ given by:

$$\tilde{G}_{25,25} = \frac{1}{G_{25,25}}; \qquad e^{2\tilde{\Phi}} = \frac{e^{2\Phi}}{G_{25,25}},$$

$$\tilde{G}_{\mu 25} = \frac{B_{\mu 25}}{G_{25,25}}; \qquad \tilde{B}_{\mu 25} = \frac{G_{\mu 25}}{G_{25,25}},$$

$$\tilde{G}_{\mu\nu} = G_{\mu\nu} - \frac{G_{\mu 25}G_{\nu 25} - B_{\mu 25}B_{\nu 25}}{G_{25,25}},$$

$$\tilde{B}_{\mu\nu} = B_{\mu\nu} - \frac{B_{\mu 25}G_{\nu 25} - G_{\mu 25}B_{\nu 25}}{G_{25,25}}, \quad (5.4)$$

where a *one loop* (not tree level) world-sheet computation (e.g. by checking the β-function equations again, or by considering the new path integral measure induced by integrating out v_a), gives the new dilaton. This fits with the fact that it couples at the next order in α' (which plays the role of $\hbar$ on the world-sheet) as discussed previously.

Of course, we can T-dualise on many (say d) independent circles, forming a torus T^d. It is not hard to deduce that one can succinctly write the resulting T-dual background as follows. If we define the $D{\times}D$ metric

$$E_{\mu\nu} = G_{\mu\nu} + B_{\mu\nu}, \qquad (5.5)$$

and if the circles are in the directions X^i, $i = 1, \ldots, d$, with the remaining directions labelled by X^a, then the dual fields are given by

$$\widetilde{E}_{ij} = E^{ij}, \qquad \widetilde{E}_{aj} = E_{ak}E^{kj}, \qquad e^{2\widetilde{\Phi}} = e^{2\Phi}\det(E^{ij}),$$
$$\widetilde{E}_{ab} = E_{ab} - E_{ai}E^{ij}E_{jb}, \qquad (5.6)$$

where $E_{ik}E^{kj} = \delta_i{}^j$ defines E^{ij} as the inverse of E_{ij}. We will find this succinct form of the $O(d, d)$ T-duality transformation very useful later on.

5.2 A first look at the D-brane world-volume action

The D-brane is a dynamical object, and as such, feels the force of gravity. In fact, it must be able to respond to the values of the various background fields in the theory. This is especially obvious if one recalls that the D-branes' location and shaped is controlled (in at least one way of describing them) by the open strings which end on them. These strings respond to the background fields in ways we have already studied (we have written world-sheet actions for them), and so should the D-branes. We must find a world-volume action describing their dynamics.

If we introduce coordinates ξ^a, $a = 0, \ldots, p$ on the brane, we can begin to write an action for the dynamics of the brane in terms of fields living on the world-volume in much the same way that we did for the string, in terms of fields living on the world-sheet. The background fields will act as generalised field-dependent couplings. As we discussed before, the fields on the brane are the embedding $X^\mu(\xi)$ and the gauge field $A_a(\xi)$. We shall ignore the latter for now and concentrate just on the embedding part. By direct analogy to the particle and string case studied in chapter 2, the action is

$$S_p = -T_p \int d^{p+1}\xi \, e^{-\Phi} \det{}^{1/2}G_{ab}, \qquad (5.7)$$

where G_{ab} is the induced metric on the brane, otherwise known as the 'pull-back' of the spacetime metric $G_{\mu\nu}$ to the brane:

$$G_{ab} \equiv \frac{\partial X^\mu}{\partial \xi^a} \frac{\partial X^\nu}{\partial \xi^b} G_{\mu\nu}. \tag{5.8}$$

T_p is the tension of the Dp-brane, which we shall discuss at length later. The dilaton dependence $e^{-\Phi} = g_{\mathrm{s}}^{-1}$ arises because this is an open string tree level action, and so this is the appropriate function of the dilaton to introduce.

N.B. The world-volume reparametrisation invariant action we have just written is in terms of the determinant of the metric. It is a common convention to leave the a, b indices dangling in writing this action and its generalisations, and we shall adopt that somewhat loose notation here. More careful authors sometimes use other symbols, like $\det^{1/2} P[G]$, where the P denotes the pull-back, and G means the metric, now properly thought of as a matrix whose determinant is to be taken. Here, the meaning of what we write using the looser notation should always be clear from the context.

Of course, this cannot be the whole story, and indeed it is clear that we shall need a richer action, since the rules of T-duality action on the background fields mean that T-dualising to a D$(p+1)$– or D$(p-1)$-brane's action will introduce a dependence on $B_{\mu\nu}$, since it mixes with components of the metric. Furthermore, there will be mixing with components of a world-volume gauge field, since some of kinetic terms for the transverse fields, $\partial_a X^m$, $m = p+1, \ldots, D-1$, implicit in the action (5.8), will become derivatives of gauge fields, $2\pi\alpha' \partial_a A_m$ according to the rules of T-duality for open strings deduced in the previous chapter. We shall construct the full T-duality respecting action in the next subsection. Before we do that, let us consider what we can learn about the tension of the D-brane from this simple action, and what we learned about the transformation of the dilaton.

The tension of the brane controls its response to outside influences which try to make it change its shape, absorb energy, etc., just as we saw for the tension of a string. We shall compute the actual value of the tension in chapter 6. Here, we are going to uncover a useful recursion relation relating the tensions of different D-branes, which follows from T-duality[76, 29]. The mass of a Dp-brane wrapped around a p-torus T^p is

$$T_p e^{-\Phi} \prod_{i=1}^{p} (2\pi R_i). \tag{5.9}$$

T-dualising on the single direction X^p and recalling the transformation (5.1) of the dilaton, we can rewrite the mass (5.9) in the dual variables:

$$T_p(2\pi\sqrt{\alpha'})e^{-\Phi'}\prod_{i=1}^{p-1}(2\pi R_i) = T_{p-1}e^{-\Phi'}\prod_{i=1}^{p-1}(2\pi R_i). \qquad (5.10)$$

Hence,

$$T_p = T_{p-1}/2\pi\sqrt{\alpha'} \qquad \Rightarrow \qquad T_p = T_{p'}(2\pi\sqrt{\alpha'})^{p'-p}, \qquad (5.11)$$

where we performed the duality recursively to deduce the general relation.

The next step is to take into account new couplings for the embedding coordinates/fields which result of other background spacetime fields like the antisymmetric tensor $B_{\mu\nu}$. This again appears as an induced tensor B_{ab} on the worldvolume, via a formula like (5.8).

It is important to notice that that there is a restriction due to spacetime gauge symmetry on the precise combination of B_{ab} and A^a which can appear in the action. The combination $B_{ab} + 2\pi\alpha'F_{ab}$ can be understood as follows. In the world-sheet sigma model action of the string, we have the usual closed string term (2.103) for B and the boundary action (2.108) for A. So the fields appear in the combination:

$$\frac{1}{2\pi\alpha'}\int_{\mathcal{M}}B + \int_{\partial\mathcal{M}}A. \qquad (5.12)$$

We have written everything in terms of differential forms, since B and A are antisymmetric. For example $\int A \equiv \int A_a d\xi^a$.

This action is invariant under the spacetime gauge transformation $\delta A = d\lambda$. However, the spacetime gauge transformation $\delta B = d\zeta$ will give a surface term which must be cancelled with the following gauge transformation of A: $\delta A = -\zeta/2\pi\alpha'$. So the combination $B + 2\pi\alpha'F$, where $F = dA$ is invariant under both symmetries; this is the combination of A and B which must appear in the action in order for spacetime gauge invariance to be preserved.

5.2.1 World-volume actions from tilted D-branes

There are many ways to deduce pieces of the world-volume action. One way is to redo the computation for Weyl invariance of the complete sigma model, including the boundary terms, which will result in the $(p+1)$-dimensional equations of motion for the world-volume fields G_{ab}, B_{ab} and A_a. One can then deduce the $p+1$-dimensional world-volume action from which those equations of motion may be derived. We will comment on this below.

Another way, hinted at in the previous subsection, is to use T-duality to build the action piece by piece. For the purposes of learning more about how the branes work, and in view of the various applications to which we will put the branes, this second way is perhaps more instructive.

Consider[38] a D2-brane extended in the X^1 and X^2 directions, and let there be a constant gauge field F_{12}. (We leave the other dimensions unspecified, so the brane could be larger by having extent in other directions. This will not affect our discussion.) We can choose a gauge in which $A_2 = X^1 F_{12}$. Now consider T-dualising along the x^2-direction. The relation (4.68) between the potential and coordinate gives

$$X'^2 = 2\pi\alpha' X^1 F_{12}, \tag{5.13}$$

This says that the resulting D1-brane is tilted at an angle*

$$\theta = \tan^{-1}(2\pi\alpha' F_{12}) \tag{5.14}$$

to the X^2-axis! This gives a geometric factor in the D1-brane world-volume action,

$$S \sim \int_{\text{D1}} ds = \int dX^1 \sqrt{1 + (\partial_1 X'^2)^2} = \int dX^1 \sqrt{1 + (2\pi\alpha' F_{12})^2}. \tag{5.15}$$

We can always boost the D-brane to be aligned with the coordinate axes and then rotate to bring $F_{\mu\nu}$ to block-diagonal form, and in this way we can reduce the problem to a product of factors like (5.15) giving a determinant:

$$S \sim \int d^D X \, \det^{1/2}(\eta_{\mu\nu} + 2\pi\alpha' F_{\mu\nu}). \tag{5.16}$$

This is the Born–Infeld action.[42]

In fact, this is the complete action (in a particular 'static' gauge which we will discuss later) for a space-filling D25-brane in flat space, and with the dilaton and antisymmetric tensor field set to zero. In the language of section 2.7, Weyl invariance of the open string sigma model (2.108) amounts to the following analogue of (2.105) for the open string sector:

$$\beta^A_\mu = \alpha' \left(\frac{1}{1 - (2\pi\alpha' F)^2} \right)^{\nu\lambda} \partial_{(\nu} F_{\lambda)\mu} = 0, \tag{5.17}$$

---

* The reader concerned about achieving irrational angles and hence densely filling the (x^1, x^2) torus should suspend disbelief until chapter 8. There, when we work in the fully consistent quantum theory of superstrings, it will be seen that the fluxes are quantised in just the right units to make this sensible.

these equations of motion follow from the action. In fact, in contrast to the Maxwell action written previously (2.107), and the closed string action (2.106), this action is true to all orders in α', although only for slowly varying field strengths; there are corrections from derivatives of $F_{\mu\nu}$.[32]

5.3 The Dirac–Born–Infeld action

We can uncover a lot of the rest of the action by simply dimensionally reducing. Starting with (5.16), where $F_{\mu\nu} = \partial_\mu A_\nu - \partial_\nu A_\mu$ as usual (we will treat the non-Abelian case later) let us assume that $D-p-1$ spatial coordinates are very small circles, small enough that we can neglect all derivatives with respect to those directions, labelled X^m, $m = p+1, \ldots, D-1$. (The uncompactified coordinates will be labelled X^a, $a = 0, \ldots, p$.) In this case, the matrix whose determinant appears in (5.16) is:

$$\begin{pmatrix} N & -A^T \\ A & M \end{pmatrix}, \tag{5.18}$$

where

$$N = \eta_{ab} + 2\pi\alpha' F_{ab}; \qquad M = \delta_{mn}; \qquad A = 2\pi\alpha' \partial_a A_m. \tag{5.19}$$

Using the fact that its determinant can be written as $|M||N + A^T M^{-1} A|$, our action becomes[56]

$$S \sim -\int d^{p+1}X \, \det^{1/2}(\eta_{ab} + \partial_a X^m \partial_b X_m + 2\pi\alpha' F_{ab}), \tag{5.20}$$

up to a numerical factor (coming from the volume of the torus we reduced on. Once again, we used the T-duality rules (4.68) to replace the gauge fields in the T-dual directions by coordinates: $2\pi\alpha' A_m = X^m$.

This is (nearly) the action for a Dp-brane and we have uncovered how to write the action for the collective coordinates X^m representing the fluctuations of the brane transverse to the world-volume. There now remains only the issue of putting in the case of non-trivial metric, $B_{\mu\nu}$ and dilaton. This is easy to guess given that which we have encountered already:

$$S_p = -T_p \int d^{p+1}\xi \, e^{-\Phi} \det^{1/2}(G_{ab} + B_{ab} + 2\pi\alpha' F_{ab}). \tag{5.21}$$

This is the Dirac–Born–Infeld Lagrangian, for arbitrary background fields. The factor of the dilaton is again a result of the fact that all of this physics arises at open string tree level, hence the factor g_s^{-1}, and the B_{ab} is in the right place because of spacetime gauge invariance. T_p and G_{ab} are in the right place to match onto the discussion we had when we computed the tension. Instead of using T-duality, we could have also deduced this action by a generalisation of the sigma model methods described earlier, and in fact this is how it was first derived in this context[34].

We have re-introduced independent coordinates ξ^a on the world-volume. Note that the actions given in equations (5.15) and (5.20) were written using a choice where we aligned the world-volume with the first $p + 1$ spacetime coordinates as $\xi^a = X^a$, leaving the $D - p - 1$ transverse coordinates called X^m. We can always do this using world-volume and spacetime diffeomorphism invariance. This choice is called the 'static gauge', and we shall use it quite a bit in these notes. Writing this out (for vanishing dilaton) using the formula (5.8) for the induced metric, for the case of $G_{\mu\nu} = \eta_{\mu\nu}$ we see that we get the action (5.20).

5.4 The action of T-duality

It is amusing[41, 51] to note that our full action obeys (as it should) the rules of T-duality which we already wrote down for our background fields. The action for the Dp-brane is built out of the determinant $|E_{ab} + 2\pi\alpha' F_{ab}|$, where the $(a, b = 0, \ldots, p)$ indices on E_{ab} mean that we have performed the pullback of $E_{\mu\nu}$ (defined in (5.5)) to the world-volume. This matrix becomes, if we T-dualise on n directions labelled by X^i and use the rules we wrote in (5.6):

$$\begin{vmatrix} E_{ab} - E_{ai}E^{ij}E_{jb} + 2\pi\alpha' F_{ab} & -E_{ak}E^{kj} - \partial_a X^i \\ E^{ik}E_{kb} + \partial_b X^i & E^{ij} \end{vmatrix}, \qquad (5.22)$$

which has determinant $|E^{ij}||E_{ab} + 2\pi\alpha' F_{ab}|$. In forming the square root, we get again the determinant needed for the definition of a T-dual DBI action, as the extra determinant $|E^{ij}|$ precisely cancels the determinant factor picked up by the dilaton under T-duality. (Recall, E^{ij} is the inverse of E_{ij}.)

Furthermore, the tension $T_{p'}$ comes out correctly, because there is a factor of $\Pi_i^n(2\pi R_i)$ from integrating over the torus directions, and a factor $\Pi_i^n(R_i/\sqrt{\alpha'})$ from converting the factor $e^{-\Phi}$, (see equation (5.1)), which fits nicely with the recursion formula (5.11) relating T_p and $T_{p'}$.

The above was done as though the directions on which we dualised were all Neumann or all Dirichlet. Clearly, we can also extrapolate to the more general case.

5.5 Non-Abelian extensions

For N D-branes the story is more complicated. The various fields on the brane representing the collective motions, A_a and X^m, become matrices valued in the adjoint. In the Abelian case, the various spacetime background fields (here denoted F_μ for the sake of argument) which can appear on the world-volume typically depend on the transverse coordinates X^m in some (possibly) non-trivial way. In the non-Abelian case, with N D-branes, the transverse coordinates are really $N \times N$ matrices, $2\pi\alpha'\Phi^m$, since they are T-dual to non-Abelian gauge fields as we learned in previous sections, and so inherit the behaviour of gauge fields (see equation (4.68)). We write them as $\Phi^m = X^m/(2\pi\alpha')$. So not only should the background fields F_μ depend on the Abelian part, but they ought to possibly depend (implicitly or explicitly) on the full non-Abelian part as $F(\Phi)_\mu$ in the action.

Furthermore, in (5.21) we have used the partial derivatives $\partial_a X^\mu$ to pull back spacetime indices μ to the world-volume indices, a, e.g. $F_a = F_\mu \partial_a X^\mu$, and so on. To make this gauge covariant in the non-Abelian case, we should pull back with the covariant derivative: $F_a = F_\mu \mathcal{D}_a X^\mu = F_\mu(\partial_a X^\mu + [A_a, X^\mu])$.

With the introduction of non-Abelian quantities in all of these places, we need to consider just how to perform a trace, in order to get a gauge invariant quantity to use for the action. Starting with the fully Neumann case (5.16), a first guess is that things generalise by performing a trace (in the fundamental of $U(N)$) of the square rooted expression. The meaning of the Tr needs to be stated, It is proposed that is means the 'symmetric' trace, denoted 'STr' which is to say that one symmetrises over gauge indices, consequently ignoring all commutators of the field strengths encountered while expanding the action[43]. (This suggestion is consistent with various studies of scattering amplitudes and also the BPS nature of various non-Abelian soliton solutions. There is still apparently some ambiguity in the definition which results in problems beyond fifth order in the field strength[44, 343].)

Once we have this action, we can then again use T-duality to deduce the form for the lower dimensional, Dp-brane actions. The point is that we can reproduce the steps of the previous analysis, but keeping commutator terms such as $[A_a, \Phi^m]$ and $[\Phi^m, \Phi^n]$. We will not reproduce those steps here, as they are similar in spirit to that which we have already done (for a complete discussion, the reader is invited to consult some of the literature[45].) The resulting action is:

$$S_p = -T_p \int d^{p+1}\xi e^{-\Phi} \mathcal{L}, \quad \text{where}$$

$$\mathcal{L} = \text{STr}\left\{ \det^{1/2}[E_{ab} + E_{ai}(Q^{-1} - \delta)^{ij}E_{jb} + 2\pi\alpha' F_{ab}] \det^{1/2}[Q^i{}_j] \right\}, (5.23)$$

where $Q^i{}_j = \delta^i{}_j + i2\pi\alpha'[\Phi^i, \Phi^k]E_{kj}$, and we have raised indices with E^{ij}.

5.6 D-branes and gauge theory

In fact, we are now in a position to compute the constant C in equation (2.107), by considering N D25-branes, which is the same as an ordinary (fully Neumann) N-valued Chan–Paton factor. Expanding the D25-brane Lagrangian (5.16) to second order in the gauge field, we get

$$-\frac{T_{25}}{4}(2\pi\alpha')^2 e^{-\Phi}\mathrm{Tr}F_{\mu\nu}F^{\mu\nu}, \tag{5.24}$$

with the trace in the fundamental representation of $U(N)$. This gives the precise numerical relation between the open and closed string couplings.

Actually, with Dirichlet and Neumann directions, performing the same expansion, and in addition noting that

$$\det[Q^i_{\ j}] = 1 - \frac{(2\pi\alpha')^2}{4}[\Phi^i, \Phi^j][\Phi^i, \Phi^j] + \cdots, \tag{5.25}$$

one can write the leading order action (5.23) as

$$S_p = -\frac{T_p(2\pi\alpha')^2}{4}\int d^{p+1}\xi\, e^{-\Phi}\mathrm{Tr}\left[F_{ab}F^{ab} + 2\mathcal{D}_a\Phi^i\mathcal{D}_a\Phi^i + [\Phi^i, \Phi^j]^2\right]\Big\}. \tag{5.26}$$

This is the dimensional reduction of the D-dimensional Yang–Mills term, displaying the non-trivial commutator for the adjoint scalars. This is an important term in many modern applications, as we shall see. Note that the $(p + 1)$-dimensional Yang–Mills coupling for the theory on the branes is

$$g_{\mathrm{YM},p}^2 = g_s T_p^{-1}(2\pi\alpha')^{-2}. \tag{5.27}$$

This is worth noting[70]. With the superstring value of T_p which we will compute later, it is used in many applications to give the correct relation between gauge theory couplings and string quantities.

5.7 BPS lumps on the world-volume

We can of course treat the Dirac–Born–Infeld action as an interesting theory in its own right, and seek for interesting solutions of it. These solutions will have both a $(p + 1)$-dimensional interpretation and a D-dimensional one.

We shall not dwell on this in great detail, but include a brief discussion here to illustrate an important point, and refer to the literature for more complete discussions.[55] More details will appear when we get to the supersymmetric case. One can derive an expression for the energy density contained in the fields on the world-volume:

$$\mathcal{E}^2 = E^a E^b F_{ca} F_{cb} + E^a E^b G_{ab} + \det(G + 2\pi\alpha' F), \qquad (5.28)$$

where here the matrix F_{ab} contains only the magnetic components (i.e. no time derivatives) and E^a are the electric components, subject to the Gauss Law constraint $\vec{\nabla} \cdot \vec{E} = 0$. Also, as before

$$G_{ab} = \eta_{ab} + \partial_a X^m \partial_b X^m, \qquad m = p+1, \ldots, D-1. \qquad (5.29)$$

Let us consider the case where we have no magnetic components and only one of the transverse fields, say X^{25}, switched on. In this case, we have

$$\mathcal{E}^2 = (1 \pm \vec{E} \cdot \vec{\nabla} X^{25})^2 + (\vec{E} \mp \vec{\nabla} X^{25})^2, \qquad (5.30)$$

and so we see that we have the Bogomol'nyi condition

$$\mathcal{E} \geq |\vec{E} \cdot \vec{\nabla} X^{25}| + 1. \qquad (5.31)$$

This condition is saturated if $\vec{E} = \pm\vec{\nabla} X^{25}$. In such a case, we have

$$\nabla^2 X^{25} = 0 \qquad \Rightarrow \qquad X^{25} = \frac{c_p}{r^{p-2}}, \qquad (5.32)$$

a harmonic solution, where c_p is a constant to be determined. The total energy (beyond that of the brane itself) is, integrating over the world-volume:

$$\begin{aligned}
E_{\text{tot}} &= \lim_{\epsilon \to \infty} T_p \int_\epsilon^\infty r^{p-1} dr d\Omega_{p-1} (\vec{\nabla} X^{25})^2 = \lim_{\epsilon \to \infty} T_p \frac{c_p^2(p-2)\Omega_{p-1}}{\epsilon^{p-2}} \\
&= \lim_{\epsilon \to \infty} T_p c_p (p-2)\Omega_{p-1} X^{25}(\epsilon), \qquad (5.33)
\end{aligned}$$

where Ω_{p-1} is the volume of the sphere S^{p-1} surrounding our point charge source, and we have cut off the divergent integral by integrating down to $r = \epsilon$. (We will save the case of $p = 1$ for later[140, 60].) Now we can choose[†] a value of the electric flux such that we get $(p-2)c_p\Omega_{p-1}T_p = (2\pi\alpha')^{-1}$. Putting this into our equation for the total energy, we see that the (divergent) energy of our configuration is:

$$E_{\text{tot}} = \frac{1}{2\pi\alpha'} X^{25}(\epsilon). \qquad (5.34)$$

What does this mean? Well, recall that $X^{25}(\xi)$ gives the transverse position of the brane in the X^{25} direction. So we see that the brane

---

[†] In the supersymmetric case, this has a physical meaning, since overall consistency of the D-brane charges set a minimum electric flux. Here, it is more arbitrary, and so we choose a value by hand to make the point we wish to illustrate.

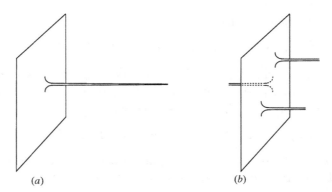

Fig. 5.1. The D-dimensional interpretation of the BIon solution. (a) It is an infinitely long spike representing a fundamental string ending on the D-brane. (b) BIons are BPS and therefore can be added together at no cost to make a multi-BIon solution.

has grown a semi-infinite spike at $r = 0$, and the base of this spike is our point charge. The interpretation of the divergent energy is simply the (infinite) length of the spike multiplied by a mass per unit length. But this mass per unit length is precisely the fundamental string tension $T = (2\pi\alpha')^{-1}$! In other words, the spike solution is the fundamental string stretched perpendicular to the brane and ending on it, forming a point electric charge, known as a 'BIon'; see figure 5.1(a). In fact, a general BIon includes the non-linear corrections to this spike solution, which we have neglected here, having only written the linearised solution.

It is a worthwhile computation to show that if test source with the same charges is placed on the brane, there is no force of attraction or repulsion between it and the source just constructed, as would happen with pure Maxwell charges. This is because our sources have in addition to electric charge, some scalar (X^{25}) charge, which can also be attractive or repulsive. In fact, the scalar charges are such that the force due to electromagnetic charges is cancelled by the force of the scalar charge, another characteristic property of these solutions, which are said to be 'Bogomol'nyi–Prasad–Sommerfield' (BPS)-saturated[61, 62]. We shall encounter solutions with this sort of behaviour a number of times in what is to follow.

Because of this property, the solution is easily generalised to include any number of BIons, at arbitrary positions, with positive and negative charges. The two choices of charge simply represents strings either leaving from, or arriving on the brane; see figure 5.1(b).

6

D-brane tension and boundary states

We have already stated that since the D-brane is a dynamical object, and couples to gravity, it should have a mass per unit volume. This tension will govern the strength of its response to outside influences which try to make it change its shape, absorb energy, etc. We have already computed a recursion relation (5.11) for the tension, whcih follows from the underlying T-duality which we used to discover D-branes in the first place.

In this chapter we shall see in detail just how to compute the value of the tension for the D-brane, and also for the orientifold plane. While the numbers that we will get will not (at face value) be as useful as the analogous quantities for the supersymmetric case, the structure of the computation is extremely important. The computation puts together many of the things that we have learned so far in a very elegant manner which lies at the heart of much of what will follow in more advanced chapters.

Along the way, we will see that D-branes can be constructed and studied in an alternative formalism known as the 'boundary state' formalism, which is essentially conformal field theory with certain sorts of boundaries included[33]. For much of what we will do, it will be a clearly equivalent way of formulating things which we also say (or have already said) based on the spacetime picture of D-branes. However, it should be noted that it is much more than just a rephrasing since it can be used to consistently formulate D-branes in many more complicated situations, even when a clear spacetime picture is not available. The method becomes even more useful in the supersymmetric situation, since it provides a natural way of constructing stable D-brane vacua of the superstring theories which do not preserve any supersymmetries, a useful starting point for exploring dualities and other non-perturbative physics in dynamical regimes which ultimately may have relevance to observable physics.

141

6.1 The D-brane tension

6.1.1 An open string partition function

Let us now compute the D-brane tension T_p. As noted previously, it is proportional to g_s^{-1}. We can in principle calculate it from the gravitational coupling to the D-brane, given by the disk with a graviton vertex operator in the interior. However, it is much easier to obtain the absolute normalisation in the following manner.

Consider two parallel Dp-branes at positions $X'^\mu = 0$ and $X'^\mu = Y^\mu$. These two objects can feel each other's presence by exchanging closed strings as shown in figure 6.1. This string graph is an annulus, with no vertex operators. It is therefore as easily calculated as our closed string one loop amplitudes done earlier in chapter 3.

In fact, this is rather like an open string partition function, since the amplitude can be thought of as an open string going in a loop. We should sum over everything that goes around in the loop. Once we have computed this, we will then change our picture of it as an open string one-loop amplitude, and look at it as a closed string amplitude for propagation between one D-brane and another. We can take a low energy limit of the result to focus on the massless closed string states which are being exchanged. Extracting the poles from graviton and dilaton exchange (we shall see that the antisymmetric tensor does not couple in this limit) then give the coupling T_p of closed string states to the D-brane.

Let us parametrise the string world-sheet as $(\sigma^2 = \tau, \sigma^1 = \sigma)$ where now τ is periodic and runs from 0 to $2\pi t$, and σ runs (as usual) from 0 to π. This vacuum graph (a cylinder) has the single modulus t, running

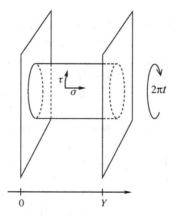

Fig. 6.1. Exchange of a closed string between two D-branes. This is equivalent to a vacuum loop of an open string with one end on each D-brane.

from 0 to ∞. If we slice horizontally, so that $\sigma^2 = \tau$ is world-sheet time, we get an open string going in a loop. If we instead slice vertically, so that σ is time, we see a single closed string propagating in the tree channel.

Notice that the world-line of the open string boundary can be regarded as a vertex connecting the vacuum to the single closed string, i.e. a one-point closed string vertex, which is a useful picture in a 'boundary state' formalism, which we will develop a bit further shortly. This diagram will occur explicitly again in many places in our treatment of this subject. String theory produces many examples where one-loop gauge/field theory results (open strings) are related to tree level geometrical/gravity results. This is all organised by diagrams of this form, and is the basis of much of the gauge theory/geometry correspondences to be discussed.

Let us consider the limit $t \to 0$ of the loop amplitude. This is the ultra-violet limit for the open string channel, since the circle of the loop is small. However, this limit is correctly interpreted as an *infrared* limit of the closed string. (This is one of the earliest 'dualities' of string theory, discussed even before it was known to be a theory of strings.) Time-slicing vertically shows that the $t \to 0$ limit is dominated by the lowest lying modes in the closed string spectrum. This all fits with the idea that there are no 'ultra-violet limits' of the moduli space which could give rise to high energy divergences. They can always be related to amplitudes which have a handle pinching off. This physics is controlled by the lightest states, or the long distance physics. (This relationship is responsible for the various 'UV/IR' relations which are a popular feature of current research[315].)

One-loop vacuum amplitudes are given by the Coleman–Weinberg [35, 36] formula, which can be thought of as the sum of the zero point energies of all the modes (see insert 6.1):

$$\mathcal{A} = V_{p+1} \int \frac{d^{p+1}k}{(2\pi)^{p+1}} \int_0^\infty \frac{dt}{2t} \sum_I e^{-2\pi\alpha' t(k^2 + M_I^2)}. \tag{6.1}$$

Here the sum I is over the physical spectrum of the string, i.e. the transverse spectrum, and the momentum k is in the $p+1$ extended directions of the D-brane world-sheet.

The mass spectrum is given by a familiar formula

$$M^2 = \frac{1}{\alpha'} \left(\sum_{n=1}^\infty \alpha^i_{-n} \alpha^i_n - 1 \right) + \frac{Y \cdot Y}{4\pi^2 \alpha'^2}, \tag{6.2}$$

where Y^m is the separation of the D-branes. The sums over the oscillator modes work just like the computations we did before (see insert 3.4 (p. 92)), giving

$$\mathcal{A} = 2V_{p+1} \int_0^\infty \frac{dt}{2t} (8\pi^2 \alpha' t)^{-\frac{(p+1)}{2}} e^{-Y \cdot Yt/2\pi\alpha'} f_1(q)^{-24}. \tag{6.3}$$

Insert 6.1.   Vacuum energy

The Coleman–Weinberg[35, 36] formula evaluates the one-loop vacuum amplitude, which is simply the logarithm of the partition function $\mathcal{A} = Z_{\text{vac}}$ for the complete theory:

$$\ln\left(Z_{\text{vac}}\right) = -\frac{1}{2}\text{Tr}\ln\left(\Box^2 + M^2\right) = -\frac{V_D}{2}\int \frac{d^D k}{(2\pi)^D}\ln\left(k^2 + M^2\right).$$

But since we can write

$$-\frac{1}{2}\ln(k^2 + M^2) = \int_0^\infty \frac{dt}{2t}e^{-(k^2+M^2)t/2},$$

we have

$$\mathcal{A} = V_D\int \frac{d^D k}{(2\pi)^D}\int_0^\infty \frac{dt}{2t}e^{-(k^2+M^2)t/2}.$$

Recall finally that $(k^2 + M^2)/2$ is just the Hamiltonian, H, which in our case is just L_0/α' (see equation (2.64)).

Here $q = e^{-2\pi t}$, and the overall factor of two is from exchanging the two ends of the string. (See insert 6.2 for news of $f_1(q)$.)

In the present case (using the asymptotics derived in insert 6.2),

$$\mathcal{A} = 2V_{p+1}\int_0^\infty \frac{dt}{2t}(8\pi^2\alpha't)^{-\frac{(p+1)}{2}}e^{-Y\cdot Yt/2\pi\alpha'}t^{12}\left(e^{2\pi/t} + 24 + \cdots\right). \tag{6.4}$$

The leading divergence is from the tachyon and is the usual bosonic string artifact not relevant to this discussion. The massless pole, from the second term, is

$$\mathcal{A}_{\text{massless}} \sim V_{p+1}\frac{24}{2^{12}}(4\pi^2\alpha')^{11-p}\pi^{(p-23)/2}\Gamma((23-p)/2)|Y|^{p-23}$$

$$= V_{p+1}\frac{24\pi}{2^{10}}(4\pi^2\alpha')^{11-p}G_{25-p}(Y), \tag{6.5}$$

where $G_d(Y)$ is the massless scalar Green's function in d dimensions:

$$G_d(Y) = \frac{\pi^{d/2}}{4}\Gamma\left(\frac{d}{2} - 1\right)\frac{1}{Y^{d-2}}. \tag{6.6}$$

Here, $d = 25 - p$, the dimension of the space transverse to the brane.

Insert 6.2. Translating closed to open

Compare our open string appearance of $f_1(q)$, for $q = e^{-2\pi t}$ with the expressions for $f_1(q)$, ($q = e^{2\pi\tau}$) defined in our closed string discussion in (4.44). Here the argument is real. The translation between definitions is done by setting $t = -\text{Im}\,\tau$. From the modular transformations (4.46), we can deduce some useful asymptotics. While the asymptotics as $t \to \infty$ are obvious, we can get the $t \to 0$ asymptotics using (4.46):

$$f_1(e^{-2\pi/s}) = \sqrt{s}\, f_1(e^{-2\pi s}), \quad f_3(e^{-2\pi/s}) = f_3(e^{-2\pi s}),$$
$$f_2(e^{-2\pi/s}) = f_4(e^{-2\pi s}).$$

6.1.2 A background field computation

We must do a a field theory calculation to work out the amplitude for the exchange of the graviton and dilaton between a pair of D-branes. Our result can the be compared to the low energy string result above to extract the value of the tension. We need propagators and couplings as per the usual field theory computation. The propagator is from the bulk action (2.106) and the couplings are from the D-brane action (5.21), but we must massage them a bit in order to find them.

In fact, we should work in the Einstein frame, since that is the appropriate frame in which to discuss mass and energy, because the dilaton and graviton don't mix there. We do this (recall equation (2.109)) by sending the metric $G_{\mu\nu}$ to $\tilde{G}_{\mu\nu} = \exp(4(\Phi_0 - \Phi)/(D-2))G_{\mu\nu}$, which gives the metric in equation (2.111). Let us also do this in the Dirac–Born–Infeld action (5.21), with the result:

$$S_p^E = -\tau_p \int d^{p+1}\xi\, e^{-\tilde{\Phi}} \det{}^{1/2}(e^{\frac{4\tilde{\Phi}}{D-2}}\tilde{G}_{ab} + B_{ab} + 2\pi\alpha' F_{ab}), \qquad (6.7)$$

where $\tilde{\Phi} = \Phi - \Phi_0$ and $\tau_p = T_p e^{-\Phi_0}$ is the physical tension of the brane; it is set by the background value, Φ_0, of the dilaton.

The next step is to linearise about a flat background, in order to extract the propagator and the vertices for our field theory. In fact, we have already discussed some of the logic of this in the introductory chapter, in section (1.2), where we came to grips with the idea of a graviton, so the reader is presumably aware that this is not really a daunting procedure.

We simply write the metric as $G_{\mu\nu} = \eta_{\mu\nu} + h_{\mu\nu}(X)$, and this time expand up to second order in $h_{\mu\nu}$. Also, if we do this with the action (6.7) as well, we see that the antisymmetric fields $B_{ab} + 2\pi\alpha' F_{ab}$ do not contribute at this order, and so we will drop them in what follows*.

Another thing which we did in section (1.2) was to fix the gauge degree of freedom (1.21) so that we would write the linearised (first order) Einstein equations in a nice gauge (1.22). We shall pick the same gauge here:

$$F_\mu \equiv \eta^{\rho\sigma}(\partial_\rho h_{\sigma\mu} - \frac{1}{2}\partial_\mu h_{\rho\sigma}) = 0, \tag{6.8}$$

and introduce the gauge choice into the Lagrangian via the addition of a gauge fixing term:

$$\mathcal{L}_{\text{fix}} = -\frac{\eta^{\mu\nu}}{4\kappa^2} F_\mu F_\nu. \tag{6.9}$$

The result for the bulk action is:

$$S_{\text{bulk}} = -\frac{1}{2\kappa^2} \int d^D X \left\{ \frac{1}{2} \left[\eta^{\mu\rho}\eta^{\nu\sigma} + \eta^{\mu\sigma}\eta^{\nu\rho} - \frac{2}{D-2}\eta^{\mu\nu}\eta^{\rho\sigma} \right] h_{\mu\nu}\partial^2 h_{\rho\sigma} \right.$$
$$\left. + \frac{4}{D-2}\tilde{\Phi}\partial^2\tilde{\Phi} \right\}, \tag{6.10}$$

and the interaction terms from the Dirac–Born–Infeld action are:

$$S_{\text{brane}} = -\tau_p \int d^{p+1}\xi \left(\left(\frac{2p-D+4}{D-2}\right) \tilde{\Phi} - \frac{1}{2}h_{aa} \right), \tag{6.11}$$

where the trace on the metric was in the $(p+1)$-dimensional world-volume of the Dp-brane.

Now it is easy to work out the momentum space propagators for the graviton and the dilaton:

$$\langle h_{\mu\nu} h_{\rho\sigma} \rangle = -\frac{2i\kappa^2}{k^2} \left[\eta_{\mu\rho}\eta_{\nu\sigma} + \eta_{\mu\sigma}\eta_{\nu\rho} - \frac{2}{D-2}\eta_{\mu\nu}\eta_{\rho\sigma} \right];$$
$$\langle \tilde{\Phi}\tilde{\Phi} \rangle = -\frac{i\kappa^2(D-2)}{4k^2}, \tag{6.12}$$

for momentum k. The reader might recognise the graviton propagator as the generalisation of the four dimensional case. If the reader has not encountered it before, the resulting form should be thought of as entirely consistent with gauge invariance for a massless spin two particle.

---

* This fits with the intuition that the D-brane should not be a source for the antisymmetric tensor field. The source for it is the fundamental closed string itself. We shall come back to this point many times much later.

All we need to do is compute two tree level Feynman diagrams, one for exchange of the dilaton and one for the exchange of the graviton, and add the result. The vertices are given in action (6.11). The result is (returning to position space)

$$
\mathcal{A}_{\text{massless}} = V_{p+1} T_p^2 \kappa_0^2 G_{25-p}(Y) \left\{ \frac{D-2}{4} \left(\frac{2p-D+4}{D-2} \right)^2 \right.
$$

$$
\left. + \frac{1}{2} \left[2(p+1) - \frac{2}{D-2}(p+1)^2 \right] \right\}
$$

$$
= \frac{D-2}{4} V_{p+1} T_p^2 \kappa_0^2 G_{25-p}(Y), \tag{6.13}
$$

and so after comparing to our result from the string theory computation (6.5) we have

$$
T_p = \frac{\sqrt{\pi}}{16\kappa_0} (4\pi^2 \alpha')^{(11-p)/2}. \tag{6.14}
$$

This agrees rather nicely with the recursion relation (5.11). We can also write it in terms of the physical value of the D-brane tension, which includes a factor of the string coupling $g_{\text{s}} = e^{\Phi_0}$,

$$
\tau_p = \frac{\sqrt{\pi}}{16\kappa} (4\pi^2 \alpha')^{(11-p)/2} \tag{6.15}
$$

where $\kappa = \kappa_0 g_{\text{s}}$, and we shall use τ_p this to denote the tension when we include the string coupling henceforth, and reserve T for situations where the string coupling is included in the background field $e^{-\Phi}$. (This will be less confusing than it sounds, since it will always be clear from the context which we mean.)

As promised, the tension τ_p of a Dp-brane is of order g_{s}^{-1}, following from the fact that the diagram connecting the brane to the closed string sector is a disc diagram, and insert 2.4 (p. 57) shows reminds us that this is of order g_{s}^{-1}. An immediate consequence of this is that they will produce non-perturbative effects of order $\exp(-1/g_{\text{s}})$ in string theory, since their action is of the same order as their mass. This is consistent with anticipated behaviour from earlier studies of toy non-perturbative string theories[100].

Formula (6.14) will not concern us much beyond these sections, since we will derive a new one for the superstring case later.

6.2 The orientifold tension

The O-plane, like the D-brane, couples to the dilaton and metric. The most direct amplitude to use to compute the tension is the same as in the previous section, but with $\mathbb{RP}^2$ in place of the disc; i.e. a crosscap replaces the boundary loop. The orientifold identifies X^m with $-X^m$ at the opposite point on the crosscap, so the crosscap is localised near one of the orientifold fixed planes. However, once again, it is easier to organise the computation in terms of a one-loop diagram, and then extract the parts we need.

6.2.1 Another open string partition function

To calculate this via vacuum graphs, the cylinder has one or both of its boundary loops replaced by crosscaps. This gives the Möbius strip and Klein bottle, respectively. To understand this, consider figure 6.2, which shows two copies of the fundamental region for the Möbius strip. The lower half is identified with the reflection of the upper, and the edges $\sigma^1 = 0, \pi$ are boundaries. Taking the lower half as the fundamental region gives the familiar representation of the Möbius strip as a strip of length $2\pi t$, with ends twisted and glued. Taking instead the left half of the figure, the line $\sigma^1 = 0$ is a boundary loop while the line $\sigma^1 = \pi/2$ is identified with itself under a shift $\sigma^2 \to \sigma^2 + 2\pi t$ plus reflection of σ^1: it is a crosscap. The same construction applies to the Klein bottle, with the right and left edges now identified. Another way to think of the Möbius strip amplitude we are going to compute here is as representing the exchange of a closed string between a D-brane and its mirror image, as shown in figure 6.3. The identification with a twist is performed on the two D-branes, turning the cylinder into a Möbius strip. The Möbius strip is given by the vacuum

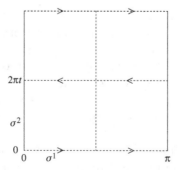

Fig. 6.2. Two copies of the fundamental region for the Möbius strip.

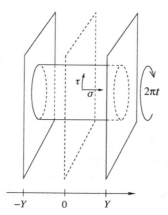

Fig. 6.3. The Möbius strip as the exchange of closed strings between a brane and its mirror image. The dotted plane is the orientifold plane.

amplitude

$$\mathcal{A}_{\mathrm{M}} = V_{p+1} \int \frac{d^{p+1}k}{(2\pi)^{p+1}} \int_0^\infty \frac{dt}{2t} \sum_i \frac{\Omega_i}{2} e^{-2\pi\alpha' t(p^2+M_I^2)}, \qquad (6.16)$$

where Ω_I is the Ω eigenvalue of state i. The oscillator contribution to Ω_I is $(-1)^n$ from equation (2.94). Actually, in the directions orthogonal to the brane and orientifold there are two additional signs in Ω_I which cancel. One is from the fact that world-sheet parity contributes an extra minus sign in the directions with Dirichlet boundary conditions (this is evident from the mode expansions we shall list later, in equations (11.1)). The other is from the fact that spacetime reflection produces an additional sign.

For the $SO(N)$ open string the Chan–Paton factors have $\frac{1}{2}N(N+1)$ even states and $\frac{1}{2}N(N-1)$ odd for a total of $+N$. For $USp(N)$ these numbers are reversed for a total of $-N$. Focus on a D-brane and its image, which correspondingly contribute ± 2. The diagonal elements, which contribute to the trace, are those where one end is on the D-brane and one on its image. The total separation is then $Y^m = 2X^m$. Then,

$$\mathcal{A}_{\mathrm{M}} = \pm V_{p+1} \int_0^\infty \frac{dt}{2t} (8\pi^2\alpha' t)^{-\frac{(p+1)}{2}} e^{-2\vec{Y}\cdot\vec{Y}t/\pi\alpha'}$$
$$\times \left[q^{-2} \prod_{k=1}^\infty (1+q^{4k-2})^{-24}(1-q^{4k})^{-24} \right].$$

The factor in braces is

$$f_3(q^2)^{-24}f_1(q^2)^{-24} = (2t)^{12}f_3(e^{-\pi/2t})^{-24}f_1(e^{-\pi/2t})^{-24}$$
$$= (2t)^{12}\left(e^{\pi/2t} - 24 + \cdots\right). \qquad (6.17)$$

One therefore finds a pole

$$\mp 2^{p-12} V_{p+1} \frac{3\pi}{2^6} (4\pi^2 \alpha')^{11-p} G_{25-p}(Y). \qquad (6.18)$$

This is to be compared with the field theory result

$$\frac{D-2}{2} V_{p+1} T_p T'_p \kappa_0^2 G_{25-p}(Y), \qquad (6.19)$$

where T'_p is the O-plane tension. A factor of two as compared to the earlier field theory calculation (6.13) comes because the spacetime boundary forces all the flux in one direction. Therefore the O-plane and D-brane tensions are related by

$$\tau'_p = \mp 2^{p-13} \tau_p. \qquad (6.20)$$

A similar calculation with the Klein bottle gives a result proportional to τ'^2_p.

Noting that there are 2^{25-p} O-planes (recall that one doubles the number every time another new direction is T-dualised, starting with e single D25-brane), the total charge of an O-plane source must be $\mp 2^{12} \tau_p$. Now, by Gauss's law, the total source must vanish because the volume of the torus T^p on which we are working is finite and of course the flux must end on sinks and sources.

So we conclude that there are $2^{(D-2)/2} = 2^{12}$ D-branes (times two for the images) and that the gauge group[37] is $SO(2^{13}) = SO(2^{D/2})$. For this group the 'tadpoles' associated with the dilaton and graviton, representing violations of the field equations, cancel at order g_s^{-1}. This has no special significance in the bosonic string due to the tachyon instability, but similar considerations will give a restriction on allowed Chan–Paton gauge groups in the superstring.

6.3 The boundary state formalism

The asymptotics (6.4) can be interpreted in terms of a sum over closed string states exchanged between the two D-branes. One can write the cylinder path integral in a Hilbert space formalism treating σ_1 rather than σ_2 as time. It then has the form

$$\langle B | e^{-(L_0 + \tilde{L}_0)\pi/t} | B \rangle \qquad (6.21)$$

where the 'boundary state' $|B\rangle$ is the closed string state created by the boundary loop.

Let us unpack this formalism a little, seeing where it all comes from. Recall that a Dp-brane is specified by the following open string boundary conditions:

$$\partial_\sigma X^\mu|_{\sigma=0,\pi} = 0, \qquad \mu = 0, \ldots, p;$$
$$X^m|_{\sigma=0,\pi} = Y^m, \qquad m = p+1, \ldots, D-1. \qquad (6.22)$$

Now we have to reinterpret this as a closed string statement. This involves exchanging τ and σ. So we write, focusing on the initial time:

$$\partial_\tau X^\mu|_{\tau=0} = 0, \qquad \mu = 0, \ldots, p;$$
$$X^m|_{\tau=0} = Y^m, \qquad m = p+1, \ldots, D-1. \qquad (6.23)$$

Recall that in the quantum theory we pass to an operator formalism, and so the conditions above should be written as an operator statement, where we are operating on some state in the Hilbert space. This defines for us then the boundary state $|B\rangle$:

$$\partial_\tau X^\mu|_{\tau=0}|B\rangle = 0, \qquad \mu = 0, \ldots, p;$$
$$(X^m|_{\tau=0} - Y^m)|B\rangle = 0, \qquad m = p+1, \ldots, D-1. \qquad (6.24)$$

As with everything we did in chapter 2, we can convert our equations above into a statement about the modes:

$$(\alpha_n^\mu + \tilde{\alpha}_{-n}^\mu)|B\rangle = 0, \qquad \mu = 0, \ldots, p;$$
$$(\alpha_n^m - \tilde{\alpha}_{-n}^m)|B\rangle = 0, \qquad m = p+1, \ldots, D-1;$$
$$p^\mu|B\rangle = 0, \qquad \mu = 0, \ldots, p;$$
$$(x^m - Y^m)|B\rangle = 0, \qquad m = p+1, \ldots, D-1. \qquad (6.25)$$

As before, we either use only $D-2$ of the oscillator modes here (ignoring $\mu = 0, 1$) or we do everything covariantly and make sure that we include the ghost sector and impose BRST invariance. We shall do the former here.

The solution to the condition above can we found by analogy with the (perhaps) familiar technology of coherent states in harmonic oscillator physics (see insert 6.3).

$$|B\rangle = \mathcal{N}_p \delta(x^m - Y^m) \left(\prod_{n=1}^\infty e^{-\frac{1}{n}\alpha_{-n}\cdot\mathcal{S}\cdot\tilde{\alpha}_{-n}} \right) |0\rangle. \qquad (6.26)$$

The object $\mathcal{S} = (\eta^{\mu\nu}, -\delta^{mn})$ is just shorthand for the fact that the dot product must be the usual Lorentz one in the directions parallel to the brane, but there is a minus sign for the transverse directions.

Insert 6.3.   The boundary state as a coherent state

Let us recall that all we are playing with are creation and annihilation operators with a slightly unusual normalisation, as noticed at the beginning of section 2.3. Working with one set of the standard operators, a and $a^\dagger$, for the left and an *independent* set $\tilde{a}$ and $\tilde{a}^\dagger$ for the right, in essence we are trying to solve the equation

$$a|b>= \mp\tilde{a}|b\rangle.$$

Now recall how coherent states are made. We have

$$[a, a^\dagger] = 1, \qquad a|0\rangle = 0,$$

and so we can define a conjugation operation which shifts a by z, by defining

$$a(z) = e^{-za^\dagger} a e^{za^\dagger}.$$

It is easy to see that $a(z) = a + z$, since by elementary differentiation and the use of the commutator, we have

$$\frac{\partial a(z)}{\partial z} = 1.$$

Therefore the state

$$|z\rangle = e^{za^\dagger}|0\rangle$$

is an eigenvalue of the annihilation operator a, since

$$a|z>= e^{za^\dagger}e^{-za^\dagger} a e^{za^\dagger}|0\rangle = e^{za^\dagger}(a + z)|0\rangle = z|z\rangle.$$

We can therefore use as a solution to our first equation above, the coherent state with the choice $z = \mp\tilde{a}^\dagger$,

$$|b\rangle = Ne^{\mp\tilde{a}^\dagger a^\dagger}|0\rangle,$$

where N is a normalisation constant.

The normalisation constant is determined by simply computing the closed string amplitude directly in this formalism. The closed string is prepared in a boundary state that corresponds to a D-brane, and it propagates for a while, ending in a similar boundary state at position $\vec{Y}$:

$$\mathcal{A} = \langle B|\Delta|B\rangle, \tag{6.27}$$

where Δ is the closed string propagator. How is this object constructed? Well, we might expect that it is essentially the inverse of $H_{cl} = 2(L_0 + \bar{L}_0 - 2)/\alpha'$, the closed string Hamiltonian, which we can easily represent as:

$$\Delta = \frac{\alpha'}{2} \int_0^1 d\rho \rho^{L_0 + \bar{L}_0 - 3},$$

and we must integrate over the modulus $\ell = -\log \rho$ of the cylinder from 0 to ∞. We must remember, however, that a physical state $|\phi\rangle$ is annihilated by $L_0 - \bar{L}_0$, and so we can modify our propagator so that it only propagates such states:

$$\Delta = \frac{\alpha'}{2} \int_0^1 d\rho \frac{d\phi}{2\pi} \int_0^{2\pi} \frac{d\phi}{2\pi} \rho^{L_0 + \bar{L}_0 - 3} e^{i\phi(L_0 - \bar{L}_0)},$$

which, after the change of variable to $z = \rho e^{i\phi}$, gives

$$\Delta = \frac{\alpha'}{4\pi} \int_{|z| \leq 1} \frac{dz d\bar{z}}{|z|^2} z^{L_0 - 1} \bar{z}^{\bar{L}_0 - 1}.$$

Computing the amplitude (6.27) by using this definition of the propagator is a straightforward exercise, similar in spirit to what we did in the open string sector. We get geometric sums over the oscillator modes resulting from traces, and integrals over the continuous quantities. If we make the choices $|z| = e^{-\pi s}$ and $dz d\bar{z} = -\pi e^{-2\pi s} ds d\phi$ for our closed string cylinder, the result is:

$$\mathcal{A} = \mathcal{N}_p^2 V_{p+1} \frac{\alpha' \pi}{2} (2\pi\alpha')^{-\frac{25-p}{2}} \int_0^\infty \frac{ds}{s} s^{-\frac{25-p}{2}} e^{-Y \cdot Y / s 2\pi\alpha'} f_1(q)^{-24}. \quad (6.28)$$

Here $q = e^{-2\pi/s}$.

Now we can compare to the open string computation, which is the result in equation (6.3). We must do a modular transformation $s = -1/t$, and using the modular transformation properties given in insert 6.2, we find exactly the open string result if we have

$$\mathcal{N}_p = \frac{T_p}{2},$$

where T_p is the brane tension (6.14) computed earlier.

This is a very useful way of formulating the whole D-brane construction. In fact, the boundary state constructed above is just a special case of a sensible conformal field theory object. It is a state that can arise in the conformal field theory with boundary. Not all boundary states have such a simple spacetime interpretation as the one we made here. We see therefore that D-branes, if interpreted simply as resulting from the introduction of open string sectors into closed string theory, have a world-sheet formulation which does not necessarily always have a spacetime interpretation as its counterpart. Similar things happen in closed string conformal field theory. There are very many conformal field theories which are perfectly good string vacua, which have no spacetime interpretation in terms of an unambiguous target space geometry. It is natural that this also be true for the open string sector.

7

Supersymmetric strings

The discussion of bosonic strings in the previous five chapters allowed us to uncover a great deal of the structure essential to understanding D-branes and other background solutions, in addition to the basic concepts used in discussing and working with critical string theory.

At the back of our mind was always the expectation that we would move on to include supersymmetry. Two of the main reasons are that we can remove the tachyon from the spectrum and that we will be able to use supersymmetry to endow many of our results with extra potency, since stability and non-renormalisation arguments will allow us to extrapolate beyond perturbation theory.

Let us set aside D-branes and T-duality for a while and use the ideas we discussed earlier to construct the supersymmetric string theories which we need to carry the discussion further. There are five such theories. Three of these are the 'superstrings', while two are the 'heterotic strings'[*].

7.1 The three basic superstring theories

7.1.1 Open superstrings: type I

Let us go back to the beginning, almost. We can generalise the bosonic string action we had earlier to include fermions. In conformal gauge it is:

$$S = \frac{1}{4\pi} \int_{\mathcal{M}} d^2\sigma \left\{ \frac{1}{\alpha'} \partial X^\mu \bar{\partial} X_\mu + \psi^\mu \bar{\partial} \psi_\mu + \tilde{\psi}^\mu \partial \tilde{\psi}_\mu \right\}, \qquad (7.1)$$

where the open string world-sheet is the strip $0 < \sigma < \pi$, $-\infty < \tau < \infty$.

---

[*] A looser and probably more sensible nomenclature is to call them all 'superstrings', but we'll choose the catch-all term to be the one we used for the title of this chapter.

> N.B. Recall that α' is the loop expansion parameter analogous to $\hbar$ on worldsheet. It is therefore natural for the fermions' kinetic terms to be normalised in this way.

We get a modification to the energy-momentum tensor from before (which we now denote as $T_{\rm B}$, since it is the bosonic part):

$$T_{\rm B}(z) = -\frac{1}{\alpha'}\partial X^\mu \partial X_\mu - \frac{1}{2}\psi^\mu \partial \psi_\mu, \tag{7.2}$$

which is now accompanied by a fermionic energy-momentum tensor:

$$T_{\rm F}(z) = i\frac{2}{\alpha'}\psi^\mu \partial X_\mu. \tag{7.3}$$

This enlarges our theory somewhat, while much of the logic of what we did in the purely bosonic story survives intact here. Now, one extremely important feature which we encountered in section 4.7 is the fact that the equations of motion admit two possible boundary conditions on the world-sheet fermions consistent with Lorentz invariance. These are denoted the 'Ramond' (R) and the 'Neveu–Schwarz' (NS) sectors:

$$\begin{aligned} \text{R:} \quad & \psi^\mu(0,\tau) = \tilde{\psi}^\mu(0,\tau) & \psi^\mu(\pi,\tau) = \tilde{\psi}^\mu(\pi,\tau) \\ \text{NS:} \quad & \psi^\mu(0,\tau) = -\tilde{\psi}^\mu(0,\tau) & \psi^\mu(\pi,\tau) = \tilde{\psi}^\mu(\pi,\tau). \end{aligned} \tag{7.4}$$

We have used the freedom to choose the boundary condition at, for example the $\sigma=\pi$ end, in order to have a $+$ sign, by redefinition of $\tilde{\psi}$. The boundary conditions and equations of motion are summarised by the 'doubling trick': take just left-moving (analytic) fields ψ^μ on the range 0 to 2π and define $\tilde{\psi}^\mu(\sigma,\tau)$ to be $\psi^\mu(2\pi - \sigma,\tau)$. These left-moving fields are periodic in the Ramond (R) sector and antiperiodic in the Neveu–Schwarz (NS).

On the complex z-plane, the NS sector fermions are half-integer moded while the R sector ones are integer, and we have:

$$\psi^\mu(z) = \sum_r \frac{\psi^\mu_r}{z^{r+1/2}}, \quad \text{where } r \in \mathbb{Z} \text{ or } r \in \mathbb{Z} + \tfrac{1}{2} \tag{7.5}$$

and canonical quantisation gives

$$\{\psi^\mu_r, \psi^\nu_s\} = \{\tilde{\psi}^\mu_r, \tilde{\psi}^\nu_s\} = \eta^{\mu\nu}\delta_{r+s}. \tag{7.6}$$

Similarly we have

$$T_{\rm B}(z) = \sum_{m=-\infty}^{\infty} \frac{L_m}{z^{m+2}} \quad \text{as before, and}$$

$$T_{\rm F}(z) = \sum_r \frac{G_r}{z^{r+3/2}}, \quad \text{where } r \in \mathbb{Z} \text{ (R) or } \mathbb{Z} + \tfrac{1}{2} \text{ (NS)}. \tag{7.7}$$

Correspondingly, the Virasoro algebra is enlarged, with the non-zero (anti)commutators being

$$[L_m, L_n] = (m - n)L_{m+n} + \frac{c}{12}(m^3 - m)\delta_{m+n}$$

$$\{G_r, G_s\} = 2L_{r+s} + \frac{c}{12}(4r^2 - 1)\delta_{r+s}$$

$$[L_m, G_r] = \frac{1}{2}(m - 2r)G_{m+r}, \tag{7.8}$$

with

$$L_m = \frac{1}{2}\sum_m : \alpha_{m-n} \cdot \alpha_m : + \frac{1}{4}\sum_r (2r - m) : \psi_{m-r} \cdot \psi_r : + a\delta_{m,0}$$

$$G_r = \sum_n \alpha_n \cdot \psi_{r-n}. \tag{7.9}$$

In the above, c is the total contribution to the conformal anomaly, which is $D + D/2$, where D is from the D bosons while $D/2$ is from the D fermions.

The values of D and a are again determined by any of the methods mentioned in the discussion of the bosonic string. For the superstring, it turns out that $D = 10$ and $a = 0$ for the R sector and $a = -1/2$ for the NS sector. This comes about because the contributions from the X^0 and X^1 directions are cancelled by the Faddeev–Popov ghosts as before, and the contributions from the ψ^0 and ψ^1 oscillators are cancelled by the superghosts. Then, the computation uses the mnemonic/formula given in equation (2.80).

$$\text{NS sector:} \quad \text{z.p.e} = 8\left(-\frac{1}{24}\right) + 8\left(-\frac{1}{48}\right) = -\frac{1}{2},$$

$$\text{R sector:} \quad \text{z.p.e} = 8\left(-\frac{1}{24}\right) + 8\left(\frac{1}{24}\right) = 0. \tag{7.10}$$

As before, there is a physical state condition imposed by annihilating with the positive modes of the (super) Virasoro generators:

$$G_r|\phi\rangle = 0, \ r > 0; \quad L_n|\phi\rangle = 0, \ n > 0; \quad (L_0 - a)|\phi\rangle = 0. \tag{7.11}$$

The L_0 constraint leads to a mass formula:

$$M^2 = \frac{1}{\alpha'}\left(\sum_{n,r} \alpha_{-n} \cdot \alpha_n + r\psi_{-r} \cdot \psi_r - a\right). \tag{7.12}$$

In the NS sector the ground state is a Lorentz singlet and is assigned odd fermion number, i.e. under the operator $(-1)^F$, it has eigenvalue -1.

In order to achieve spacetime supersymmetry, the spectrum is projected on to states with even fermion number. This is called the 'GSO projection'[71], and for our purposes, it is enough to simply state that this obtains spacetime supersymmetry, as we will show at the massless level. A more complete treatment – which gets it right for all mass levels – is contained in the full superconformal field theory. The GSO projection there is a statement about locality with the gravitino vertex operator. Yet another way to think of its origin is as a requirement of modular invariance.

Since the open string tachyon clearly has $(-1)^F = -1$, it is removed from the spectrum by GSO. This is our first achievement, and justifies our earlier practice of ignoring the tachyon's appearance in the bosonic spectrum in what has gone before. From what we will do for the rest of the this book, the tachyon will largely remain in the wings, but it (and other tachyons) do have a role to play, since they are often a signal that the vacuum wants to move to a (perhaps) more interesting place.

Massless particle states in ten dimensions are classified by their $SO(8)$ representation under Lorentz rotations, that leave the momentum invariant: $SO(8)$ is the 'little group' of $SO(1,9)$. The lowest lying surviving states in the NS sector are the eight transverse polarisations of the massless open string photon, A^μ, made by exciting the ψ oscillators:

$$\psi^\mu_{-1/2}|k\rangle, \qquad M^2 = 0. \tag{7.13}$$

These states clearly form the vector of $SO(8)$. They have $(-)^F = 1$ and so survive GSO.

In the R sector the ground state energy always vanishes because the world-sheet bosons and their superconformal partners have the same moding. The Ramond vacuum has a 32-fold degeneracy, since the ψ^μ_0 take ground states into ground states. The ground states form a representation of the ten dimensional Dirac matrix algebra

$$\{\psi^\mu_0, \psi^\nu_0\} = \eta^{\mu\nu}. \tag{7.14}$$

(Note the similarity with the standard Γ-matrix algebra, $\{\Gamma^\mu, \Gamma^\nu\} = 2\eta^{\mu\nu}$. We see that $\psi^\mu_0 \equiv \Gamma^\mu/\sqrt{2}$.)

For this representation, it is useful to choose this basis:

$$d^\pm_i = \frac{1}{\sqrt{2}} \left(\psi^{2i}_0 \pm i\psi^{2i+1}_0 \right) \qquad i = 1, \ldots, 4$$

$$d^\pm_0 = \frac{1}{\sqrt{2}} \left(\psi^1_0 \mp \psi^0_0 \right). \tag{7.15}$$

In this basis, the Clifford algebra takes the form

$$\{d^+_i, d^-_j\} = \delta_{ij}. \tag{7.16}$$

The $d_i^{\pm}$, $i = 0, \ldots, 4$ act as creation and annihilation operators, generating the $2^{10/2} = 32$ Ramond ground states. Denote these states

$$|s_0, s_1, s_2, s_3, s_4\rangle = |\mathbf{s}\rangle \qquad (7.17)$$

where each of the s_i takes the values $\pm\frac{1}{2}$, and where

$$d_i^- |-\tfrac{1}{2}, -\tfrac{1}{2}, -\tfrac{1}{2}, -\tfrac{1}{2}, -\tfrac{1}{2}\rangle = 0 \qquad (7.18)$$

while d_i^+ raises s_i from $-\frac{1}{2}$ to $\frac{1}{2}$. This notation has physical meaning: the fermionic part of the ten dimensional Lorentz generators is

$$S^{\mu\nu} = -\frac{i}{2} \sum_{r \in \mathbf{Z}+\kappa} [\psi^{\mu}_{-r}, \psi^{\nu}_r], \qquad (7.19)$$

(recall equation (2.124)). The states (7.17) above are eigenstates of $S_0 = iS^{01}$, $S_i = S^{2i,2i+1}$, with s_i the corresponding eigenvalues. Since by construction the Lorentz generators (7.19) always flip an even number of s_i, the Dirac representation **32** decomposes into a **16** with an even number of $-\frac{1}{2}$s and **16′** with an odd number.

The physical state conditions (7.11), on these ground states, reduce to $G_0 = (2\alpha')^{1/2} p_\mu \psi^\mu$. (Note that $G_0^2 \sim L_0$.) Let us pick the (massless) frame $p^0 = p^1$. This becomes

$$G_0 = \alpha'^{1/2} p_1 \Gamma_0 (1 - \Gamma_0 \Gamma_1) = 2\alpha'^{1/2} p_1 \Gamma_0 \left(\tfrac{1}{2} - S_0\right), \qquad (7.20)$$

which means that $s_0 = \frac{1}{2}$, giving a 16-fold degeneracy for the *physical* Ramond vacuum. This is a representation of $SO(8)$ which decomposes into $\mathbf{8_s}$ with an even number of $-\frac{1}{2}$s and $\mathbf{8_c}$ with an odd number. One is in the **16** and the **16′**, but the two choices, **16** or **16′**, are physically equivalent, differing only by a spacetime parity redefinition, which would therefore swap the $\mathbf{8_s}$ and the $\mathbf{8_c}$.

In the R sector the GSO projection amounts to requiring

$$\sum_{i=1}^{4} s_i = 0 \quad (\mathrm{mod}\ 2), \qquad (7.21)$$

picking out the $\mathbf{8_s}$. Of course, it is just a convention that we associated an even number of $\frac{1}{2}$s with the $\mathbf{8_s}$; a physically equivalent discussion with things the other way around would have resulted in $\mathbf{8_c}$. The difference between these two is only meaningful when they are both present, and at this stage we only have one copy, so either is as good as the other.

The ground state spectrum is then $\mathbf{8_v} \oplus \mathbf{8_s}$, a vector multiplet of $D = 10$, $\mathcal{N} = 1$ spacetime supersymmetry. Including Chan–Paton factors gives again a $U(N)$ gauge theory in the oriented theory and $SO(N)$ or $USp(N)$

in the unoriented. This completes our tree-level construction of the open superstring theory.

Of course, we are not finished, since this theory is (on its own) inconsistent for many reasons. One such reason (there are many others) is that it is anomalous. Both gauge invariance and coordinate invariance have anomalies arising because it is a chiral theory: e.g. the fermion $\mathbf{8_s}$ has a specific chirality in spacetime. The gauge and gravitational anomalies are very useful probes of the consistency of any theory. These show up quantum inconsistencies of the theory resulting in the failure of gauge invariance and general coordinate invariance, and hence must be absent. See insert 7.1 for more on anomalies.

Another reason we will see that the theory is inconsistent is that, as we learned in chapter 4, the theory is equivalent to some number of space-filling D9-branes in spacetime, and it will turn out later that these are positive electric sources of a particular 10-form field in the theory. The field equation for this field asks that all of its sources must simply vanish, and so we must have a negative source of this same field in order to cancel the D9-branes' contribution. This will lead us to the closed string sector i.e. one-loop, the same level at which we see the anomaly.

Let us study some closed strings. We will find three of interest here. Two of them will stand in their own right, with two ten dimensional supersymmetries, while the third will have half of that, and will be anomalous. This latter will be the closed string sector we need to supplement the open string we made here, curing its one-loop anomalies.

7.1.2 Closed superstrings: type II

Just as we saw before, the closed string spectrum is the product of two copies of the open string spectrum, with right- and left-moving levels matched. In the open string the two choices for the GSO projection were equivalent, but in the closed string there are two inequivalent choices, since we have to pick two copies to make a closed string.

Taking the same projection on both sides gives the 'type IIB' case, while taking them opposite gives 'type IIA'. These lead to the massless sectors

$$\text{Type IIA:} \quad (\mathbf{8_v} \oplus \mathbf{8_s}) \otimes (\mathbf{8_v} \oplus \mathbf{8_c})$$
$$\text{Type IIB:} \quad (\mathbf{8_v} \oplus \mathbf{8_s}) \otimes (\mathbf{8_v} \oplus \mathbf{8_s}). \tag{7.22}$$

Let us expand out these products to see the resulting Lorentz ($SO(8)$) content. In the NS–NS sector, this is

$$\mathbf{8_v} \otimes \mathbf{8_v} = \Phi \oplus B_{\mu\nu} \oplus G_{\mu\nu} = \mathbf{1} \oplus \mathbf{28} \oplus \mathbf{35}. \tag{7.23}$$

Insert 7.1.   Gauge and gravitational anomalies

The beauty of the anomaly is that it is both a UV and an IR tool: UV since it represents the failure to be able to find a consistent regulator at the quantum level and IR since it cares only about the massless sector of the theory: Any potentially anomalous variations for the *effective* action $\Gamma = \ln Z$ should be written as the variation of a local term which allows it to be cancelled by adding a local counterterm. Massive fields always give effectively local terms at long distance.

An anomaly in D dimensions arises from complex representations of the Lorentz group which include chiral fermions in general but also bosonic representations if $D = 4k + 2$, e.g. the rank $2k + 1$ (anti)self-dual tensor. The anomalies are controlled by the so-called 'hexagon' diagram which generalises the (perhaps more familiar) triangle of four dimensional field theory or a square in six dimensions.

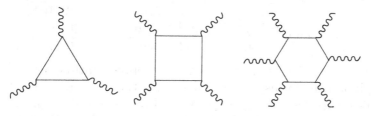

The external legs are either gauge bosons, gravitons, or a mixture. We shall not spend any time on the details[53], but simply state that consistency demands that the structure of the anomaly,

$$\delta \ln Z = \frac{i}{(2\pi)^{D/2}} \int \hat{I}_D,$$

is in terms of a D-form $\hat{I}_D$, polynomial in traces of even powers of the field strength two-forms $F = dA + A^2$ and $R = d\omega + \omega^2$. (Recall section 2.8.) It is naturally related to a $(D+2)$-form polynomial $\hat{I}_{D+2}$ which is gauge invariant and written as an exact form $\hat{I}_{D+2} = d\hat{I}_{D+1}$. The latter is not gauge invariant, but its variation is another exact form: $\delta \hat{I}_{D+1} = d\hat{I}_D$. A key example of this is the Chern–Simons three-form, which is discussed in insert 7.3, p. 167. See also insert 7.2 on p. 162 for explicit expressions in dimensions $D = 4k + 2$. We shall see that the anomalies are a useful check of the consistency of string spectra that we construct in various dimensions.

Insert 7.2.   A list of anomaly polynomials

It is useful to list here some anomaly polynomials for later use. In ten dimensions, the contributions to the polynomial come from three sorts of field, the spinors $\mathbf{8}_{s,c}$, the gravitinos $\mathbf{56}_{c,s}$, and the fifth rank antisymmetric tensor field strength with its self-dual and anti-self-dual parts. The anomalies for each pair within each sort are equal and opposite in sign, i.e. $\hat{I}_{12}^{\mathbf{8}_s} = -\hat{I}_{12}^{\mathbf{8}_c}$, etc., and we have:

$$\hat{I}_{12}^{\mathbf{8}_s} = -\frac{\mathrm{Tr}(F^6)}{1440}$$
$$+ \frac{\mathrm{Tr}(F^4)\mathrm{tr}(R^2)}{2304} - \frac{\mathrm{Tr}(F^2)\mathrm{tr}(R^4)}{23040} - \frac{\mathrm{Tr}(F^2)[\mathrm{tr}(R^2)]^2}{18432}$$
$$+ \frac{n\mathrm{tr}(R^6)}{725760} + \frac{n\mathrm{tr}(R^4)\mathrm{tr}(R^2)}{552960} + \frac{n[\mathrm{tr}(R^2)]^3}{1327104};$$

$$\hat{I}_{12}^{\mathbf{56}_c} = -495\frac{\mathrm{tr}(R^6)}{725760} + 225\frac{\mathrm{tr}(R^4)\mathrm{tr}(R^2)}{552960} - 63\frac{[\mathrm{tr}(R^2)]^3}{1327104};$$

$$\hat{I}_{12}^{\mathbf{35}+} = +992\frac{\mathrm{tr}(R^6)}{725760} - 448\frac{\mathrm{tr}(R^4)\mathrm{tr}(R^2)}{552960} + 128\frac{[\mathrm{tr}(R^2)]^3}{1327104},$$

and n is the dimension of the gauge representation under which the spinor transforms, for which we use the trace denoted Tr. We also have suppressed the use of $\wedge$, for brevity. For $D = 6$, there are anomaly eight-forms. We denote the various fields by their transformation properties of the $D = 6$ little group $SO(4) \sim SU(2) \times SU(2)$:

$$\hat{I}_8^{(\mathbf{1,2})} = +\frac{\mathrm{Tr}(F^4)}{24} - \frac{\mathrm{Tr}(F^2)\mathrm{tr}(R^2)}{96} + \frac{n\mathrm{tr}(R^4)}{5760} + \frac{n[\mathrm{tr}(R^2)]^2}{4608};$$

$$\hat{I}_8^{(\mathbf{3,2})} = +245\frac{\mathrm{tr}(R^4)}{5760} - 43\frac{[\mathrm{tr}(R^2)]^2}{4608};$$

$$\hat{I}_8^{(\mathbf{3,1})} = +28\frac{\mathrm{tr}(R^4)}{5760} - 8\frac{[\mathrm{tr}(R^2)]^2}{4608}.$$

Note that the first two are for complex fermions. For real fermions, one must divide by two. For completeness, for $D = 2$ we list the three analogous anomaly four-forms:

$$\hat{I}_4^{1/2} = \frac{n\mathrm{tr}(R^2)}{48} - \frac{\mathrm{Tr}(F^2)}{2}, \qquad \hat{I}_4^{3/2} = -23\frac{\mathrm{tr}(R^2)}{48}, \qquad \hat{I}_4^0 = \frac{\mathrm{tr}(R^2)}{48}.$$

It is amusing to note that the anomaly polynomials can be written in terms of geometrical characteristic classes. This should be kept at the back of the mind for a bit later, in section 9.5.

In the R–R sector, the IIA and IIB spectra are respectively

$$\mathbf{8_s} \otimes \mathbf{8_c} = [1] \oplus [3] = \mathbf{8_v} \oplus \mathbf{56_t}$$
$$\mathbf{8_s} \otimes \mathbf{8_s} = [0] \oplus [2] \oplus [4]_+ = \mathbf{1} \oplus \mathbf{28} \oplus \mathbf{35_+}. \tag{7.24}$$

Here $[n]$ denotes the n-times antisymmetrised representation of $SO(8)$, and $[4]_+$ is self-dual. Note that the representations $[n]$ and $[8-n]$ are the same, as they are related by contraction with the eight dimensional ϵ-tensor. The NS–NS and R–R spectra together form the bosonic components of $D = 10$ IIA (nonchiral) and IIB (chiral) supergravity respectively; We will write their effective actions shortly.

In the NS–R and R–NS sectors are the products

$$\mathbf{8_v} \otimes \mathbf{8_c} = \mathbf{8_s} \oplus \mathbf{56_c}$$
$$\mathbf{8_v} \otimes \mathbf{8_s} = \mathbf{8_c} \oplus \mathbf{56_s}. \tag{7.25}$$

The $\mathbf{56_{s,c}}$ are gravitinos. Their vertex operators are made roughly by tensoring a NS field ψ^μ with a vertex operator $\mathcal{V}_\alpha = e^{-\varphi/2}\mathbf{S}_\alpha$, where the latter is a 'spin field', made by bosonising the d_is of equation (7.15) and building:

$$\mathbf{S} = \exp\left[i\sum_{i=0}^{4} s_i H^i\right]; \quad d_i = e^{\pm iH^i}. \tag{7.26}$$

(The factor $e^{-\varphi/2}$ is the bosonisation (see section 4.7) of the Faddeev–Popov ghosts (see insert 3.2), about which we will have nothing more to say here.) The resulting full gravitino vertex operators, which correctly have one vector and one spinor index, are two fields of weight $(0,1)$ and $(1,0)$, respectively, depending upon whether ψ^μ comes from the left or right. These are therefore holomorphic and anti-holomorphic world-sheet currents, and the symmetry associated to them in spacetime is the supersymmetry. In the IIA theory the two gravitinos (and supercharges) have opposite chirality, and in the IIB the same.

Consider the vertex operators for the R–R states[1]. This will involve a product of spin fields[74], one from the left and one from the right. These again decompose into antisymmetric tensors, now of $SO(9,1)$:

$$V = \mathcal{V}_\alpha \mathcal{V}_\beta (\Gamma^{[\mu_1} \ldots \Gamma^{\mu_n]} C)_{\alpha\beta} G_{[\mu_1 \cdots \mu_n]}(X) \tag{7.27}$$

with C the charge conjugation matrix. In the IIA theory the product is $\mathbf{16} \otimes \mathbf{16'}$ giving even n (with $n \cong 10 - n$) and in the IIB theory it is $\mathbf{16} \otimes \mathbf{16}$ giving odd n. As in the bosonic case, the classical equations of motion follow from the physical state conditions, which at the massless level reduce to $G_0 \cdot V = \tilde{G}_0 \cdot V = 0$. The relevant part of G_0 is just $p_\mu \psi_0^\mu$ and similarly for $\tilde{G}_0$. The p_μ act by differentiation on G, while ψ_0^μ

acts on the spin fields as it does on the corresponding ground states: as multiplication by Γ^μ. Noting the identity

$$\Gamma^\nu \Gamma^{[\mu_1} \ldots \Gamma^{\mu_n]} = \Gamma^{[\nu} \ldots \Gamma^{\mu_n]} + \left(\delta^{\nu\mu_1} \Gamma^{[\mu_2} \ldots \Gamma^{\mu_n]} + \text{perms} \right) \qquad (7.28)$$

and similarly for right multiplication, the physical state conditions become

$$dG = 0 \qquad\qquad d^*G = 0. \qquad (7.29)$$

These are the Bianchi identity and field equation for an antisymmetric tensor field strength. This is in accord with the representations found: in the IIA theory we have odd-rank tensors of $SO(8)$ but even-rank tensors of $SO(9,1)$ (and reversed in the IIB), the extra index being contracted with the momentum to form the field strength. It also follows that R–R amplitudes involving elementary strings vanish at zero momentum, so strings do not carry R–R charges[†].

As an aside, when the dilaton background is nontrivial, the Ramond generators have a term $\Phi_{,\mu}\partial\psi^\mu$, and the Bianchi identity and field strength pick up terms proportional to $d\Phi \wedge G$ and $d\Phi \wedge {}^*G$. The Bianchi identity is non-standard, so G is not of the form dC. Defining $G' = e^{-\Phi}G$ removes the extra term from both the Bianchi identity and field strength. The field G' is thus decoupled from the dilaton. In terms of the action, the fields G in the vertex operators appear with the usual closed string $e^{-2\Phi}$ but with non-standard dilaton gradient terms. The fields we are calling G' (which in fact are the usual fields used in the literature, and so we will drop the prime symbol in the sequel) have a dilaton-independent action.

The type IIB theory is chiral since it has different numbers of left moving fermions from right-moving. Furthermore, there is a self-dual R–R tensor. These structures in principle produce gravitational anomalies, and it is one of the miracles (from the point of view of the low energy theory) of string theory that the massless spectrum is in fact anomaly free. There is a delicate cancellation between the anomalies for the $\mathbf{8}_c$ and for the $\mathbf{56}_s$ and the $\mathbf{35}_+$. The reader should check this by using the anomaly polynomials in insert 7.2, (of course, put $n = 1$ and $F = 0$) to see that

$$-2\hat{I}_{12}^{\mathbf{8}_s} + 2\hat{I}_{12}^{\mathbf{56}_c} + \hat{I}_{12}^{\mathbf{35}_+} = 0, \qquad (7.30)$$

which is in fact miraculous, as previously stated[339].

[†] The reader might wish to think of this as analogous to the discovery that a moving electric point source generates a magnetic field, but of course is not a basic magnetic monopole source.

7.1.3 Type I from type IIB, the prototype orientifold

As we saw in the bosonic case, we can construct an unoriented theory by projecting onto states invariant under world-sheet parity, Ω. In order to get a consistent theory, we must of course project a theory which is invariant under Ω to start with. Since the left and right moving sectors have the same GSO projection for type IIB, it is invariant under Ω, so we can again form an unoriented theory by gauging. We cannot gauge Ω in type IIA to get a consistent theory, but see later.

Projecting onto $\Omega = +1$ interchanges left-moving and right-moving oscillators and so one linear combination of the R–NS and NS–R gravitinos survives, so there can be only one supersymmetry remaining. In the NS–NS sector, the dilaton and graviton are symmetric under Ω and survive, while the antisymmetric tensor is odd and is projected out. In the R–R sector, by counting we can see that the $\mathbf{1}$ and $\mathbf{35}_+$ are in the symmetric product of $\mathbf{8_s} \otimes \mathbf{8_s}$ while the $\mathbf{28}$ is in the antisymmetric. The R–R state is the product of right- and left-moving fermions, so there is an extra minus in the exchange. Therefore it is the $\mathbf{28}$ that survives. The bosonic massless sector is thus $\mathbf{1} \oplus \mathbf{28} \oplus \mathbf{35}$, and together with the surviving gravitino, this give us the $D = 10$ $N = 1$ supergravity multiplet.

Sadly, this supergravity is in fact anomalous. The delicate balance (7.30) between the anomalies from the various chiral sectors, which we noted previously, vanishes since one each of the $\mathbf{8}_c$ and $\mathbf{56}_s$, and the $\mathbf{35}_+$, have been projected out. Nothing can save the theory unless there is an additional sector to cancel the anomaly.[107]

This sector turns out to be $N = 1$ supersymmetric Yang–Mills theory, with gauge group $SO(32)$ or $E_8 \times E_8$. Happily, we already know at least one place to find the first choice: We can use the low-energy (massless) sector of $SO(32)$ unoriented open superstring theory. This fits nicely, since as we have seen before, at one loop open strings couple to closed strings. We will not be able to get gauge group $E_8 \times E_8$ from perturbative open string theory (Chan–Paton factors can't make this sort of group), but we will see shortly that there is another way of getting this group, but from a closed string theory.

The total anomaly is that of the gravitino, dilatino and the gaugino, the latter being charged in the adjoint of the gauge group:

$$I_{12} = -\hat{I}_{12}^{\mathbf{8}_s}(R) + \hat{I}_{12}^{\mathbf{56}_c}(R) + \hat{I}_{12}^{\mathbf{8}_s}(F, R). \tag{7.31}$$

Using the polynomials given in insert 7.2, it should be easily seen that there is an irreducible term

$$(n - 496)\frac{\mathrm{tr}(R^6)}{725760}, \tag{7.32}$$

which must simply vanish, and so n, the dimension of the group, must be 496. Since $SO(32)$ and $E_8 \times E_8$ both have this dimension, this is encouraging. That the rest of the anomaly cancels is a very delicate and important story which deserves some attention. We will do that in the next section.

Finishing the present discussion, in the language we learned in section 4.11, we put a single (space-filling) O9-plane into type IIB theory, making the type IIB theory into the unoriented $N = 1$ closed string theory. This is anomalous, but we can cancel the resulting anomalies by adding 16 D9-branes.

Another way of putting it is that (as we shall see) the O9-plane has 16 units of C_{10} charge, which cancels that of 16 D9-branes, satisfying the equations of motion for that field.

We have just constructed our first (and in fact, the simplest) example of an 'orientifolding' of a superstring theory to get another. More complicated orientifolds may be constructed by gauging combinations of Ω with other discrete symmetries of a given string theory which form an 'orientifold group' G_Ω under which the theory is invariant[28]. Generically, there will be the requirement to cancel anomalies by the addition of open string sectors (i.e. D-branes), which results in consistent new string theory with some spacetime gauge group carried by the D-branes. In fact, these projections give rise to gauge groups containing any of $U(n)$, $USp(n)$ factors, and not just $SO(n)$ sectors.

7.1.4 The Green–Schwarz mechanism

Let us finish showing that the anomalies of $\mathcal{N} = 1$, $D = 10$ supergravity coupled to Yang–Mills do vanish for the groups $SO(32)$ and $E_8 \times E_8$. We have already shown above that the dimension of the group must be $n = 496$. Some algebra shows that that the rest of the anomaly (7.31), for *this* value of n can be written suggestively as:

$$I_{12}^{(n=496)} = \frac{1}{3 \times 2^8} Y_4 X_8 \tag{7.33}$$
$$- \frac{1}{1440} \left(\mathrm{Tr}_{\mathrm{adj}}(F^6) - \frac{\mathrm{Tr}_{\mathrm{adj}}(F^2)\mathrm{Tr}_{\mathrm{adj}}(F^4)}{48} + \frac{[\mathrm{Tr}_{\mathrm{adj}}(F^2)]^3}{14400} \right),$$

where

$$Y_4 = \mathrm{tr}(R^2) - \frac{1}{30}\mathrm{Tr}_{\mathrm{adj}}(F^2), \tag{7.34}$$
$$X_8 = \frac{\mathrm{Tr}_{\mathrm{adj}}(F^4)}{3} - \frac{[\mathrm{Tr}_{\mathrm{adj}}(F^2)]^2}{900} - \frac{\mathrm{Tr}_{\mathrm{adj}}(F^2)\mathrm{tr}(R^2)}{30} +$$
$$+ \mathrm{tr}(R^4) + \frac{[\mathrm{tr}(R^2)]^2}{4}.$$

Insert 7.3.   The Chern–Simons three-form

The Chern–Simons three-form is a very important structure which will appear in a number of places, and it is worth pausing a while to consider its properties. Recall from insert 2.5 that we can write the gauge potential, and the field strength as Lie Algebra–valued forms: $A = t^a A^a_\mu dx^\mu$, where the t^a are generators of the Lie algebra. We can write the Yang–Mills field strength as a matrix-valued two-form, $F = t^a F^a_{\mu\nu} dx^\mu \wedge dx^\nu$. We can define the Chern–Simons three-form as

$$\omega_{3Y} = \text{Tr}\left(A \wedge F - \frac{1}{3}A \wedge A \wedge A\right) = \text{Tr}\left(A \wedge dA + \frac{2}{3}A \wedge A \wedge A\right).$$

One interesting thing about this object is that we can write:

$$d\omega_{3Y} = \text{Tr}\left(F \wedge F\right).$$

Furthermore, under a gauge transformation $\delta A = d\Lambda + [A, \Lambda]$:

$$\delta\omega_{3Y} = \text{Tr}(d\Lambda dA) = d\omega_2, \quad \omega_2 = \text{Tr}(\Lambda dA).$$

So its gauge variation, while not vanishing, is an exact three-form. Note that there is a similar structure in the pure geometry sector. From section 2.8, we recall that the potential analogous to A is the spin connection one-form $\omega^a{}_b = \omega^a{}_{b\mu} dx^\mu$, with a and b being Minkowski indices in the space tangent to the point x^μ in spacetime and so ω is an $SO(D-1, 1)$ matrix in the fundamental representation. The curvature is a two-form $R^a{}_b = d\omega^a{}_b + \omega^a{}_c \wedge \omega^c{}_b = R^a{}_{b\mu\nu} dx^\mu \wedge dx^\nu$, and the gauge transformation is now $\delta\omega = d\Theta + [\omega, \Theta]$. We can define:

$$\omega_{3L} = \text{tr}\left(\omega \wedge d\omega + \frac{2}{3}\omega \wedge \omega \wedge \omega\right),$$

with similar properties to ω_{3Y}, above. Here tr means trace on the indices a, b.

On the face of it, it does not really seem possible that this can be cancelled, since the the gaugino carries gauge charge and nothing else does, and so there are a lot of gauge quantities which simply stand on their own. This seems hopeless because we have so far restricted ourselves to quantum anomalies arising from the gauge and gravitational sector. If we include the rank two R–R potential $C_{(2)}$ in a cunning way, we can generate a mechanism for cancelling the anomaly. Consider the interaction

$$S_{\text{GS}} = \frac{1}{3 \times 2^6 (2\pi)^5 \alpha'} \int C_{(2)} \wedge X_8. \tag{7.35}$$

It is invariant under the usual gauge transformations

$$\delta A = d\Lambda + [A, \Lambda]; \quad \delta\omega = d\Theta + [\omega, \Theta], \tag{7.36}$$

since it is constructed out of the field strengths F and R. It is also invariant under the two-form potential's standard transformation $\delta C_{(2)} = d\lambda$. Let us however give $C_{(2)}$ another gauge transformation rule. While A and ω transform under (7.36), let it transform as:

$$\delta C_{(2)} = \frac{\alpha'}{4} \left(\frac{1}{30} \text{Tr}(\Lambda F) - \text{tr}(\Theta R) \right). \tag{7.37}$$

Then the variation of the action does not vanish, and is:

$$\delta S_{\text{GS}} = \frac{1}{3 \times 2^8 (2\pi)^5} \int \left[\frac{1}{30} \text{Tr}(\Lambda F) - \text{tr}(\Theta R) \right] \wedge X_8.$$

However, using the properties of the Chern–Simons three-form discussed in insert 7.3, this classical variation can be written as descending *via* the consistency chain in insert 7.1 from precisely the 12-form polynomial given in the first line of equation (7.34), but with a minus sign. Therefore we cancel that offending term with this classical modification of the transformation of $C_{(2)}$. Later on, when we write the supergravity action for this field in the type I model, we will use the modified field strength:

$$\widetilde{G}^{(3)} = dC^{(2)} - \frac{\alpha'}{4} \left[\frac{1}{30} \omega_{3\text{Y}}(A) - \omega_{3\text{L}}(\Omega) \right], \tag{7.38}$$

where because of the transformation properties of the Chern–Simons three-form (see insert 7.3), $\widetilde{G}^{(3)}$ is gauge invariant under the new transformation rule (7.37).

N.B. It is worth noting here that this is a quite subtle mechanism. We are cancelling the anomaly generated by a one loop diagram with a tree-level graph. It is easy to see what the tree level diagram is. The kinetic term for the modified field strength will have its square appearing, and so looking at its definition (7.38), we see that there is a vertex coupling $C_{(2)}$ to two gauge bosons or to two gravitons. There is another vertex that comes from the interaction (7.35) which couples $C_{(2)}$ to four particles, pairs of gravitons and pairs of gauge bosons, or a mixture. So the tree level diagram in figure 7.1 can mix with the hexagon anomaly of insert 7.1.

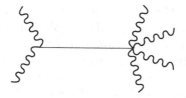

Fig. 7.1. The tree which cures the $\mathcal{N} = 1$ $D = 10$ anomalies. A two-form field is exchanged.

Somehow, the terms in the second line must cancel amongst themselves. Miraculously, they do for a number of groups, $SO(32)$ and $E_8 \times E_8$ included. For the first group, it follows from the fact that for the group $SO(n)$, we can write:

$$\mathrm{Tr}_{\mathrm{adj}}(t^6) = (n - 32)\mathrm{Tr}_\mathrm{f}(t^6) + 15\mathrm{Tr}_\mathrm{f}(t^2)\mathrm{Tr}_\mathrm{f}(t^4);$$
$$\mathrm{Tr}_{\mathrm{adj}}(t^4) = (n - 8)\mathrm{Tr}_\mathrm{f}(t^4) + 3\mathrm{Tr}_\mathrm{f}(t^2)\mathrm{Tr}_\mathrm{f}(t^2);$$
$$\mathrm{Tr}_{\mathrm{adj}}(t^2) = (n - 2)\mathrm{Tr}_\mathrm{f}(t^2), \tag{7.39}$$

where the subscript 'f' denotes the fundamental representation. For E_8, we have that

$$\mathrm{Tr}_{\mathrm{adj}}(t^6) = \frac{1}{7200}[\mathrm{Tr}_\mathrm{f}(t^2)]^3,$$
$$\mathrm{Tr}_{\mathrm{adj}}(t^4) = \frac{1}{100}[\mathrm{Tr}_\mathrm{f}(t^2)]^2. \tag{7.40}$$

In checking these (which of course the reader will do) one should combine the traces as $\mathrm{Tr}_{G_1 \times G_2} = \mathrm{Tr}_{G_1} + \mathrm{Tr}_{G_2}$, etc.

Overall, the results[107] of this subsection are quite remarkable, and generated a lot of excitement which we now call the First Superstring Revolution. This excitement was of course justified, since the discovery of the mechanism revealed that there were consistent superstring theories with considerably intricate structures with promise for making contact with the physics that we see in Nature.

7.2 The two basic heterotic string theories

In addition to the three superstring theories briefly constructed above, there are actually two more supersymmetric string theories which live in ten dimensions. In addition, they have non-Abelian spacetime gauge symmetry, and they are also free of tachyons. These are the 'Heterotic Strings'[20]. The fact that they are chiral, have fermions and non-Abelian gauge symmetry meant that they were considered extremely attractive as starting points for constructing 'realistic' phenomenology based on string theory. It is in fact remarkable that one can come tantalisingly close to naturally realising many of the features of the Standard Model of particle physics by starting with, say, the $E_8 \times E_8$ Heterotic String, while remaining entirely in the perturbative regime. This was the focus of much of the First Superstring Revolution. Getting many of the harder questions right led to the search for non-perturbative physics, which ultimately led us to the Second Superstring Revolution, and the realisation that all of the other string theories were just as important too, because of duality.

One of the more striking things about the heterotic strings, from the point of view of what we have done so far, is the fact that they have non-Abelian gauge symmetry and are still closed strings. The $SO(32)$ of the type I string theory comes from Chan–Paton factors at the ends of the open string, or in the language we now use, from 16 coincident D9-branes.

We saw a big hint of what is needed to get spacetime gauge symmetry in the heterotic string in chapter 4. Upon compactifying bosonic string theory on a circle, at a special radius of the circle, an enhanced $SU(2)_L \times SU(2)_R$ gauge symmetry arose. From the two dimensional world-sheet point of view, this was a special case of a current algebra, which we uncovered further in section 4.6. We can take two key things away from that chapter for use here. The first is that we can generalise this to a larger non-Abelian gauge group if we use more bosons, although this would seem to force us to have many compact directions. The second is that there were identical and *independent* structures coming from the left and the right to give this result. So we can take, say, the left hand side of the construction and work with it, to produce a single copy of the non-Abelian gauge group in spacetime.

This latter observation is the origin of the word 'heterotic' which comes from 'heterosis'. The theory is a hybrid of two very different constructions on the left and the right. Let us take the right hand side to be a copy of the right hand side of the superstrings we constructed previously, and so we use only the right hand side of the action given in equation (7.1) (with closed string boundary conditions). Then the usual consistency checks give that the critical dimension is of course ten, as before: the central charge (conformal anomaly) is $-26 + 11 = 15$ from the conformal and superconformal ghosts. This is cancelled by ten bosons and their superpartners since they contribute to the anomaly an amount $10 \times 1 + 10 \times \frac{1}{2} = 15$. The left hand side is in fact a purely bosonic string, and so the anomaly is cancelled to zero by the -26 from the conformal ghosts and there must be the equivalent of 26 bosonic degrees of freedom, contributing 26×1 to the anomaly.

How can the theory make sense as a ten dimensional theory? The answer to this question is just what gives the non-Abelian gauge symmetry. Sixteen of the bosons are periodic, and so may be thought of as compactified on a torus $T^{16} \simeq (S^1)^{16}$ with very specific properties. Those properties are such that the generic $U(1)^{16}$ one might have expected from such a toroidal compactification is enhanced to one of two special rank 16 gauge groups: $SO(32)$, or $E_8 \times E_8$, via the very mechanism we saw in chapter 4: the torus is 'self-dual'. The remaining ten non-compact bosons on the left combine with the ten on the right to make the usual ten spacetime coordinates, on which the usual ten dimensional Lorentz group $SO(1, 9)$ acts.

7.2.1 $SO(32)$ and $E_8 \times E_8$ from self-dual lattices

The requirements are simple to state. We are required to have a sixteen dimensional lattice, according to the above discussion, and so we can apply the results of chapter 4, but there is a crucial difference. Recalling what we learned there, we see that since we only have a left-moving component to this lattice, we do not have the Lorenzian signature which arose there, but only a *Euclidean* signature. But all of the other conditions apply: it must be even, in order to build gauge bosons as vertex operators, and it must be self-dual, to ensure modular invariance.

The answer turns out to be quite simple. There are only two choices, since even self-dual Euclidean lattices are very rare (They only exist when the dimension is a multiple of eight). For sixteen dimensions, there is either $\Gamma_8 \times \Gamma_8$ or Γ_{16}. The lattice Γ_8 is the collection of points:

$$(n_1, n_2, \ldots, n_8) \quad \text{or} \quad (n_1 + \tfrac{1}{2}, n_2 + \tfrac{1}{2}, \ldots, n_8 + \tfrac{1}{2}), \quad \sum_i n_i \in 2\mathbb{Z},$$

with $\sum_i n_i^2 = 2$. The integer lattice points are actually the root lattice of $SO(16)$, with which the 120 dimensional adjoint representation is made. The half-integer points construct the spinor representation of $SO(16)$. A bit of thought shows that it is just like the construction we made of the spinor representations of $SO(8)$ previously; the entries are only $\pm\tfrac{1}{2}$ in eight different slots, with only an even number of minus signs appearing, which again gives a squared length of two. There are $2^7 = 128$ possibilities, which is the dimension of the spinor representation. The total dimension of the represetnation we can make is $120 + 128 = 248$ which is the dimension of E_8. The sixteen dimensional lattice is made as the obvious tensor product of two copies of this, giving gauge group $E_8 \times E_8$, which is 496 dimensional.

The lattice Γ_{16} is extremely similar, in that it is:

$$(n_1, n_2, \ldots, n_{16}) \quad \text{or} \quad (n_1 + \tfrac{1}{2}, n_2 + \tfrac{1}{2}, \ldots, n_{16} + \tfrac{1}{2}), \quad \sum_i n_i \in 2\mathbb{Z},$$

with $\sum_i n_i^2 = 2$. Again, we see that the integer points make the root lattice of $SO(32)$, but there is more. There is a spinor representation of $SO(32)$, but it is clear that since $16 \times 1/4 = 4$, the squared length is twice as large as it need to be to make a massless vector, and so the gauge bosons remain from the adjoint of $SO(32)$, which is 496 dimensional. In fact, the full structure is more than $SO(32)$, because of this spinor representation. It is not quite the cover, which is $Spin(32)$ because the conjugate spinor and the vector representations are missing. It is instead written as $Spin(32)/\mathbb{Z}_2$. In fact, $SO(32)$ in the quotient of $Spin(32)$ by another $\mathbb{Z}_2$.

Actually, before concluding, we should note that there is an alternative construction to this one using left–moving fermions instead of bosons. This is easily arrived at from here using what we learned about fermionisation in section 4.7. From there, we learn that we can trade in each of the left-moving bosons here for *two* left–moving Majorana–Weyl fermions, giving a fermionic construction with 32 fermions $\boldsymbol{\Psi}^i$. The construction divides the fermions into the NS and R sectors as before, which correspond to the integer and half-integer lattice sites in the above discussion. The difference between the two heterotic strings is whether the fermions are split into two sets with independent boundary conditions (giving $E_8 \times E_8$) or if they have all the same boundary conditions ($SO(32)$). In this approach, there is a GSO projection, which in fact throws out a tachyon, etc. Notice that in the R sector, the zero modes of the 32 $\boldsymbol{\Psi}^i$ will generate a spinor and conjugate spinor $\mathbf{2^{31}} \oplus \mathbf{2^{31}}$ of $SO(32)$ for much the same reasons as we saw a $\mathbf{16} \oplus \overline{\mathbf{16}}$ in the construction of the superstring. Just as there, a GSO projection arises in the construction, which throws out the conjugate spinor, leaving the sole massive spinor we saw arise in the direct lattice approach.

7.2.2 The massless spectrum

In the case we must consider here, we can borrow a lot of what we learned in section 4.5 with hardly any adornment. We have sixteen compact left-moving bosons, X^i, which, together with the allowed momenta P^i, define a lattice Γ. The difference between this lattice and the ones we considered in section 4.5 is that there is no second part coming from a family of right-moving momenta, and hence it is only half the expected dimension, and with a purely Euclidean signature. This sixteen dimensional lattice must again be self-dual and even. This amounts to the requirement of modular invariance, just as before. More directly, we can see what effect this has on the low-lying parts of the spectrum.

Recall that the NS and R sector of the right hand side has zero point energy equal to $-1/2$ and 0, respectively. Recall that we then make, after the GSO projection, the vector $\mathbf{8}_v$, and its superpartner the spinor $\mathbf{8}_s$ from these two sectors. On the left hand side, we have the structure of the bosonic string, with zero point energy -1. There is no GSO projection on this side, and so potentially we have the tachyon, $|0\rangle$, the familiar massless states $\alpha^\mu_{-1}|0\rangle$, and the current algebra elements $J^a_{-1}|0\rangle$. These must be tensored together with the right hand side's states, but we must be aware that the level–matching condition is modified. To work out what it is we must take the difference between the correctly normalised *ten dimensional* M^2 operators on each side. We must also recall that in making the ten dimensional M^2 operator, we are left with a remainder, the contribution to the internal momentum $\alpha' p_L^2/4$. The result is:

$$\frac{\alpha' p_L^2}{4} + N - 1 = \tilde{N} - \begin{cases} -\frac{1}{2} \\ 0 \end{cases},$$

where the choice corresponds to the NS or R sectors.

Now we can see how the tachyon is projected out of the theory, even without a GSO projection on the left. The GSO on the right has thrown out the tachyon there, and so we start with $\tilde{N} = \frac{1}{2}$ there. The left tachyon is $N = 0$, but this is not allowed, and we must have the even condition $\alpha' p_L^2/2 = 2$ which corresponds to switching on a current J^a_{-1}, making a massless state. If we do not have this state excited, then we can also make a massless state with $N = 1$, corresponding to $\alpha^\mu_{-1}|0\rangle$.

The massless states we can make by tensoring left and right, respecting level–matching are actually familiar. In the NS–NS sector, we have $\alpha^\mu_{-1}\psi^\nu_{-1/2}|0\rangle$, which is the graviton, $G_{\mu\nu}$ antisymmetric tensor $B_{\mu\nu}$ and dilaton Φ in the usual way. We also have $J^a_{-1}\psi^\mu_{-1/2}|0\rangle$, which gives an $E_8 \times E_8$ or $SO(32)$ gauge boson, $A^{\mu a}$. In the NS–R sector, we have $\alpha^\mu_{-1}|0\rangle_\alpha$ which is the gravitino, ψ^μ_α. Finally, we have $J^a_{-1}|0\rangle_\alpha$, which is the superpartner of the gauge boson, λ^a_α. In the language we used earlier, we can write the left hand representations under $SO(8) \times G$ (where G is $SO(32)$ or $E_8 \times E_8$) as $(\mathbf{8}_v, \mathbf{1})$ or $(\mathbf{1}, \mathbf{496})$. Then the tensoring is

$$(\mathbf{8}_v, \mathbf{1}) \otimes (\mathbf{8}_v + \mathbf{8}_s) = (\mathbf{1}, \mathbf{1}) + (\mathbf{35}, \mathbf{1}) + (\mathbf{28}, \mathbf{1}) + (\mathbf{56}_s, \mathbf{1}) + (\mathbf{8}_s, \mathbf{1}),$$
$$(\mathbf{1}, \mathbf{496}) \otimes (\mathbf{8}_v + \mathbf{8}_s) = (\mathbf{8}_v, \mathbf{496}) + (\mathbf{8}_s, \mathbf{496}).$$

So we see that we have again obtained the $\mathcal{N} = 1$ supergravity multiplet, coupled to a massless vector. The effective theory which must result at low energy must have the same gravity sector, but since the gauge fields arise at closed string tree level, their Lagrangian must have a dilaton coupling $e^{2\Phi}$, instead of e^{Φ} for the open string where the gauge fields arise at open string tree level.

7.3 The ten dimensional supergravities

Just as we saw in the case of the bosonic string, we can truncate consistently to focus on the massless sector of the string theories, by focusing on low energy limit $\alpha' \to 0$. Also as before, the dynamics can be summarised in terms of a low energy effective (field theory) action for these fields, commonly referred to as 'supergravity'.

The bosonic part of the low energy action for the type IIA string theory in ten dimensions may be written (cf. equation (2.106)) as (the wedge product is understood)[1, 5, 75]:

$$S_{\text{IIA}} = \frac{1}{2\kappa_0^2} \int d^{10}x (-G)^{1/2} \left\{ e^{-2\Phi} \left[R + 4(\nabla\Phi)^2 - \frac{1}{12}(H^{(3)})^2 \right] \right. $$
$$\left. - \frac{1}{4}(G^{(2)})^2 - \frac{1}{48}(G^{(4)})^2 \right\} - \frac{1}{4\kappa_0^2} \int B^{(2)} dC^{(3)} dC^{(3)}. \tag{7.41}$$

As before $G_{\mu\nu}$ is the metric in string frame, Φ is the dilaton, $H^{(3)} = dB^{(2)}$ is the field strength of the NS–NS two form, while the Ramond–Ramond field strengths are $G^{(2)} = dC^{(1)}$ and $G^{(4)} = dC^{(3)} + H^{(3)} \wedge C^{(1)\ddagger}$.

For the bosonic part in the case of type IIB, we have:

$$S_{\text{IIB}} = \frac{1}{2\kappa_0^2} \int d^{10}x (-G)^{1/2} \left\{ e^{-2\Phi} \left[R + 4(\nabla\Phi)^2 - \frac{1}{12}(H^{(3)})^2 \right] \right. $$
$$\left. - \frac{1}{12}(G^{(3)} + C^{(0)}H^{(3)})^2 - \frac{1}{2}(dC^{(0)})^2 - \frac{1}{480}(G^{(5)})^2 \right\} $$
$$+ \frac{1}{4\kappa_0^2} \int \left(C^{(4)} + \frac{1}{2}B^{(2)} C^{(2)} \right) G^{(3)} H^{(3)}. \tag{7.42}$$

Now, $G^{(3)} = dC^{(2)}$ and $G^{(5)} = dC^{(4)} + H^{(3)}C^{(2)}$ are R–R field strengths, and $C^{(0)}$ is the R–R scalar. (Note that we have canonical normalisations for the kinetic terms of forms: there is a prefactor of the inverse of $-2 \times p!$ for a p-form field strength.) There is a small complication due to the fact that we require the R–R four form $C^{(4)}$ to be self-dual, or we will have too many degrees of freedom. We write the action here and remind ourselves to always impose the self-duality constraint on its field strength $F^{(5)} = dC^{(4)}$ by hand in the equations of motion: $F^{(5)} = {}^*F^{(5)}$.

---

‡ This can be derived by dimensional reduction from the structurally simpler eleven dimensional supergravity action, presented in chapter 12, but at this stage, this relation is a merely formal one. We shall see a dynamical connection later.

Equation (2.109) tells us that, in ten dimensions, we must use

$$\widetilde{G}_{\mu\nu} = e^{(\Phi_0 - \Phi)/2} G_{\mu\nu} \qquad (7.43)$$

to convert these actions to the Einstein frame. As before (see discussion below equation(2.111)), Newton's constant will be set by

$$2\kappa^2 \equiv 2\kappa_0^2 g_s^2 = 16\pi G_N = (2\pi)^7 \alpha'^4 g_s^2, \qquad (7.44)$$

where the latter equality can be established by (for example) direct examination of the results of a graviton scattering computation. We will see that it gives a very natural normalisation for the masses and charges of the various branes in the theory. Also g_s is set by the asymptotic value of the dilaton at infinity: $g_s \equiv e^{\Phi_0}$.

Those were the actions for the ten dimensional supergravities with thirty-two supercharges. Let us consider those with sixteen supercharges. For the bosonic part of type I, we can construct it by dropping the fields which are odd under Ω and then adding the gauge sector, plus a number of cross terms which result from cancelling anomalies, as we discussed in subsection 7.1.3:

$$S_I = \frac{1}{2\kappa_0^2} \int d^{10}x (-G)^{1/2} \left\{ e^{-2\Phi} \left[R + 4(\nabla\Phi)^2 \right] \right.$$
$$\left. - \frac{1}{12} (\widetilde{G}^{(3)})^2 - \frac{\alpha'}{8} e^{-\Phi} \mathrm{Tr}(F^{(2)})^2 \right\}. \quad (7.45)$$

Here, $\widetilde{G}^{(3)}$ is a modified field strength for the two-form potential, defined in equation (7.38). Recall that this modification followed from the requirement of cancellation of the anomaly via the Green–Schwarz mechanism.

We can generate the heterotic low-energy action using a curiosity which will be meaningful later. Notice that a simple redefinition of fields:

$$G_{\mu\nu}(\text{type I}) = e^{-\Phi} G_{\mu\nu}(\text{heterotic})$$
$$\Phi(\text{type I}) = -\Phi(\text{heterotic})$$
$$\widetilde{G}^{(3)}(\text{type I}) = \widetilde{H}^{(3)}(\text{heterotic})$$
$$A_\mu(\text{type I}) = A_\mu(\text{heterotic}), \qquad (7.46)$$

takes one from the type I Lagrangian to:

$$S_H = \frac{1}{2\kappa_0^2} \int d^{10}x (-G)^{1/2} e^{-2\Phi} \left\{ R + 4(\nabla\Phi)^2 \right.$$
$$\left. - \frac{1}{12} (\widetilde{H}^{(3)})^2 - \frac{\alpha'}{8} \mathrm{Tr}(F^{(2)})^2 \right\}, \quad (7.47)$$

where (renaming $C^{(2)} \to B^{(2)}$)

$$\widetilde{H}^{(3)} = dB^{(2)} - \frac{\alpha'}{4}\left[\frac{1}{30}\omega_{3Y}(A) - \omega_{3L}(\Omega)\right]. \tag{7.48}$$

This is the low energy effective Lagrangian for the heterotic string theories. Note that in (7.47), α' is measured in heterotic units of length.

We can immediately see two key features about these theories. The first was anticipated earlier: their Lagrangian for the gauge fields have a dilaton coupling $e^{-2\Phi}$, since they arise at closed string tree level, instead of $e^{-\Phi}$ for the open string where where the gauge fields arise at open string tree level. The second observation is that since from equation (7.46) the dilaton relations tell us that $g_s(\text{type I}) = g_s^{-1}(\text{heterotic})$, there is a non-perturbative connection between these two theories, although they are radically different in perturbation theory. We are indeed *forced* to consider these theories when we study the type I string in the limit of strong coupling.

7.4 Heterotic toroidal compactifications

Much later, it will be of interest to study simple compactifications of the heterotic strings, and the simplest result from placing them on tori[174, 175]. Our interest here is not in low energy particle physics phenomenology, as this would require us to compactify on more complicated spaces to break the large amount of supersymmetry and gauge symmetry. Instead, we shall see that it is quite instructive, on the one hand, and on the other hand, studying various superstring compactifications with D-brane sectors taken into account will produce vacua which are in fact strong/weak coupling dual to heterotic strings on tori. This is another remarkable consequence of duality which forces us to consider the heterotic strings even though they cannot have D-brane sectors.

Actually, there is not much to do. From our work in section 7.2 and from that in section 4.5, it is easy to see what the conditions for the consistency of a heterotic toroidal compactification must be. Placing some of the ten dimensions on a torus T^d will give us the possibility of having windings, and right-moving momenta. In addition, the gauge group can be broken by introducing Wilson lines (see insert 4.4 and section 4.9.1) on the torus for the gauge fields A^μ. This latter choice breaks the gauge group to the maximal Abelian subgroup, which is $U(1)^{16}$.

The compactification simply enlarges our basic sixteen dimensional Euclidean lattice from $\Gamma_8 \oplus \Gamma_8$ or Γ_{16} by two dimensions of Lorentzian signature $(1, 1)$ for each additional compact direction, for the reasons we already discussed in section 4.5. So we end up with a lattice with signature $(16 + d, d)$, on which there must be an action of $O(d, 16 + d)$ generating the lattices. Again, we will have that there is a physical equivalence between some of these lattices, because physics only depends on p_L^2 and p_R^2, and further, there will be the discrete equivalences corresponding to the action of a T-duality group, which is $O(d, 16 + d, \mathbb{Z})$.

The required lattices are completely classified, as a mathematical exercise. In summary, the space of inequivalent toroidal compactifications turns out to be:

$$\mathcal{M}_{T^d} = [O(d) \times O(d + 16)] \backslash O(d, d + 16) / O(d, d + 16, \mathbb{Z}). \qquad (7.49)$$

Notice, after a quick computation, that the dimension of this space is $d^2 + 16d$. So in addition to the fields $G_{\mu\nu}$, $B_{\mu\nu}$ and Φ, we have that number of extra massless scalars in the $\mathcal{N} = 2$, $D = 6$ low energy theory. The first part of the result comes, as before from the available constant components, G_{mn} and B_{mn}, of the internal metric and antisymmetric tensor on T^d. The remaining part comes from the sixteen generic constant internal gauge bosons (the Wilson lines), A_m for each circle.

Let us compute what the generic gauge group of this compactified model is. There is of course the $U(1)^{16}$ from the original current algebra sector. In addition, there is a $U(1) \times U(1)$ coming from each compact dimension, since we have Kaluza–Klein reduction of the metric and antisymmetric tensor. Therefore, the generic gauge group is $U(1)^{16+2d}$.

To get something less generic, we must tune some moduli to special points. Of course, we can choose to switch off some of the Wilson lines, getting non-Abelian gauge groups from the current algebra sector, restoring an $E_8 \times E_8 \times U(1)^{2d}$ or $SO(32) \times U(1)^{2d}$ gauge symmetry. We also have the possibility of enhancing the Kaluza–Klein factor by tuning the torus to special points. We simply need to make states of the form $\exp(ik_L \cdot X_L)\psi^\mu_{-1/2}|0\rangle$, where we can have left-moving momenta of $\alpha' p_L^2/2 = 2$ (we are referring to the components of p_L which are in the torus T^d). This will give any of the A–D–E series of gauge groups up to a rank $2d$ in this sector.

The reader will have noticed that we only gave one family of lattices for each dimension d of the torus. We did not have one choice for the $E_8 \times E_8$ string and another for the $SO(32)$ string. In other words, as soon as we compactify one heterotic string on a circle, we find that we could have arrived at the same spectrum by compactifying the other heterotic string on a circle. This is of course T-duality. It is worth examining further, and we do this in section 8.1.3.

7.5 Superstring toroidal compactification

The placement of the superstrings on tori is at face value rather less interesting than the heterotic case, and so we will not spend much time on it here, although will return to it later when we revisit T-duality, and again when we study U-duality in section 12.7.

Imagine that we compactify one of our superstring theories on the torus T^d. We simply ask that d of the directions are periodic with some chosen radius, as we did in section 4.5 for the bosonic string. This does not not affect any of our discussion of supercharges, etc., and we simply have a $(10 - d)$-dimensional theory with the same amount of supersymmetry as the ten dimensional theory which we started with. As discussed in section 4.4, there is a large $O(d, d, \mathbb{Z})$ pattern of T-duality groups available to us. There are also Kaluza–Klein gauge groups $U(1)^{2d}$ coming from the internal components of the graviton and the antisymmetric tensor. In addition, there are Kaluza–Klein gauge groups coming from the possibility of some of the R–R sector antisymmetric tensors having internal indices. Note that there aren't the associated enhanced gauge symmetries present at special radii, since the appropriate objects which would have arisen in a current algebra, J^a_{-1}, do not give masses states in spacetime, and in any case level matching would have forbidden them from being properly paired with $\psi^\mu_{-1/2}$ to give a spacetime vector.

To examine the possibilities, it is probably best to study a specific example, and we do the case of placing the type IIA string theory on T^5.

Let us first count the gauge fields. This can be worked out simply by counting the number of ways of wrapping the metric and the various p-form potentials (with p odd) in the theory on the five circles of the T^5 to give a one-form in the remaining five non-compact directions. From the NS–NS sector there are five Kaluza–Klein gauge bosons and five gauge bosons from the antisymmetric tensor. There are 16 gauge bosons from the dimensional reduction of the various R–R forms: the breakdown is 10+5+1 from the forms $C^{(3)}$, $C^{(5)}$ and $C^{(1)}$, respectively, since, for example, there are ten independent ways of making two out of the three indices of $C^{(3)}$ be any two out of the five internal directions, and so on. Finally, in five dimensions, one can form a two form field strength from the Hodge dual $^*H$ of the three-form field strength of the NS–NS $B_{\mu\nu}$, thus defining another gauge field.

So the gauge group is generically $U(1)^{27}$. There are in fact a number of massless fields corresponding to moduli representing inequivalent sizes and shapes for the T^5. We can count them easily. We have the $5^2 = 25$ components coming from the graviton and antisymmetric tensor field. From the R–R sector there is only one way of getting a scalar from $C^{(5)}$, and five and ten ways from $C^{(1)}$ and $C^{(3)}$, respectively. This gives 41 moduli. Along with the dilaton, this gives a total of 42 scalars for this compactification.

By now, the reader should be able to construct the very same five dimensional spectrum but starting with the type IIB string and placing it on T^5. This is a useful exercise in preparation for later. The same phenomenon will happen with any torus, T^d. Thus we begin to uncover the fact that the type IIA and type IIB string theories are (T-dual) equivalent to each other when placed on circles. We shall examine this in more detail in section 8.1, showing that the equivalence is exact.

The full T-duality group is actually $O(5,5;\mathbb{Z})$. It acts on the different sectors independently, as it ought to. For example, for the gauge fields, it mixes the first ten NS–NS gauge fields among themselves, and the 16 R–R gauge fields among themselves, and leaves the final NS–NS field invariant. Notice that the fields fill out sensible representations of $O(5,5;\mathbb{Z})$. Thinking of the group as roughly $SO(10)$, those familiar with numerology from grand unification might recognise that the sectors are transforming as the **10**, **16**, and **1**.

A little further knowledge will lead to questions about the fact that **10** $\oplus$ **16** $\oplus$ **1** is the decomposition of the **27** (the fundamental representation) of the group E_6, but we should leave this for a later time, when we come to discuss U-duality in section 12.7.

7.6 A superstring orbifold: discovering the K3 manifold

Before we go any further, let us briefly revisit the idea of strings propagating on an orbifold, and take it a bit further. Imagine that we compactify one of our closed string theories on the four torus, T^4. Let us take the simple case where there the torus is simply the product of four circles, S^1, each with radius R. Let us choose that the four directions (say) x^6, x^7, x^8 and x^9 are periodic with period $2\pi R$. The resulting six dimensional theory has $\mathcal{N} = 4$ supersymmetry.

Let us orbifold the theory by the $\mathbb{Z}_2$ group which has the action

$$\mathbf{R}: \quad x^6, x^7, x^8, x^9 \rightarrow -x^6, -x^7, -x^8, -x^9, \qquad (7.50)$$

which is clearly a good symmetry to divide by. We can choose to let $\mathbf{R}$ be embedded in the $SU(2)_{\mathrm{L}}$ which acts on the $\mathbb{R}^4$ (see insert 7.4). This will leave an $SU(2)_{\mathrm{R}}$ which descends to the six dimensions as a global symmetry. It is in fact the R–symmetry of the remaining $D = 6$, $\mathcal{N} = 2$ model. We shall use this convention a number of times in what is to come.

Insert 7.4. $SU(2)_\mathrm{L}$ versus $SU(2)_\mathrm{R}$

It is well worth pausing here to note a nice way of writing things, for later use. The space $\mathbb{R}^4$ with coordinates $(x_6, x_7, x_8, x_9) = (\tau, x, y, z)$ has an obvious $SO(4)$ symmetry. Note that $SO(4) \sim SU(2)_\mathrm{L} \times SU(2)_\mathrm{R}$, where the 'L' and 'R' labels denote left and right. What is the meaning of this? To see it, present two new sets of coordinates. Write $\mathbb{R}^4$ with a radial coordinate $r = (\tau^2 + x^2 + y^2 + z^2)^{1/2}$, and Euler angles on an S^3 (r, θ, ϕ, ψ), where $0 < \theta < \pi$, $0 < \phi < 2\pi$, $0 < \psi < 4\pi$. The metric is:

$$ds^2 = d\tau^2 + dx^2 + dy^2 + dz^2 = dr^2 + \frac{r^2}{4}(d\theta^2 + d\phi^2 + d\psi^2 + 2\cos\theta d\psi d\phi).$$

Further define an element $g \in SU(2)$: $g = (\tau\mathbf{1} - i\vec{\tau} \cdot \vec{x})/r$ for Pauli matrices τ_i (given, *e.g.* in equation (13.1), where they're called σ_i):

$$g = \frac{1}{r}\begin{pmatrix} \tau + iz & -y + ix \\ y + ix & \tau - iz \end{pmatrix} = \begin{pmatrix} e^{\frac{i}{2}(\phi+\psi)}\cos\frac{\theta}{2} & -e^{\frac{i}{2}(\phi-\psi)}\sin\frac{\theta}{2} \\ e^{-\frac{i}{2}(\phi-\psi)}\sin\frac{\theta}{2} & e^{-\frac{i}{2}(\phi+\psi)}\cos\frac{\theta}{2} \end{pmatrix}.$$

There are natural independent actions of $h \in SU(2)$ on this on the left, $g \to hg$, or on the right, $g \to gh$. It is really useful to extract certain natural 'Maurer–Cartan' one-forms from this. They are $\sigma_a = -i\mathrm{Tr}(\tau_a g^{-1}dg)$ and are clearly invariant under the $SU(2)_\mathrm{L}$. The $\bar{\sigma}_a = -i\mathrm{Tr}(\tau_a dg g^{-1})$ are $SU(2)_\mathrm{R}$ invariant. Explicitly:

$$2\sigma_1 = -\sin\psi d\theta + \cos\psi\sin\theta d\phi;$$
$$2\sigma_2 = \cos\psi d\theta + \sin\psi\sin\theta d\phi; \quad 2\sigma_3 = d\psi + \cos\theta d\phi,$$

and they satisfy $d\sigma_a = \epsilon_{abc}\sigma_b \wedge \sigma_c$. Note also that $4(\sigma_1^2 + \sigma_2^2)$ is the standard round unit radius S^2 metric, while $\sigma_1^2 + \sigma_2^2 + \sigma_3^2$ gives the same for S^3. (The $\bar{\sigma}_i$ can be obtained by sending $\psi \leftrightarrow \phi$.) Now, our metric on $\mathbb{R}^4$ can be written as $ds^2 = dr^2 + r^2(\sigma_1^2 + \sigma_2^2 + \sigma_3^2)$.

7.6.1 The orbifold spectrum

We can construct the resulting six dimensional spectrum by first working out (say) the left-moving spectrum, seeing how it transforms under **R** and then tensoring with another copy from the right in order to construct the closed string spectrum.

Let us now introduce a bit of notation which will be useful in the future. Use the label x^m, $m = 6, 7, 8, 9$ for the orbifolded directions, and use x^μ,

$\mu = 0, \ldots, 5$, for the remaining. Let us also note that the ten dimensional Lorentz group is decomposed as

$$SO(1,9) \supset SO(1,5) \times SO(4).$$

We shall label the transformation properties of our massless states in the theory under the $SU(2) \times SU(2) = SO(4)$ little group. Just as we did before, it will be useful in the Ramond sector to choose a labelling of the states which refers to the rotations in the planes (x^0, x^1), (x^2, x^3), etc., as eigenstates $s_0, s_1, \ldots, s_4$ of the operator S^{01}, S^{23}, etc., (see equations (7.17) and (7.19) and surrounding discussion).

With this in mind, we can list the states on the left that survive the GSO projection.

sector	state	**R** charge	$SO(4)$ charge
NS	$\psi^{\mu}_{-\frac{1}{2}}|0;k\rangle$	$+$	$(\mathbf{2},\mathbf{2})$
	$\psi^{m}_{-\frac{1}{2}}|0;k\rangle$	$-$	$4(\mathbf{1},\mathbf{1})$
R	$|s_1 s_2 s_3 s_4\rangle$; $s_1 = +s_2$, $s_3 = -s_4$	$+$	$2(\mathbf{2},\mathbf{1})$
	$|s_1 s_2 s_3 s_4\rangle$; $s_1 = -s_2$, $s_3 = +s_4$	$-$	$2(\mathbf{1},\mathbf{2})$

Crucially, we should also examine the 'twisted sectors' which will arise, in order to make sure that we get a modular invariant theory. The big difference here is that in the twisted sector, the moding of the fields in the x^m directions is shifted. For example, the bosons are now half-integer moded. We have to recompute the zero point energies in each sector in order to see how to get massless states (see (2.80)):

$$\text{NS sector:} \quad 4\left(-\frac{1}{24}\right) + 4\left(-\frac{1}{48}\right) + 4\left(\frac{1}{48}\right) + 4\left(\frac{1}{24}\right) = 0,$$

$$\text{R sector:} \quad 4\left(-\frac{1}{24}\right) + 4\left(\frac{1}{24}\right) + 4\left(\frac{1}{48}\right) + 4\left(-\frac{1}{48}\right) = 0. \quad (7.51)$$

This is amusing; both the Ramond and NS sectors have zero vacuum energy, and so the integer moded sectors will give us degenerate vacua. We see that it is only states $|s_1 s_2\rangle$ which contribute from the R sector (since they are half-integer moded in the x^m directions) and the NS sector, since it is integer moded in the x^m directions, has states $|s_3 s_4\rangle$.

> N.B. It is worth seeing in equation (7.51) how we achieved this ability to make a massless field in this case. The single twisted sector ground state in the bosonic orbifold theory with energy 1/48, was multiplied by four since there are four such orbifolded directions. Combining this with the contribution from the four unorbifolded directions produced just the energy needed to cancel the contribution from the fermions.

The states and their charges are as follows (after imposing GSO).

sector	state	$\mathbf{R}$ charge	$SO(4)$ charge
NS	$\lvert s_3 s_4 \rangle; \ s_3 = -s_4$	$+$	$\mathbf{2(1,1)}$
R	$\lvert s_1 s_2 \rangle; \ s_1 = -s_2$	$-$	$\mathbf{(1,2)}$

Now we are ready to tensor. Recall that we could have taken the opposite GSO choice here to get a left moving with the identical spectrum, but with the swap $\mathbf{(1,2)} \leftrightarrow \mathbf{(2,1)}$. Again we have two choices: tensor together two identical GSO choices, or two opposite. In fact, since six dimensional supersymmetries are chiral, and the orbifold will keep only two of the four we started with, we can write these choices as $(0,2)$ or $(1,1)$ supersymmetry, resulting from type IIB or IIA on K3. It is useful to tabulate the result for the bosonic spectra for the untwisted sector.

sector	$SO(4)$ charge
NS–NS	$\mathbf{(3,3)} + \mathbf{(1,3)} + \mathbf{(3,1)} + \mathbf{(1,1)}$ $10\mathbf{(1,1)} + 6\mathbf{(1,1)}$
R–R (IIB)	$2\mathbf{(3,1)} + 4\mathbf{(1,1)}$ $2\mathbf{(1,3)} + 4\mathbf{(1,1)}$
R–R (IIA)	$4\mathbf{(2,2)}$ $4\mathbf{(2,2)}$

This is the result for the twisted sector.

sector	$SO(4)$ charge
NS–NS	$3\mathbf{(1,1)} + \mathbf{(1,1)}$
R–R (IIB) R–R (IIA)	$\mathbf{(1,3)} + \mathbf{(1,1)}$ $\mathbf{(2,2)}$

Recall now that we have two twisted sectors for each orbifolded circle, and hence there are 16 twisted sectors in all, for $T^4/\mathbb{Z}_2$. Therefore, to make the complete model, we must take sixteen copies of the content of the twisted sector table above.

Now let identify the various pieces of the spectrum. The gravity multiplet $G_{\mu\nu} + B_{\mu\nu} + \Phi$ is in fact the first line of our untwisted sector table, coming from the NS–NS sector, as expected. The field B can be seen to be broken into its self-dual and anti-self-dual parts $B_{\mu\nu}^+$ and $B_{\mu\nu}^-$, transforming as $(\mathbf{1},\mathbf{3})$ and $(\mathbf{3},\mathbf{1})$. There are sixteen other scalar fields, $((\mathbf{1},\mathbf{1}))$, from the untwisted NS–NS sector. The twisted sector NS–NS sector has 4×16 scalars. Not including the dilaton, there are 80 scalars in total from the NS–NS sector.

Turning to the R–R sectors, we must consider the cases of IIA and IIB separately. For type IIA, there are eight one-forms (vectors, $(\mathbf{2},\mathbf{2})$) from the untwisted sector and 16 from the twisted, giving a total of 24 vectors, and have a generic gauge group $U(1)^{24}$.

For type IIB, the untwisted R–R sector contains three self-dual and three anti-self-dual tensors, while there are an additional 16 self-dual tensors $(\mathbf{1},\mathbf{3})$. We therefore have 19 self-dual $C_{\mu\nu}^+$ and three anti-self-dual $C_{\mu\nu}^-$. There are also eight scalars from the untwisted R–R sector and 16 scalars from the twisted R–R sector. In fact, including the dilaton, there are 105 scalars in total for the type IIB case.

7.6.2 Another miraculous anomaly cancellation

This type IIB spectrum is chiral, as already mentioned, and in view of what we studied in earlier sections, the reader must be wondering whether or not it is anomaly-free. It actually is, and it is a worthwhile exercise to check this, using the polynomials in insert 7.2.

The cancellation is so splendid that we cannot resist explaining it in detail here. To do so we should be careful to understand the $\mathcal{N} = 2$ multiplet structure properly. A sensible non-gravitational multiplet has the same number of bosonic degrees of freedom as fermionic, and so it is possible to readily write out the available ones given what we have already seen. (Or we could simply finish the tensoring done in the last section, doing the NS–R and R–NS parts to get the fermions.) Either way, table 7.1 has the multiplets listed.

The 16 components of the supergravity bosonic multiplet is accompanied by two copies of the 16 components making up a gravitino and a dilatino. These two copies are the same chirality for type IIB and opposite for type IIA.

The next thing to do is to repackage the spectrum we identified earlier in terms of these multiplets. First, notice that the supergravity multiplet has one $(\mathbf{1},\mathbf{1})$, four $(\mathbf{2},\mathbf{1})$s and one $(\mathbf{1},\mathbf{3})$. With four other scalars, we can make a full tensor multiplet. (The other $(\mathbf{3},\mathbf{1})$, which is an anti-self-dual piece makes up the rest of $B_{\mu\nu}$.) That gives us 19 complete self dual

Table 7.1.   *The structure of the $\mathcal{N} = 2$ multiplets in $D = 6$*

multiplet	bosons	fermions
vector	**(2,2)+4(1,1)**	**2(1,2)+2(2,1)**
SD tensor	**(1,3)+5(1,1)**	**4(2,1)**
ASD tensor	**(3,1)+5(1,1)**	**4(2,1)**
supergravity	$(\mathbf{3},\mathbf{3}) + (\mathbf{3},\mathbf{1}) + (\mathbf{1},\mathbf{3}) + (\mathbf{1},\mathbf{1})$	$2(\mathbf{3},\mathbf{2}) + 2(\mathbf{2},\mathbf{1})$ or $2(\mathbf{2},\mathbf{3}) + 2(\mathbf{1},\mathbf{2})$

tensor multiplets in total and two complete anti-self-dual ones since the last one is not complete. Since there are five scalars in a tensor multiplet this accounts for the 105 scalars that we have.

So we can study the anomaly now, knowing what (anti-)self-dual tensors, and fermions we have. Consulting insert 7.2 (p. 162), we note that the polynomials listed for the fermions are for complex fermions, and so we must divide them by two to get the ones appropriate for the real components we have counted in the orbifolding. Putting it together according to what we have said above for the content of the spectrum, we have:

$$19\hat{I}_8^{(1,3)} + 19 \times 4\hat{I}_8^{(2,1)} + 2\hat{I}_8^{(3,1)} + 2 \times 4\hat{I}_8^{(2,1)} + 2\hat{I}_8^{(3,2)} + \hat{I}_8^{(3,1)} = 0, \quad (7.52)$$

where we have listed, respectively, the contribution of the 19 self-dual tensors, the two anti-self-dual tensors, the two gravitinos, and the remaining piece of the supergravity multiplet. That this combination of polynomials vanishes is amazing[109].

7.6.3 The K3 manifold

Quite remarkably, there is a geometrical interpretation of all of those data presented in the previous subsections in terms of compactifying type II string theory on a smooth manifold. The manifold is K3. It is a four dimensional manifold containing 22 independent two-cycles, which are topologically two-spheres more properly described as the complex surface $\mathbb{CP}^1$ (see insert 16.1), in this context. Correspondingly the space of two-forms which can be integrated over these two cycles is 22 dimensional. So we can choose a basis for this space. Nineteen of them are self-dual and three of them are anti-self-dual, in fact. The space of metrics on K3 is in fact parametrised by 58 numbers.

In compactifying the type II superstrings on K3, the ten dimensional gravity multiplet and the other R–R fields gives rise to six dimensional fields by direct dimensional reduction, while the components of the fields in the K3 give other fields. The six dimensional gravity multiplet arises by direct reduction from the NS–NS sector, while 58 scalars arise, parametrising the 58 dimensional space of K3 metrics which the internal parts of the metric, G_{mn}, can choose. Correspondingly, there are 22 scalars arising from the 19+3 ways of placing the internal components of the antisymmetric tensor, B_{mn} on the manifold. A commonly used terminology is that the form has been 'wrapped' on the 22 two-cycles to give 22 scalars.

In the R–R sector of type IIB, there is one scalar in ten dimensions, which directly reduces to a scalar in six. There is a two-form, which produces 22 scalars, in the same way as the NS–NS two-form did. The self-dual four-form can be integrated over the 22 two cycles to give 22 two forms in six dimensions, 19 of them self-dual and three anti-self-dual. Finally, there is an extra scalar from wrapping the four-form entirely on K3. This is precisely the spectrum of fields which we computed directly in the type IIB orbifold.

Alternatively, while the NS–NS sector of type IIA gives rise to the same fields as before, there is in the R–R sector a one-form, three-form and five-form. The one-form directly reduces to a one-form in six dimensions. The three-form gives rise to 22 one-forms in six dimensions while the five-form gives rise to a single one-form. We therefore have 24 one-forms (generically carrying a $U(1)$ gauge symmetry) in six dimensions. This also completes the smooth description of the type IIA on K3 spectrum, which we computed directly in the orbifold limit. See insert 7.5 for a significant comment on this spectrum.

7.6.4 Blowing up the orbifold

The connection between the orbifold and the smooth K3 manifold is as follows[78]: K3 does indeed have a geometrical limit which is $T^4/\mathbb{Z}_2$, and it can be arrived at by tuning enough parameters, which corresponds here to choosing the vev's of the various scalar fields. Starting with the $T^4/\mathbb{Z}_2$, there are 16 fixed points which look locally like $\mathbb{R}^4/\mathbb{Z}^2$, a singular point of infinite curvature. It is easy to see where the 58 geometric parameters of the K3 metric come from in this case. Ten of them are just the symmetric G_{mn} constant components, on the internal directions. This is enough to specify a torus T^4, since the hypercube of the lattice in $\mathbb{R}^4$ is specified by the ten angles between its unit vectors, $\mathbf{e}^m \cdot \mathbf{e}^n$. Meanwhile each of the 16 fixed points has three scalars associated to its metric geometry. (The remaining fixed point NS–NS scalar in the table is from the field B, about which we will have more to say later.)

The three metric scalars can be tuned to resolve or 'blow-up' the fixed point, and smooth it out into the $\mathbb{CP}^1$ which we mentioned earlier. (This accounts for 16 of the two-cycles. The other six correspond to the six $\mathbb{Z}_2$

Insert 7.5.   Anticipating a string–string duality in $D = 6$

We have seen that for type IIA we have an $\mathcal{N} = 2$, $D = 6$ supergravity with 80 additional scalars and 24 gauge bosons with a generic gauge group $U(1)^{24}$. The attentive reader will have noticed an apparent coincidence between the result for the spectrum of type IIA on K3 and another six dimensional spectrum which we obtained earlier. That was the spectrum of the heterotic string compactified on T^4, obtained in section 7.4 (put $d = 4$ in the results there). The moduli space of compactifications is in fact

$$O(20, 4, \mathbb{Z}) \backslash O(20, 4) / [O(20) \times O(4)]$$

on both sides. We have seen where this comes from on the heterotic side. On the type IIA side it arises too. Start with the known

$$O(19, 3, \mathbb{Z}) \backslash O(19, 3) / [O(19) \times O(3)]$$

for the standard moduli space of K3s (you should check that this has 57 parameters; there is an additional one for the volume). It acts on the 19 self-dual and three anti-self-dual two-cycles. This classical geometry is supplemented by stringy geometry arising from $B_{\mu\nu}$, which can have fluxes on the 22 two-cycles, giving the missing 22 parameters. We will not prove here that the moduli space is precisely as above, and hence the same as globally and locally as the heterotic one, but it will become apparent later in chapters 12 and 16.

Perturbatively, the coincidence of the spectra must be an accident. The two string theories in $D = 10$ are extremely dissimilar. One has twice the supersymmetry of the other and is simpler, having no large gauge group, while the other is chiral. We place the simpler theory on a complicated space (K3) and the more complex theory on a simple space T^4 and result in the same spectrum. The theories cannot be T-dual since the map would have to mix things which are unrelated by properties of circles. The only duality possible would have to go beyond perturbation theory. This is what we shall see later in chapter 16. Note also that there is something missing. At special points in the heterotic moduli space we have seen that it is possible to get large enhanced non-Abelian gauge groups. There is no sign of that here in how we have described the type IIA string theory using conformal field theory. In fact, we shall see how to go beyond conformal field theory and describe these special points using D-branes in chapter 13.

invariant forms $dX^m \wedge dX^n$ on the four-torus.) The smooth space has a known metric, the 'Eguchi–Hanson' metric[84], which is *locally* asymptotic to $\mathbb{R}^4$ (like the singular space) but with a global $\mathbb{Z}_2$ identification. Its metric is:

$$ds^2 = \left(1 - \left(\frac{a}{r}\right)^4\right)^{-1} dr^2 + r^2 \left(1 - \left(\frac{a}{r}\right)^4\right) \sigma_3^2 + r^2(\sigma_1^2 + \sigma_2^2), \quad (7.53)$$

where the σ_i are defined in terms of the S^3 Euler angles (θ, ϕ, ψ) in insert 7.4. From there we learn that $4(\sigma_1^2 + \sigma_2^2) = d\theta^2 + \sin^2\theta d\phi^2$. The point $r = a$ is an example of a 'bolt' singularity. Near there, the space is topologically $\mathbb{R}^2_{r\psi} \times S^2_{\theta\phi}$, with the S^2 of radius $a/2$, and the singularity is a coordinate one provided ψ has period 2π. (See insert 7.6, (p. 188).) Since on S^3, ψ would have period 4π, the space at infinity is $S^3/\mathbb{Z}_2$, just like an $\mathbb{R}^4/\mathbb{Z}_2$ fixed point. For small enough a, the Eguchi–Hanson space can be neatly slotted into the space left after cutting out the neighbourhood of the fixed point. The bolt is in fact the $\mathbb{CP}^1$ of the blow-up mentioned earlier. The parameter a controls the size of the $\mathbb{CP}^1$, while the other two parameters correspond to how the $\mathbb{R}^2$ (say) is oriented in $\mathbb{R}^4$.

The Eguchi–Hanson space is the simplest example of an 'Asymptotically Locally Euclidean' (ALE) space, which K3 can always be tuned to resemble locally. These spaces are classified[85] according to their identification at infinity, which can be any discrete subgroup[86], Γ, of the $SU(2)$ which acts on the S^3 at infinity, to give S^3/Γ. These subgroups have been characterised by McKay[87], and have an A–D–E classification which we shall study more in chapter 13. The metrics on the A–series are known explicitly as the Gibbons–Hawking metrics[91], which we shall display later, and Eguchi–Hanson is in fact the simplest of this series, corresponding[92] to A_1. We shall later use a D-brane as a probe of string theory on a $\mathbb{R}^4/\mathbb{Z}_2$ orbifold, an example which will show that the string theory correctly recovers all of the metric data (7.53) of these fixed points, and not just the algebraic data we have seen here.

For completeness, let us compute one more thing about K3 using this description. The Euler characteristic, in this situation, can be written in two ways[82]

$$\chi(K3) = \frac{1}{32\pi^2} \int_{K3} \sqrt{g} \left(R_{abcd}R^{abcd} - 4R_{ab}R^{ab} + R^2\right)$$

$$= \frac{1}{32\pi^2} \int_{K3} \sqrt{g}\epsilon_{abcd} R^{ab} R^{cd}$$

$$= -\frac{1}{16\pi^2} \int_{K3} \text{Tr} R \wedge R = 24. \quad (7.54)$$

Even though no explicit metric for K3 has been written, we can compute χ as follows[80, 82]. If we take a manifold M, divide by some group G, remove

---

Insert 7.6.   A closer look at the Eguchi–Hanson space

Let us establish some of the properties claimed in the main body of the text, while uncovering a useful technique. The S^3s in the metric (7.53) are the natural 3D 'orbits' of the $SU(2)$ action. The S^2 of (θ, ϕ) is a special 2D 'invariant submanifold'. To examine the potential singularity at $r = a$, look *near* $r = a$. Choose, if you will, $r = a + \varepsilon$ for small ε, and:

$$ds^2 = \frac{a}{4\varepsilon}\left[d\varepsilon^2 + \frac{16\varepsilon^2}{4}(d\psi + \cos\theta d\phi)^2\right] + \frac{1}{4}(a^2 + 2a\varepsilon)d\Omega_2^2,$$

which as $\varepsilon \to 0$ is obviously topologically looking locally like $\mathbb{R}^2_{\varepsilon,\psi} \times S^2_{\theta,\phi}$, where the S^2 is of radius $a/2$. (Globally, there is a fibred structure due to the $d\psi d\phi$ cross term.) Incidentally, this is perhaps the quickest way to see that the Euler number or 'Euler charachteristic' of the space has to be equal to that of an S^2, which is two. There is a potential 'bolt' singularity at $r = a$. It is a true singularity for arbitrary choices of periodicity $\Delta\psi$ of ψ, since there is a conical deficit angle in the plane. In other words, we have to ensure that as we get to the origin of the plane, $\varepsilon = 0$, the ψ-circles have circumference 2π, no more or less. Infinitesimally, we make those measures with the metric, and so the condition is:

$$2\pi = \lim_{\varepsilon \to 0}\left(\frac{d(\sqrt{a}\varepsilon^{1/2})\Delta\psi}{d\varepsilon\sqrt{(a/4)}\varepsilon^{-1/2}}\right),$$

which gives $\Delta\psi = 2\pi$. So in fact, we must spoil our S^3 which was a nice orbit of the $SU(2)$ isometry, by performing an $\mathbb{Z}_2$ identification on ψ, giving it half its usual period. In this way, the 'bolt' singularity $r = a$ is just a harmless artifact of coordinates[83, 82]. Also, we are left with an $SO(3) = SU(2)/\mathbb{Z}_2$ isometry of the metric. The space at infinity is $S^3/\mathbb{Z}_2$.

---

some fixed point set F, and add in some set of new manifolds N, one at each point of F, the Euler characteristic of the new manifold is

$$\chi = \frac{\chi(M) - \chi(F)}{|G|} + \chi(N). \tag{7.55}$$

Here, $G = \mathbf{R} \equiv \mathbb{Z}_2$, and the Euler characteristic of the Eguchi–Hanson space is equal to two, from insert 7.6 (p. 188). That of a point is one, and of the torus is zero. We therefore get

$$\chi(K3) = -\frac{16}{2} + 16 \times 2 = 24, \tag{7.56}$$

which will be of considerable use later on.

So we have constructed the consistent, supersymmetric string propagation on the K3 manifold, using orbifold techniques. We shall use this manifold to illustrate a number of beautiful properties of D-branes and string theory in the rest of these lectures.

7.6.5 Some other K3 orbifolds

We can construct K3 in its orbifold limits using other $\mathbb{Z}_N$ group actions. We begin with the space $\mathbb{R}^4 \equiv \mathbb{C}^2$, with complex coordinates $z^1 = x^6 + ix^7$ and $z^2 = x^8 + ix^9$, upon which we make the identifications $z^i \sim z^i + 1 \sim z^i + i$, for $N=2$ or 4, and $z^i \sim z^i + 1 \sim z^i + \exp(\pi i/3)$ for $N=3$ or 6. These lattices define for us the torus T^4, upon which the discrete rotations $\mathbb{Z}_N$, acts naturally as

$$(z^1, z^2) \rightarrow (\beta z^1, \beta^{-1} z^2), \tag{7.57}$$

for $\beta = \exp(2\pi i/N)$.

We may therefore define a new space by identifying points under the action of $\mathbb{Z}_N$. This is the orbifold $T^4/\mathbb{Z}_N$, which is a smooth surface except at fixed points, which are invariant under $\mathbb{Z}_N$ or some non-trivial subgroup of it. For $N \in \{2, 3, 4, 6\}$, this procedure produces a family of compact spaces which are also orbifold limits of the K3 surface.

The smooth K3 manifold is obtained from these limits by blowing up the orbifold points, removing each of the points and replacing it by a smooth space, just as we did in the previous section. The neighbourhood of a fixed point is $\mathbb{R}^4/\mathbb{Z}_M$, where $N \geq M \in \{2, 3, 4, 6\}$, which is the asymptotic region of the A–series ALE space with which we replace the excised point. Note that the Euler characteristic of the A_n ALE space is $n + 1$.

Let us denote the generator of $\mathbb{Z}_N$ by α_N The group elements are then the powers α_N^m, for $m \in \{0, 1, \ldots, N - 1\}$. In fact the number, F_M, of points fixed under the $\mathbb{Z}_M$ subgroup of $\mathbb{Z}_N$, (generated by $\alpha_N^{N/M}$) is simply $F_M = 16 \sin^4 \frac{\pi}{M}$, where M is a divisor of N.

For $T^4/\mathbb{Z}_2$, as we have already seen, we have 16 points fixed under the action of α_2, each of which are replaced by the A_1 ALE space in order to resolve to smooth K3. For $T^4/\mathbb{Z}_3$ there are nine fixed points of α_3, which are each replaced by the A_2 ALE space to make the blow-up.

From formula (7.55), we get

$$\chi(K3) = -\frac{9}{3} + 9 \times 3 = 24.$$

The case $T^4/\mathbb{Z}_4$ has 16 fixed points. Four of them are fixed under the action of α_4, while the other 12 are only fixed under α_4^2. Under α_4, these 12 $\mathbb{Z}_2$ points transform as six doublets. Consequently, the blow-up is carried out by first constructing the $\mathbb{Z}_4$-invariant region by identifying these pairs of fixed points. One can then replace each of the original four $\mathbb{Z}_4$ fixed points by an A_3 ALE space and the six pairs by an A_1. From formula (7.55), we get

$$\chi(K3) = -\frac{16}{4} + 4 \times 4 + 6 \times 2 = 24.$$

For $T^4/\mathbb{Z}_6$ the situation is similar. There are 24 fixed points altogether. There is only one point fixed under α_6. It is replaced by the A_5 ALE space to make the blow-up. There are eight points fixed under the $\mathbb{Z}_3$ subgroup, generated by α_6^2, which transform as doublets under the action of α_6. They are therefore replaced by four copies of the A_2 ALE space. There are 15 points fixed under α_6^3, which transform as triplets under the action of α_6. Consequently, they are replaced by five copies of the A_1 space in performing the blow-up surgery. Once again, we get the correct value of the Euler number:

$$\chi(K3) = -\frac{24}{6} + 5 \times 2 + 4 \times 3 + 1 \times 6 = 24.$$

We can go a lot further and recover other geometric properties of the K3 in each case. For example, as we shall see later in chapter 13, the A_n ALE space is generically like $n + 1$ $\mathbb{CP}^1$s (i.e. S^2s) intersecting in a particular pattern. There is in fact a self-dual cycle associated to n of these. So its contribution to the K3s count of $(19, 3)$ cycles is $(n, 0)$. It s

Table 7.2. *Recovering some properties of the K3 geometry in orbifold limits*

case	T^4 parameters	ALE parameters	T^4 forms	ALE forms
$\mathbb{Z}_2$	10	$16 \times 3 = 48$	(3,3)	$16 \times (1,0)$
$\mathbb{Z}_3$	4	$18 \times 3 = 54$	(1,3)	$9 \times (2,0)$
$\mathbb{Z}_4$	4	$18 \times 3 = 54$	(1,3)	$6 \times (1,0) + 4 \times (3,0)$
$\mathbb{Z}_6$	4	$18 \times 3 = 54$	(1,3)	$(5,0) + 5 \times (1,0) + 4 \times (2,0)$

useful to combine this with the contribution from the torus to compute the result for K3, and table 7.2 has a list of the arithmetic in each case. The origin of the 58 metric parameters can similarly be computed, using the fact that some come from the torus and some from the parameters (three for each $\mathbb{CP}^1$ in fact) of the ALE spaces. This is also given in table 7.2. We've listed the $\mathbb{Z}_2$ case which we already computed in the previous subsection. Notice that it is in some sense more special than the others. In both forms and metric parameters, the bare torus contributes more than in the other cases. This is because it is more symmetric than the others. This is traceable to the fact that the T^4 is written naturally in terms of the complex parameters $z_1 = x_6 + ix_7$ and $z_2 = x_8 + ix_9$, and the form of the action on it is given by equation (7.57). It is only for $\mathbb{Z}_2$ that $\beta = 1/\beta$, and thus there is more symmetry between the x^ms.

Therefore of the 6 forms (made from $dx^m \wedge dx^n$) and 10 scalars one can make, only four survive in each non-$\mathbb{Z}_2$ case. (This can be worked out most easily by working directly with z_1 and z_2. Then the forms are $dz_1 \wedge dz_2$, $d\bar{z}_1 \wedge d\bar{z}_2$, etc., but, for example, $dz_1 \wedge d\bar{z}_2$ is clearly not invariant since it transforms as β^2.)

7.6.6 Anticipating D-manifolds

We've just made some traditional superstring compactifications by including in the internal space the pure geometry of K3, resulting in a six dimensional vacuum. Later we will see that it is possible to construct a whole new class of string 'compactification' vacua by including D-branes in the spectrum in such a way that their contribution to spacetime anomalies, etc., combines with that of the pure geometry in a way that is crucial to the consistency of the model. This gives the idea of a 'D-manifold'[116].

An analogue of the orbifold method for making these supersymmetric vacua is the generalised 'orientifold' construction already mentioned. There are constructions of 'K3 orientifolds' which follow the ideas presented in this section, combined with D-brane orbifold techniques to be developed in chapter 14[131]. We shall also encounter K3 in its orbifold limits in chapter 16, where we use our knowledge gained here to explore properties of remarkable non-perturbative type IIB vacua made using F-theory. D-branes will be present there too, but in a somewhat different way.

8

Supersymmetric strings and T-duality

8.1 T-duality of supersymmetric strings

We noticed in section 7.5, when considering the low energy spectrum of the type II superstrings compactified on tori, that there is an equivalence between them. We saw much the same things happen for the heterotic strings in section 7.4 too. This is of course T-duality, as we should examine it further here and check that it is the familiar exact equivalence. Just as in the case of bosonic strings, doing this when there are open string sectors present will uncover D-branes of various dimensions.

8.1.1 T-duality of type II superstrings

T-duality on the closed oriented Type II theories has a somewhat more interesting effect than in the bosonic case[12, 8]. Consider compactifying a single coordinate X^9, of radius R. In the $R \to \infty$ limit the momenta are $p_R^9 = p_L^9$, while in the $R \to 0$ limit $p_R^9 = -p_L^9$. Both theories are $SO(9,1)$ invariant but under *different* $SO(9,1)$s. T-duality, as a right-handed parity transformation (see (4.18)), reverses the sign of the right-moving $X^9(\bar{z})$; therefore by superconformal invariance it does so on $\psi^9(\bar{z})$. Separate the Lorentz generators into their left- and right-moving parts $M^{\mu\nu} + \widetilde{M}^{\mu\nu}$. Duality reverses all terms in $\widetilde{M}^{\mu 9}$, so the $\mu 9$ Lorentz generators of the T-dual theory are $M^{\mu 9} - \widetilde{M}^{\mu 9}$. In particular this reverses the sign of the helicity $\tilde{s}_4$ and so switches the chirality on the right-moving side. If one starts in the IIA theory, with opposite chiralities, the $R \to 0$ theory has the same chirality on both sides and is the $R \to \infty$ limit of the IIB theory, and vice versa. In short, T-duality, as a one-sided spacetime parity operation, reverses the relative chiralities of the right- and left-moving ground states. The same is true if one dualises on any odd number of dimensions, whilst dualising on an even number returns the original type II theory.

Since the IIA and IIB theories have different R–R fields, T_9 duality must transform one set into the other. The action of duality on the spin fields is of the form

$$S_\alpha(z) \to S_\alpha(z), \qquad \tilde{S}_\alpha(\bar{z}) \to P_9 \tilde{S}_\alpha(\bar{z}) \tag{8.1}$$

for some matrix P_9, the parity transformation (nine-reflection) on the spinors. In order for this to be consistent with the action $\tilde{\psi}^9 \to -\tilde{\psi}^9$, P_9 must anticommute with Γ^9 and commute with the remaining Γ^μ. Thus $P_9 = \Gamma^9 \Gamma^{11}$ (the phase of P_9 is determined, up to sign, by hermiticity of the spin field). Now consider the effect on the R–R vertex operators (7.27). The Γ^{11} just contributes a sign, because the spin fields have definite chirality. Then by the Γ-matrix identity (7.28), the effect is to add a 9-index to G if none is present, or to remove one if it is. The effect on the potential C ($G = dC$) is the same. Take as an example the type IIA vector C_μ. The component C_9 maps to the IIB scalar C, while the $\mu \neq 9$ components map to $C_{\mu 9}$. The remaining components of $C_{\mu\nu}$ come from $C_{\mu\nu 9}$, and so on.

Of course, these relations should be translated into rules for T-dualising the spacetime fields in the supergravity actions (7.41) and (7.42). The NS–NS sector fields' transformations are the same as those shown in equations (5.4),(5.6), while for the R–R potentials[77]:

$$\tilde{C}^{(n)}_{\mu\cdots\nu\alpha 9} = C^{(n-1)}_{\mu\cdots\nu\alpha} - (n-1)\frac{C^{(n-1)}_{[\mu\cdots\nu|9}G_{|\alpha]9}}{G_{99}} \tag{8.2}$$

$$\tilde{C}^{(n)}_{\mu\cdots\nu\alpha\beta} = C^{(n+1)}_{\mu\cdots\nu\alpha\beta 9} + nC^{(n-1)}_{[\mu\cdots\nu\alpha}B_{\beta]9} + n(n-1)\frac{C^{(n-1)}_{[\mu\cdots\nu|9}B_{|\alpha|9}G_{|\beta]9}}{G_{99}}.$$

8.1.2 T-duality of type I superstrings

Just as in the case of the bosonic string, the action of T-duality in the open and unoriented open superstring theory produces D-branes and orientifold planes. Having done it once (say on X^9 with radius R), we get a T_9-dual theory on the line interval $S^1/\mathbb{Z}_2$, where $\mathbb{Z}_2$ acts as the reflection $X^9 \to -X^9$. The S^1 has radius $R' = \alpha'/R$). There are 16 D8-branes and their mirror images (coming from the 16 D9-branes), together with two orientifold O8-planes located at $X^9 = 0, \pi R'$. This is called the 'type I'' theory (and sometimes the 'type IA' theory, and then the usual open string is 'type IB'), about which we will have more to say later as well.

Starting with the type IB theory, i.e. 16 D9-branes and one O9-plane, we can carry this out n times on n directions, giving us 16 D$(9-n)$ and their mirror images through 2^n O$(9-n)$-planes arranged on the hypercube of fixed points of $T^n/\mathbb{Z}_2$, where the $\mathbb{Z}_2$ acts as a reflection in the n directions. If n is odd, the bulk theory away from the planes and branes is type IIA string theory, while we are back in type IIB otherwise.

Let us focus here on a single D-brane, taking a limit in which the other D-branes and the O-planes are very far away and can be ignored. Away from the D-brane, only closed strings propagate. The local physics is that of the type II theory, with two gravitinos. This is true even though we began with the unoriented type I theory which has only a single gravitino. The point is that the closed string begins with two gravitinos, one with the spacetime supersymmetry on the right-moving side of the world-sheet and one on the left. The orientation projection of the type I theory leaves one linear combination of these. However in the T-dual theory, the orientation projection does not constrain the local state of the string, but relates it to the state of the (distant) image gravitino. Locally there are two independent gravitinos, with equal chiralities if n, (the number of dimensions on which we dualised) is even and opposite if n is odd.

This is all summarised nicely by saying that while the type I string theory comes from projecting the type IIB theory by Ω, the T-dual string theories come from projecting type II string theory compactified on the torus T^n by $\Omega \prod_m [R_m(-1)^F]$, where the product over m is over all the n directions, and R_m is a reflection in the mth direction. This is indeed a symmetry of the theory and hence a good symmetry with which to project. So we have that T-duality takes the orientifold groups into one another:

$$\{\Omega\} \leftrightarrow \{1, \Omega \prod_m [R_m(-1)^F]\}. \tag{8.3}$$

This is a rather trivial example of an orientifold group, since it takes type II strings on the torus T^n and simply gives a theory which is simply related to type I string theory on T^n by n T-dualities. Nevertheless, it is illustrative of the general constructions of orientifold backgrounds made by using more complicated orientifold groups. This is a useful piece of technology for constructing string backgrounds with interesting gauge groups, with fewer symmetries, as a starting point for phenomenological applications.

8.1.3 T-duality for the heterotic strings

As we noticed in section 7.4, there is a T-duality equivalence between the heterotic strings once we compactify on a circle. Let us uncover it carefully.

We can begin by compactifying the $SO(32)$ string on a circle of radius R, with Wilson line:

$$A_9^i = \frac{1}{2\pi R}\text{diag}\left\{\frac{1}{2}, \ldots \frac{1}{2}, 0, \ldots, 0\right\}, \tag{8.4}$$

with eight $\frac{1}{2}$s and eight 0s breaking down the gauge group to $SO(16) \times SO(16)$. We can compute the mass spectrum of the nine dimensional theory which results from this reduction, in the presence of the Wilson line. This is no harder than the computations which we did in chapter 4. The Wilson line simply shifts the contribution to the spectrum coming from the p_L^i momenta. We can focus on the sector which is uncharged under the gauge group, i.e. we switch off the p_L^i. The mass formula is:

$$p_{\substack{L\\R}} = \frac{(n + 2m)}{R} \pm \frac{2mR}{\alpha'},$$

where we see that the allowed windings (coming in units of two) are controlled by the integer m, and the momenta are controlled by m and n in the combination $n + 2m$.

We could instead have started from the $E_8 \times E_8$ string on a circle of radius R', with Wilson line

$$A_9^i = \frac{1}{2\pi R'}\text{diag}\{1, 0 \ldots 0, 1, 0, \ldots, 0\}, \tag{8.5}$$

again in two equal blocks of eight. This also breaks down the gauge group to $SO(16) \times SO(16)$. A computation of the spectrum of the neutral states gives:

$$p'_{\substack{L\\R}} = \frac{(n' + 2m')}{R'} \pm \frac{2m'R'}{\alpha'},$$

for integers n' and m'. We see that if we exchange $n + 2m$ with m' and m with $n' + 2m'$ then the spectrum is invariant if we do the right handed parity identification $p_L \leftrightarrow p'_L$, $p_R \leftrightarrow -p'_R$, provided that the circles' radii are inversely related as $R' = \alpha'/(2R)$.

We shall see that this relation will result in some very remarkable connections between non-perturbative string vacua much later, in chapters 12 and 16.

8.2 D-branes as BPS solitons

Let us return to the type II strings, and the D-branes which we can place in them. While there is type II string theory in the bulk (i.e. away from the branes and orientifolds), notice that the open string boundary conditions are invariant under only one supersymmetry. In the original type I theory, the left-moving world-sheet current for spacetime supersymmetry $j_\alpha(z)$ flows into the boundary and the right-moving current $\tilde{j}_\alpha(\bar{z})$ flows out, so only the total charge $Q_\alpha + \tilde{Q}_\alpha$ of the left- and right-movers is conserved. Under T-duality this becomes

$$Q_\alpha + (\textstyle\prod_m P_m)\, \tilde{Q}_\alpha, \tag{8.6}$$

where the product of reflections P_m runs over all the dualised dimensions, that is, over all directions orthogonal to the D-brane. Closed strings couple to open, so the general amplitude has only one linearly realised supersymmetry. That is, the vacuum without D-branes is invariant under $N = 2$ supersymmetry, but the state containing the D-brane is invariant under only $N = 1$: *it is a BPS state*[265, 93].

BPS states must carry conserved charges. In the present case there is only one set of charges with the correct Lorentz properties, namely the antisymmetric R–R charges. The world volume of a p-brane naturally couples to a $(p + 1)$-form potential $C_{(p+1)}$, which has a $(p + 2)$-form field strength $G_{(p+2)}$. This identification can also be made from the $g_{\rm s}^{-1}$ behaviour of the D-brane tension: this is the behaviour of an R–R soliton[94, 96] as will be developed further later.

The IIA theory has Dp-branes for $p = 0$, 2, 4, 6, and 8. The vertex operators (7.27) describe field strengths of all even ranks from zero to ten. The n-form and $(10 - n)$-form field strengths are Hodge dual to one another*, so a p-brane and $(6 - p)$-brane are sources for the same field, but one magnetic and one electric. The field equation for the ten-form field strength allows no propagating states, but the field can still have a physically significant energy density [265, 97, 98].

The IIB theory has Dp-branes for $p = -1$, 1, 3, 5, 7, and 9. The vertex operators (7.27) describe field strengths of all odd ranks from one to nine, appropriate to couple to all but the nine-brane. The nine-brane does couple to a non-trivial *potential,* as we will see below.

A (-1)-brane is a Dirichlet instanton, defined by Dirichlet conditions in the time direction as well as all spatial directions[99]. Of course, it is not clear that T-duality in the time direction has any meaning, but one can argue for the presence of (-1)-branes as follows. Given zero-branes in the IIA theory, there should be virtual zero-brane world-lines that wind in a purely spatial direction. Such world-lines are required by quantum mechanics, but note that they are essentially instantons, being localised in time. A T-duality in the winding direction then gives a (-1)-brane. One

---

* This works at the level of vertex operators via a Γ-matrix identity.

of the first clues to the relevance of D-branes[25], was the observation that D-instantons, having action g_s^{-1}, would contribute effects of order e^{-1/g_s} as expected from the behaviour of large orders of string perturbation theory[100].

The D-brane, unlike the fundamental string, carries R–R charge. This is consistent with the fact that they are BPS states, and so there must be a conserved charge. A more careful argument, involving the R–R vertex operators, can be used to show that they *must* couple thus, and furthermore that fundamental strings cannot carry R–R charges (see also insert 8.1).

8.3 The D-brane charge and tension

The discussion of section 5.3 will supply us with the world-volume action (5.21) for the bosonic excitations of the D-branes in this supersymmetric context. Now that we have seen that Dp-branes are BPS states, and couple to R–R sector $(p+1)$-form potential, we ought to compute the values of their charges and tensions.

Focusing on the R–R sector for now, supplementing the spacetime supergravity action with the D-brane action we must have at least (recall that the dilaton will not appear here, and also that we cannot write this for $p = 3$):

$$S = -\frac{1}{2\kappa_0^2} \int G_{(p+2)}{}^*G_{(p+2)} + \mu_p \int_{\mathcal{M}_{p+1}} C_{(p+1)}, \qquad (8.7)$$

where μ_p is the charge of the Dp-brane under the $(p+1)$-form $C_{(p+1)}$. $\mathcal{M}_{p+1}$ is the world-volume of the Dp-brane.

Now the same vacuum cylinder diagram as in the bosonic string, as we did in chapter 6. With the fermionic sectors, our trace must include a sum over the NS and R sectors, and furthermore must include the GSO projection onto even fermion number. Formally, therefore, the amplitude looks like[265]:

$$A = \int_0^\infty \frac{dt}{2t} \text{Tr}_{\text{NS+R}} \left\{ \frac{1 + (-1)^F}{2} e^{-2\pi t L_0} \right\}. \qquad (8.8)$$

Performing the traces over the open superstring spectrum gives

$$A = 2V_{p+1} \int \frac{dt}{2t} (8\pi^2\alpha't)^{-(p+1)/2} e^{-t\frac{Y^2}{2\pi\alpha'}}$$

$$\frac{1}{2} f_1^{-8}(q) \left\{ -f_2(q)^8 + f_3(q)^8 - f_4(q)^8 \right\}, \qquad (8.9)$$

where again $q = e^{-2\pi t}$, and we are using the definitions given in chapter 4, when we computed partition functions of various sorts. Insert 14.1, p. 327, uncovers more of the properties of the f-functions.

Insert 8.1.　A summary of forms and branes

Common to both type IIA and IIB are the NS–NS sector fields

$$\Phi, G_{\mu\nu}, B_{\mu\nu}.$$

The latter is a rank two antisymmetric tensor potential, and we have seen that the fundamental closed string couples to it electrically by the coupling

$$\nu_1 \int_{\mathcal{M}_2} B_{(2)},$$

where $\nu_1 = (2\pi\alpha')^{-1}$, $\mathcal{M}_2$ is the world sheet, with coordinates ξ^a, $a = 1, 1$. $B_{(2)} = B_{ab}d\xi^a d\xi^b$, and B_{ab} is the pullback of $B_{\mu\nu}$ via (5.8). By ten dimensional Hodge duality, we can also construct a six form potential $B_{(6)}$, by the relation $dB_{(6)} = *dB_{(2)}$. There is a natural electric coupling $\nu_5 \int_{\mathcal{M}_6} B_{(6)}$, to the world-volume $\mathcal{M}_6$ of a five dimensional extended object. This NS–NS charged object, which is commonly called the 'NS5-brane' is the magnetic dual of the fundamental string[72, 73]. It is in fact, in the ten dimensional sense, the monopole of the $U(1)$ associated to $B_{(2)}$. We shall be forced to discuss it by strong coupling considerations in section 12.3.

The string theory has other potentials, from the R–R sector:

$$\text{type IIA}: \quad C_{(1)}, C_{(3)}, C_{(5)}, C_{(7)}$$
$$\text{type IIB}: \quad C_{(0)}, C_{(2)}, C_{(4)}, C_{(6)}, C_{(8)}$$

where in each case the last two are Hodge duals of the first two, and $C_{(4)}$ is self dual. (A p-form potential and a rank q-form potential are Hodge dual to one another in D dimensions if $p + q = D - 2$.)

Dp-branes are the basic p-dimensional extended sources which couple to all of these via an electric coupling of the form:

$$\mu_p \int_{\mathcal{M}_{p+1}} C_{(p+1)}$$

to their $p + 1$-dimensional world volumes $\mathcal{M}_{p+1}$.

The three terms in the braces come from the open string R sector with $\frac{1}{2}$ in the trace, from the NS sector with $\frac{1}{2}$ in the trace, and the NS sector with $\frac{1}{2}(-1)^F$ in the trace; the R sector with $\frac{1}{2}(-1)^F$ gives no net contribution. In fact, these three terms sum to zero by Jacobi's abstruse identity (*'aequatio identico satis abstrusa'*, see insert 14.2, p. 328) as they ought to since the open string spectrum is supersymmetric, and we are computing a vacuum diagram.

What does this result mean? Recall that this vacuum diagram also represents the exchange of closed strings between two identical branes. the result $\mathcal{A} = 0$ is simply a restatement of the fact that D-branes are BPS states: the net forces from the NS–NS and R–R exchanges cancel. $\mathcal{A} = 0$ has a useful structure, nonetheless, and we can learn more by identifying the separate NS–NS and R–R pieces. This is easy, if we look at the diagram afresh in terms of closed string: In the terms with $(-1)^F$, the world-sheet fermions are *periodic* around the cylinder thus correspond to R–R exchange. Meanwhile the terms without $(-1)^F$ have *antiperiodic* fermions and are therefore NS–NS exchange.

Obtaining the $t \to 0$ behaviour as before (use the limits in insert 6.2 (p. 145)) gives

$$
\begin{aligned}
\mathcal{A}_{\mathrm{NS}} = -\mathcal{A}_{\mathrm{R}} &\sim \frac{1}{2} V_{p+1} \int \frac{dt}{t} (2\pi t)^{-(p+1)/2} (t/2\pi\alpha')^4 e^{-t\frac{Y^2}{8\pi^2\alpha'^2}} \\
&= V_{p+1} 2\pi (4\pi^2\alpha')^{3-p} G_{9-p}(Y^2).
\end{aligned}
\tag{8.10}
$$

Comparing with field theory calculations runs just as it did in chapter 6, with the result[265]:

$$
2\kappa_0^2 \mu_p^2 = 2\kappa^2 \tau_p^2 = 2\pi (4\pi^2\alpha')^{3-p}.
\tag{8.11}
$$

Finally, using the explicit expression (7.44) for κ in terms of string theory quantities, we get an extremely simple form for the charge:

$$
\mu_p = (2\pi)^{-p} \alpha'^{-\frac{(p+1)}{2}}, \quad \text{and} \quad \tau_p = g_{\mathrm{s}}^{-1} \mu_p.
\tag{8.12}
$$

(For consistency with the discussion in the bosonic case, we shall still use the symbol T_p to mean $\tau_p g_{\mathrm{s}}$, in situations where we write the action with the dilaton present. It will be understood then that $e^{-\Phi}$ contains the required factor of g_{s}^{-1}.)

It is worth updating our bosonic formula (5.27) for the coupling of the Yang–Mills theory which appears on the world-volume of Dp-branes with our superstring result above, to give:

$$
g_{\mathrm{YM},p}^2 = \tau_p^{-1} (2\pi\alpha')^{-2} = (2\pi)^{p-2} \alpha'^{(p-3)/2},
\tag{8.13}
$$

a formula we will use a lot in what is to follow.

Note that our formula for the tension (8.12) gives for the D1-brane

$$
\tau_1 = \frac{1}{2\pi\alpha' g_{\mathrm{s}}},
\tag{8.14}
$$

which sets the ratios of the tension of the fundamental string, $\tau_1^F \equiv T = (2\pi\alpha')^{-1}$, and the D–string to be simply the string coupling g_s. This is a very elegant normalisation and is quite natural.

D-branes that are not parallel feel a net force since the cancellation is no longer exact. In the extreme case, where one of the D-branes is rotated by π, the coupling to the dilaton and graviton is unchanged but the coupling to the R–R tensor is reversed in sign. So the two terms in the cylinder amplitude add, instead of cancelling, as Jacobi cannot help us. The result is:

$$\mathcal{A} = V_{p+1} \int \frac{dt}{t} (2\pi t)^{-(p+1)/2} e^{-t(Y^2 - 2\pi\alpha')/8\pi^2\alpha'^2} f(t), \qquad (8.15)$$

where $f(t)$ approaches zero as $t \to 0$. Differentiating this with respect to Y to extract the force per unit world-volume, we get

$$F(Y) = Y \int \frac{dt}{t} (2\pi t)^{-(p+3)/2} e^{-t(Y^2 - 2\pi\alpha')/8\pi^2\alpha'^2} f(t). \qquad (8.16)$$

The point to notice here is that the force diverges as $Y^2 \to 2\pi\alpha'$. This is significant. One would expect a divergence, of course, since the two oppositely charged objects are on their way to annihilating[101]. The interesting feature it that the divergence begins when their separation is of order the string length. This is where the physics of light fundamental strings stretching between the two branes begins to take over. Notice that the argument of the exponential is tU^2, where $U = Y/(2\alpha')$ is the energy of the lightest open string connecting the branes. A scale like U will appear again, as it is a useful guide to new variables to D-brane physics at 'substringy' distances[102, 103, 104] in the limit where α' and Y go to zero.

8.4 The orientifold charge and tension

Orientifold planes also break half the supersymmetry and are R–R and NS–NS sources. In the original type I theory the orientation projection keeps only the linear combination $Q_\alpha + \tilde{Q}_\alpha$. In the T-dualised theory this becomes $Q_\alpha + (\prod_m P_m)\tilde{Q}_\alpha$ just as for the D-branes. The force between an orientifold plane and a D-brane can be obtained from the Möbius strip as in the bosonic case; again the total is zero and can be separated into NS–NS and R–R exchanges. The result is similar to the bosonic result (6.18),

$$\mu_p' = \mp 2^{p-5}\mu_p, \qquad \tau_p' = \mp 2^{p-5}\tau_p, \qquad (8.17)$$

where the plus sign is correlated with $SO(n)$ groups and the minus with $USp(n)$. Since there are 2^{9-p} orientifold planes, the total O-plane charge is $\mp 16\mu_p$, and the total fixed-plane tension is $\mp 16\tau_p$.

8.5 Type I from type IIB, revisited

A non-zero total tension represents a source for the graviton and dilaton, for which the response is simply a time dependence of these background fields[105]. A non-zero total R–R source is more serious, since this would mean that the field equations are inconsistent: there is a violation of Gauss's Law, as R–R flux lines have no place to go in the compact space T^{9-p}. So our result tells us that on T^{9-p}, we need exactly 16 D-branes, with the SO projection, in order to cancel the R–R $G_{(p+2)}$ form charge. This gives the T-dual of $SO(32)$, completing our simple orientifold story.

The spacetime anomalies for $G \neq SO(32)$ (see also section 7.1.3) are thus accompanied by a divergence[107] in the full string theory, as promised, with inconsistent field equations in the R–R sector: as in field theory, the anomaly is related to the ultra-violet limit of a (open string) loop graph. But this ultraviolet limit of the annulus/cylinder ($t \to \infty$) is in fact the infrared limit of the closed string tree graph, and the anomaly comes from this infrared divergence. From the world-sheet point of view, as we have seen in the bosonic case, inconsistency of the field equations indicates that there is a conformal anomaly that cannot be cancelled. This is associated to the presence of a 'tadpole' which is simply an amplitude for creating quanta out of the vacuum with a one-point function, which is a sickness of the theory which must be cured.

The prototype of all of this is the original $D = 10$ type I theory[31]. The N D9-branes and single O9-plane couple to an R–R ten-form, and we can write its action formally as

$$(32 \mp N)\frac{\mu_{10}}{2} \int C_{10}. \tag{8.18}$$

The field equation from varying C_{10} is just $G = SO(32)$.

8.6 Dirac charge quantisation

We are of course studying a quantum theory, and so the presence of both magnetic and electric sources of various potentials in the theory should give some cause for concern. We should check that the values of the charges are consistent with the appropriate generalisation of[114] the Dirac quantisation condition. The field strengths to which a Dp-brane and D$(6 - p)$-brane couple are dual to one another, $G_{(p+2)} = *G_{(8-p)}$.

We can integrate the field strength $*G_{(p+2)}$ on an $(8 - p)$-sphere surrounding a Dp-brane, and using the action (8.7), we find a total flux $\Phi = \mu_p$. We can write $*G_{(p+2)} = G_{(8-p)} = dC_{(7-p)}$ everywhere except on a Dirac 'string' (see also insert 9.2; here it is really a sheet), at the end of which lives the $D(6-p)$ 'monopole'. Then

$$\Phi = \frac{1}{2\kappa_0^2} \int_{S_{8-p}} *G_{(p+2)} = \frac{1}{2\kappa_0^2} \int_{S_{7-p}} C_{(7-p)}, \qquad (8.19)$$

where we perform the last integral on a small sphere surrounding the Dirac string. A $(6-p)$-brane circling the string picks up a phase $e^{i\mu_{6-p}\Phi}$. The condition that the string be invisible is

$$\mu_{6-p}\Phi = 2\kappa_0^2 \mu_{6-p}\mu_p = 2\pi n. \qquad (8.20)$$

The D-branes' charges (8.11) satisfy this condition with the minimum quantum $n = 1$.

While this argument does not apply directly to the case $p = 3$, as the self-dual five-form field strength has no covariant action, the result follows by the T-duality recursion relation (5.11) and the BPS property.

8.7 D-branes in type I

As we saw in section 7.1.3, the only R–R potentials available in type I theory are the two-form and its dual, the 6-form, and so we can have D1-branes in the theory, and D5-branes, which are electromagnetic duals of each other. The overall 16 d9-branes carry an $SO(32)$ gauge group, as we have seen from many points of view. Let us remind ourselves of how this gauge group came about, since there are important subtleties of which we should be mindful[132].

The action of Ω has representation γ_Ω, which acts on the Chan–Paton indices, as discussed in chapter 4:

$$\Omega: \quad |\psi, ij\rangle \longrightarrow (\gamma_\Omega)_{ii'} |\Omega\psi, j'i'\rangle (\gamma_\Omega^{-1})_{j'j},$$

where ψ represents the vertex operator which makes the state in question, and $\Omega\psi$ is the action of Ω on it. The reader should recall that we transposed the indices because Ω exchanges the endpoints of the string. We can consider the square of Ω:

$$\Omega^2: \quad |\psi, ij\rangle \longrightarrow \left[\gamma_\Omega(\gamma_\Omega^T)^{-1}\right]_{ii'} |\psi, i'j'\rangle \left[\gamma_\Omega^T \gamma_\Omega\right]_{j'j}, \qquad (8.21)$$

and so we see that we have the choice

$$\gamma_\Omega^T = \pm\gamma_\Omega.$$

If γ_Ω is symmetric, the with n branes we can write it as $\mathbf{I}_{2n}$, the $2n \times 2n$ identity matrix. Since the 99 open string vertex operator is $\partial_t X^\mu$, it has (as we have seen a lot in chapter 4) $\Omega = -1$. Therefore we do have the symmetric choice since, as we tacitly assumed in equation (8.21) $\Omega^2 = 1$, and so we conclude that the Chan–Paton wavefunction is antisymmetric. Since $n = 16$, we have gauge group $SO(32)$.

If γ_Ω was antisymmetric, then we could have written it as

$$\gamma_\Omega = \begin{pmatrix} 0 & i\mathbf{I}_n \\ -i\mathbf{I}_n & 0 \end{pmatrix},$$

and we would have been able to have gauge group $USp(2n)$. In fact, we shall have to make this choice for D5-branes. Let us see why. Let us place the D5-branes so that they are pointlike in the directions X^m, $m = 6, 7, 8, 9$, and aligned in the directions X^μ, $\mu = 0, 1, \ldots, 5$.

Consider the 5–5 sector, i.e. strings beginning and ending on D5-branes. Again we have $\Omega = -1$ for the vectors $\partial_t X^\mu$, and the opposite sign for the transverse scalars $\partial_n X^m$. In general, other sectors can have different mode expansions. Generically the mode for a fermion is ψ_r and Ω acts on this as $\pm(-1)^r = \pm e^{i\pi r}$ (see chapter 11 for more discussion of these possible modings). In the NS sector they are half-integer and since GSO requires them to act in pairs in vertex operators, their individual $\pm i$s give $\Omega = \pm 1$, with a similar result in the R sector by supersymmetry.

The 59 sector is more subtle[132]. The X^m are now half-integer moded and the ψ^m are integer moded. The ground states of the latter therefore form a representation of the Clifford algebra and we can bosonise them into a spin field, as we did in chapter 7 in a similar situation: $e^{iH_3} \sim \psi^6 + i\psi^7$, and $e^{iH_4} \sim \psi^8 + i\psi^9$. In fact, the vertex operator (the part of it relevant to this discussion) in that sector is

$$V_{59} \sim e^{i(H_3 + H_4)/2}.$$

Now consider the square of this operator. It has parts which are either in the 55 sector or the 99 sector, and is of the form

$$V_{59}^2 \sim e^{i(H_3 + H_4)} \sim (\psi^6 + i\psi^7)_{-1/2}(\psi^8 + i\psi^9)_{-1/2}|0\rangle.$$

So it has $\Omega = -1$, since each $\psi_{-1/2}$ gives $\pm i$. So $\Omega^2 = -1$ for V_{59} for consistency.

Returning to our problem of the choices to make for the Chan–Paton factors we see that we have an extra sign in equation (8.21), and so must choose the antisymmetric condition $\gamma_\Omega^T = -\gamma_\Omega$. Therefore, in type I string theory, n D5-branes have gauge group $USp(2n)$. Notice that this means that a single one has $SU(2)$, and the Chan–Paton wavefunction can be chosen as the Pauli matrices. The Chan–Paton wavefunction for the scalars for transverse motion must simply be δ^{ij}, since we have another sign. This simply means that the two D5-branes (corresponding to the two index choices) are forced to move with each other as one unit.

Notice that this fits rather nicely with our charge quantisation computation of the previous section[132]. The orientifold projection will halve the force between D1-branes and between D5-branes in the charge calculation, and so their effective charges would be reduced by $\sqrt{2}$, violating the Dirac quantisation condition by a factor of a half. However, the fact that the D5-branes are forced to move as a pair restores a factor of two in the quantisation condition, and so we learn that D-branes are still the smallest consistent charge carries of the R–R sector.

We can augment the argument above for Dp branes in type I in general, and obtain[132]

$$\Omega^2 = (\pm i)^{\frac{9-p}{2}}.$$

For $p = 3$ and $p = 7$, we see that simply gives an inconsistency, which is itself consistent with the fact that there is no R–R four-forms or eight-form for a stable D3-brane or D7-brane to couple to. For $p = 1$ we recover the naively expected result that they have an $SO(2n)$ gauge group.

In chapter 14 we shall see that when we combine the orientifold action with other spacetime orbifold symmetries, we can recover extra phase factors by means analogous to what we have uncovered here in order to discover other choices for D5- and D9-brane gauge groups.

9
World-volume curvature couplings

We've now seen that we can construct D-branes which, in superstring theory, have important extra properties. Much of what we have learned about them in the bosonic theory is still true here of course, a key result being that the world-volume dynamics is governed by the dynamics of open strings, etc. Still relevant is the Dirac–Born–Infeld action (equation (5.21)) for the coupling to the background NS–NS fields,

$$S_{\mathrm{DBI}} = -\tau_p \int_{\mathcal{M}_{p+1}} d^{p+1}\xi \, e^{-\Phi} \det^{1/2}(G_{ab} + B_{ab} + 2\pi\alpha' F_{ab}), \qquad (9.1)$$

and the non-Abelian extensions mentioned later in chapter 5.

As we have seen in the previous chapter, for the R–R sector, they are sources of $C_{(p+1)}$. We therefore also have the Wess–Zumino-like term

$$S_{\mathrm{WZ}} = \mu_p \int_{\mathcal{M}_{p+1}} C_{(p+1)}. \qquad (9.2)$$

Perhaps not surprisingly, there are other terms of great importance, and this chapter will uncover a number of them. In fact, there are many ways of deducing that there *must* be other terms, and one way is to use the fact that D-branes turn into each other under T-duality.

9.1 Tilted D-branes and branes within branes

There are additional terms in the action involving the D-brane gauge field. Again these can be determined from T-duality. Consider, as an example, a D1-brane in the 1–2 plane. The action is

$$\mu_1 \int dx^0 \, dx^1 \left(C_{01} + \partial_1 X^2 C_{02} \right). \qquad (9.3)$$

Under a T-duality in the x^2-direction this becomes

$$\mu_2 \int dx^0 dx^1 dx^2 \, (C_{012} + 2\pi\alpha' F_{12}C_0). \tag{9.4}$$

We have used the T-transformation of the C fields as discussed in section 8.1.1, and also the recursion relation (5.11) between D-brane tensions.

This has an interesting interpretation. As we saw before in section 5.2.1, a Dp-brane tilted at an angle θ is equivalent to a D$(p+1)$-brane with a constant gauge field of strength $F = (1/2\pi\alpha')\tan\theta$. Now we see that there is additional structure: the flux of the gauge field couples to the R–R potential $C^{(p)}$. In other words, the flux acts as a source for a D$(p-1)$-brane living in the world-volume of the D$(p+1)$-brane. In fact, given that the flux comes from an integral over the whole world-volume, we cannot localise the smaller brane at a particular place in the world-volume: it is 'smeared' or 'dissolved' in the world-volume.

In fact, we shall see when we come to study supersymmetric combinations of D-branes that supersymmetry requires the D0-brane to be completely smeared inside the D2-brane. It is clear here how it manages this, by being simply T-dual to a tilted D1-brane. We shall see many consequences of this later.

9.2 Anomalous gauge couplings

The T-duality argument of the previous section can be generalised to discover more terms in the action, but we shall take another route to discover such terms, exploiting some important physics in which we already have invested considerable time.

Let us return to the type I string theory, and the curious fact that we had to employ the Green–Schwarz mechanism (see section 7.1.4, where we mixed a classical and a quantum anomaly in order to achieve consistency). Focusing on the gauge sector alone for the moment, the classical coupling which we wrote in equation (7.35) implies a mixture of the two-form $C_{(2)}$ with gauge field strengths:

$$S = \frac{1}{3 \times 2^6 (2\pi)^5 \alpha'} \int C_{(2)} \left(\frac{\mathrm{Tr}_{\mathrm{adj}}(F^4)}{3} - \frac{[\mathrm{Tr}_{\mathrm{adj}}(F^2)]^2}{900} \right). \tag{9.5}$$

We can think of this as an interaction on the world-volume of the D9-branes showing a coupling to a D1-brane, analogous to that which we saw for a D0-brane inside a D2-brane in equation (9.4). This might seem a bit of a stretch, but let us write it in a different way:

$$
S = \mu_9 \int \frac{(2\pi\alpha')^4}{3 \times 2^6} C_{(2)} \left(\frac{\mathrm{Tr}_{\mathrm{adj}}(F^4)}{3} - \frac{[\mathrm{Tr}_{\mathrm{adj}}(F^2)]^2}{900} \right)
$$

$$
= \mu_9 \int \frac{(2\pi\alpha')^4}{4!} C_{(2)} \mathrm{Tr}(F^4), \tag{9.6}
$$

where, crucially, in the last line we have used the properties (7.39) of the traces for $SO(32)$ to rewrite things in terms of the trace in the fundamental.

Another exhibit we would like to consider is the kinetic term for the modified three-form field strength, $\widetilde{G}_{(3)}$, which is

$$
S = -\frac{1}{4\kappa_0^2} \int \widetilde{G}_{(3)} \wedge^* \widetilde{G}_{(3)}. \tag{9.7}
$$

Since $d\omega_{3Y} = \mathrm{Tr}(F \wedge F)$ and $d\omega_{3L} = \mathrm{Tr}(R \wedge R)$, this gives, after integrating by parts and, dropping the parts with R for now:

$$
S = \frac{\alpha'}{4\kappa^2} \int C_{(6)} \wedge \left(\frac{1}{30} \mathrm{Tr}_{\mathrm{adj}}(F \wedge F) \right)
$$

$$
= \mu_9 \int \frac{(2\pi\alpha')^2}{2} C_{(6)} \wedge (\mathrm{Tr}(F \wedge F)) \tag{9.8}
$$

again, we have converted the traces using (7.39), we've used the relation (7.44) for κ_0 and we've recalled the definition (7.38).

Upon consideration of the three examples (9.4), (9.6), and (9.8), it should be apparent that a pattern is forming. The full answer for the gauge sector is the result[118, 119]

$$
\mu_p \int_{\mathcal{M}_{p+1}} \left[\sum_p C_{(p+1)} \right] \wedge \mathrm{Tr}\, e^{2\pi\alpha' F + B}, \tag{9.9}
$$

(We have included non-trivial B on the basis of the argument given in section 5.2.) So far, the gauge trace (which is in the fundamental) has the obvious meaning. We note that there is the possibility that in the full non-Abelian situation, the C can depend on *non-commuting* transverse fields X^i, and so we need something more general. We will return to this later. The expansion of the integrand (9.9) involves forms of various rank; the notation means that the integral picks out precisely the terms whose rank is $(p + 1)$, the dimension of the Dp-brane's world-volume.

Looking at the first non-trivial term in the expansion of the exponential in the action we see that there is the term that we studied above corresponding to the dissolution of a D$(p - 2)$-brane into the sub two-plane in the Dp-brane's world volume formed by the axes X^i and X^j, if field strength components F_{ij} are turned on.

At the next order, we have a term which is quadratic in F which we could rewrite as:

$$S = \frac{\mu_{p-4}}{8\pi^2} \int C_{(p-3)} \wedge \text{Tr}(F \wedge F). \tag{9.10}$$

We have used the fact that $\mu_{p-4}/\mu_p = (2\pi\sqrt{\alpha'})^4$. Recall that there are non-Abelian field configurations called 'instantons' for which the quantity $\int \text{Tr}(F \wedge F)/8\pi^2$ gives integer values. (See, for example, insert 9.4.) Interestingly, we see that if we excite an instanton configuration on a four dimensional sub-space of the Dp-brane's world-volume, it is equivalent to precisely one unit of D$(p-4)$-brane charge, which is remarkable.

In trying to understand what might be the justification (other than T-duality) for writing the full result (9.9) for all branes so readily, the reader might recognise something familiar about the object we built the action out of. The quantity $\exp(iF/(2\pi))$, using a perhaps more familiar normalisation, generates polynomials of the Chern classes of the gauge bundle of which F is the curvature. It is called the Chern character. In the Abelian case we first studied, we had non-vanishing first Chern class $\text{Tr}F/(2\pi)$, which after integrating over the manifold, gives a number which is in fact quantised. For the non-Abelian case, the second Chern class $\text{Tr}(F \wedge F)/(8\pi^2)$ computes the integer known as the instanton number, and so on.

These numbers, being integers, are topological invariants of the gauge bundle. By the latter, we mean the fibre bundle of the gauge group over the world-volume, for which the gauge field A is a connection.

A fibre bundle is a rule for assigning a copy of a certain space (the fibre: in this case, the gauge group G) to every point of another space (the base: here, the world-volume). The most obvious case of this is simply a product of two manifolds (since one can be taken as the base and then the product places a copy of the other at every point of the base), but this is awfully trivial. More interesting is to have only a product space locally. Then, the whole structure of the bundle is given by a collection of such local products glued together in an overlapping way, together with a set of transition functions which tell one how to translate from one local patch to another on the overlap. In the case of a gauge theory, this is all familiar. The transition rule is simply a G gauge transformation, and we are allowed to use the term 'vector bundle' in this case. For the connection or gauge field this is: $A \to gAg^{-1} + gdg^{-1}$. So the gauge field is not globally defined. Perhaps the most familiar gauge bundle is the monopole bundle corresponding to a Dirac monopole. See insert 9.1.

Insert 9.1. The Dirac monopole as a gauge bundle

A gauge bundle is sometimes called a principal fibre bundle. Perhaps everybody's favourite gauge bundle is the Dirac monopole. Take a sphere S^2 as our base. We will fibre a circle over it. Recall that S^2 cannot be described by a global set of coordinates, but we can use two, the Northern and the Southern hemisphere, with overlap in the vicinity of the Equator. Put standard polar coordinates (θ, ϕ) on S^2, where $\theta = \pi/2$ is the Equator. Put an angular coordinate $e^{i\psi}$ on the circle. We will use ϕ_+ in the North and ψ_- in the South.
So our bundle is a copy of two patches which are locally $S^2 \times S^1$,

$$+\text{Patch}: \quad \{\theta, \phi, e^{i\psi_+}\}; \qquad -\text{Patch}: \quad \{\theta, \phi, e^{i\psi_-}\},$$

together with a transition function which relates them.

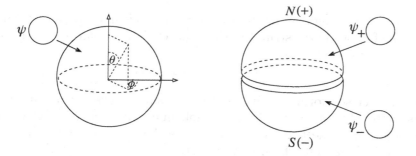

The relation between the two can be chosen to be

$$e^{i\psi_-} = e^{in\phi}e^{i\psi_+},$$

where n is an integer, since as we go around the equator, $\phi \to \phi + 2\pi$, the gluing together of the fibres must still make sense.
The boring case $n = 0$ is sensible, but it simply gives the trivial bundle $S^2 \times S^1$. The case $n = 1$ is the familiar Hopf fibration, which describes the manifold S^3 as a circle bundle over S^2. It is a Dirac monopole of unit charge. Higher values of n give charge n monopoles. The integer n is characteristic of the bundle. It is in fact (minus) the integral of the first Chern class.
The reader who found this a little dry might turn straight to insert 9.2 where we describe the connection on the bundle and compute the first Chern class explicitly.

Insert 9.2.   The first Chern class of the Dirac bundle

Following what we did in insert 9.1, we can uncover more features, which will be useful later on. A natural choice for the connection one-form (gauge potential) in each patch is simply

$$+\text{Patch}: \quad A_+ + d\psi_+; \qquad -\text{Patch}: \quad A_- + d\psi_-,$$

so that the transition function defined in insert 9.1 allows us to connect the two patches, defining the standard $U(1)$ gauge transformation

$$A_+ = A_- + n\phi.$$

Here are the gauge potentials which are standard in this example:

$$A_\pm = n\frac{(\pm 1 - \cos\theta)}{2}d\phi,$$

which, while being regular almost everywhere, clearly have a singularity (the famous Dirac string) in the $\mp$ patch. The curvature two-form is simply

$$F = dA = \frac{n}{2}\sin\theta d\theta \wedge d\phi.$$

This is a closed form, but it is not exact, since there is not a unique answer to what A can be over the whole manifold. According to what we describe in the text, we can compute the first Chern number by integrating the first Chern class to get:

$$\int_{S^2}\frac{F}{2\pi} = \int_+\frac{F}{2\pi} + \int_-\frac{F}{2\pi} = n.$$

9.3 Characteristic classes and invariant polynomials

The topology of a particular fibration can be computed by working out just the right information about its collection of transition functions. For a gauge bundle, the field strength or curvature two-form $F = dA + A \wedge A$ is a nice object with which to go and count, since it is globally defined over the whole base manifold. When the group is Abelian, $F = dA$ and so $dF = 0$. If the bundle is not trivial, then we can't write F as dA everywhere and so F is closed but not exact. Then F is said to be an element of the cohomology group $H^2(\mathcal{B}, \mathbb{R})$ of the base, which we'll call $\mathcal{B}$. The first Chern class $F/2\pi$ defines an integer when integrated over $\mathcal{B}$,

---

Insert 9.3.   The Yang–Mills instanton as a gauge bundle

A favourite non-Abelian example[120] is the $SU(2)$ Yang–Mills instanton. The base is S^4, with coordinates (r, θ, ϕ, ψ), which is $\mathbb{R}^4$ with the point at infinity added. A metric on it for radius $\rho/2$ is:

$$ds^2 = \left(1 + \frac{r^2}{\rho^2}\right)^{-2} \left(dr^2 + r^2(\sigma_1^2 + \sigma_2^2 + \sigma_3^2)\right).$$

The gauge group (fibre) is $G = SU(2)$, which also happens to be the manifold S^3. By analogy with what we saw in insert 9.1, we can divide the S^4 into Northern and Southern hemispheres. The equator is in fact an S^3 and that is where we define the overlap region. Recall that there is a natural $SU(2)$ favoured writing of the coordinates, defining an element $h(\theta, \phi, \psi) \in SU(2)$ as in insert 7.4. We can define similar Euler angles (α, β, γ) as coordinates on the fibre g, for the North (+) and South (−) patches, giving:

+Patch : $\{\theta, \phi, \psi, \alpha_+, \beta_+, \gamma_+\}$;   −Patch : $\{\theta, \phi, \psi, \alpha_-, \beta_-, \gamma_-\}$.

Our transition functions at the equator, taking us from the North to the South fibres are again parametrised by an integer, k:

$$g(\alpha_+, \beta_+, \gamma_+) = h^k(\theta, \phi, \psi)g(\alpha_-, \beta_-, \gamma_-).$$

Again $k = 0$ is trivial. The case $k = 1$ is the Hopf fibration of S^7 as an S^3 over S^4. It is the one instanton solution. Other k are the multi-instantons. Also k will give the second Chern class of the bundle.

---

telling us to which topological class F belongs; this integer is a topological invariant.

For the non-Abelian case, F is no longer closed, and so we don't have the first Chern class. However, the quantity $\text{Tr}(F \wedge F)$ is closed, since as we know from insert 7.3 (p. 167), it is actually $d\omega_{3Y}$. So if the Chern–Simons three-form ω_{3Y} is not globally defined, we have a non-trivial bundle, and $\text{Tr}(F \wedge F)$, being closed but not exact, defines an element of the cohomology group $H^4(\mathcal{B}, \mathbb{R})$. The second Chern class $\text{Tr}(F \wedge F)/8\pi^2$ integrated over $\mathcal{B}$ gives an integer which says to which topological class F belongs. See insert 9.4.

As we have said above, D-branes appear to compute certain topological features of the gauge bundle on their world-volumes, corresponding here

---

Insert 9.4.   The BPST one-instanton connection

Just as with the Dirac monopole case, we can write the connection 1-form for each patch:

+Patch : $g_+^{-1}A_+g_+ + g_+^{-1}dg_+$;   −Patch : $g_-^{-1}A_-g_- + g_-^{-1}dg_-$,

so that the transition function defined in insert 9.3 allows us to connect the two patches with a gauge transformation

$$A_+ = h^k A_- h^{-k} + h^k dh^{-k}.$$

The $k = 1$ solution can be written quite simply:

$$A_+ = \frac{r^2}{r^2 + \rho^2} h^{-1}dh = \frac{r^2}{r^2 + \rho^2} i\tau_n \sigma_n,$$

where the σ_n are the left-invariant one-forms. This solution is smooth everywhere except at a singularity at $r = 0$. The South pole solution is obtained by gauge transformation:

$$A_- = hA_+h^{-1} + hdh^{-1} = -\frac{\rho^2}{\rho^2 + r^2} dhh^{-1} = \frac{\rho^2}{\rho^2 + r^2} i\tau_n \bar{\sigma}_n,$$

where the $\bar{\sigma}_n$ are the right-invariant one-forms. This solution is singular at $r = \infty$. The curvature two-form is best described using the veilbiens $\{e^0, e^1, e^2, e^3\} = (1 + r/\rho)^{-2}\{dr, r\sigma_1, r\sigma_2, r\sigma_3\}$:

$$F_+ = dA + A \wedge A = i\tau_k \frac{2}{\rho^2}\left(e^0 \wedge e^k + \frac{1}{2}\epsilon_{kij}e^i \wedge e^j\right).$$

Of course, $F_- = hF_+h^{-1}$. It is worth checking that this solution is self dual, i.e. $^*F = F$, with anti-self duality made by $\sigma_n \leftrightarrow \bar{\sigma}_n$. The instanton number is (minus) the second Chern class integrated over the S^4:

$$k = -\frac{1}{8\pi^2}\int_{S_4} \text{Tr}(F \wedge F) = \frac{1}{8\pi^2}\left(\frac{48}{\rho^4}\right)\int_{S_4} e^0 \wedge e^1 \wedge e^2 \wedge e^3 = 1,$$

where in the latter we have used that the volume of the S^4 is $\pi^2\rho^4/6$. Here, ρ has the interpretation as the 'core size' of the instanton.

to the Chern classes of the cohomology. As we shall see, they compute other topological numbers as well, and so let us pause to appreciate a little of the tools that they employ, in order to better be able to put them to work for us.

The first and second Chern classes shall be denoted $c_1(F)$ and $c_2(F)$ and so on, $c_j(F)$ for the jth Chern class. Let us call the gauge group G, and keep in mind $U(n)$ (we will make appropriate modifications to our statements to include $O(n)$ later). We'd like to know how to compute the $c_j(F)$. The remarkable thing is that they arise from forming polynomials in F which are invariant under G. Forget that F is a two-form for now, and just think of it as an $n \times n$ matrix. The $c_j(F)$ are found by expanding a natural invariant expression in F as a series in t:

$$\det\left(t\mathbf{I} + \frac{iF}{2\pi}\right) = \sum_{j=0}^{n} c_{n-j}(F)t^j. \tag{9.11}$$

(Here, we use the i in F to keep the expression real, since $U(N)$ generators are anti-Hermitean.) The great thing about this is that there is an excellent trick for finding explicit expressions for the c_js which will allow us to manipulate them and relate them to other quantities. Assume that the matrix $iF/2\pi$ has been diagonalised. Call this diagonal matrix X, with n distinct non-vanishing eigenvalues x_i, $i = 1, \ldots, n$. Then we have

$$\det(t\mathbf{I} + X) = \prod_{i=1}^{n}(t + x_i) = \sum_{j=0}^{n} c_{n-j}(x)t^j, \tag{9.12}$$

and we find by explicit computation that the c_js are symmetric polynomials:

$$c_0 = 1, \quad c_1 = \sum_{i}^{n} x_i, \quad c_2 = \sum_{i_1 < i_2}^{n} x_{i_1} x_{i_2}, \quad \ldots$$

$$c_j = \sum_{i_1 < i_2 < \cdots i_j}^{n} x_{i_1} x_{i_2} \ldots x_{i_j}, \quad c_n = x_1 x_2 \ldots x_n. \tag{9.13}$$

Now rewrite the expressions on the eigenvalues back as matrix expressions in terms of X, and then replace X by $iF/2\pi$, to get:

$$c_0(F) = 1, \quad c_1(F) = \frac{i}{2\pi} \mathrm{Tr} F,$$

$$c_2(F) = \frac{1}{2}\left(\frac{i}{2\pi}\right)^2 [\mathrm{Tr} F \wedge \mathrm{Tr} F - \mathrm{Tr}(F \wedge F)],$$

$$c_n(F) = \left(\frac{i}{2\pi}\right)^n \det F. \tag{9.14}$$

In the case of $SU(N)$, the generators are traceless, and so

$$c_2(F) = \frac{1}{8\pi^2} \text{Tr}(F \wedge F),$$

the expression we saw before. The $c_j(F)$ are rank $2j$ forms, so of course, the largest one that gives a meaningful quantity is the one for which $\dim(\mathcal{B}) = 2j$.

The natural object which D-branes seem to have on their world-volume is in fact the Chern character, $ch(F) = \text{Tr}\exp(iF/2\pi)$. This computes a specific combination of the Chern classes, and we can compute this by using our symmetric polynomial expressions in (9.13). Working with the diagonal X again we have

$$ch(x) = \sum_i e^{x_i} = \sum_i \left(1 + x_i + \frac{x_i^2}{2} + \cdots\right)$$

$$= n + c_1 + \frac{1}{2}(c_1^2 - 2c_2) + \cdots, \quad \text{and so we have:}$$

$$ch(F) = n + c_1(F) + \frac{1}{2}(c_1^2(F) - 2c_2(F)) + \cdots. \tag{9.15}$$

The Chern character is to be thought of as an important generating function of the Chern classes and in fact it is a powerful tool, in that it is well behaved in the sense that for bundle E and a bundle F, the relations

$$ch(E \oplus F) = ch(E) + ch(F) \quad \text{and} \quad ch(E \otimes F) = ch(E) \wedge ch(F) \tag{9.16}$$

are true. This is part of an important technology to doing 'algebra' on bundles allowing one to perform operations which compare them to each other, etc.

For the case $G = O(n)$, the characteristic classes are called Pontryagin classes. We may think of the bundle as a real vector bundle. Now we have

$$\det\left(t\mathbf{I} + \frac{F}{2\pi}\right) = \sum_{j=0}^{n} p_{n-j}(F)t^j. \tag{9.17}$$

Again, consider having diagonalised to X. We can't quite diagonalise, but can get it into the block diagonal form:

$$X = \begin{pmatrix} 0 & x_1 & & & \\ -x_1 & 0 & & & \\ & & 0 & x_2 & \\ & & -x_2 & 0 & \\ & & & & \ddots \end{pmatrix}. \tag{9.18}$$

Now we have the relation:

$$\det\left(t\mathbf{I} + X\right) = \det\left(t\mathbf{I} + X^T\right) = \det\left(t\mathbf{I} - X\right),$$

and so we see that the $p_j(F)$ must be even in F. A bit of work similar to that which we did above for the Chern classes gives:

$$p_1(F) = -\frac{1}{2}\left(\frac{1}{2\pi}\right)^2 \mathrm{Tr}F^2,$$

$$p_2(F) = \frac{1}{8}\left(\frac{1}{2\pi}\right)^4\left[(\mathrm{Tr}F^2)^2 - 2\mathrm{Tr}F^4\right], \ldots, \text{etc.},$$

$$p_{[\frac{n}{2}]}(F) = \left(\frac{1}{2\pi}\right)^n \det F, \tag{9.19}$$

where $[n/2] = n/2$ if n is even or $(n-1)/2$ otherwise.

Now an important case of orthogonal groups is of course the tangent bundle to a manifold of dimension n. Using the veilbiens formalism of section 2.8, the structure group is $O(n)$. The two-form to use is the curvature two-form R. Then we have, e.g.

$$p_1(R) = -\frac{1}{8\pi^2}\mathrm{Tr}R \wedge R. \tag{9.20}$$

The Euler class is naturally defined here too. For an orientable even dimensional $n = 2k$ manifold M, the Euler class class $e(M)$ is defined via

$$e(X)e(X) = p_k(X).$$

We write X here and not the two-form R, since we would have a $4k$-form which vanishes on M. However, $e(R)$ makes sense as a form since its rank is n, which is the dimension of M. For an example, see insert 9.5.

Two other remarkable generating functions of importance are the $\hat{A}$ ('A-roof') or Dirac genus:

$$\hat{A} = \prod_{j=1}^{n}\frac{x_j/2}{\sinh x_j/2} = \prod_{j=1}^{n}\left(1 + \sum_{n\geq 1}(-1)^n\frac{2^{2n} - 2}{(2n)!}B_n x_j^{2n}\right)$$

$$= 1 - \frac{1}{24}p_1 + \frac{1}{5760}(7p_1^2 - 4p_2) + \cdots, \tag{9.21}$$

and the Hirzebruch $\hat{\mathcal{L}}$-polynomial

$$\hat{\mathcal{L}} = \prod_{j=1}^{n}\frac{x_j}{\tanh x_j} = \prod_{j=1}^{n}\left(1 + \sum_{n\geq 1}(-1)^{n-1}\frac{2^{2n}}{(2n)!}B_n x_j^{2n}\right)$$

$$= 1 + \frac{1}{3}p_1 + \frac{1}{45}(7p_2 - p_1^2) + \cdots, \tag{9.22}$$

Insert 9.5.   The Euler number of the sphere

Lets test this out for the two-sphere S^2. Using the formalism of section 2.8, the curvature two-form can be computed as $R_{\theta\phi} = \sin\theta d\theta \wedge d\phi$. Then we can compute

$$p_1(S^2) = -\frac{1}{8\pi^2}\mathrm{Tr}R \wedge R = \left(\frac{1}{2\pi}\sin\theta d\theta \wedge d\phi\right)^2.$$

So we see that

$$e(S^2) = \frac{1}{2\pi}\sin\theta d\theta \wedge d\phi.$$

The integral of this from over the manifold given the Euler number:

$$\chi = \int_{S^2} e(S^2) = 2,$$

a result we know well and have used extensively.

---

where the B_n are the Bernoulli numbers, $B_1 = 1/6, B_2 = 1/30, B_3 = 1/42, \ldots$. These are very important characteristics as well, and again have useful algebraic properties for facilitating the calculus of vector bundles along the lines given by equations (9.16). As we shall see, they also play a very natural role in our story here.

9.4 Anomalous curvature couplings

So we seem to have wandered away from our story somewhat, but in fact we are getting closer to a big part of the answer. If the above formula (9.9) is true, then D-branes evidently know how to compute the topological properties of the gauge bundle associated to their world-volumes. This is in fact a hint of a deeper mathematical structure underlying the structure of D-branes and their charge, and we shall see it again later.

There is another strong hint of what is going on based on the fact that we began to deduce much of this structure using the terms we discovered were needed to cancel anomalies. So far we have only looked at the terms involving the curvature of the gauge bundle, and not the geometry of the brane itself which might have non-trivial R associated to its tangent bundle. Indeed, since the gauge curvature terms came from anomalies, we might expect that the tangent bundle curvatures do too. Since these are so closely related, one might expect that there is a very succinct formula for those couplings as well. Let us look at the anomaly terms again. The key terms are the curvature terms in (7.35) and the curvature terms arising from the modification (7.38) of the field strength of $C_{(3)}$ to achieve gauge invariance. The same deduction we made to arrive at (9.8) will lead us to $\text{Tr}R^2$ terms coupling to $C_{(6)}$. Also, if we convert to the fundamental representation, we can see that there is a term

$$-\frac{1}{3 \times 2^6 (2\pi)^5} \int C_{(2)} \text{Tr}F^2 \text{tr}R^2.$$

This mixed anomaly type term can be generated in a number of ways, but a natural guess[110, 111, 112] (motivated by remarks we shall make shortly) is that there is a $\sqrt{\hat{\mathcal{A}}}$ term on the world volume, multiplying the Chern characteristic. In fact, the precise term, written for all branes, is:

$$\mu_p \int_{\mathcal{M}_{p+1}} \sum_i C_{(i)} \left[e^{2\pi\alpha'F+B} \right] \sqrt{\hat{\mathcal{A}}(4\pi^2\alpha'R)}. \tag{9.23}$$

Working with this expression, using the precise form given in (9.21) will get the mixed term precisely right, but the $C_{(6)} \text{tr}R^2$ will not have the right coefficient, and also the remaining fourth order terms coupling to $C_{(2)}$ are incorrect, after comparing the result to (7.35).

The reason why they are not correct is because there is another crucial contribution which we have not included. There is an orientifold O9-plane of charge $-32\mu_9$ as well. As we saw, it is crucial in the anomaly cancellation story of the previous chapter and it must be included here for precisely the same reasons. While it does not couple to the $SO(32)$ gauge fields (open strings), it certainly has every right to couple to gravity, and hence source curvature terms involving R. Again, as will be clear shortly, the precise term for Op-planes of this type is [125]:

$$\tilde{\mu}_p \int_{\mathcal{M}_{p+1}} \sum_i C_{(i)} \sqrt{\hat{\mathcal{L}}(\pi^2\alpha'R)}, \tag{9.24}$$

where $\hat{\mathcal{L}}(R)$ is defined above in equation (9.22). Remarkably, expanding this out will repair the pure curvature terms so as to give all of the correct terms in X_8 to reproduce (7.35), and the $C_{(6)}$ coupling is precisely:

$$S = \mu_9 \int \frac{(2\pi\alpha')^2}{2} C_{(6)} \wedge (\text{Tr}(F \wedge F) - \text{Tr}R \wedge R)$$

$$= \mu_9 \int \frac{(2\pi\alpha')^2}{2} C_{(6)} \wedge Y_4. \tag{9.25}$$

Beyond just type I, it is worth noting that the $R \wedge R$ term will play an important role on the world volumes of branes. It can be written in the form:

$$\frac{\mu_p (4\pi^2 \alpha')^2}{48} \int_{\mathcal{M}_{p+1}} C_{(p-3)} \wedge p_1(R). \tag{9.26}$$

By straightforward analogy with what we have already observed about instantons, another way to get a D$(p-4)$-brane inside the world-volume of a Dp-brane is to wrap the brane on a four dimensional surface of non-zero $p_1(R)$. Indeed, as we saw in equation (7.54), the K3 surface has $p_1 = -2\chi = -48$, and so wrapping a Dp-brane on K3 gives D$(p-4)$-brane charge of precisely -1. This will be important to us later[115, 121].

9.5 A relation to anomalies

There is one last amusing fact that we should notice, which will make it very clear that the curvature couplings that we have written above are natural for branes and O-planes of all dimensionalities. The point is that the curvature terms just don't accidentally resemble the anomaly polynomials we saw before, but are built out of the very objects which can be used to generate the anomaly polynomials that we listed in insert 7.2.

In fact, we can use them to generate anomaly polynomials for dimension $D = 4k + 2$. We can pick out the appropriate powers of the curvature two forms by using the substitution

$$\sum_{i=1}^{2k+1} x_i^{2m} = \frac{1}{2}(-1)^m \mathrm{tr} R^{2m}.$$

Then in fact the polynomial $\hat{I}_{1/2}$ is given by the $\hat{A}$ genus:

$$\begin{aligned}
\hat{I}_{1/2} = \hat{A} &= \prod_{j=1}^{2k+1} \frac{x_j/2}{\sinh x_j/2} \\
&= \prod_{j=1}^{2k+1} \left(1 + \frac{y_j^2}{3!} + \frac{y_j^4}{5!} + \cdots \right)^{-1} \\
&= \prod_{j=1}^{2k+1} \left(1 - \frac{1}{6}y_j^2 + \frac{7}{360}y_j^4 - \frac{31}{15120}y_j^6 + \cdots \right) \\
&= 1 - \frac{1}{6}\mathcal{Y}_2 + \frac{1}{180}\mathcal{Y}_4 + \frac{1}{72}\mathcal{Y}_2^2 \\
&\quad - \frac{1}{2835}\mathcal{Y}_6 - \frac{1}{1080}\mathcal{Y}_2\mathcal{Y}_4 - \frac{1}{1296}\mathcal{Y}_2^3 + \cdots
\end{aligned} \tag{9.27}$$

where

$$\mathcal{Y}_{2m} = \sum_{i=1}^{2k+1} y_i^{2m} = \frac{1}{2}\left(-\frac{1}{4}\right)^m \operatorname{tr}R^{2m}.$$

The trick is then to simply pick out the piece of the expansion which fits the dimension of interest, remembering that the desired polynomial is of rank $D+2$. So for example, picking out the order 12 terms will give precisely the 12-form polynomial in insert 7.2, etc.

The gravitino polynomials come about in a similar way. In fact,

$$I_{3/2} = I_{1/2}\left(-1 + 2\sum_{j=1}^{2k+1}\cosh x_i\right)$$

$$= I_{1/2}\left(D - 1 + 4\mathcal{Y}_2 + \frac{4}{3}\mathcal{Y}_4 + \frac{8}{45}\mathcal{Y}_6 + \cdots\right). \qquad (9.28)$$

Also, the polynomials for the antisymmetric tensor come from

$$I_A = -\frac{1}{8}\hat{\mathcal{L}}(R) = -\frac{1}{8}\sum_{j=1}^{2k+1}\frac{x_j}{\tanh x_j}$$

$$= -\frac{1}{8} - \frac{1}{6}\mathcal{Y}_2 + \left(\frac{7}{45} - \frac{1}{9}\mathcal{Y}_2^2\right)$$

$$+ \frac{1}{2835}\left(-496\mathcal{Y}_6 + 588\mathcal{Y}_2\mathcal{Y}_4 - 140\mathcal{Y}_2^3 + \cdots\right). \qquad (9.29)$$

Finally, it is easy to work out the anomaly polynomial for a charged fermion. One simply multiplies by the Chern character:

$$I_{1/2}(F, R) = \operatorname{Tr}e^{iF}I_{1/2}(R). \qquad (9.30)$$

Now it is perhaps clearer what must be going on[111, 112]. The D-branes and O-planes, and any intersections between them all define sub-spacetimes of the ten dimensional spacetime, where potentially anomalous theories live. This is natural, since as we have already learned, and shall explore much more, there are massless fields of various sorts living on them, possibly charged under any gauge group they might carry.

As the world-volume intersections may be thought of as embedded in the full ten dimensional theory, there is a mechanism for cancelling the anomaly which generalises that which we have already encountered. For example, since the Dp-brane is also a source for the R–R sector field $G^{(p+2)}$, it modifies it according to

$$G_{(p+2)} = dC_{(p+1)} - \mu_p\delta(x_0)\ldots\delta(x_p)dx_0 \wedge \ldots dx_p\mathcal{F}(R, F), \qquad (9.31)$$

where the delta functions are chosen to localise the contribution to the world-volume of the brane, extended in the directions $x_0, x_1, \ldots, x_p$. Also μ_p is the Dp-brane (or Op-plane) charge, and the polynomial $\mathcal{F}$ must be chosen so that the classically anomalous variation $\delta C_{(p+1)}$ required to keep $G^{(p+2)}$ gauge invariant can cancel the anomaly on the branes' intersection. Following this argument to its logical conclusion, and using the previously mentioned fact that the possible anomalies are described in terms of the characteristic classes $\exp(iF)$, $\hat{A}(R)$ and $\hat{\mathcal{L}}(R)$, reveals that $\mathcal{F}$ takes the form of the couplings that we have already written. We see that the Green–Schwarz mechanism from type I is an example of something much more general, involving the various geometrical objects which can appear embedded in the theory, and not just the fundamental string itself.

Arguments along these lines also uncover the feature that the normal bundle also contributes to the curvature couplings as well. The full expressions, for completeness, are:

$$\mu_p \int_{\mathcal{M}_{p+1}} \sum_i C_{(i)} \left[e^{2\pi\alpha' F + B} \right] \sqrt{\frac{\hat{A}(4\pi^2\alpha' R_{\mathrm{T}})}{\hat{A}(4\pi^2\alpha' R_{\mathrm{N}})}}, \qquad (9.32)$$

and

$$\tilde{\mu}_p \int_{\mathcal{M}_{p+1}} \sum_i C_{(i)} \sqrt{\frac{\hat{\mathcal{L}}(\pi^2\alpha' R_{\mathrm{T}})}{\hat{\mathcal{L}}(\pi^2\alpha' R_{\mathrm{N}})}}, \qquad (9.33)$$

where the subscripts 'T', 'N' denote curvatures of the tangent and the normal frame, respectively.

9.6 D-branes and K-theory

In fact, the sort of argument above is an independent check on the precise normalisation of the D-brane charges, which we worked out by direct computation in previous sections. As already said before, the close relation to the topology of the gauge and tangent bundles of the branes suggests a connection with tools which might uncover a deeper classification. This tool is called '*K-theory*'. K-theory should be thought of as a calculus for working out subtle topological differences between vector bundles, and as such makes a natural physical appearance here[113, 18].

This is because there is a means of constructing a D-brane by a mechanism known as 'tachyon condensation' on the world-volume of higher dimensional branes. Recall that in chapter 8 we observed that a Dp-brane and an anti-Dp-brane will annihilate. Indeed, there is a tachyon in the spectrum of $p\bar{p}$ strings. Let us make the number of branes be N, and the number of anti-branes be $\bar{N}$. Then the tachyon is charged under the gauge group $U(N) \times U(\bar{N})$. The idea is that a suitable choice of the tachyon can give rise to topology which must survive even if all of the parent branes annihilate away. For example, if the tachyon field is given a topologically stable kink (see insert 1.4, p. 18) as a function of one of the dimensions inside the brane, then there will be a $p-1$ dimensional structure left over, to be identified with a D$(p-1)$-brane. This idea is the key to seeing how to classify D-branes, by constructing all branes in this way.

Most importantly, we have two gauge bundles, that of the Dp-branes, which we might call E, and that of the D$\bar{p}$-branes, called F. To classify the possible D-branes which can exist in the world volume, one must classify all such bundles, defining as equivalent all pairs which can be reached by brane creation or annihilation: If some number of Dp-branes annihilate with D$\bar{p}$-branes (or if the reverse happens, i.e. creation of Dp-D$\bar{p}$ pairs), the pair (E, F) changes to $(E \oplus G, F \oplus G)$, where G is the gauge bundle associated to the new branes, identical in each set. These two pairs of bundles are equivalent. The group of distinct such pairs is (roughly) the object called $K(X)$, where X is the spacetime that the branes fill (the base of the gauge bundles). Physically distinct pairs have non-trivial differences in their tachyon configurations which would correspond to different D-branes after complete annihilation had taken place. So K-theory, defined in this way, is a sort of more subtle or advanced cohomology which goes beyond the more familiar sort of cohomology we encounter daily.

The technology of K-theory is beyond that which we have room for here, but it should be clear from what we have seen in this chapter that it is quite natural, since the world-volume couplings of the charge of the branes is via the most natural objects with which one would want to perform sensible operations on the gauge bundles of the branes like addition and subtraction: the characteristic classes, $\exp(iF)$, $\hat{\mathcal{A}}(R)$ and $\hat{\mathcal{L}}(R)$. Actually, this might have been enough to simply get the result that D-brane charges were classified by cohomology. That it is in fact K-theory (which can compute differences between bundles that cohomology alone would miss) is probably related to a very important physical fact about the underlying theory which will be more manifest one day.

9.7 Further non-Abelian extensions

One can use T-duality to go a bit further and deduce a number of non-Abelian extensions of the action, being mindful of the sort of complications mentioned at the beginning of section 5.5. In the absence of geometrical curvature terms it turns out to be[51, 52]:

$$\mu_p \int_{p-\text{brane}} \text{Tr} \left(\left[e^{2\pi\alpha' i_\Phi i_\Phi} \sum_p C_{(p+1)} \right] e^{2\pi\alpha' F + B} \right). \qquad (9.34)$$

Here, we ascribe the same meaning to the gauge trace as we did previously (see section 5.5). The meaning of i_X is as the 'interior product' in the direction given by the vector Φ^i, which produces a form of one degree fewer in rank. For example, on a two form $C_{(2)}(\Phi) = (1/2)C_{ij}(\Phi)dX^i dX^j$, we have

$$\mathbf{i}_\Phi C_{(2)} = \Phi^i C_{ij}(\Phi)dX^j; \quad \mathbf{i}_\Phi \mathbf{i}_\Phi C_{(2)}(\Phi) = \Phi^j \Phi^i C_{ij}(\Phi) = \frac{1}{2}[\Phi^i, \Phi^j]C_{ij}(\Phi),$$
$$(9.35)$$

where we see that the result of acting twice is non-vanishing when we allow for the non-Abelian case, with C having a non-trivial dependence on Φ. We shall see this action work for us to produce interesting physics later.

9.8 Further curvature couplings

We deduced geometrical curvature couplings to the R–R potentials a few subsections ago. In particular, such couplings induce the charge of lower p branes by wrapping larger branes on topologically non-trivial surfaces.

In fact, as we saw before, if we wrap a Dp-brane on K3, there is induced precisely -1 units of charge of a D$(p-4)$-brane. This means that the charge of the effective $(p-4)$-dimensional object is

$$\mu = \mu_p V_{\text{K3}} - \mu_{p-4}, \qquad (9.36)$$

where V_{K3} is the volume of the K3. However, we can go further and notice that since this is a BPS object of the six dimensional $\mathcal{N} = 2$ string theory obtained by compactifying on K3, we should expect that it has a tension which is

$$\tau = \tau_p V_{\text{K3}} - \tau_{p-4} = g_s^{-1}\mu. \qquad (9.37)$$

If this is indeed so, then there must be a means by which the curvature of K3 induces a shift in the tension in the world-volume action. Since the part of the action which refers to the tension is the Dirac–Born–Infeld action, we deduce that there must be a set of curvature couplings for that part of the action as well. Some of them are given by the following[122, 128]:

$$S = -\tau_p \int d^{p+1}\xi \ e^{-\Phi}\det^{1/2}(G_{ab} + \mathcal{F}_{ab})\left(1 - \frac{1}{3 \times 2^8\pi^2} \times\right.$$

$$\left.\left(\mathcal{R}_{abcd}\mathcal{R}^{abcd} - \mathcal{R}_{\alpha\beta ab}R^{\alpha\beta ab} + 2\hat{\mathcal{R}}_{\alpha\beta}\hat{\mathcal{R}}^{\alpha\beta} - 2\hat{\mathcal{R}}_{ab}\hat{\mathcal{R}}^{ab}\right) + O(\alpha'^4)\right),$$

$$(9.38)$$

where $\mathcal{R}_{abcd} = (4\pi^2\alpha')R_{abcd}$, etc., and a, b, c, d are the usual tangent space indices running along the brane's world-volume, while α, β are normal indices, running transverse to the world-volume.

Some explanation is needed. Recall that the embedding of the brane into D-dimensional spacetime is achieved with the functions $X^\mu(\xi^a)$, $(a = 0, \ldots, p; \mu = 0, \ldots, D-1)$ and the pull-back of a spacetime field F_μ is performed by soaking up spacetime indices μ with the local 'tangent frame' vectors $\partial_a X^\mu$, to give $F_a = F_\mu \partial_a X^\mu$. There is another frame, the 'normal frame', with basis vectors ζ^μ_α, $(\alpha = p+1, \ldots, D-1)$. Orthonormality gives $\zeta^\mu_\alpha\zeta^\nu_\beta G_{\mu\nu} = \delta_{\alpha\beta}$ and also we have $\zeta^\mu_\alpha\partial_a X^\nu G_{\mu\nu} = 0$.

We can pull back the spacetime Riemann tensor $R_{\mu\nu\kappa\lambda}$ in a number of ways, using these different frames, as can be seen in the action. $\hat{R}$ with two indices are objects which were constructed by contraction of the *pulled-back* fields. They are *not* the pull-back of the bulk Ricci tensor, which vanishes at this order of string perturbation theory anyway.

In fact, for the case of K3, it is Ricci flat and everything with normal space indices vanishes and so we get only $R_{abcd}R^{abcd}$ appearing, which alone computes the result (7.54) for us, and so after integrating over K3, the action becomes:

$$S = -\int d^{p-3}\xi \ e^{-\Phi}\left[\tau_p V_{K3} - \tau_{p-4}\right]\det^{1/2}(G_{ab} + \mathcal{F}_{ab}), \qquad (9.39)$$

where again we have used the recursion relation between the D-brane tensions. So we see that we have correctly reproduced the shift in the tension that we expected on general grounds for the effective D$(p-4)$-brane. We will use this action later.

10

The geometry of D-branes

As we have seen, branes of various sorts are solutions of string theory which are localised to some extent, and have well-defined mass and charge per unit volume. Since these masses and charges are measured at infinity, meaning that the branes are sources of fields from the massless sector, we might expect that they must be actually be solutions of the low energy equations of motion: the gravity sector and other fields such as the various antisymmetric tensor fields, and possibly the dilaton. These field configurations can be thought of as representing interesting backgrounds in which the string can propagate. It has become increasingly important in many recent research areas to consider the details of such solutions, and we shall begin exploring this highly developed technology in the present chapter.

10.1 A look at black holes in four dimensions

Before we launch into a description of the solutions associated to branes, it is a good idea to start with something more familiar in order to gain some intuition about how the solutions work. We will start in four dimensions with a familiar system: Einstein's gravity coupled to Maxwell's electromagnetism. The more advanced reader may wish to skip directly to section 10.2 if the following is too elementary, but beware, since we shall be uncovering and emphasising probably less familiar features in order to prepare for analogous properties of branes in higher dimensions.

10.1.1 A brief study of the Einstein–Maxwell system

Let us consider the Einstein–Hilbert action for gravity coupled to the Maxwell system:

$$S = \frac{1}{16\pi G} \int d^4x \, (-g)^{1/2} [R - G F_{\mu\nu} F^{\mu\nu}], \qquad (10.1)$$

where
$$F_{\mu\nu} = \partial_\mu A_\nu - \partial_\nu A_\mu.$$
The equations of motion for this system resulting from varying with respect to $g_{\mu\nu}$ are of course:
$$R_{\mu\nu} - \frac{1}{2}g_{\mu\nu}R = 8\pi G T_{\mu\nu}, \tag{10.2}$$
where
$$T_{\mu\nu} = \frac{1}{4\pi}\left(g^{\gamma\delta}F_{\mu\gamma}F_{\nu\delta} - \frac{1}{4}g_{\mu\nu}F_{\gamma\delta}F^{\gamma\delta}\right). \tag{10.3}$$
A particularly interesting spherically symmetric solution of this system, (see insert 10.1) representing a source of mass M and electric charge Q is, for the metric:
$$ds^2 = -\left(1 - \frac{2MG}{r} + \frac{Q^2}{r^2}\right)dt^2 + \left(1 - \frac{2MG}{r} + \frac{Q^2}{r^2}\right)^{-1}dr^2 + r^2 d\Omega_2^2, \tag{10.4}$$
where $d\Omega_2^2 \equiv d\theta^2 + \sin^2\theta d\phi^2$, is the metric on a round S^2 in standard polar coordinates, and
$$F_{tr} = E_r(r) = -F_{rt}, \quad E_r(r) = \frac{Q}{r^2} \quad \text{or} \quad F = \frac{Q}{r^2}dt \wedge dr.$$
Let us note some of the key properties of these solutions.

10.1.2 Basic properties of Schwarzschild

We begin with the case $Q = 0$, an empty-space solution (i.e. a solution of pure Einstein gravity), which is the Schwarzschild solution. The first thing to take note of is that the solution has various obvious symmetries. Notice that the metric components do not depend on t or ϕ. So there is a pair of symmetries coming from invariance under translations in these coordinates. In other words, the solution is static, and symmetric about the ϕ axis. Well, of course it is manifestly spherically symmetric as well. In a more sophisticated language, we would say that there are 'Killing vectors' $\mathbf{k}$, of this solution satisfying
$$\nabla_\mu \mathbf{k}_\nu + \nabla_\nu \mathbf{k}_\mu = 0, \tag{10.7}$$
where ∇_μ is the covariant derivative. Our two obvious ones are:
$$\xi^\mu = \left(\frac{\partial}{\partial t}\right)^\mu = (1, 0, 0, 0); \tag{10.8}$$
$$\eta^\mu = \left(\frac{\partial}{\partial\phi}\right)^\mu = (0, 0, 0, 1), \tag{10.9}$$

Insert 10.1.   Checking the Reissner–Nordström solution

It is worthwhile listing some of the objects that the diligent reader would have computed if checking by hand that equation (10.4) is a solution. They will be useful later. The non-vanishing components of the 'affine' or 'metric' connection are:

$$\Gamma^t_{tr} = \frac{M\,r - Q^2}{r\,(r^2 - 2\,M\,r + Q^2)}; \quad \Gamma^r_{tt} = \frac{(r^2 - 2\,M\,r + Q^2)\,(M\,r - Q^2)}{r^5};$$

$$\Gamma^r_{rr} = -\frac{M\,r - Q^2}{r\,(r^2 - 2\,M\,r + Q^2)}; \quad \Gamma^r_{\theta\theta} = -\frac{r^2 - 2\,M\,r + Q^2}{r};$$

$$\Gamma^r_{\phi\phi} = -\frac{(r^2 - 2\,M\,r + Q^2)\sin^2\theta}{r};$$

$$\Gamma^\theta_{\phi\phi} = -\sin\theta\,\cos\theta; \quad \Gamma^\phi_{r\phi} = \frac{1}{r}; \quad \Gamma^\phi_{\theta\phi} = \frac{\cos\theta}{\sin\theta}; \quad \Gamma^\theta_{\phi\theta} = \frac{1}{r}, \quad (10.5)$$

remembering that it is symmetric in its lower components. Taking some more derivatives to make the Riemann–Christoffel tensor, and then contracting gives the non-vanishing components of the Ricci tensor:

$$R_{tt} = \frac{(r^2 - 2\,M\,r + Q^2)\,Q^2}{r^6}; \quad R_{rr} = -\frac{Q^2}{r^2\,(r^2 - 2\,M\,r + Q^2)};$$

$$R_{\theta\theta} = \frac{Q^2}{r^2}; \quad R_{\phi\phi} = \frac{\sin^2\theta\,Q^2}{r^2}, \quad (10.6)$$

from which it is easy to see that its trace, the Ricci scalar R, actually vanishes. Computing the stress tensor gives the result that $T_{\mu\nu} = R_{\mu\nu}/8\pi$, proving that it is a solution.

---

in an obvious notation*. To see the full spherical symmetry, it is in fact better to change variables to the 'isotropic coordinates', so called because it makes the spatial part of the metric conformal to flat space, which means that all distances measured are rescaled by an overall factor, but the locally measured angles between vectors are preserved. Changing to

---

* Here, and in many other places, we will use the fact that in curved spacetime it is very useful to define vectors as differential operators.

a new coordinate ρ defined by

$$r = \rho \left(1 + \frac{M}{2\rho}\right)^2,$$

the metric becomes[†]

$$ds^2 = -\frac{\left(1 - \frac{M}{2\rho}\right)^2}{\left(1 + \frac{M}{2\rho}\right)^2} dt^2 + \left(1 + \frac{M}{2\rho}\right)^4 \left(dx^2 + dy^2 + dz^2\right), \qquad (10.10)$$

where $\rho^2 = x^2 + y^2 + z^2$. Then the Killing vectors corresponding to spherical symmetry are

$$\mathbf{L}_3 = x\frac{\partial}{\partial y} - y\frac{\partial}{\partial x}, \quad \mathbf{L}_1 = y\frac{\partial}{\partial z} - z\frac{\partial}{\partial y}, \quad \mathbf{L}_2 = z\frac{\partial}{\partial x} - x\frac{\partial}{\partial z}.$$

One can check that they satisfy the A_1 (i.e. $SO(3)$) Lie algebra: $[\mathbf{L}_i, \mathbf{L}_j] = \epsilon_{ijk}\mathbf{L}_k$. It is worth knowing that the existence of Killing vectors guarantees certain important properties of the solutions, helping to exhibit certain conserved physical quantities. For example, $\partial/\partial t$, being timelike, ensures that the geometry is static, since Killing's equation results in $\partial g_{\mu\nu}/\partial t = 0$.

Recall that a vector (or more properly a vector field in curved spacetime) define a curve, by being the tangent to it at every point. In fact, along a curve generated by a Killing vector $\mathbf{k}$, the combination $\mathbf{u} \cdot \mathbf{k}$ is a conserved quantity, which will be useful later on. Notice that ξ and η, as defined above, define for us (respectively) a conserved energy and angular momentum per unit rest mass.

Now, it is of course a familiar feature of the solution that the spherical surface $r = r_{\mathrm{H}} = 2M$ is an horizon, since we can see that, for example, g_{tt} vanishes there. While looking at the vanishing of g_{tt} is a quick way of reading off the location horizon, for the general geometry (10.4), it is misleading in general. We should characterise it as follows:

The spherical surface at radius $r = R$ has a unit normal vector to it, $\mathbf{n}$, given by (see insert 10.2)

$$n^\mu = \frac{1}{\sqrt{|g_{rr}|}} \left(\frac{\partial}{\partial r}\right)^\mu. \qquad (10.11)$$

In fact, the norm $n^2 = n^\mu n_\mu$ takes the value $+1$ for $r > r_{\mathrm{H}}$ and -1 for $r < r_{\mathrm{H}}$, while for $r = r_{\mathrm{H}}$, it is zero. So the spherical surface corresponding to the horizon is a 'null hypersurface'. For $r > r_{\mathrm{H}}$, had we

---

[†] It is worth checking that this can be done for non-zero Q also, solving for the new radial coordinate via $(r^2 - 2Mr + Q^2)^{-1/2} dr = \rho^{-1} d\rho$. More generally, any spherically symmetric solution can be written in isotropic form, with sufficient effort.

approached this spacetime in a spaceship, we can blast our rockets and avoid the horizon if we choose, so any hypersurface this side of it is time-like, while any hypersurface the other side of it is spacelike, since we have to encounter them. Why do we have to encounter them? Well, looking back at the metric we see that in fact the role of t and r have exchanged roles for $r < r_{\mathrm{H}}$. This is because it is now the coefficient of dr^2 which is negative, and so it is really a *time* coordinate. So once we are in the region $r < r_{\mathrm{H}}$, all smaller values of r are in the inevitable future. The 'singularity' at $r = 0$ is a special case of one the inevitable spacelike hypersurfaces, so it is in our future as soon as we cross the horizon. In other words, Schwarzschild has a spacelike singularity, which is an important fact.

10.1.3 Basic properties of Reissner–Nordstrom

Let us consider the case of $Q \neq 0$, the charged black hole geometry. The set of spacelike Killing vectors representing spherical symmetry is similar to the case we had before, and there is again a timelike Killing vector arising form the t-invariance of the metric components, showing that the solution is static. When we come to look at the horizon structure, things get interesting. There are two, since there are two places where the hypersurface normal in equation (10.11) can go null:

$$r_{\pm} = M \pm \sqrt{M^2 - Q^2}.$$

It should be clear that there is a singularity at $r = 0$ again. Very interestingly, we can can see by looking at $\mathbf{n}$ that the singularity is *timelike*, and so it is in fact avoidable with sufficient effort, if one were moving in the geometry.

We have tacitly assumed that $M \geq Q$, or there will be no horizons, and the singularity at $r = 0$ will be a 'naked singularity', which is not allowed by the cosmic censors, it is believed[292]. That this is a strict and physical bound makes a lot of sense when we study this solution further, especially in a supersymmetric context, which we should do next.

10.1.4 Extremality, supersymmetry, and the BPS condition

There is a very important special case arising when we saturate the lower bound on M, making it equal to Q. Then we see that both horizons coincide at $r = Q$. Let us change coordinates to $R = r - Q$, giving:

$$ds^2 = -\frac{R^2}{(R+Q)^2}dt^2 + \frac{(R+Q)^2}{R^2}[dR^2 + R^2(d\theta^2 + \sin^2\theta d\phi^2)], \quad (10.15)$$

Insert 10.2. A little hypersurface technology

Let us formulate the idea of hypersurfaces within the parent geometry a bit more generally. This is a natural thing to consider in a text emphasising branes as hypersurfaces, and it shall be very useful to us later. Our spacetime M has coordinates x^μ, and a metric $G_{\mu\nu}$. A general hypersurface Σ within M deserves its own coordinates ξ^a, and so it is specified by an equation of the form $f(x^\mu(\xi^a)) = 0$. We have already met that there is natural metric induced on Σ, which is the 'pull-back' of the spacetime metric:

$$G_{ab} = \frac{\partial x^\mu}{\partial \xi^a} \frac{\partial x^\nu}{\partial \xi^b} G_{\mu\nu},$$

and we can define other useful quantities too. For example, the unit vector normal to this hypersurface is then specified as

$$n_\mu^\pm = \pm\sigma \frac{\partial f}{\partial x^\mu}, \quad \text{where} \quad \sigma = \left| G^{\mu\nu} \frac{\partial f}{\partial x^\mu} \frac{\partial f}{\partial x^\nu} \right|^{-1/2}. \quad (10.12)$$

In the simple case where Σ is, say, a spherical hypersurface of radius R, of one dimension fewer than M (with radial coordinate r), the equation specifying Σ is just $f = r - R = 0$. We can use the remaining angular coordinates of M as coordinates on Σ. Now, $\partial f/\partial r = 1$, giving (note the contravariant index):

$$n^\mu = \pm \frac{1}{\sqrt{|G_{rr}|}} \left(\frac{\partial}{\partial r} \right)^\mu.$$

A final useful thing we shall need is the extrinsic curvature or 'second fundamental form' of the surface, which is given by the pull-back of the covariant derivative of the normal vector:

$$K_{ab} = \frac{\partial x^\mu}{\partial \xi^a} \frac{\partial x^\nu}{\partial \xi^b} \nabla_\mu n_\nu = -n_\mu \left(\frac{\partial^2 x^\mu}{\partial \xi^a \partial \xi^b} + \Gamma^\mu_{\nu\rho} \frac{\partial x^\nu}{\partial \xi^a} \frac{\partial x^\rho}{\partial \xi^b} \right). \quad (10.13)$$

Like the induced metric, this is a tensor in the spacetime Σ. This might seem to be a daunting expression, but (like many things) it simplifies a lot in simple symmetric cases. So in our spherical example, using $r = R$, and the coordinates $\xi^a = x^a$, we get the simple expression:

$$K_{ab}^\pm = \frac{1}{2} n_\pm^\mu \frac{\partial G_{ab}}{\partial x^\mu}. \quad (10.14)$$

and the reader should notice that the metric is in a very special isotropic form. It is worth emphasising that the whole solution has a nice form, and can be written as:

$$ds^2 = -e^{2U}dt^2 + e^{-2U}(dR^2 + R^2 d\Omega_2^2);$$

$$A = -\left(e^{-U} - 1\right)dt, \quad \text{where} \quad e^{-U} = 1 + \frac{Q}{R}. \qquad (10.16)$$

This special form and generalisations of it (involving higher dimensions, extended objects, and the presence of other fields) will appear many times in what we study later, and so this is a good place to admire it properly before things get more complicated.

A very important reason why the extremal Reissner–Nordström solution is quite special is because it behaves very much like a BPS object, where $M \geq Q$ is the BPS bound. This is worth looking at very carefully, since it is an important theme that we have already visited, and we shall see many times again. To see the BPS properties, we can think of our Einstein–Maxwell action as the bosonic part of an $\mathcal{N} = 2$ supersymmetric theory of gravity. $\mathcal{N} = 2$ supergravity in four dimensions has three important types of massless multiplet. The gravity multiplet itself contains the graviton, two gravitinos and a vector called the graviphoton. So the bosonic content of our Einstein–Maxwell theory matches this nicely. We need only include a pair of spin $\frac{3}{2}$ ('*Rarita–Schwinger*') fields Ψ to play the role of the gravitino. The other two multiplets are the massless vector multiplet which contains a vector, a scalar and two spin $\frac{1}{2}$ particles, and the hypermultiplet which contains two spin $\frac{1}{2}$ particles and four scalars. The supersymmetry variations take bosonic fields into fermionic ones and vice versa, and the algebra can be written as:

$$\{Q_\alpha^i, Q_\beta^{\dagger j}\} = 2\gamma_{\alpha\beta}^\mu P_\mu \delta^{ij},$$

$$\{Q_\alpha^i, Q_\beta^j\} = 2\epsilon_{\alpha\beta} Z^{ij}, \qquad (10.17)$$

where the supercharges are written as Weyl spinors Q_α^j, ($\alpha = 1, 2$, $i = 1, 2$), with $Q_\alpha^{\dagger j}$ being the Hermitian conjugate. The quantity Z^{ij} is anti-symmetric, and commutes with everything else in the algebra. It is the *central charge*. Let us consider massive representations of the superalgebra. We can choose a basis in which $P_\mu = (M, 0, 0, 0)$. The little group is $SO(3)$. Writing the Z eigenvalue as simply $Z^{12} = Z$, we get

$$\{Q_\alpha^i, Q_\beta^{\dagger j}\} = M\delta^{ij},$$

$$\{Q_\alpha^i, Q_\beta^j\} = \epsilon_{\alpha\beta} |Z| \epsilon^{ij},$$

which, after taking linear combinations, we can write in terms of two families of fermionic creation and annihilation operators, a_α, a_α^+ and b_α, b_α^+:

$$\{a_\alpha, a_\beta^+\} = (M + |Z|)\delta^{ij},$$
$$\{b_\alpha, b_\beta^+\} = (M - |Z|)\delta^{ij}.$$

We can build representations of the algebra by starting with a Lorentz representation of some $SO(3)$ spin, s. We can write a ground state $|s\rangle$, which is defined as being annihilated by a_α^+ and b_α^+, and then we can proceed make 2^4 states by acting with the a_α and b_α. For example, starting with spin 1, one can make a massive vector multiplet whose content is the sum of the vector and hypermultiplet above. This the generic 'long' massive multiplet[63].

Since we must make unitary representations, the left hand side of the algebra above must be positive, and so we find that there is a bound

$$M \geq |Z|. \tag{10.18}$$

The only way to saturate this bound is if the state is annihilated by the b_α^+s, which is to say the state is invariant under half of the supersymmetry algebra. Then we only have the a_αs acting to make our multiplets and they are half the size. These are the special 'short' massive multiplets[63, 64]. There is a vector and a hyper of the same content mentioned above for the massless case, except that these can have any mass M.

The key point about extremal Reissner–Nordström is that it is part of a short hypermultiplet[65, 69]. This comes about in two stages. First, it has no fermion fields, and so the variation of all of the bosonic fields vanish when evaluated on this solution. This would be true for any old bosonic solution, of course. The remaining property is of course that the fermionic variations vanish for some choice of infinitessimal spinor ϵ_α generating the variation. Of course, it must be that only some of the spinors do this, otherwise we would be in a trivial situation. Setting the variation of the gravitino to zero, asks that there exists a spinor which solves:

$$D_\mu \epsilon_\alpha - \frac{1}{4} F_{\nu\kappa}^- \Gamma^\nu \Gamma^\kappa \Gamma_\mu \epsilon_\alpha = 0, \tag{10.19}$$

where $F_{\mu\nu}^\pm = \frac{1}{2}(F_{\mu\nu} \pm i^* F_{\mu\nu})$, and recall from equation (2.125) that the covariant derivative on the spinor involves the spin connection, $\omega^{ab}{}_\mu$

$$D_\mu \epsilon_\alpha \equiv \left(\delta_{\alpha\beta}\partial_\mu + \frac{1}{8}\omega^{ab}{}_\mu [\Gamma_a, \Gamma_b]_{\alpha\beta}\right)\epsilon^\beta.$$

This is asking for the existence of a *'covariantly constant'*, or *'Killing'*, spinor. It is a useful exercise to show that there are indeed such spinors. In fact, the problem reduces to just one differential equation which is satisfied everywhere by half of the available spinor components, matching the result above that half of the supersymmetries annihilate the solution.

In terms of the mass and the charge, things match as well. The graviphoton embedded in the gravity multiplet is a $U(1)$ gauge field whose charge is in fact the central charge. (There are gauge symmetries associated to the central charge operator which are local symmetries in supergravity[67].) So in fact, Z is the integral of the field strength two-form: $Z = (\int F)/4\pi = Q$, in the normalisation we are using. This matches with the property of our black hole solution.

10.1.5 Multiple black holes and multicentre solutions

It is important to note that there is a simple generalisation of the extremal solution to a case representing N distinct black holes of the same type:

$$ds^2 = -e^{2U}dt^2 + e^{-2U}(dR^2 + R^2 d\Omega_2^2);$$

$$A = -\left(e^{-U} - 1\right) dt, \quad \text{where} \quad e^{-U} = 1 + \sum_{i=1}^{N} \frac{q_i}{|\vec{R} - \vec{R_i}|},$$

(10.20)

where, in this 'multicentre' solution, $\vec{R_i}$ is a three-vector giving the location of the centre of the ith black hole with mass $m_i = q_i$. The total charge sourced by the whole configuration is, by Gauss's Law, simply $Q = \sum_{i=1}^{n} q_i$, which, by the BPS bound, is also equal to the total mass. This implicitly tells us that there is also a no-force condition applying to our black holes, since the total mass-energy is simply the sum of the individual mass-energies – there is no binding energy, coming from work against interaction forces.

The quickest way to see that this form arises as a solution is to rewrite the equation for the present Killing spinor as a condition on the solution written in the form in the first line of equation (10.20). We can do it for the slightly more general form where $dR^2 + R^2 d\Omega_2^2$ is replaced by $d\vec{x} \cdot d\vec{x}$. The resulting equation is simply that the e^{-U} be an *harmonic* function on the transverse space $\mathbb{R}^3$, for which after normalising it to be unity at infinity, we can choose for it to be written in the multicentre form. These are in general known as the Majumdar–Papapetrou solutions[66], and the spherical cases we've been looking at here are a special subclass corresponding to Reissner–Nordström.

10.1.6 Near horizon geometry and an infinite throat

It is particularly interesting to look closely at the horizon of the charged black hole in the extremal limit. Let us look at equation (10.15) in the neighbourhood of $R = 0$, the horizon, where we have

$$ds^2 = -\frac{R^2}{Q^2}dt^2 + \frac{Q^2}{R^2}dR^2 + Q^2(d\theta^2 + \sin^2\theta d\phi^2). \qquad (10.21)$$

The spatial part of the solution has degenerated into the product of an infinitely long tube or 'throat' of topology $\mathbb{R} \times S^2$ with fixed radius set by the charge. The whole geometry, called the 'Bertotti–Robinson' universe is actually $AdS_2 \times S^2$, a two dimensional 'anti-de Sitter' spacetime being the (t, R) part. Anti-de Sitter spacetime is the most symmetric 'vacuum' solution to *two dimensional* Einstein's equations with a negative cosmological constant. This pleasingly simple near-horizon geometry is a sign of something more general which will occur in all its glory in chapter 18 and so it is worthwhile understanding the toy example presented here, and also worthwhile digressing on solutions of Einstein's equations in the presence of cosmological constant, for later use.

This has special meaning for the supersymmetric discussion above as well. At infinity, the solution is of course flat space, which has all eight of the maximum set of available Killing spinors. At arbitrary radius, there are four, as mentioned above. It turns out that the Bertotti–Robinson geometry also has eight Killing spinors, and so is also a maximally suppersymmetric vacuum of the theory, just like flat space. In this sense we see that the extremal Reissner–Nordström solution is akin to a soliton[65], since it behaves as an interpolating solution between two vacua (see insert 1.4). Much the same thing will be true for some of the extremal brane solutions which we shall encounter later[68].

10.1.7 Cosmological constant; de Sitter and anti-de Sitter

In General Relativity, the Einstein tensor $G_{\mu\nu} \equiv R_{\mu\nu} - \frac{1}{2}g_{\mu\nu}R$ is arrived at by asking that the field equations be written in terms of the unique symmetric, rank two covariantly conserved object constructed out of the metric and its derivatives which has Minkowski space as a vacuum solution. If we wish to relax that final condition somewhat, we have a slightly more general choice. Of course, the metric itself is a symmetric rank two tensor, and since $\nabla_\mu g^{\mu\nu} = 0$, so it is also a candidate. We can add it in with an arbitrary constant, to give

$$R_{\mu\nu} - \frac{1}{2}g_{\mu\nu}R - \Lambda g_{\mu\nu} = 8\pi G T_{\mu\nu}, \qquad (10.22)$$

for which the generalisation of the Einstein–Hilbert Lagrangian is

$$\mathcal{L} = (-g)^{1/2}(R - 2\Lambda).$$

Recall from General Relativity[292] the form of the stress tensor for a perfect fluid of scalar density and pressure ρ and p:

$$T^{\mu\nu} = (\rho + p)u^\mu u^\nu + pg^{\mu\nu}. \tag{10.23}$$

We see that the 'cosmological constant' Λ acts like an intrinsic universal pressure. $\Lambda > 0$ is a cosmological repulsion, while $\Lambda < 0$ is an attraction.

While Minkowski space is no longer a solution, there are highly symmetric solutions analogous to it in the presence of non-zero Λ. Actually, the type of solutions we are looking for are called 'maximally symmetric' and satisfy the condition

$$R_{\lambda\mu\kappa\nu} = \mp\frac{2}{\ell^2}\left(g_{\lambda\kappa}g_{\mu\nu} - g_{\lambda\nu}g_{\kappa\mu}\right), \quad \text{or} \quad R_{\mu\nu} = \pm\frac{2\Lambda}{(D-2)}g_{\mu\nu}$$

$$\text{where} \quad \ell^2 = -\frac{(D-1)(D-2)}{2\Lambda}. \tag{10.24}$$

Already familiar are the signature $(+ + + \cdots)$ spaces which satisfy equation (10.24) with the plus sign, the round spheres S^D. In fact, for signature $(- + + \cdots)$ the spaces of interest here may be written as:

$$ds^2 = -\left(1 - \pm\frac{r^2}{\ell^2}\right)dt^2 + \left(1 - \pm\frac{r^2}{\ell^2}\right)^{-1}dr^2 + r^2 d\Omega^2_{D-2}, \tag{10.25}$$

where $d\Omega^2_{D-2}$ is the metric on a unit round $D - 2$ sphere.

The cosmological constant sets a length scale, ℓ. The larger the cosmological constant, the smaller the scale. The limit $r \ll \ell$ therefore returns us locally to Minkowski space, since if we fall below the length scale set by Λ, we simply do not notice, locally, that we have a cosmological constant, Λ. For $r \simeq \ell$ or greater we cannot ignore the effect of the cosmological constant.

10.1.8 de-Sitter spacetime and the sphere

For instance, notice that for the case of the plus sign, de Sitter space, there is an horizon at $r = \ell$. Since r cannot exceed ℓ, we might as well write $r = \ell \sin\theta$. A little algebra shows that, if we analytically continue time via $it = \ell\psi$, we get the metric

$$ds^2 = \ell^2(d\theta^2 + \cos^2\theta \, d\psi^2 + \sin^2\theta \, d\Omega^2_{D-2}), \tag{10.26}$$

which is the metric on a round sphere S^D, with radius ℓ, if ψ and θ have the appropriate periodicities.

10.1.9 Anti-de Sitter in various coordinate systems

The case of anti-de Sitter, the minus sign, we can instead take $r = \ell \sinh \rho$, and get

$$ds^2 = -\cosh^2\rho\, dt^2 + \ell^2 d\rho^2 + \ell^2 \sinh^2 \rho\, d\Omega^2_{D-2}, \tag{10.27}$$

which is a useful form which we will see much later. Notice that we can view this as an analytic continuation of the metric of the sphere S^D, given in equation (10.26).

There is a useful form of the metric to present which can be thought of as the $r \gg \ell$ limit. In this case, drop the 1 from $(1+r^2/\ell^2)$, and work with local coordinates. So write $\ell^2 d\Omega^2_{D-2}$, the metric on the S^{D-2} of radius ℓ embedded in $\mathbb{R}^{D-1}$ in Cartesian coordinates

$$\ell^2 d\Omega^2_{D-2} = dx_1^2 + dx_2^2 + \cdots + dx_{D-2}^2 + \frac{(x_1 dx_1 + x_2 dx_2 + \cdots + x_{D-2} dx_{D-2})^2}{x_{D-1}^2},$$

where $x_{D-1}^2 = \ell^2 - \sum_{i=1}^{D-1} x_i^2$. Then we can work in the local neighbourhood of $x_i \sim 0$, $x_{D-1} \sim \ell$, giving

$$\ell^2 d\Omega^2_{D-2} \simeq dx_1^2 + dx_2^2 + \cdots + dx_{D-2}^2.$$

Choosing these local coordinates is equivalent to the large radius limit of the sphere, and the rest of the geometry therefore takes the form:

$$ds^2 = \frac{r^2}{\ell^2}\left(-dt^2 + dx_1^2 + \cdots + dx_{D-2}^2\right) + \ell^2 \frac{dr^2}{r^2}, \tag{10.28}$$

which is known as the 'Poincaré' form of the metric, which arose already as part of the throat (10.21) of the Reissner–Nordstrom solution, and it shall arise again later. The radial coordinate R used there should be compared to r here, and the infinite line $\mathbb{R}$ coordinatised by t should be compared to the $\mathbb{R}^{D-1}$ coordinatised by $(t, x_1, \ldots, x_{D-2})$. Notice that the metric on that subspace (obtained by radial slices of constant r) is actually that of $D-1$ dimensional Minkowski, a fact which will be important for us later. The horizon at $R = 0$ compares to an horizon at $r = 0$ here, which is an important clue as to where anti-de Sitter will arise in later sections and chapters.

Actually, we can write another metric for AdS as follows:

$$ds^2 = -\left(-1 + \frac{r^2}{\ell^2}\right) dt^2 + \left(-1 + \frac{r^2}{\ell^2}\right)^{-1} dr^2 + r^2 d\Xi^2_{D-2},$$

where $d\Xi^2_{D-2}$ is the 'unit' metric on a $D-2$ dimensional hyperbolic space $\mathbb{H}^{D-2}$. This metric can be obtained by analytically continuing $d\Omega^2_{D-2}$. For this case, the radial slices are $\mathbb{H}^{D-2} \times \mathbb{R}$ instead of $D-1$ Minkowski space for the previous form (10.28) or $S^{D-2} \times \mathbb{R}$ for the form in equation (10.25). Just as before, we can do a hyperbolic change to a new coordinate $r = \ell \cosh \rho$, and get

$$ds^2 = -\sinh^2 \rho \, dt^2 + \ell^2 d\rho^2 + \ell^2 \cosh^2 \rho \, d\Xi^2_{D-2}).$$

In summary, we have AdS$_D$ in the following metrics:

$$ds^2 = -\left(k + \frac{r^2}{\ell^2}\right) dt^2 + \frac{dr^2}{\left(k + \frac{r^2}{\ell^2}\right)} + \frac{r^2}{\ell^2} d\Sigma^2_{k,D-2}, \qquad (10.29)$$

where the $(D-2)$-dimensional metric $d\Sigma^2_{k,D-2}$ is

$$d\Sigma^2_{k,D-2} = \begin{cases} \ell^2 d\Omega^2_{D-2} & \text{for } k = +1 \\ \sum_{i=1}^{D-2} dx_i^2 & \text{for } k = 0 \\ \ell^2 d\Xi^2_{D-2} & \text{for } k = -1, \end{cases} \qquad (10.30)$$

The $k = 0$ form can be thought of as the local physics in all three cases.

Anti-de Sitter space in D dimensions has an $SO(2, D-1)$ isometry, of which a subgroup $SO(1,1) \times ISO(1, D-2)$ is manifest as

$$(t, u, x_1, \ldots, x_{D-2}) \longrightarrow (\lambda t, \lambda^{-1} u, \lambda x_1, \ldots, \lambda x_{D-2}),$$

for the first factor, and the Poincaré group (i.e. Lorentz boosts and translations) acting on the Minkowski part. The group $SO(2, D-1)$ is the conformal group in $D-1$-dimensional Minkowski space, and the $SO(1,1)$ is the dilation part of it. The reader may recall that we met this group all the way back in chapter 3, and its appearance here will be given physical significance in terms of a duality in chapter 18.

10.1.10 Anti-de Sitter as a hyperbolic slice

It is worth noting that AdS$_D$ has a very natural geometrical representation. Start with the $(D+1)$-dimensional spacetime with signature $(-, -, +, +, \cdots)$, with metric:

$$ds^2 = -dX_0^2 - dX_D^2 + \sum_{i=1}^{D-1} dX_i^2. \qquad (10.31)$$

Notice that the isometry group of this homogeneous and isotropic spacetime is $SO(2, D-1)$. Now consider the hyperboloid within this spacetime, given by the equation

$$X_0^2 + X_D^2 - \sum_{i=1}^{D-1} X_i^2 = \ell^2.$$

A solution of this equation is

$$X_0 = \ell \cosh \rho \cos \tau / \ell, \quad X_D = \ell \cosh \rho \sin \tau / \ell, \quad X_i = \ell \Omega_i \sinh \rho,$$

where the angles Ω_i are chosen such that $\sum_{i=1}^{D-1} \Omega_i = 1$. We can substitute this solution into the metric (10.31) in order to find the metric on this hyperboloid, and we find the global AdS_D metric given in equation (10.27). With $0 \leq \tau \leq 2\pi$ and $0 \leq \rho$, our solution covers the entire hyperboloid once, and this is why these are called the 'global' coordinates on AdS. The time τ is usually taken not as a circle (which gives closed timelike curves) but on the real line, $-\infty \leq \tau + \infty$ giving the universal cover of the hyperboloid.

Another solution to the hyperboloid equation is:

$$X_0 = \frac{1}{2r}\left(1 + r^2(\ell^2 + \vec{x}^2 - t^2)\right), \quad X_D = rt$$

$$X_{D-1} = \frac{1}{2r}\left(1 - r^2(\ell^2 - \vec{x}^2 + t^2)\right), \quad X_i = rx_i,$$

which defines coordinates which cover a half of the hyperboloid. The resulting metric after substitution into equation (10.31) is the Poincaré form exhibited in equation (10.28). These are the 'local' coordinates.

10.1.11 Revisiting the extremal solution

How did constant curvature spaces, and negative cosmological constant become relevant to the Reissner–Nordström solution near the horizon at extremality? Well, it is worth examining the Ricci tensor in the extremal limit, in the coordinate $R = r - Q$, in the neighbourhood of the horizon $r = Q$:

$$R_{tt} = \frac{R^2}{Q^4}; \quad R_{rr} = -\frac{1}{R^2}; \quad R_{\theta\theta} = 1; \quad R_{\phi\phi} = \sin^2\theta, \tag{10.32}$$

and so we see that, upon comparing to equation (10.21):

$$R_{\mu\nu} = -\frac{1}{Q^2}g_{\mu\nu}; \quad \text{for } \mu, \nu = t \text{ or } r;$$

$$R_{\mu\nu} = +\frac{1}{Q^2}g_{\mu\nu}; \quad \text{for } \mu, \nu = \theta \text{ or } \phi. \tag{10.33}$$

Since the Maxwell stress tensor essentially obeys the same relations, giving something proportional to the metric tensor, it can be seen that the flux due to the charge carried by the hole is what is responsible for supplying the effective cosmological constant. It is worth noting that we could have formulated the same sort of features in terms of magnetic fields. In that case, we would have traded in the electric two form components for magnetic components $F = Q\epsilon_2$, where $\epsilon_2 = \sin\theta d\theta \wedge d\phi$ is the volume form of S^2. In this form, the decomposition of the throat solution by dualising the electric source into a magnetic source will generalise into something called the 'Freund–Rubin' ansatz in higher dimensional supergravity[19].

10.2 The geometry of D-branes

Now let us return to the full ten dimensional equations of motion of the type IIA and type IIB supergravity equations (7.41) and (7.42), where we have additional fields coming from the R–R sector and the NS–NS sector.

10.2.1 A family of 'p-brane' solutions

There is an interesting family of ten dimensional solutions, which source gravity, the dilaton, and the R–R potentials, and can be written as follows[94, 95]:

$$dS^2 = Z_p^{-1/2}(r)\left(-K(r)dt^2 + \sum_{i=1}^{p} dx_i^2\right) + Z_p^{1/2}(r)\left(\frac{dr^2}{K(r)} + r^2 d\Omega_{8-p}^2\right),$$
(10.34)

where $d\Omega_{8-p}^2$ is the metric on a unit round S^{8-p} sphere, and

$$Z_p(r) = 1 + \alpha_p \left(\frac{r_p}{r}\right)^{7-p},$$

$$K(r) = 1 - \left(\frac{r_H}{r}\right)^{7-p},$$

$$e^{2\Phi} = g_s^2 Z_p(r)^{\frac{(3-p)}{2}},$$

$$C_{(p+1)} = g_s^{-1}\left[Z_p(r)^{-1} - 1\right] dx^0 \wedge \cdots \wedge dx^p.$$
(10.35)

In the above

$$r_p^{7-p} = d_p(2\pi)^{p-2} g_s N \alpha'^{(7-p)/2}, \quad d_p = 2^{7-2p} \pi^{\frac{9-3p}{2}} \Gamma\left(\frac{7-p}{2}\right),$$

$$\alpha_p = \sqrt{1 + \left(\frac{r_H^{7-p}}{2r_p^{7-p}}\right)^2} - \frac{r_H^{7-p}}{2r_p^{7-p}}.$$
(10.36)

One should not be intimidated by the form of these solutions. They represent p-dimensional extended objects called 'p-branes', and as such, are localised in the $9 - p$ directions transverse to them. Since we have rotational symmetry in those directions, we can use polar coordinates with a radial coordinate r, and the angles on an $(8 - p)$-sphere. The branes are aligned along the $(x^1, x^2, \ldots, x^p)$ directions, and move in time, so they have a $(p + 1)$ dimensional world volume, with geometry $\mathbb{R}^{p+1}$, generalising the worldline of the black hole solutions we studied earlier. It is useful to observe how the solution is split between the transverse and parallel coordinates and then look at, say, the Schwarzschild or Reissner–Nordström solution (10.4) and see that the analogue of this is happening in that solution too. There, the world-volume is replaced by a simple world-line, the space $\mathbb{R}$ coordinatised by t. The rest of the solution concerns the transverse part of the spacetime. Since there is rotational symmetry it has a simple presentation in terms of the radius r and the two angles on the round S^2. From our analysis of the black hole solutions, it should be clear that these solutions have an horizon at radius $r = r_\mathrm{H}$, and a singularity at $r = 0$.

10.2.2 *The boost form of solution*

Actually there is another way of writing the solution which is instructive and useful for later. We could instead write:

$$Z_p(r) = 1 + \alpha_p \left(\frac{r_p}{r_\mathrm{H}}\right)^{7-p} \left(\frac{r_\mathrm{H}}{r}\right)^{7-p} = 1 + \sinh^2 \beta_p \left(\frac{r_\mathrm{H}}{r}\right)^{7-p},$$

where, given the nice form of α_p in equation (10.36), we can write

$$\alpha_p \left(\frac{r_p}{r_\mathrm{H}}\right)^{7-p} = \sqrt{\frac{1}{4} + \left(\frac{r_p^{7-p}}{r_\mathrm{H}^{7-p}}\right)^2} - \frac{1}{2} = \sinh^2 \beta_p,$$

and hence

$$\alpha_p = \tanh \beta_p; \qquad \cosh^2 \beta_p = \frac{1}{2} + \sqrt{\frac{1}{4} + \left(\frac{r_p^{7-p}}{r_\mathrm{H}^{7-p}}\right)^2}.$$

The tension and charge can be written in terms of these nicely as:

$$\tau_p = \left(\frac{r_\mathrm{H}}{r_p}\right)^{7-p} \frac{N}{g_\mathrm{s}} \left(\frac{1}{7 - p} + \cosh^2 \beta_p\right),$$

$$Q_p = N \left(\frac{r_\mathrm{H}}{r_p}\right)^{7-p} \sinh \beta_p \cosh \beta_p = N. \qquad (10.37)$$

So we see that in fact the solutions above are normalised such that they carry N units of the basic D-brane R–R charge μ_p, where N is an integer. Observe that the mass is larger than the charge, in a manner analogous to the Reissner–Nordström solution.

Notice that when the parameter β_p goes to zero, the solution simplifies drastically, becoming uncharged. The function Z_p becomes unity, the dilaton becomes constant, and the solution simply becomes a $(10 - p)$-dimensional Schwarzschild black hole, with horizon at $r = r_{\mathrm{H}}$, times the space $\mathbb{R}^p$.

10.2.3 The extremal limit and coincident D-branes

Just like in the case of the charged black hole solution, there are extremal limits of these solutions. The extremal cases are BPS solutions of the ten dimensional supersymmetry algebra, as we shall see. For now, the similarity with the detailed case study of Reissner–Nordström black holes in earlier sections should be borne in mind, although there are differences which will become apparent shortly. The extremal limit is simply $\alpha_p = 1$, where the solutions are:

$$
\begin{aligned}
ds^2 &= H_p^{-1/2}\eta_{\mu\nu}dx^\mu dx^\nu + H_p^{1/2}dx^i dx^i, \\
e^{2\Phi} &= g_{\mathrm{s}}^2 H_p^{\frac{(3-p)}{2}}, \\
C_{(p+1)} &= -(H_p^{-1} - 1)g_{\mathrm{s}}^{-1}dx^0 \wedge \cdots \wedge dx^p,
\end{aligned} \tag{10.38}
$$

where $\mu = 0, \ldots, p$, and $i = p+1, \ldots, 9$, and the harmonic function H_p is

$$
H_p = 1 + \left(\frac{r_p}{r}\right)^{7-p}, \tag{10.39}
$$

where r_p is still given in equation (10.36). In the boost form mentioned at the end of the last subsection, it is the limit of infinite boost, $\beta_p \to \infty$, combined with sending the horizon parameter r_{H} to zero while holding fixed the combination $r_{\mathrm{H}}^{(7-p)}e^{2\beta_p}/4 = r_p^{7-p}$.

It is worth comparing this to the form in equation (10.16), where the extremal black hole is written in isotropic form analogous to what we have here. Furthermore, it should be clear that there is a multicentre generalisation of this solution, where we write for the harmonic function

$$
H_p = 1 + \sum_{i=1}^{N} \frac{r_p^{7-p}}{|\vec{r} - \vec{r}_i|^{7-p}}. \tag{10.40}
$$

This represents N different branes located at arbitrary positions given by the vectors $\vec{r}_i$. A clear sign that the solution is a BPS object made of lots of smaller such objects is the fact that the mass computed for this solution is just the sum of the individual masses and is equal to the total charge. There is no binding energy since the interaction forces are zero.

It is clear that in all cases (except $p = 3$) the horizon, located at $r = 0$, is a singular place of zero area, since the radius of the S^{8-p} vanishes there. In the $p = 3$ case, however, the inverse quartic power of r appearing in the harmonic function means that the square root yields a cancellation between the vanishing of the horizon size and the divergence of the metric, leaving an horizon of finite size $r_3^{1/2} = \alpha'(4\pi g_s N)^{1/2}$. Some simple algebra shows that the geometry is simply $\text{AdS}_5 \times S^5$, with the sizes of each factor set by $r_3^{1/2}$. The dilaton is constant, and the R–R field is $F_{(5)} = dC_{(4)} + *dC_{(4)}$, where $dC_{(4)} = r_3 \epsilon_{(5)}$ where $\epsilon_{(5)}$ is the volume form on S^5.

Note again the sharp analogy with the case of Reissner–Nordström. The appearance of this simple smooth near-horizon geometry is interesting, and we will explore this much later, in chapter 18.

More complicated supergravity solutions preserving fewer supersymmetries (in the extremal case) can be made by combining these simple solutions in various ways, by intersecting them with each other, boosting them to finite momentum, and by wrapping, and/or warping them on compact geometries. This allows for the construction of finite area horizon solutions, corresponding to R–R charged Reissner–Nordström black holes, and generalisations thereof. We shall in fact do this in chapter 17.

These solutions are R–R charged with N units of Dp-brane charge, but we have already established to all orders in string perturbation theory that Dp-branes actually are the *basic sources* of the R–R fields. It is natural to suppose that there is a connection between these two families of objects: perhaps the solution (10.38) is 'made of D-branes' in the sense that it is actually the field due to N Dp-branes, all located at $r = 0$. This is precisely how we are to make sense of this solution as a supergravity soliton solution. We *must* do so, since (except for $p = 3$ as we have seen) the solution is actually singular at $r = 0$, and so one might have simply discarded them as pathological, since solitons 'ought to be smooth'. However, string duality, which we shall encounter in chapter 12, *forces* us to consider them, since smooth NS–NS solitons of various extended sizes (which can be made by wrapping or warping NS5-branes (see section 12.3 for their entry into our story) in an arbitrary compactification) are mapped[165] into these R–R solitons under it, generalising what we have already seen in ten dimensions. With the understanding that there are D-branes 'at their core', which fits with the fact that they are R–R charged, they make sense of the whole spectrum of extended solitons in string theory.

Let us build up the logic of how they can be related to D-branes. Recall that the form of the action of the ten dimensional supergravity with NS–NS and R–R field strengths H and G respectively is, roughly:

$$S = \int d^{10}x \sqrt{-g} \left[e^{-2\Phi} \left(R - H^2 \right) - G^2 \right]. \tag{10.41}$$

There is a balance between the dilaton dependence of the NS–NS and gravitational parts, and so the mass of a soliton solution[95] carrying NS–NS charge (like the NS5-brane) scales like the action: $T_{\mathrm{NS}} \sim e^{-2\Phi} \sim g_{\mathrm{s}}^{-2}$. An R–R charged soliton has, on the other hand, a mass which goes like the geometric mean of the dilaton dependence of the R–R and gravitational parts: $T_{\mathrm{R}} \sim e^{-\Phi} \sim g_{\mathrm{s}}^{-1}$. This is just the behaviour we saw for the tension of the Dp-brane, computed in string perturbation theory, treating them as boundary conditions.

We have so far treated Dp-branes as point-like (in their transverse dimensions) in an otherwise flat spacetime. We were able to study an arbitrary number of them by placing the appropriate Chan–Paton factors into amplitudes. However, the solutions (10.38) have non-trivial spacetime curvature, and is only asymptotically flat. How are these two descriptions related?

Well, for every Dp-brane which is added to a situation, another boundary is added to the problem, and so a typical string diagram has a factor $g_{\mathrm{s}}N$ since every boundary brings in a factor g_{s} and there is the trace over the N Chan–Paton factors. So open string perturbation theory is good as long as $g_{\mathrm{s}}N < 1$. Notice that this is the regime where the supergravity solution (10.38) fails to be valid, since the typical squared curvature invariant behaves as

$$R^2 \sim \left(\frac{r_p}{r} \right)^{7-p} \sim g_{\mathrm{s}}N \left(\frac{\sqrt{\alpha'}}{r} \right)^{7-p}.$$

On the other hand, for $g_{\mathrm{s}}N > 1$, the supergravity solution has its curvature weakened, and can be considered as a workable solution. This regime is where the open string perturbation theory, on the other hand, breaks down.

So we have a fruitful complementarity between the two descriptions. In particular, since we only derived the supergravity equations of motion in string perturbation theory, i.e. $g_{\mathrm{s}} < 1$, for most computations, we can work with the supergravity solution with the interpretation that N is very large, such that the curvatures are small. Alternatively, if one restricts oneself to studying only the BPS sector, then one can work with arbitrary N, and extrapolate results – computed with the D-brane description for small

g_s – to the large g_s regime (since there are often non-renormalisation theorems which apply), where they can be related to properties of the non-trivial curved solutions. This is the basis of the successful statistical enumeration of the entropy of black holes, for cases where the solutions (10.38) are used to construct R–R charged black holes. We shall do this in chapter 17.

In summary, for a large enough number of coincident D-branes or for strong enough string coupling, one cannot consider them as points in flat space: they deform the spacetime according to the geometry given in equation (10.38). Given that D-branes are also described very well at low energy by gauge theories, this gives plenty of scope for finding a complementarity between descriptions of non-trivially curved geometry and of gauge theory. This is the basis of what might be called 'gauge theory/geometry' correspondences. In some cases, when certain conditions are satisfied, there is a complete decoupling of the supergravity description from that of the gauge theory, signalling a complete *duality* between the two. This is the basis of the AdS/CFT correspondence, which we shall come to in chapter 18.

10.3 Probing p-brane geometry with Dp-branes

In the previous section, we argued that the spacetime geometry given by equations (10.38) represents the spacetime fields produced by N Dp-branes. We noted that as a reliable solution to supergravity, the product $g_s N$ ought be be large enough that the curvatures are small. This corresponds to either having N small and g_s large, or vice versa. Since we are good at studying situations with g_s small, we can safely try to see if it makes sense to make N large.

10.3.1 Thought experiment: building p with Dp

One way to imagine that this spacetime solution came about at weak coupling was that we built it by bringing in N Dp-branes, one by one, from infinity. If this is to be a sensible process, we must study whether it is really possible to do this. Imagine that we have been building the geometry for a while, bringing up one brane at a time from $r = \infty$ to $r = 0$. Let us now imagine bringing the next brane up, in the background fields created by all the other N branes. Since the branes share p common directions where there is no structure to the background fields, we can ignore those directions and see that the problem reduces to the motion of a test particle in the transverse $9 - p$ spatial directions. What is the mass of this particle, and what is the effective potential in which it moves?

We can answer this sort of question using the toolbox which combines the fact that at low energy we know the world-volume action of the D-brane, describing how it interacts with the background fields with the fact that the probe brane is a heavy object which can examine many distance scales in the theory[106].

10.3.2 Effective Lagrangian from the world-volume action

We can find the answers to all of the above questions by deriving an effective Lagrangian for the problem which results from the world-volume action of the brane. We can exploit the fact that we have spacetime Lorentz transformations and world-volume reparametrisations at our disposal to choose the work in the 'static gauge'. In this gauge, we align the world-volume coordinates, ξ^a, of the brane with the spacetime coordinates such that:

$$
\begin{aligned}
\xi^0 &= x^0 = t; \\
\xi^i &= x^i; \quad i = 1, \ldots, p, \\
\xi^m &= \xi^m(t); \quad m = p+1, \ldots, 9.
\end{aligned}
\tag{10.42}
$$

The Dirac–Born–Infeld part of the action (5.21) requires the insertion of the induced metric derived from the metric in question. In static gauge, it is easy to see that the induced metric becomes:

$$
[G]_{ab} = \begin{pmatrix}
G_{00} + \sum_{mn} G_{mn} v_m v_n & 0 & 0 & \cdots & 0 \\
0 & & G_{11} & 0 & \cdots & \vdots \\
\vdots & & \vdots & \ddots & \vdots & \vdots \\
\vdots & & \vdots & & \ddots & \vdots \\
0 & & 0 & 0 & \cdots & G_{pp}
\end{pmatrix},
\tag{10.43}
$$

where $v_m \equiv dx^m/d\xi^0 = \dot{x}^m$.

In our particular case of a simple diagonal metric, the determinant turns out as

$$
\det[-G_{ab}] = H_p^{-\frac{(p+1)}{2}} \left(1 - H_p \sum_{m=p+1}^{9} v_m^2\right) = H_p^{-\frac{(p+1)}{2}} \left(1 - H_p v^2\right).
\tag{10.44}
$$

The Wess–Zumino term representing the electric coupling of the brane is, in this gauge:

$$
\mu_p \int C_{(p+1)} = \mu_p \int d^{p+1}\xi \; \varepsilon^{a_0 a_1 \cdots a_p} [C_{(p+1)}]_{\mu_0 \mu_1 \cdots \mu_p} \frac{\partial x^{\mu_0}}{\partial \xi^{a_0}} \frac{\partial x^{\mu_1}}{\partial \xi^{a_1}} \cdots \frac{\partial x^{\mu_p}}{\partial \xi^{a_p}}
$$

$$
= \mu_p V_p \int dt \left[H_p^{-1} - 1\right] g_s^{-1},
\tag{10.45}
$$

where $V_p = \int d^p x$, the spatial world-volume of the brane. Now, we are going to work in the approximation that we bring the branes *slowly* up the the main stack of branes so we keep the velocity small enough such that only terms up to quadratic order in v are kept in our computation. We can therefore the expand the square root of our determinant, and putting it all together (not forgetting the crucial insertion of the background functional dependence of the dilaton from (10.38)) we get that the action is:

$$S = \mu_p V_p \int dt \left(-g_s^{-1} H_p^{-1} + \frac{1}{2g_s} v^2 + g_s^{-1} H_p^{-1} - g_s^{-1} \right)$$
$$= \int dt \mathcal{L} = \int dt \left(\frac{1}{2} m_p v^2 - m_p \right), \tag{10.46}$$

which is just a Lagrangian for a free particle moving in a constant potential, (which we can set to zero) where $m_p = \tau_p V_p$ is the mass of the particle.

This result has a number of interesting interpretations. The first is simply that we have successfully demonstrated that our procedure of 'building' our geometry (10.38) by successively bringing branes up from infinity to it, one at a time, makes sense. There is no non-trivial potential in the effective Lagrangian for this process, so there is no force required to do this; correspondingly there is no binding energy needed to make this system.

That there is no force is simply a restatement of the fact that these branes are BPS states, all of the same species. This manifests itself here as the fact that the R–R charge is equal to the tension (with a factor of $1/g_s$), saturating the BPS bound. It is this fact which ensured the cancellation between the r-dependent parts in (10.46) which would have otherwise resulted in a non-trivial potential $U(r)$. (Note that the cancellation that we saw only happens at order v^2 – the slow probe limit. Beyond that order, the BPS condition is violated, since it really only applies to statics.)

10.3.3 A metric on moduli space

All of this is pertinent to the world-volume field theory as well. Recall that there is a $U(N)$ $(p + 1)$-dimensional gauge theory on a family of N Dp-branes. Recall furthermore that there is a sector of the theory which consists of a family of $(9 - p)$ scalars, Φ^m, in the adjoint. Geometrically, these are the collective coordinates for motions of the branes transverse to their world-volumes. Classical background values for the fields, (defining vacua about which we would then do perturbation theory) are equivalent to data about how the branes are distributed in this transverse space. Well, we have just confirmed that there is a 'moduli space' of inequivalent vacua of the theory corresponding to the fact that one can give a vacuum expectation value to a component of a Φ^m, representing a brane moving away from the clump of N branes. That there is no potential translates into the fact that we can place the brane anywhere in this transverse clump, and it will stay there.

It is also worth noting that this metric on the moduli space is *flat*; treating the fields Φ^m as coordinates on the space $\mathbb{R}^{9-p}$, we see (from the fact that the velocity squared term in (10.46) appears as $v^2 = \delta_{mn}v^m v^n$) that the metric seen by the probe is simply

$$ds^2 \sim \delta_{mn} d\Phi^m d\Phi^n. \tag{10.47}$$

This flatness is a consequence of the high amount of supersymmetry (16 supercharges). For the case of D3-branes (whether or not they are in the $\text{AdS}_5 \times S^5$ limit, to be described later), this result translates into the fact there that there is no running of the gauge coupling g_{YM}^2 of the superconformal gauge theory on the brane, (since in this example, and in the case of eight supercharges, supersymmetry relates the coupling to the kinetic term). This is read off from the prefactor $g_{\text{YM}}^{-2} = \tau_3 (2\pi\alpha')^2 = (2\pi g_{\text{s}})^{-1}$ in the metric. The supersymmetry ensures that any corrections which could have been generated are zero. We shall later see less trivial versions, where we have nontrivial metrics in the case of eight supercharges and even four supercharges. Before we do that, we have to go back to studying D-branes as boundary conditions, in order to see how to put together multiple D-branes, and branes of different types.

10.4 T-duality and supergravity solutions

In principle, nothing stops us from studying the action of T-duality on the Dp-branes, now starting with their representation as a supergravity solution, and correspondingly using the background field T-duality rules given in equation (5.4) for the NS–NS sector, and equations (8.2) for the R–R sector. One should expect to get the supergravity solution of a D$(p+1)$-brane or D$(p-1)$-brane, depending upon whether one T-dualised in a direction containing the Dp-brane's world-volume or not. This expectation is indeed borne out to some extent, but we must be careful. Let us discuss the subtlety by example.

10.4.1 D$(p+1)$ from Dp

Start with the case of T-dualising in a direction transverse to a Dp-brane, lying in directions $X^1, \ldots, X^p$. What this really means, recall, is that we must place the branes on a circle of radius R, and find an equivalent representation for the system on a dual circle of radius $R' = \alpha'/R$. We can represent this as an infinite array of identical branes on the line with coordinate X^{p+1}, a distance $2\pi R$ apart, identifying $X^{p+1} \sim X^{p+1} + 2\pi R$. We can easily write a supergravity solution for this, since the branes are BPS, and so the multibrane harmonic function in equation (10.40) can be employed here. Let us write the radius in the directions transverse to the Dp-brane in terms of X^{p+1} and a radius in the remaining directions:

$$r^2 = (X^{p+1})^2 + (X^{p+2})^2 + \cdots + (X^{9-p})^2 = \hat{r}^2 + (X^{p+1})^2,$$

in terms of which the appropriate harmonic function including all of the images is:

$$H_p^{\text{array}} = 1 + \sum_{n=-\infty}^{+\infty} \frac{r_p^{7-p}}{|\hat{r}^2 + (X^{p+1} - 2\pi nR)^2|^{(7-p)/2}}. \tag{10.48}$$

If the circle's radius is very small, then the sum in the above can be replaced by an integral, to a good approximation, since the difference between each term in the sum is small. Defining a new variable u via $\hat{r}u = 2n\pi R - X^{p+1}$, we get:

$$H_p^{\text{array}} \sim 1 + \frac{r_p^{7-p}}{2\pi R} \frac{1}{\hat{r}^{6-p}} \int_{-\infty}^{\infty} \frac{du}{(1+u^2)^{(7-p)/2}}, \tag{10.49}$$

where we have used $\hat{r}\delta u = 2\pi R\delta n$ to get the measure right. The integral is:

$$\int_{-\infty}^{\infty} \frac{du}{(1+u^2)^{(7-p)/2}} = \frac{\sqrt{\pi}\Gamma\left[\frac{1}{2}(6-p)\right]}{\Gamma\left[\frac{1}{2}(7-p)\right]},$$

and so looking at the definition of the constant r_p^{7-p} given in equation (10.36), we see that

$$H_p^{\text{array}} \sim H_{p+1} = 1 + \frac{\sqrt{\alpha'}r_{p+1}^{7-(p+1)}}{R} \frac{1}{\hat{r}^{7-(p+1)}}, \tag{10.50}$$

which is the correct form of the harmonic function for a D$(p+1)$-brane.

We should check normalisations here. If we had started with a single brane on the array, i.e. with $N = 1$, then we get the new number of branes as $\tilde{N} = \sqrt{\alpha'}/R$. So if $R = \sqrt{\alpha'}$, then we have the correct normalisation for a single brane on the dual side also. Better perhaps is to have $N = R/\sqrt{\alpha'}$, giving a single $\tilde{N} = 1$ as the T-dual. This has the interpretation in the original theory as $R/\sqrt{\alpha'}$ for each $2\pi R$ of length, or $2\pi\sqrt{\alpha'}$ branes per unit length.

We can work on the full Dp-brane metric with the T-duality rules (5.4), treating X^{p+1} as the isometry direction. Following the rules through, we see that the transformation will invert the metric function $G_{p+1,p+1}$, which will indeed convert the metric for a p-brane to that of a $(p+1)$-brane. So the new dilaton is, according to the rules in equation (5.4),

$$e^{2\tilde{\Phi}} = \frac{e^{2\Phi}}{G_{p+1,p+1}} = \frac{e^{2\Phi}}{H_p^{1/2}} = g_s^2 H_p^{\frac{(3-(p+1))}{2}},$$

which after replacing H_p by H_p^{array}, which becomes H_{p+1} as we have shown above, gives the dilaton for the D$(p+1)$-brane supergravity solution. Similarly, equations (8.2) give the correct R–R potential.

This works very well because it is easy to soften the power of r which appears in the denominator of the harmonic function, as needed for a larger brane.

10.4.2   D$(p-1)$ from Dp

Harder to get is the increase of the power of r in the dependence of the harmonic function, which we would need for a D$(p-1)$-brane, if we T-dualised in a world-volume direction, say X^p. Clearly the powers of the harmonic function itself will in the metric, dilaton and R–R potential, using the rules (5.4) and (8.2). The problem is that we would get

$$H_p = 1 + \frac{r_p^{7-p}}{r^{7-p}}. \tag{10.51}$$

This is not really what we want. We can, however, interpret this as the result of 'smearing' the brane in the direction X^p, i.e. the result of integrating a uniform density of branes (with the correct $1/r^{(8-p)}$ behaviour) over X^p. This will indeed yield the behaviour given in (10.51). We shall encounter such smeared solutions, or 'brane distributions' in later chapters.

11

Multiple D-branes and bound states

In chapter 5, we saw a number of interesting terms arise in the Dp-brane world-volume action which had interpretations as smaller branes. For example, a $U(1)$ flux was a $D(p-2)$-brane fully delocalised in the world-volume, while for the non-Abelian case, we saw a $D(p-4)$-brane arise as an instanton in the world-volume gauge theory. Interestingly, while the latter breaks half of the supersymmetry again, as it ought to, the former is still half BPS, since it is T-dual to a tilted $D(p+1)$-brane.

It is worthwhile trying to understand this better back in the basic description using boundary conditions and open string sectors, and this is the first goal of this chapter. After that, we'll have a closer look at the nature of the BPS bound and the superalgebra, and study various key illustrative examples.

11.1 Dp and Dp' from boundary conditions

Let us consider two D-branes, Dp and Dp', each parallel to the coordinate axes. (We can of course have D-branes at angles[129], but we will not consider this here.) An open string can have both ends on the same D-brane or one on each. The $p-p$ and $p'-p'$ spectra are the same as before, but the $p-p'$ strings are new if $p \neq p'$. Since we are taking the D-branes to be parallel to the coordinate axes, there are four possible sets of boundary conditions for each spatial coordinate X^i of the open string, namely NN (Neumann at both ends), DD, ND, and DN. What really will matter is the number ν of ND plus DN coordinates. A T-duality can switch NN and DD, or ND and DN, but ν is invariant. Of course ν is even because we only have p even or p odd in a given theory in order to have a chance of preserving supersymmetry.

249

The respective mode expansions are

$$\text{NN:} \quad X^\mu(z,\bar{z}) = x^\mu - i\alpha' p^\mu \ln(z\bar{z}) + i\sqrt{\frac{\alpha'}{2}} \sum_{m\neq 0} \frac{\alpha_m^\mu}{m}(z^{-m} + \bar{z}^{-m}),$$

$$\text{DN, ND:} \quad X^\mu(z,\bar{z}) = i\sqrt{\frac{\alpha'}{2}} \sum_{r\in\mathbb{Z}+1/2} \frac{\alpha_r^\mu}{r}(z^{-r} \pm \bar{z}^{-r}),$$

$$\text{DD:} \quad X^\mu(z,\bar{z}) = -i\frac{\delta X^\mu}{2\pi} \ln(z/\bar{z}) + i\sqrt{\frac{\alpha'}{2}} \sum_{m\neq 0} \frac{\alpha_m^\mu}{m}(z^{-m} - \bar{z}^{-m}).$$

$$(11.1)$$

In particular, the DN and ND coordinates have half-integer moding. The fermions have the same moding in the Ramond sector (by definition) and opposite in the Neveu–Schwarz sector. The string zero point energy is 0 in the R sector as always, and using (2.80) we get:

$$(8-\nu)\left(-\frac{1}{24}-\frac{1}{48}\right) + \nu\left(\frac{1}{48}+\frac{1}{24}\right) = -\frac{1}{2}+\frac{\nu}{8} \qquad (11.2)$$

in the NS sector.

The oscillators can raise the level in half-integer units, so only for ν a multiple of four is degeneracy between the R and NS sectors possible. Indeed, it is in this case that the Dp–Dp′ system is supersymmetric. We can see this directly. As discussed in sections 8.1.1 and 8.2, a D-brane leaves unbroken the supersymmetries

$$Q_\alpha + P\tilde{Q}_\alpha, \qquad (11.3)$$

where P acts as a reflection in the direction transverse to the D-brane. With a second D-brane, the only unbroken supersymmetries will be those that are also of the form

$$Q_\alpha + P'\tilde{Q}_\alpha = Q_\alpha + P(P^{-1}P')\tilde{Q}_\alpha, \qquad (11.4)$$

with P' the reflection transverse to the second D-brane. Then the unbroken supersymmetries correspond to the $+1$ eigenvalues of $P^{-1}P'$. In DD and NN directions this is trivial, while in DN and ND directions it is a net parity transformation. Since the number ν of such dimensions is even, we can pair them as we did in section 7.1.1, and write $P^{-1}P'$ as a product of rotations by π,

$$e^{i\pi(J_1+\cdots+J_{\nu/2})}. \qquad (11.5)$$

In a spinor representation, each $e^{i\pi J}$ has eigenvalues $\pm i$, so there will be unbroken supersymmetry only if ν is a multiple of four as found above*.

For example, type I theory, besides the D9-branes, will have D1-branes and D5-branes. This is consistent with the fact that the only R–R field strengths are the three-form and its Hodge-dual seven-form. The D5-brane is required to have two Chan–Paton degrees of freedom (which can be thought of as images under Ω) and so an $SU(2)$ gauge group[130, 132].

When $\nu = 0$, $P^{-1}P' = 1$ identically and there is a full ten-dimensional spinor of supersymmetries. This is the same as for the original type I theory, to which it is T-dual. In $D = 4$ units, this is $\mathcal{N} = 4$, or sixteen supercharges. For $\nu = 4$ or $\nu = 8$ there is $D = 4$ $\mathcal{N} = 2$ supersymmetry.

Let us now study the spectrum for $\nu = 4$, saving $\nu = 8$ for later. Sometimes it is useful to draw a quick table showing where the branes are located. Here is one for the (9,5) system, where the D5-brane is pointlike in the x^6, x^7, x^8, x^9 directions and the D9-brane is (of course) extended everywhere.

	x^0	x^1	x^2	x^3	x^4	x^5	x^6	x^7	x^8	x^9
D9	—	—	—	—	—	—	—	—	—	—
D5	—	—	—	—	—	—	•	•	•	•

A dash under x^i means that the brane is extended in that direction, while a dot means that it is pointlike there.

Continuing with our analysis, we see that the NS zero-point energy is zero. There are four periodic world-sheet fermions ψ^i, namely those in the ND directions. The four zero modes generate $2^{4/2}$ or four ground states, of which two survive the GSO projection. In the R sector the zero-point energy is also zero; there are four periodic transverse ψ, from the NN and DD directions not counting the directions $\mu = 0, 1$. Again these generate four ground states of which two survive the GSO projection. The full content of the $p - p'$ system is then is half of an $N = 2$ hypermultiplet. The other half comes from the $p' - p$ states, obtained from the orientation reversed strings: these are distinct because for $\nu \neq 0$ the ends are always on different D-branes.

Let us write the action for the bosonic $p - p'$ fields χ^A, starting with $(p, p') = (9, 5)$. Here A is a doublet index under the $SU(2)_R$ of the $N = 2$ algebra. The field χ^A has charges $(+1, -1)$ under the $U(1) \times U(1)$ gauge theories on the branes, since one end leaves, and the other arrives. The

---

* We will see that there are supersymmetric *bound states* when $\nu = 2$.

minimally coupled action is then

$$\int d^6\xi \left(\sum_{a=0}^{5} |(\partial_a + iA_a - iA'_a)\chi|^2 + \left(\frac{1}{4g^2_{\mathrm{YM},p}} + \frac{1}{4g^2_{\mathrm{YM},p'}} \right) \sum_{I=1}^{3} (\chi^\dagger \tau^I \chi)^2 \right),$$

$$(11.6)$$

with A_a and A'_a the brane gauge fields, $g_{\mathrm{YM},p}$ and $g_{\mathrm{YM},p'}$ the effective Yang–Mills couplings (8.13), and τ^I the Pauli matrices. The second term is from the $N = 2$ D-terms for the two gauge fields. It can also be written as a commutator $\mathrm{Tr}\,[\phi^i, \phi^j]^2$ for appropriately chosen fields ϕ^i, showing that its form is controlled by the dimensional reduction of an F^2 pure Yang–Mills term. See section 13.1 for more on this.

The integral is over the five-brane world-volume, which lies in the nine-brane world-volume. Under T-dualities in any of the ND directions, one obtains $(p, p') = (8, 6), (7, 7), (6, 8),$ or $(5, 9)$, but the intersection of the branes remains $(5+1)$-dimensional and the $p - p'$ strings live on the intersection with action (11.6). In the present case the D-term is non-vanishing only for $\chi^A = 0$, though more generally (say when there are several coincident p and p'-branes), there will be additional massless charged fields and flat directions arise.

Under T-dualities in r NN directions, one obtains $(p, p') = (9-r, 5-r)$. The action becomes

$$\int d^{6-r}\xi \left(\sum_{a=0}^{5-r} |(\partial_a + iA_a - iA'_a)\chi|^2 + \frac{\chi^\dagger \chi}{(2\pi\alpha')^2} \sum_{a=6-r}^{5} (X_a - X'_a)^2 \right.$$

$$\left. + \left(\frac{1}{4g^2_{\mathrm{YM},p}} + \frac{1}{4g^2_{\mathrm{YM}p'}} \right) \sum_{i=1}^{3} (\chi^\dagger \tau^I \chi)^2 \right).$$

$$(11.7)$$

The second term, proportional to the separation of the branes, is from the tension of the stretched string.

11.2 The BPS bound for the Dp–Dp' system

The ten dimensional $\mathcal{N} = 2$ supersymmetry algebra (in a Majorana basis) is

$$\{Q_\alpha, Q_\beta\} = 2(\Gamma^0\Gamma^\mu)_{\alpha\beta}(P_\mu + Q^{\mathrm{NS}}_\mu/2\pi\alpha')$$

$$\{\tilde{Q}_\alpha, \tilde{Q}_\beta\} = 2(\Gamma^0\Gamma^\mu)_{\alpha\beta}(P_\mu - Q^{\mathrm{NS}}_\mu/2\pi\alpha')$$

$$\{Q_\alpha, \tilde{Q}_\beta\} = 2\sum_p \frac{\tau_p}{p!}(\Gamma^0\Gamma^{m_1}\ldots\Gamma^{m_p})_{\alpha\beta}Q^{\mathrm{R}}_{m_1\ldots m_p}.$$

$$(11.8)$$

Here Q^{NS} is the charge to which the NS–NS two-form couples, it is essentially the winding of a fundamental string stretched along $\mathcal{M}_1$:

$$Q_\mu^{\mathrm{NS}} \equiv \frac{Q^{\mathrm{NS}}}{v_1} \int_{\mathcal{M}_1} dX^\mu, \quad \text{with} \quad Q^{\mathrm{NS}} = \frac{1}{\mathrm{Vol}\, S^7} \int_{S^7} e^{-2\Phi} \, {}^*H^{(3)} \quad (11.9)$$

and the charge Q^{NS} is normalised to one per unit spatial world-volume, $v_1 = L$, the length of the string. It is obtained by integrating over the S^7 which surrounds the string. The Q^{R} are the R–R charges, defined as a generalisation of winding on the space $\mathcal{M}_p$:

$$Q_{\mu_1 \dots \mu_p}^{\mathrm{R}} \equiv \frac{Q_p^{\mathrm{R}}}{v_p} \int_{\mathcal{M}_p} dX^{\mu_1} \wedge \dots dX^{\mu_p}, \quad \text{with} \quad Q_p^{\mathrm{R}} = \frac{1}{\mathrm{Vol}\, S^{8-p}} \int_{S^{8-p}} {}^*G^{(p+2)}.$$
$$(11.10)$$

The sum in (11.8) runs over all orderings of indices, and we divide by $p!$. Of course, p is even for IIA or odd for IIB. The R–R charges appear in the product of the right- and left-moving supersymmetries, since the corresponding vertex operators are a product of spin fields, while the NS–NS charges appear in right–right and left–left combinations of supercharges.

As an example of how this all works, consider an object of length L, with the charges of p fundamental strings ('F-strings', for short) and q D1-branes ('D-strings') in the IIB theory, at rest and aligned along the direction X^1. The anticommutator implies

$$\frac{1}{2}\left\{ \begin{bmatrix} Q_\alpha \\ \tilde{Q}_\alpha \end{bmatrix}, \begin{bmatrix} Q_\beta & \tilde{Q}_\beta \end{bmatrix} \right\} = \begin{bmatrix} 1 & 0 \\ 0 & 1 \end{bmatrix} M\delta_{\alpha\beta} + \begin{bmatrix} p & q/g_\mathrm{s} \\ q/g_\mathrm{s} & -p \end{bmatrix} \frac{L(\Gamma^0\Gamma^1)_{\alpha\beta}}{2\pi\alpha'}.$$
$$(11.11)$$

The eigenvalues of $\Gamma^0\Gamma^1$ are ± 1 so those of the right hand side are $M \pm L(p^2 + q^2/g^2)^{1/2}/2\pi\alpha'$. The left side is a positive matrix, and so we get the 'BPS bound' on the tension [133]

$$\frac{M}{L} \geq \frac{\sqrt{p^2 + q^2/g_\mathrm{s}^2}}{2\pi\alpha'} \equiv \tau_{p,q}. \quad (11.12)$$

Quite pleasingly, this is saturated by the fundamental string, $(p,q) = (1,0)$, and by the D-string, $(p,q) = (0,1)$.

It is not too hard to extend this to a system with the quantum numbers of Dirichlet p and p' branes. The result for ν a multiple of four is

$$M \geq \tau_p v_p + \tau_{p'} v_{p'} \quad (11.13)$$

and for ν even but not a multiple of four is [†]

$$M \geq \sqrt{\tau_p^2 v_p^2 + \tau_{p'}^2 v_{p'}^2}. \quad (11.14)$$

[†] The difference between the two cases comes from the relative sign of $\Gamma^M(\Gamma^{M'})^T$ and $\Gamma^{M'}(\Gamma^M)^T$.

The branes are wrapped on tori of volumes v_p and v'_p in order to make the masses finite.

The results (11.13) and (11.14) are consistent with the earlier results on supersymmetry breaking. For ν a multiple of four, a separated p-brane and p'-brane do indeed saturate the bound (11.13). For ν not a multiple of four, they do not saturate the bound (11.14) and cannot be supersymmetric.

11.3 Bound states of fundamental strings and D-strings

Consider a parallel D1-brane (D-string) and a fundamental string (F-string) lying along X^1. The total tension

$$\tau_{D1} + \tau_{F1} = \frac{g_s^{-1} + 1}{2\pi\alpha'} \tag{11.15}$$

exceeds the BPS bound (11.12) and so this configuration is not supersymmetric. However, it can lower its energy[26] as shown in figure 11.1. The F-string breaks, its endpoints attached to the D-string. The endpoints can then move off to infinity, leaving only the D-string behind. Of course, the D-string must now carry the charge of the F-string as well. This comes about because the F-string endpoints are charged under the D-string gauge field, so a flux runs between them; this flux remains at the end. Thus the final D-string carries both the NS–NS and R–R two-form charges. The flux is of order g_s, its energy density is of order g_s, and so the final tension is $(g_s^{-1} + O(g_s))/2\pi\alpha'$. This is below the tension of the separated strings and of the same form as the BPS bound (11.12) for a $(1,1)$ string. A more detailed calculation shows that the final tension saturates

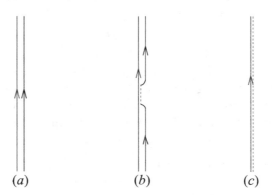

(a) (b) (c)

Fig. 11.1. (a) A parallel D-string and F-string, which is not supersymmetric. (b) The F-string breaks, its ends attaching to the D-string, resulting in (c) the final supersymmetric state, a D-string with flux.

the bound[118], so the state is supersymmetric. In effect, the F-string has dissolved into the D-string, leaving flux behind.

We can see quite readily that this is a supersymmetric situation using T-duality. We can choose a gauge in which the electric flux is $F_{01} = \dot{A}_1$. T-dualising along the x^1 direction, we ought to get a D0-brane, which we do, except that it is moving with constant velocity, since we get $\dot{X}^1 = 2\pi\alpha'\dot{A}_1$. This clearly has the same supersymmetry as a stationary D0-brane, having been simply boosted.

To calculate the number of BPS states we should put the strings in a box of length L to make the spectrum discrete. For the $(1, 0)$ F-string, the usual quantisation of the ground state gives eight bosonic and eight fermionic states moving in each direction for $16^2 = 256$ in all. This is the ultrashort representation of supersymmetry: half the 32 generators annihilate the BPS state and the other half generate $2^8 = 256$ states. The same is true of the $(0, 1)$ D-string and the $(1, 1)$ bound state just found, as will be clear from the later duality discussion of the D-string.

It is worth noting that the $(1, 0)$ F-string leaves unbroken half the supersymmetry and the $(0, 1)$ D-string leaves unbroken a different half of the supersymmetry. The $(1, 1)$ bound state leaves unbroken not the intersection of the two (which is empty), but yet a different half. The unbroken symmetries are linear combinations of the unbroken and broken supersymmetries of the D-string.

All the above extends immediately to p F-strings and one D-string, forming a supersymmetric $(p, 1)$ bound state. The more general case of p F-strings and q D-strings is more complicated. The gauge dynamics are now non-Abelian, the interactions are strong in the infrared, and no explicit solution is known. When p and q have a common factor, the BPS bound makes any bound state only neutrally stable against falling apart into subsystems. To avoid this complication let p and q be relatively prime, so any supersymmetric state is discretely below the continuum of separated states. This allows the Hamiltonian to be deformed to a simpler supersymmetric Hamiltonian whose supersymmetric states can be determined explicitly, and again there is one ultrashort representation, 256 states. It is left to the reader to consult the literature[26, 1] for the details.

11.4 The three-string junction

Let us consider further the BPS saturated formula derived and studied in the two previous subsections, and write it as follows:

$$\tau_{p,q} = \sqrt{(p\tau_{1,0})^2 + (q\tau_{0,1})^2}. \tag{11.16}$$

An obvious solution to this is

$$\tau_{p,q} \sin \alpha = q\tau_{0,1}, \quad \tau_{p,q} \cos \alpha = p\tau_{1,0}. \tag{11.17}$$

with $\tan \alpha = q/(pg_s)$. Recall that these are tensions of strings, and therefore we can interpret the equations (11.17) as balance conditions for the components of forces. In fact, it is the required balance for three strings[137, 135], and we draw the case of $p = q = 1$ in figure 11.2.

Is this at all consistent with what we already know? The answer is yes. An F-string is allowed to end on a D-string by definition, and a (1,1) string is produced, due to flux conservation, as we discussed above. The issue here is just how we see that there is bending. The first thing to notice is that the angle α goes to $\pi/2$ in the limit of zero string coupling, and so the D-string appears in that case to run straight. This had better be true, since it is then clear that we simply were allowed to ignore the bending in our previous weakly coupled string analysis. (This study of the bending of branes beyond zero coupling has important consequences for the study of one-loop gauge theory data[139]. We shall study some of this later on.)

Parenthetically, it is nice to see that in the limit of infinite string coupling, α goes to zero. The diagram is better interpreted as a D-string ending on an F-string with no resulting bending. This fits nicely with the fact that the D- and F-strings exchange roles under the strong/weak coupling duality ('S-duality') of the type IIB string theory, as we shall see in chapter 12.

When we wrote the linearised BIon equations in section 5.7, we ignored the 1+1 dimensional case. Let us now include that part of the story here

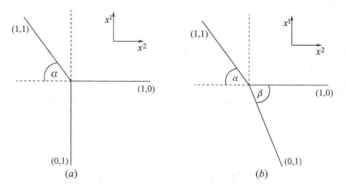

Fig. 11.2. (*a*) When an F-string ends on a D-string it causes it to bend at an angle set by the string coupling. On the other side of the junction is a (1,1) string. This is in fact a BPS state. (*b*) Switching on some amount of the R–R scalar can vary the other angle, as shown.

as a 1+1 dimensional gauge theory discussion. There is a flux F_{01} on the world-volume, and the end of the F-string is an electric source. Given that there is only one spatial dimension, the F-string creates a discontinuity on the flux, such that e.g.[140, 60]:

$$F_{01} = \begin{cases} g_s, & x_1 > 0 \\ 0, & x_1 < 0 \end{cases},$$ (11.18)

so we can choose a gauge such that

$$A_0 = \begin{cases} g_s x^1, & x_1 > 0 \\ 0, & x_1 < 0 \end{cases}.$$ (11.19)

Just as in section 5.7, this is BPS if one of the eight scalars Φ^m is also switched on so that

$$\Phi^2(x^1) = A_0.$$ (11.20)

How do we interpret this? Since $(2\pi\alpha')\Phi^2$ represents the x^2 position of the D-string, we see that for $x^1 < 0$ the D-string is lying along the x^1 axis, while for $x^1 > 0$, it lies on a line forming an angle $\tan^{-1}(1/g_s)$ with the x^1, axis.

Recall the T_1-dual picture we mentioned in the previous section, where we saw that the flux on the D-string (making the (1,1) string) is equivalent to a D0-brane moving with velocity $(2\pi\alpha')F_{01}$. Now we see that the D0-brane loses its velocity at $x^1 = 0$. This is fine, since the apparent impulse is accounted for by the momentum carried by the F-string in the T-dual picture. (One has to tilt the diagram in order to T-dualise along the (1,1) string in order to see that there is F-string momentum.)

Since we have seen many times that the presence of flux on the world-volume of a Dp-brane is equivalent to having a dissolved D$(p-2)$-brane, i.e. non-zero $C_{(p-1)}$ source, we can modify the flux on the $x^1 < 0$ part of the string this way by turning on the R–R scalar C_0. This means that $\Phi^2(x^1)$ will be linear there too, and so the angle β between the D- and F-strings can be varied too (see figure 11.2(b)). It is interesting to derive the balance conditions from this, and then convert it into a modified tension formula, but we will not do that here[140].

It is not hard to imagine that given the presence we have already deduced of a general (p, q) string in the theory that there are three-string junctions to be made out of any three strings such that the (p, q)-charges add up correctly, giving a condition on the angles at which they can meet. This is harder to do in the full non-Abelian gauge theories on their world-volumes, but in fact a complete formula can be derived using the underlying $SL(2, \mathbb{Z})$ symmetry of the type IIB string theory. We will have more to say about this symmetry later.

General three-string junctions have been shown to be important in a number of applications, and there is a large literature on the subject which we are unfortunately not able to review here.

11.5 Aspects of D-brane bound states

Bound states of p-branes and p'-branes have many applications. Some of them will appear in our later lectures, and so it is worth listing some of the results here. Here we focus on $p' = 0$, since we can always reach it from a general (p, p') using T-duality.

11.5.1 0–0 bound states

The BPS bound for the quantum numbers of n 0-branes is $n\tau_0$, so any bound state will be at the edge of the continuum. What we would like to know is if there is actually a true bound state wave function, i.e. a wavefunction which is normalisable. To make the bound state counting well defined, compactify one direction and give the system momentum m/R with m and n relatively prime[141]. The bound state now lies discretely below the continuum, because the momentum cannot be shared evenly among unbound subsystems.

This bound state problem is T-dual to the one considered in section 11.3. Taking the T-dual, the n D0-branes become D1-branes, while the momentum becomes winding number, corresponding to m F-strings. There is therefore one ultrashort multiplet of supersymmetric states when m and n are relatively prime[141]. This bound state should still be present back in infinite volume, since one can take R to be large compared to the size of the bound state. There is a danger that the size of the wavefunction we have just implicitly found might simply grow with R such that as $R \rightarrow \infty$ it becomes non-normalisable again. More careful analysis is needed to show this. It is sufficient to say here that the bound states for arbitrary numbers of D0-branes are needed for the consistency of string duality, so this is an important problem. Some strong arguments have been presented in the literature ($n = 2$ is proven), but the general case is not yet proven[142].

11.5.2 0–2 bound states

Now the BPS bound (expression (11.14)) puts any bound state discretely below the continuum. One can see a hint of a bound state forming by noticing that for a coincident D0-brane and D2-brane the NS 0–2 string has a negative zero-point energy (11.2) and so a tachyon (which survives the GSO projection), indicating an instability towards forming something else. In fact the bound state (one short representation) is easily described: the D0-brane dissolves in the D2-brane, leaving flux, as we have seen numerous times. The brane R–R action (expression (9.9)) contains the coupling $C_{(1)}F$, so with the flux the D2-brane also carries the D0-brane charge[143]. There is also one short multiplet for n D0-branes. This same bound state is always present when $\nu = 2$.

11.5.3 0–4 bound states

The BPS bound (11.13) makes any bound state marginally stable, so the problem is made well-defined as in the 0–0 case by compactifying and adding momentum[144]. The interactions in the action (11.7) are relevant in the infrared so this is again a hard problem, but as before it can be deformed into a solvable supersymmetric system. Again there is one multiplet of bound states[144]. Now, though, the bound state is invariant only under $\frac{1}{4}$ of the original supersymmetry, the intersection of the supersymmetries of the D0-brane and of the D4-brane. The bound states then lie in a short (but not ultrashort) multiplet of 2^{12} states.

For two D0-branes and one D4-brane, one gets the correct count as follows[145]. Think of the case that the volume of the D4-brane is large. The 16 supersymmetries broken by the D4-brane generate 256 states that are delocalised on the D4-brane. The eight supersymmetries unbroken by the D4-brane and broken by the D0-brane generate 16 states (half bosonic and half fermionic), localised on the D0-brane. The total number is the product 2^{12}. Now count the number of ways two D0-branes can be put into their 16 states on the D4-brane: there are eight states with both D0-branes in the same (bosonic) state and $\frac{1}{2}16 \cdot 15$ states with the D-branes in different states, for a total of $8 \cdot 16$ states. But in addition, the two-branes can bind, and there are again 16 states where the bound state binds to the D4-brane. The total, tensoring again with the D4-brane ground states, is $9 \cdot 16 \cdot 256$.

For n D0-branes and one D4-brane, the degeneracy D_n is given by the generating functional [145] (see insert 3.4, p. 92):

$$\sum_{n=0}^{\infty} q^n D_n = 256 \prod_{k=1}^{\infty} \left(\frac{1 + q^k}{1 - q^k} \right)^8, \qquad (11.21)$$

where the term k in the product comes from bound states of k D0-branes then bound to the D4-brane. Some discussion of the D0–D4 bound state, and related issues, can be found in the references[146].

11.5.4   0–6 bound states

The relevant bound is (11.14) and again any bound state would be below
the continuum. The NS zero-point energy for 0–6 strings is positive, so
there is no sign of decay. One can give D0-brane charge to the D6-brane
by turning on flux, but there is no way to do this and saturate the BPS
bound. So it appears that there are *no* supersymmetric bound states.
Notice that, unlike the 0–2 case, the 0–6 interaction is repulsive, both at
short distance and at long.

11.5.5   0–8 bound states

The case of the D8-brane is special, since it is rather big. It is a domain
wall, because there is only one spatial dimension transverse to it. In fact,
the D8-brane on its own is not really a consistent object. Trying to put
it into type IIA runs into trouble, since the string coupling blows up a
finite distance from it on either side because of the nature of its coupling
to the dilaton. To stop this happening, one has to introduce a pair of
O8-planes, one on each side, because they (for SO groups) have negative
charge (-8 times that of the D8-brane) and can soak up the dilaton. We
therefore should have 16 D8-branes for consistency, and so we end up in
the type I$'$ theory, the T-dual of type I. The bound state problem is now
quite different, and certain details of it pertain to the strong coupling
limit of certain string theories, and their 'matrix'[157] formulation[147, 148].
We shall revisit this in section 12.5.

12

Strong coupling and string duality

One of the most striking results of the mid-1990s was the realisation that all of the superstring theories are in fact dual to one another at strong coupling[149]. This also brought eleven dimensional supergravity into the picture and started the search for M-theory, the dynamical theory within which all of those theories would fit as various effective descriptions of perturbative limits. All of this is referred to as the 'Second Superstring Revolution'. Every revolution is supposed to have a hero or heroes. We shall consider branes to be cast in that particular role, since they (and D-branes especially) supplied the truly damning evidence of the strong coupling fate of the various string theories.

We shall discuss aspects of this in the present section. We simply study the properties of D-branes in the various string theories, and then trust to that fact that as they are BPS states, many of these properties will survive at strong coupling.

12.1 Type IIB/type IIB duality

12.1.1 D1-brane collective coordinates

Let us first study the D1-brane. This will be appropriate to the study of type IIB and the type I string by Ω-projection. Its collective dynamics as a BPS soliton moving in flat ten dimensions is captured by the 1+1 dimensional world-volume theory, with 16 or 8 supercharges, depending upon the theory we are in. (See figure 12.1(a).)

It is worth first setting up a notation and examining the global symmetries. Let us put the D1-brane to lie along the x^1 direction, as we will do many times in what is to come. This arrangement of branes breaks the

261

Lorentz group up as follows:

$$SO(1,9) \supset SO(1,1)_{01} \times SO(8)_{2-9}. \tag{12.1}$$

Accordingly, the supercharges decompose under (12.1) as

$$\mathbf{16} = \mathbf{8}_+ \oplus \mathbf{8}_- \tag{12.2}$$

where $\pm$ subscripts denote a chirality with respect to $SO(1,1)$.

For the 1–1 strings, there are eight Dirichlet–Dirichlet (DD) directions, the Neveu–Schwarz (NS) sector has zero point energy $-1/2$. The massless excitations form vectors and scalars in the 1+1 dimensional model. For the vectors, the Neumann–Neumann (NN) directions give a gauge field A^μ. Now, the gauge field has no local dynamics, so the only contentful bosonic excitations are the transverse fluctuations. These come from the eight Dirichlet–Dirichlet (DD) directions x^m, $m = 2, \ldots, 9$, and are

$$\phi^m(x^0, x^1): \quad \lambda_\phi \psi^m_{-\frac{1}{2}}|0\rangle. \tag{12.3}$$

The fermionic states ξ from the Ramond (R) sector (with zero point energy 0, as always) are built on the vacua formed by the zero modes ψ^i_0, $i=0,\ldots,9$. This gives the initial $\mathbf{16}$. The GSO projection acts on the vacuum in this sector as:

$$(-1)^F = e^{i\pi(S_0+S_1+S_2+S_3+S_4)}. \tag{12.4}$$

A left- or right-moving state obeys $\Gamma^0\Gamma^1\xi_\pm = \pm\xi_\pm$, and so the projection onto $(-1)^F\xi=\xi$ says that left- and right-moving states are odd and (respectively) even under $\Gamma^2\ldots\Gamma^9$, which is to say that they are either in the $\mathbf{8}_s$ or the $\mathbf{8}_c$. So we see that the GSO projection simply correlates world-sheet chirality with spacetime chirality: ξ_- is in the $\mathbf{8}_c$ of $SO(8)$ and ξ_+ is in the $\mathbf{8}_s$.

So we have seen that for a D1-brane in type IIB string theory, the right-moving spinors are in the $\mathbf{8}_s$ of $SO(8)$, and the left-moving spinors in the $\mathbf{8}_c$. These are the same as the fluctuations of a fundamental IIB string, in static gauge[26], and here spacetime supersymmetry is manifest. (It is in 'Green–Schwarz' form[108].) There, the supersymmetries Q_α and $\tilde{Q}_\alpha$ have the same chirality. Half of each spinor annihilates the F-string and the other half generates fluctuations. Since the supersymmetries have the same $SO(9,1)$ chirality, the $SO(8)$ chirality is correlated with the direction of motion.

So far we have been using the string metric. We can switch to the Einstein metric, $g^{(E)}_{\mu\nu} = e^{-\Phi/2}g^{(S)}_{\mu\nu}$, since in this case gravitational action has no dependence on the dilaton, and so it is invariant under duality. The tensions in this frame are:

$$\text{F-string: } g_{\text{s}}^{1/2}/2\pi\alpha'$$
$$\text{D-string: } g_{\text{s}}^{-1/2}/2\pi\alpha'. \tag{12.5}$$

Since these are BPS states, we are able to trust these formulae at arbitrary values of g_{s}.

Let us see what interpretation we can make of these formulae. At weak coupling the D-string is heavy and the F-string tension is the lightest scale in the theory. At strong coupling, however, the D-string is the lightest object in the theory (a dimensional argument shows that the lowest dimensional branes have the lowest scale[150]), and it is natural to believe that the theory can be reinterpreted as a theory of weakly coupled D-strings, with $g_{\text{s}}' = g_{\text{s}}^{-1}$. One cannot prove this without a non-perturbative definition of the theory, but quantising the light D-string implies a large number of the states that would be found in the dual theory, and self-duality of the IIB theory seems by far the simplest interpretation – given that physics below the Planck energy is described by some specific string theory, it seems likely that there is a unique extension to higher energies. This agrees with the duality deduced from the low energy action and other considerations[149, 164]. In particular, the NS–NS and R–R two-form potentials, to which the D- and F-strings respectively couple, are interchanged by this duality.

This duality also explains our remark about the strong and weak coupling limits of the three string junction depicted in figure 11.2. The roles of the D- and F-strings are swapped in the $g_{\text{s}} \to 0, \infty$ limits, which fits with the two limiting values $\alpha \to \pi/2, 0$.

12.1.2 S-duality and $SL(2, \mathbb{Z})$

The full duality group of the $D = 10$ type IIB theory is expected to be $SL(2, \mathbb{Z})$[151, 153]. This relates the fundamental string not only to the R–R string but to a whole set of strings with the quantum numbers of p F-strings and q D-strings for p and q relatively prime[133]. The bound states found in section 11.3 are just what is required for $SL(2, \mathbb{Z})$ duality[26]. As the coupling and the R–R scalar are varied, each of these strings becomes light at the appropriate point in moduli space. We shall study this further in section 16.1, on the way to uncovering 'F-theory', a tool for generating very complicated type IIB backgrounds by geometrising the $SL(2, \mathbb{Z})$ symmetry.

12.2 $SO(32)$ **Type I/heterotic duality**

12.2.1 D1-brane collective coordinates

Let us now consider the D1-brane in the type I theory. We must modify our previous analysis in two ways. First, we must project onto Ω-even states.

As in section 2.6, the $U(1)$ gauge field A is in fact projected out, since ∂_t is odd under Ω. The normal derivative ∂_n, is even under Ω, and hence the Φ^m survive. Turning to the fermions, we see that Ω acts as $e^{i\pi(S_1+S_2+S_3+S_4)}$ and so the left-moving $\mathbf{8_c}$ is projected out and the right-moving $\mathbf{8_s}$ survives.

Recall that D9-branes must be introduced after doing the Ω projection of the type IIB string theory. These are the $SO(32)$ Chan–Paton factors. This means that we must also include the massless fluctuations due to strings with one end on the D1-brane and the other on a D9-brane (see figure 12.1(b)). The zero point energy in the NS sector for these states is $1/2$, and so there is way to make a massless state. The R sector has zero point energy zero, as usual, and the ground states come from excitations in the x^0, x^1 direction, since it is in the NN sector that the modes are integer. The GSO projection $(-)^F = \Gamma^0\Gamma^1$ will project out one of these, λ_-, while the right-moving one will remain. The Ω projection simply relates 1–9 strings to 9–1 strings, and so places no constraint on them. Finally, we should note that the 1–9 strings, as they have one end on a D9-brane, transform as vectors of $SO(32)$.

Now, by the argument that we saw in the case of the type IIB string, we should deduce that this string becomes a light fundamental string in some dual string theory at strong coupling. We have seen such a string before in section 7.2. It is the 'heterotic' string, which has $(0,1)$ spacetime supersymmetry, and a left-moving family of 32 fermions transforming as the $\mathbf{32}$ of $SO(32)$. They carry a current algebra which realises the

(a) (b)

Fig. 12.1. D1-branes (a) in type IIB theory its fluctuations are described by 1–1 strings; (b) in type I string theory, there are additional contributions from 1–9 strings.

$SO(32)$ as a spacetime gauge symmetry. The other ten dimensional heterotic string, with gauge group $E_8 \times E_8$, has a strong coupling limit which we will examine shortly, using the fact that upon compactifying on a circle, the two heterotic string theories are perturbatively related by T-duality (see section 8.1.3)[173, 174].

We have obtained the $SO(32)$ string here with spacetime supersymmetry and with a left-moving current algebra $SO(32)$ in fermionic form[162]. As we learned in section 7.2, we can bosonise these into the 16 chiral bosons which we then used to construct the heterotic string in the first instance. This also fits rather well with the fact that we had already noticed that we could have deduced that such a string theory might exist just by looking at the supergravity sector in section 7.3. This is just how type I/heterotic duality was deduced first[153, 164] and then D-brane constructions were used to test it more sharply[162]. We shall see that considerations of the strong coupling limit of various other string theories will again point to the existence of the heterotic string. We have already seen hints of that in chapter 7, as discussed in insert 7.5. Of course, the heterotic strings were discovered by direct perturbative construction, but it is amusing to thing that, in another world, they may be discovered by string duality.

We end with a brief remark about some further details that we shall not pursue. Recall that it was mentioned at the end of section 7.2, the fermionic $SO(32)$ current algebra requires a GSO projection. By considering a closed D1-brane we see that the Ω projection removes the $U(1)$ gauge field, but in fact allows a discrete gauge symmetry: a holonomy ± 1 around the D1-brane. This discrete gauge symmetry is the GSO projection, and we should sum over all consistent possibilities. The heterotic strings have spinor representations of $SO(32)$, and we need to be able to make them in the Type I theory, in order for duality to be correct. In the R sector of the discrete D1-brane gauge theory, the 1–9 strings are periodic. The zero modes of the fields Ψ^i, representing the massless 1–9 strings, satisfy the Clifford algebra $\{\Psi^i_0, \Psi^j_0\} = \delta^{ij}$, for $i, j = 1, \ldots, 32$, and so just as for the fundamental heterotic string we get spinors $\mathbf{2^{31}} \oplus \mathbf{\overline{2^{31}}}$. One of them is removed by the discrete gauge symmetry to match the spectrum with a single massive spinor which we uncovered directly using lattices in section 7.2.1.

12.3 Dual branes from 10D string–string duality

There is an instructive way to see how the D-string tension turns into that of an F-string. In terms of supergravity fields, part of the duality transformation (7.46) involves

$$G_{\mu\nu} \to e^{-\tilde{\Phi}}\tilde{G}_{\mu\nu}, \qquad \Phi \to -\tilde{\Phi}, \tag{12.6}$$

where the quantities on the right, with tildes, are in the dual theory. This means that in addition to $g_s = \tilde{g}_s^{-1}$, for the relation of the string coupling to the dual string coupling, there is also a redefinition of the string length, via

$$\alpha' = \tilde{g}_s^{-1}\tilde{\alpha}',$$

which is the same as

$$\alpha' g_s^{-1} = \tilde{\alpha}'.$$

Starting with the D-string tension, these relations give:

$$\tau_1 = \frac{1}{2\pi\alpha' g_s} \to \frac{1}{2\pi\tilde{\alpha}'} = \tau_1^{\mathrm{F}},$$

precisely the tension of the fundamental string in the dual string theory, measured in the correct units of length.

One might understandably ask the question about the fate of other branes under S-dualities[165]. For the type IIB's D3-brane:

$$\tau_3 = \frac{1}{(2\pi)^3 \alpha'^2 g_s} \to \frac{1}{(2\pi)^3 \tilde{\alpha}'^2 \tilde{g}_s} = \tau_3,$$

showing that the dual object is again a D3-brane. For the D5-brane, in either type IIB or type I theory:

$$\tau_5 = \frac{1}{(2\pi)^5 \alpha'^3 g_s} \to \frac{1}{(2\pi)^5 \tilde{\alpha}'^3 \tilde{g}_s^2} = \tau_5^{\mathrm{F}}.$$

This is the tension of a fivebrane which is *not* a D5-brane. This is interesting, since for both dualities, the R–R two-form $C^{(2)}$ is exchanged for the NS–NS two-form $B^{(2)}$, and so this fivebrane is magnetically charged under the latter. It is in fact the magnetic dual of the fundamental string. Its g_s^{-2} behaviour identifies it as a soliton of the NS–NS sector.

So we conclude that there exists in both the type IIB and $SO(32)$ heterotic theories such a brane, and in fact such a brane can be constructed directly as a soliton solution. They should perhaps be called 'F5-branes', since they are magnetic duals to fundamental strings or 'F1-branes', but this name never stuck. They go by various names like 'NS5-brane', since they are made of NS–NS sector fields, or 'solitonic fivebrane', and so on. As they are constructed completely out of closed string fields, T-duality along a direction parallel to the brane does not change its dimensionality, as would happen for a D-brane. We conclude therefore that they also exist in the T-dual type IIA and $E_8 \times E_8$ string theories. Let us study them a bit further.

12.3.1 The heterotic NS-fivebrane

For the heterotic cases, the soliton solution also involves a background gauge field, which is in fact an instanton. This follows from the fact that in type I string theory, the D5-brane is an instanton of the D9-brane gauge fields as we saw with dramatic success in section 9.2. We shall have even more to say about this later, when we uncover more properties of how to probe branes with branes. As we saw there and in chapter 7, through equation (7.38), $\mathrm{Tr}(F \wedge F)$ and $\mathrm{tr}(R \wedge R)$ both magnetically source the two-form potential $C_{(2)}$, since by taking one derivative:

$$d\tilde{G}^{(3)} = -\frac{\alpha'}{4}\left[\mathrm{Tr}F^2 - \mathrm{tr}R^2\right].$$

By strong/weak coupling duality, this must be the case for the NS–NS two form $B_{(2)}$. To leading order in α', we can make a solution of the heterotic low energy equation of motion with these clues quite easily as follows. Take for example an $SU(2)$ instanton (the very object described in insert 9.4 when we reminded ourselves about non-trivial second Chern class) embedded in an $SU(2)$ subgroup of the $SO(4)$ in the natural decomposition: $SO(32) \supset SO(28) \times SO(4)$. As we said, this will source some dH, which in turn will source the metric and the dilaton. In fact, to leading order in α', the corrections to the metric away from flat space will not give any contribution to $\mathrm{tr}(R \wedge F)$, which has more derivatives than $\mathrm{Tr}(F \wedge F)$, and is therefore subleading in this discussion. The result should be an object which is localised in $\mathbb{R}^4$ with a finite core size (the 'dressed' instanton), and translationally invariant in the remaining $5+1$ directions. This deserves to be called a fivebrane. A solution realising this logic can be found, and it can be written as[72, 73]:

$$ds^2 = -dt^2 + (dx^1)^2 + \cdots + (dx^5)^2 + e^{2\Phi}\left(dr^r + r^2 d\Omega_3^2\right)$$

$$e^{2\Phi} = g_s^2\left(1 + \alpha'\frac{(r^2 + 2\rho^2)}{(r^2 + \rho^2)^2} + O(\alpha'^2)\right), \quad H_{\mu\nu\lambda} = -\epsilon_{\mu\nu\lambda}{}^\sigma \partial_\sigma \Phi$$

$$A_\mu = \left(\frac{r^2}{r^2 + \rho^2}\right)g^{-1}\partial_\mu g, \quad g = \frac{1}{r}\begin{pmatrix} x^6 + ix^7 & x^8 + ix^9 \\ x^8 - ix^9 & x^6 - ix^7 \end{pmatrix}, \quad (12.7)$$

showing its structure as an $SU(2)$ instanton localised in x^6, x^7, x^8, x^9, with core size ρ. As before, r^2 is the radial coordinate, and $d\Omega_3^2$ is a metric on a round S^3.

Once we have deduced the existence of this object in the $SO(32)$ heterotic string, it is straightforward to see that it must exist in the $E_8 \times E_8$ heterotic string too. We simply compactify on a circle in a world-volume direction where there is no structure at all. Shrinking it away takes us to the other heterotic theory, with an NS5-brane of precisely the same sort of structure as above. Alternatively, we could have just constructed the fivebrane directly using the ideas above without appealing to T-duality at all.

12.3.2 The type IIA and type IIB NS5-brane

As already stated, similar reasoning leads one to deduce that there must be an NS5-brane in type II string theory*. We can deduce its supergravity fields by using the ten dimensional S-duality transformations to convert the case $p = 5$ of equations (10.38), (10.39), to give[72, 73]:

$$ds^2 = -dt^2 + (dx^1)^2 + \cdots + (dx^5)^2 + \tilde{Z}_5 \left(dr^r + r^2 d\Omega_3^2 \right)$$

$$e^{2\Phi} = g_s^2 \tilde{Z}_5 = g_s^2 \left(1 + \frac{\alpha' N}{r^2} \right),$$

$$B_{(6)} = (\tilde{Z}_5^{-1} - 1)g_s dx^0 \wedge \cdots \wedge dx^5. \tag{12.8}$$

This solution has N units of the basic magnetic charge of $B_{(2)}$, and is a point in x^6, x^7, x^8, x^9. Note that the same sort of transformation will give a solution for the fields around a fundamental IIB string, by starting with the $p = 1$ case of (10.38)[163, 164]. We shall do this in chapter 16.

For the same reasons as for the heterotic string, once we have made an NS5-brane for the type IIB string, it is easy to see that we can use T-duality along a world-volume direction (where the solution is trivial) in order to make one in the type IIA string theory as well.

A feature worth considering is the world-volume theory describing the low energy collective motions of these type II branes. This can be worked out directly, and string duality is consistent with the answer: from the duality, we can immediately deduce that the type IIB's NS5-brane must have a vector multiplet, just like the D5-brane. Also as with D5-branes, there is enhanced $SU(N)$ gauge symmetry when N coincide[160], the extra massless states being supplied by light D1-branes stretched between them. (See figure 12.2.) The vector multiplet can be read off from table 7.1 as $(\mathbf{2}, \mathbf{2}) + 4(\mathbf{1}, \mathbf{1}) + 2(\mathbf{1}, \mathbf{2}) + 2(\mathbf{2}, \mathbf{1})$. There are four scalars, which are the four transverse positions of the brane in ten dimensions. The fermionic content can be seen to be manifestly non-chiral giving a $(1, 1)$ supersymmetry on the world-volume.

---

* In the older literature, it is sometimes called a 'symmetric fivebrane', after its left-right symmetric σ-model description, in constrast to that of the heterotic NS5-brane.

NS5-branes

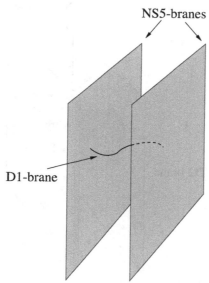

D1-brane

Fig. 12.2. D1-branes stretched between NS5-branes in type IIB string theory will give extra massless vectors when the NS5-branes coincide.

For the type IIA NS5-brane, things are different. Following the T-duality route mentioned above, it can be seen that the brane actually must have a chiral (0,2) supersymmetry. So it cannot have a vector multiplet any more, and instead there is a six dimensional tensor multiplet on the brane. So there is a two-form potential instead of a one-form potential, which is interesting. The tensor multiplet can be read off from table 7.1 as $(\mathbf{1},\mathbf{3}) + 5(\mathbf{1},\mathbf{1}) + 4(\mathbf{2},\mathbf{1})$, with a manifestly chiral fermionic content. There are now *five* scalars, which is suggestive, since in their interpretation as collective coordinates for transverse motions of the brane, there is an implication of an *eleventh* direction. This extra direction will become even more manifest in section 12.4.

There is an obvious $U(1)$ gauge symmetry under the transformation $B^+_{(2)} \to B^+_{(2)} + d\Lambda_{(1)}$, and the question arises as to whether there is a non-Abelian generalisation of this when many branes coincide. On the D-brane side of things, it is clear how to construct the extra massless states as open strings stretched between the branes whose lengths can shrink to zero size in the limit. Here, there is a similar, but less well-understood phenomenon. The tensor potential on the world-volume is naturally sourced by six dimensional strings, which are in fact the ends of open D2-branes ending on the NS5-branes. The mass or tension of these strings is set by the amount that the D2-branes are stretched between

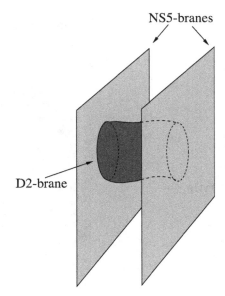

Fig. 12.3. D2-branes stretched between NS5-branes in type IIA string theory will give extra massless self-dual tensors when the NS5-branes coincide.

two NS-branes, by precise analogy with the D-brane case. So we are led to the interesting case that there are tensionless strings when many NS5-branes coincide, forming a generalised enhanced gauge tensor multiplet. (See figure 12.3.) These strings are not very well understood, it must be said. They are not sources of a gravity multiplet, and they appear not to be weakly coupled in any sense that is understood well enough to develop an intrinsic perturbation theory for them[†].

However, the theory that they imply for the branes is apparently well-defined. The information about how it works fits well with the dualities to better understood things, as we have seen here, and as we will see later when we shall say a little more about it in chapter 18, since it can be indirectly defined using the AdS/CFT correspondence. It should be noted that we do not have to use D-branes or duality to deduce a number of the features mentioned above for the world-volume theories on the NS5-branes. That there is either a (1,1) vector multiplet or a (0,2) tensor multiplet was first uncovered by direct analysis of the collective dynamics of the NS5-branes as supergravity solitons in the type II theories[159].

[†] The cogniscenti will refer to theories of non-Abelian 'gerbes' at this point. The reader should know that these are not small furry pets, but well-defined mathematical objects. They are (reportedly) a generalisation of the connection on a vector bundle, appropriate to two-form gauge fields[90].

12.4 Type IIA/M-theory duality

Let us turn our attention to the type IIA theory and see if at strong coupling we can see signs of a duality to a useful weakly coupled theory. In doing this we will find that there are even stranger dualities than just a string–string duality (which is strange and beautiful enough as it is!), but in fact a duality which points us firmly in the direction of the unexplored and the unknown.

12.4.1 A closer look at D0-branes

Notice that, in the IIA theory, the D0-brane has a mass $\tau_0 = \alpha'^{-1/2} g_s$, as measured in the string metric. As $g_s \to \infty$, this mass becomes light, and eventually becomes the lightest scale in the theory, lighter even than that of the fundamental string itself.

We can trust the extrapolation of the mass formula done in this way because the D0-brane is a BPS object, and so the formula is protected from, for example, levelling off to some still not-too-light scale by loop corrections, etc. So we are being shown new features of the theory here, and it would be nice to make sense of them. Notice that in addition, we have seen in section 11.5 that n D0-branes have a single supersymmetric bound state with mass $n\tau_0$. So in fact, these are genuine physical particles, charged under the $U(1)$ of the R–R one-form $C_{(1)}$, and forming an evenly spaced tower of mass states which is become light as we go further to strong coupling. How are we to make sense of this in ten dimensional string theory?

In fact, the spectrum we just described is characteristic of the appearance of an additional dimension[152, 153], where the momentum (Kaluza–Klein) states have masses n/R and form a continuum is $R \to \infty$. Here, $R = \alpha'^{1/2} g_s$, so weak coupling is small R and the theory is effectively ten dimensional, while strong coupling is large R, and the theory is eleven dimensional. We saw such Kaluza–Klein behaviour in section 4.2. The charge of the nth Kaluza–Klein particle corresponds to n units of momentum $1/R$ in the hidden dimension. In this case, this $U(1)$ is the R–R one form of type IIA, and so we interpret D0-brane charge as eleven dimensional momentum.

12.4.2 Eleven dimensional supergravity

In this way, we are led to consider eleven dimensional supergravity as the strong coupling limit of the type IIA string. This is only for *low energy*, of course, and the issue of the complete description of the short distance physics at strong coupling to complete the 'M-theory', is yet to be settled.

It cannot be simply eleven dimensional supergravity, since that theory (like all purely field theories of gravity) is ill-defined at short distances. A most widely examined proposal for the structure of the short distance physics is 'Matrix Theory'[157], and we shall briefly discuss it in chapter 16.

In the absence of a short distance theory, we have to make do with the low-energy effective theory, which is a graviton, and antisymmetric 3-form tensor gauge field $A_{(3)}$, and their superpartners. Notice that this theory has the same number of bosonic and fermionic components as the type II theory. Take type IIA and note that the NS–NS sector has 64 bosonic components as does the R–R sector, giving a total of 128. Now count the number of physical components of a graviton, together with a three-form in eleven dimensions. The answer is $9 \times 10/2 - 1 = 44$ for the graviton and $9 \times 8 \times 7/(3 \times 2) = 84$ for the three-form. The superpartners constitute the same number of fermionic degrees of freedom, of course, giving an $\mathcal{N} = 1$ supersymmetry in eleven dimensions, equivalent to 32 supercharges. In fact, a common trick to be found in many discussions for remembering how to write the type IIA Lagrangian[5] is simply to dimensionally reduce the eleven dimensional supergravity Lagrangian. Now we see that a physical reason lies behind it. The bosonic part of the action is:

$$S_{\text{IID}} = \frac{1}{2\kappa_{11}^2} \int d^{11}x \sqrt{-G} \left(R - \frac{1}{48}(F_{(4)})^2 \right) - \frac{1}{12\kappa_{11}^2} \int A_{(3)} \wedge F_{(4)} \wedge F_{(4)},$$
(12.9)

and we shall work out $2\kappa_{11}^2 = 16\pi G_{\text{N}}^{11}$ shortly.

To relate the type IIA string coupling to the size of the eleventh dimension we need to compare the respective Einstein–Hilbert actions[153], ignoring the rest of the actions for now:

$$\frac{1}{2\kappa_0^2 g_s^2} \int d^{10}x \sqrt{-G_s} R_s = \frac{2\pi R}{2\kappa_{11}^2} \int d^{10}x \sqrt{-G_{11}} R_{11}.$$
(12.10)

The string and eleven dimensional supergravity metrics are equal up to an overall rescaling,

$$G_{s\mu\nu} = \zeta^2 G_{11\mu\nu}$$
(12.11)

and so $\zeta^8 = 2\pi R \kappa_0^2 g_s^2 / \kappa_{11}^2$. The respective masses are related $n/R = m_{11} = \zeta m_s = n\zeta\tau_0$ or $R = \alpha'^{1/2} g_s / \zeta$. Combining these with the result (7.44) for κ_0, we obtain

$$\zeta = g_s^{1/3} \left[2^{7/9} \pi^{8/9} \alpha' \kappa_{11}^{-2/9} \right]$$
(12.12)

and the radius in eleven dimensional units is:

$$R = g_s^{2/3} \left[2^{-7/9} \pi^{-8/9} \kappa_{11}^{2/9} \right].$$
(12.13)

In order to emphasise the basic structure we hide in braces numerical factors and factors of κ_{11} and α'. The latter factors are determined by dimensional analysis, with κ_{11} having units of (11D supergravity length$^{9/2}$) and α' (string theory length^2). We are free to set $\zeta = 1$, using the same metric and units in M-theory as in string theory. In this case

$$\kappa_{11}^2 = g_{\rm s}^3 \left[2^7 \pi^8 \alpha'^{9/2} \right], \qquad \text{and then} \quad R = g_{\rm s} \ell_{\rm s}. \qquad (12.14)$$

The reason for not always doing so is that when we have a series of dualities, as below, there will be different string metrics. For completeness, let us note that if we define Newton's constant via $2\kappa_{11}^2 = 16\pi G_{\rm N}^{11}$, then we have:

$$\kappa_{11}^2 = 2^7 \pi^8 \ell_{\rm p}^9; \qquad \ell_{\rm p} = g_{\rm s}^{1/3} \sqrt{\alpha'} = g_{\rm s}^{1/3} \ell_{\rm s}. \qquad (12.15)$$

See insert 12.1 for more about the Kaluza–Klein reduction.

12.5 $E_8 \times E_8$ heterotic string/M-theory duality

We have deduced the duals of four of the five ten dimensional string theories. Let us study the final one, the $E_8 \times E_8$ heterotic string, which is T-dual to the $SO(32)$ string[173, 174].

Compactify on a large radius $R_{\rm HA}$ and turn on a Wilson line which breaks $E_8 \times E_8$ to $SO(16) \times SO(16)$. As we learned in section 8.1.3, this is T-dual to the $SO(32)$ heterotic string, again with a Wilson line breaking the group to $SO(16) \times SO(16)$. The couplings and radii are related

$$R_{\rm HB} = \frac{\ell_{\rm s}^2}{R_{\rm HA}},$$

$$g_{\rm s,HB} = g_{\rm s,HA} \frac{\ell_{\rm s}}{R_{\rm HA}}. \qquad (12.16)$$

Now use type I/heterotic duality to write this as a type I theory with[153]

$$R_{\rm IB} = g_{\rm s,HB}^{-1/2} R_{\rm HB} = g_{\rm s,HA}^{-1/2} \frac{\ell_{\rm s}^{3/2}}{R_{\rm HA}^{1/2}},$$

$$g_{\rm s,IB} = g_{\rm s,HB}^{-1} = g_{\rm s,HA}^{-1} \frac{R_{\rm HA}}{\ell_{\rm s}}. \qquad (12.17)$$

The radius is very small, so it is useful to make another T-duality, to the 'type I′' or 'type IA' theory. The compact dimension is then a segment of

---

Insert 12.1.   Kaluza–Klein relations

It is amusing to work out the relationship between the metric in Einstein frame, and the metric and scalar in one dimension fewer, in either Einstein or another frame. The general case might be useful, so we will work it out, bearing in mind that for the eleven to ten case, the scalar is the type IIA dilaton Φ, but in other cases it is simply an additional modulus (there may already be a dilaton). We want to get to D dimensions, reducing on x^D, and the higher dimensional metric shall be written in Kaluza–Klein form:

$$G_{MN}^{(D+1)}dx^M dx^N = e^{2\alpha\Phi}\left(G_{\mu\nu}^{(D)}dx^\mu dx^\nu + e^{2\Phi}(dx^D + A_\mu dx^\mu)^2\right),$$

where we have included the possibility that we will have to do a rescaling to change frames in the lower dimensions, by multiplying by $e^{2\alpha\Phi}$. This results in the Ricci scalar of the new metric being multiplied by $e^{-2\alpha\Phi}$. The determinant of the original metric becomes $e^{2\alpha(D+1)\Phi}e^{2\Phi}\det[-G^{(D)}]$, and so the reduced action is

$$\int d^D x \, e^{\alpha(D+1)\Phi}e^\Phi \det^{\frac{1}{2}}[-G^{(D)}]e^{-2\alpha\Phi}R^{(D)}.$$

The total power of e^Φ which appears is $\alpha(D+1) + 1 - 2\alpha = 1 + \alpha(D-1)$. So now we can dial up whatever frame we desire. String frame would have an $e^{-2\Phi}$, and so we get $1 + \alpha(D-1) = -2$, i.e. $\alpha = -3/(D-1)$. For the case the $D = 10$, $\alpha = -1/3$ and this means that

$$G_{MN}^{(11)}dx^M dx^N = e^{-\frac{2}{3}\Phi}G_{\mu\nu}^{(10)}dx^\mu dx^\nu + e^{\frac{4}{3}\Phi}(dx^{10} + A_\mu dx^\mu)^2.$$

---

length πR_{IA} with eight D8-branes and O8-planes at each end, and

$$R_{\mathrm{IA}} = \frac{\ell_s^2}{R_{\mathrm{IB}}} = g_{s,\mathrm{HA}}^{1/2}R_{\mathrm{HA}}^{1/2}\ell_s^{1/2},$$

$$g_{s,\mathrm{IA}} = g_{s,\mathrm{IB}}\frac{\ell_s}{R_{\mathrm{IB}}\sqrt{2}} = g_{s,\mathrm{HA}}^{-1/2}\frac{R_{\mathrm{HA}}^{3/2}}{\ell_s^{3/2}\sqrt{2}}. \tag{12.18}$$

It is worth drawing a picture of this arrangement, and it is displayed in figure 12.4. Notice that since the charge of an O8-plane is precisely that of eight D8-branes, the charge of the R–R sector is locally cancelled at

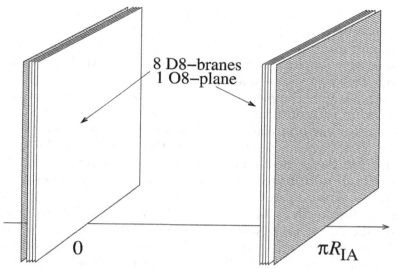

Fig. 12.4. The type IA configuration of two groups of eight D8-branes and O8-planes resulting from a $SO(16) \times SO(16)$ Wilson line.

each end. There is therefore no R–R flux in the interior of the interval and so crucially, we see that the physics between the ends of the segment is given locally by the IIA string. Now we can take $R_{\mathrm{HA}} \to \infty$ to recover the original ten dimensional theory (in particular the Wilson line is irrelevant and the original $E_8 \times E_8$ restored). Both the radius and the coupling of the type IA theory become large. Since the bulk physics is locally that of the IIA string[‡], the strongly coupled limit is eleven dimensional. Taking into account the transformations (12.11) and (12.13), the radii of the two compact dimensions in M-theory units are

$$R_9 = \zeta_{\mathrm{IA}}^{-1} R_{\mathrm{IA}} = g_{\mathrm{s}}^{2/3} \left[2^{-11/18} \pi^{-8/9} \kappa_{11}^{2/9} \right]$$
$$R_{10} = g_{\mathrm{s,IA}}^{2/3} \left[2^{-7/9} \pi^{-8/9} \kappa_{11}^{2/9} \right] = g_{\mathrm{s,HA}}^{-1/3} R_{\mathrm{HA}} \left[2^{-10/9} \pi^{-8/9} \alpha'^{-1/2} \kappa_{11}^{2/9} \right].$$
$$(12.19)$$

Again, had we chosen $\zeta_{\mathrm{IA}} = 1$, we would have

$$R_{10} = R_{\mathrm{HA}} 2^{-1/3}, \quad R_9 = g_{\mathrm{s}} \ell_{\mathrm{s}} 2^{1/6}. \qquad (12.20)$$

[‡] Notice that this is not the case if the D8-branes are placed in a more general arrangement where the charges are not cancelled locally. For such arrangements, the dilaton and R–R nine-form is allowed to vary piecewise linearly between neighbouring D8-branes. The supergravity between the branes is the 'massive' supergravity considered by Romans[97]. This is a very interesting topic in its own right, which we shall not have room to touch upon here. A review of some aspects, with references, is given in the bibliography[181].

As $R \to \infty$, $R_{10} \to \infty$ also, while R_9 remains fixed and (for g_s large) large compared to the Planck scale. This suggests that in the strongly coupled limit of the ten dimensional $E_8 \times E_8$ heterotic string an eleventh dimension again appears: it is a line segment of length R_9, with one E_8 ninebrane factor on each endpoint[169].

We have not fully completed the argument, since we only have argued for $SO(16)$ at each end. One way to see how the E_8 arises is to start from the other end and place eleven dimensional supergravity on a line segment. This theory is anomalous, but the anomaly can be cancelled by having 248 vector fields on each ten dimensional boundary[169]. So the **120** of $SO(16)$ is evidently joined by 128 new massless states at strong coupling. As we saw in section 7.2 in the decomposition of E_8 to $SO(16)$, the adjoint breaks up as $\mathbf{248} = \mathbf{120} \oplus \mathbf{128}$, where the **128** is the spinor representation of $SO(16)$. Now we see why we could not construct this in perturbative type IA string theory. Spinor representations of orthogonal groups cannot be made with Chan–Paton factors. However, we can see these states as massive D0–D8 bound states, T-dual to the D1–D9 spinors we were able to make in the $SO(32)$ case in section 12.2. Now, with $SO(16)$ at each end, we can make precisely the pair of **128**s we need.

12.6 M2-branes and M5-branes

12.6.1 *Supergravity solutions*

Just as in the other supergravities, we can make extended objects in the theory. The most natural one to consider first, given what we have displayed as the content of the theory is one which carries the charge of the higher rank gauge field, $A_{(3)}$. This is a two dimensional brane (a membrane) which we shall call the '*M2-brane*', and the solution is[166]:

$$ds^2 = f_3^{-2/3} \left(-dt^2 + (dx^1)^2 + (dx^2)^2 \right) + f_3^{1/3}(dr^2 + r^2 d\Omega_7^2)$$

$$f_3 = \left(1 + \frac{\pi N \ell_p^3}{r^3} \right), \qquad A_{(3)} = f_3^{-1} dt \wedge dx^1 \wedge dx^2, \quad (12.21)$$

where the eleven dimensional Planck length ℓ_p is given by equation (12.15).

By eleven dimensional Hodge duality, it is easy to see that there is another natural object, a fivebrane which is magnetically dual to the M2-brane, called the '*M5-brane*'[167]:

$$ds^2 = f_5^{-1/3} \left(-dt^2 + (dx^1)^2 + \cdots + (dx^5)^2 \right) + f_5^{2/3}(dr^2 + r^2 d\Omega_4^2)$$

$$f_5 = \left(1 + \frac{32\pi^2 N \ell_p^6}{r^6} \right), \qquad A_{(6)} = f_5^{-1} dt \wedge dx^1 \wedge \cdots \wedge dx^5. \quad (12.22)$$

The tensions of the single (i.e. $N = 1$) M2- and M5-branes of eleven dimensional supergravity are:

$$\tau_2^{\mathrm{M}} = (2\pi)^{-2}\ell_{\mathrm{p}}^{-3}; \qquad \tau_5^{\mathrm{M}} = (2\pi)^{-5}\ell_{\mathrm{p}}^{-6}. \tag{12.23}$$

The product of the M-branes' tensions gives

$$\tau_2^{\mathrm{M}}\tau_5^{\mathrm{M}} = 2\pi(2\pi)^{-8}\ell_{\mathrm{p}}^{-9} = \frac{2\pi}{2\kappa_{11}^2} \tag{12.24}$$

and so is the minimum allowed by the quantum theory, in close analogy with what we know for D-branes from equation (8.20).

12.6.2 From D-branes and NS5-branes to M-branes and back

It is interesting to track the eleven dimensional origin of the various branes of the IIA theory[154]. The D0-branes are, as we saw above, are Kaluza–Klein states[153]. The F1-branes, the IIA strings themselves, are wrapped M2-branes of M-theory. The D2-branes are M2-branes transverse to the eleventh dimension X^{10}. The D4-branes are M5-theory wrapped on X^{10}, while the NS5-branes are M5-branes transverse[§] to X^{10}. The D6-branes, being the magnetic duals of the D0-branes, are Kaluza–Klein monopoles[168, 152] (we shall see this directly later in section 15.2). As mentioned before, the D8-branes have a more complicated fate. To recapitulate, the point is that the D8-branes cause the dilaton to diverge within a finite distance[162], and must therefore be a finite distance from an orientifold plane, which is essentially a boundary of spacetime as we saw in section 4.11. As the coupling grows, the distance to the divergence and the boundary necessarily shrinks, so that they disappear into it in the strong coupling limit: they become part of the gauge dynamics of the nine dimensional boundary of M-theory[169], used to make the $E_8 \times E_8$ heterotic string, as discussed in more detail above. This raises the issue of the strong coupling limit of orientifolds in general. There are various results in the literature, but since the issues are complicated, and because the techniques used are largely strongly coupled field theory deductions, which take us well beyond the scope of this book, except for an O6-plane in section 15.3 and the O7-plane in sections 16.1.11 and 16.1.12, we will have to refer the reader to the literature[235].

One can see further indication of the eleventh dimension in the world-volume dynamics of the various branes. We have already seen this in

---

[§] The reader might like to check, using the Kaluza–Klein relations given at the end of insert 12.1, that the D2-brane and NS5-brane metrics can be obtained from the M2- and M5-brane metrics and vice versa by reduction or the reverse, 'oxidation'.

section 12.3.2 where we saw that the type IIA NS5-brane has a chiral tensor multiplet on its world-volume, the five scalars of which are indicative of an eleven dimensional origin. We saw in the above that this is really a precursor of the fact that it lifts to the M5-brane with the same world-volume tensor multiplet, when type IIA goes to strong coupling. The world-volume theory is believed to be a 5+1 dimensional fixed point theory (see insert 3.1). Consider as another example the D2-brane. In $2+1$ dimensions, the vector field on the brane is dual to a scalar, through Hodge duality of the field strength, $*F_2 = d\phi$. This scalar is the eleventh embedding dimension[155]. It joins the other seven scalars already defining the collective modes for transverse motion to show that there are *eight* transverse dimensions. Carrying out the duality in detail, the D2-brane action is therefore found to have a hidden eleven dimensional Lorentz invariance. We shall see this feature in certain probe computations later on in section 15.2. So we learn that the M2-brane, which it becomes, has a 2+1 dimensional theory with eight scalars on its world-volume. The existence of this theory may be inferred in purely field theory terms as being an infra-red fixed point (see insert 3.1) of the 2+1 dimensional gauge theory[180].

12.7 U-duality

A very interesting feature of string duality is the enlargement of the non-perturbative duality group under further toroidal compactification. There is a lot to cover, and it is somewhat orthogonal to most of what we want to do for the rest of the book, so we will err on the side of brevity (for a change). The example of the type II string on a five-torus T^5 is useful, since it is the setting for the simplest black hole state counting that we'll study in chapter 17, and we have already started discussing it in section 7.5.

12.7.1 *Type II strings on T^5 and $E_{6(6)}$*

As we saw in section 7.5, the T-duality group is $O(5, 5, \mathbb{Z})$. The 27 gauge fields split into 10+16+1 where the middle set have their origin in the R–R sector and the rest are NS–NS sector fields. The $O(5, 5; \mathbb{Z})$ representations here correspond directly to the **10**, **16**, and **1** of $SO(10)$. There are also 42 scalars.

The crucial point here is that there is a larger symmetry group of the supergravity, which is in fact $E_{6,(6)}$. It generalises the $SL(2, \mathbb{R})$ $(SU(1,1))$ S-duality group of the type IIB string in ten dimensions. In that case there are two scalars, the dilaton Φ and the R–R scalar $C_{(0)}$, and they take values on the coset space

$$\frac{SL(2,\mathbb{R})}{U(1)} \simeq \frac{SU(1,1)}{SO(2)}.$$

The low energy supergravity theory for this compactification has a continuous symmetry, $E_{6(6)}$ which is a non-compact version[176] of E_6. (See insert 12.2.)

The gauge bosons are in the **27** of $E_{6(6)}(\mathbb{Z})$, which is the same as the **27** of $E_{6(6)}$. The decomposition under $SO(10) \sim O(5,5;\mathbb{Z})$ is familiar from grand unified model building: **27** → **10** + **16** + **1**. Another generalisation is that the 42 scalars live on the coset

$$\frac{E_{6,(6)}}{USp(8)}.$$

In the light of string duality, just as the various branes in type IIB string theory formed physical realisations of multiplets of $SL(2,\mathbb{Z})$, so do the branes here. A discrete subgroup $E_{6(6)}(\mathbb{Z})$ is the 'U-duality' symmetry. The particle excitations carrying the **10** charges are just the Kaluza–Klein and winding strings. The U-duality requires also states in the **16**. These are just the various ways of wrapping Dp-branes to give D-particles (10 for D2, 5 for D4 and 1 for D0). Finally, the state carrying the **1** charge is the NS5-brane, wrapped entirely on the T^5.

In fact, the U-duality group for the type II strings on T^d is $E_{d+1,(d+1)}$, where for $d = 4,3,2,1,0$ we have that the definition of the appropriate E-group is $SO(5,5), SL(5), SL(2) \times SL(3), SL(2) \times \mathbb{R}_+, SL(2)$. These groups can be seen with similar embedding of Dynkin diagrams to what we have done in insert 12.2.

12.7.2 U-duality and bound states

It is interesting to see how some of the bound state results from chapter 11 fit the predictions of U-duality. We will generate U-transformations as a combination of $T_{mn\cdots p}$, which is a T-duality in the indicated directions, and S, the IIB strong/weak coupling transformation. The former switches between N and D boundary conditions and between momentum and winding number in the indicated directions. The latter interchanges the NS–NS and R–R 2-duals but leaves the R–R four-dual invariant, and acts correspondingly on the solitons carrying these charges. We denote by D$_{mn\cdots p}$ a D-brane extended in the indicated directions, and similarly for F$_m$ a fundamental string and p_m a momentum-carrying BPS state.

The first duality chain is

$$(D_9, F_9) \overset{T_{78}}{\to} (D_{789}, F_9) \overset{S}{\to} (D_{789}, D_9) \overset{T_9}{\to} (D_{78}, D_\emptyset).$$

Insert 12.2.   Origins of $E_{6,(6)}$ and other U-duality Groups

One way of seeing roughly where $E_{6,(6)}$ comes from is as follows: The naive symmetry resulting from a T^5 compactification would be $SL(5, \mathbb{R})$, the generalisation of the $SL(2, \mathbb{R})$ of the T^2 to the higher dimensional torus. There are two things which enlarge this somewhat. The first is an enlargement to $SL(6, \mathbb{R})$, which ought to be expected, since the type IIB string already has an $SL(2, \mathbb{R})$ in ten dimensions. This implies the existence of an an extra circle, enlarging the naive torus from T^5 to T^6. This is of course something we have already discovered in section 12.4: at strong coupling, the type IIA string sees an extra circle. Below ten dimensions, T-duality puts both type II strings on the same footing, and so it is most efficient to simply think of the problem as M-theory (at least in its eleven dimensional supergravity limit) compactified on a T^6. Another enlargement is due to T-duality. As we have learned, the full T-duality group is $O(5,5,\mathbb{Z})$, and so we should expect a classical enlargement of the naive $SL(5, \mathbb{R})$ to $O(5,5)$. That $E_{6,(6)}$ contains these two enlargements can be seen quite efficiently in the following Dynkin diagrams[181].

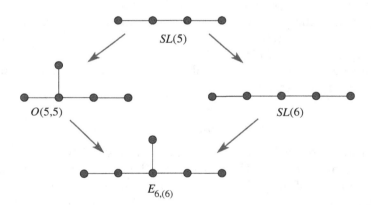

(Actually, the above embedding is not unique, but we are not attempting a proof here; we are simply showing that $E_{6,(6)}$ as is not unreasonable, given what we already know.) The notation $E_{6,(6)}$ means that by analytic continuation of some of the generators, we make a non-compact version of the group (much as in the same way as we get $SL(2, \mathbb{R})$ from $SU(2)$). The maximal number of generators for which this is possible is the relevant case here.

(The last symbol denotes a D0-brane, which is of course not extended anywhere.) Thus the D-string–F-string bound state is U-dual to the 0–2 bound state, as previously indicated in sections 11.3 and 11.5.

The second chain is

$$(D_{6789}, D_{\emptyset}) \xrightarrow{T_6} (D_{789}, D_6) \xrightarrow{S} (D_{789}, F_6) \xrightarrow{T_{6789}} (D_6, p_6) \xrightarrow{S} (F_6, p_6).$$

The bound states of n D0-branes and m D4-branes are thus U-dual to fundamental string states with momentum n and winding number m. The bound state degeneracy (11.21) for $m = 1$ precisely matches the fundamental string degeneracy[177, 144, 178].

For $m > 1$ the same form (11.21) should hold but with $n \to mn$. This is believed to be the case, but the analysis (which requires the instanton picture described in the next section) does not seem to be complete[178].

A related issue is the question of branes ending on other branes[179], and we shall see more of this later. An F-string can of course end on a D-string, so from the first duality chain it follows that a Dp-brane can end on a D$(p + 2)$-brane. The key issue is whether the coupling between spacetime forms and world-brane fields allows the source to be conserved, as with the NS–NS two-dual source in figure 11.1. Similar arguments can be applied to the extended objects in M-theory[179, 143].

13

D-branes and geometry I

In previous chapters we became increasingly aware of the intimate relation of D-branes to both spacetime geometry and to gauge theory, via the collective description of their low energy dynamics. In fact, we have already seen that we can reinterpret many aspects of the spacetime geometry in which the brane moves by reference to the vevs of scalars in the world-volume gauge theory. In this chapter we explore this in much more detail, by using D-branes to probe a number of string theory backgrounds, and find that they allow us to get a new handle on quite detailed properties of the geometry. In addition, we will find that D-branes can take on the properties of a variety of familiar objects, such as monopoles and instantons, depending upon the situation.

13.1 D-branes as probes of ALE spaces

One of the beautiful results which we uncovered soon after constructing the type II strings was that we can 'blow-up' the 16 fixed points of the $T^4/\mathbb{Z}_2$ 'orbifold compactification' to recover string propagation on the smooth hyper-Kähler manifold K3. (We had a lot of fun with this in section 7.6.) Strictly speaking, we only recovered the algebraic data of the K3 manifold this way, and it seemed plausible that the full metric geometry of the space is recovered, but how can we see this directly?

We can recover the metric data by using a brane as a short distance 'probe' of the geometry. This is a powerful technique, which has many useful applications as we shall see in numerous examples as we proceed.

13.1.1 Basic setup and a quiver gauge theory

Let us focus on a single orbifold fixed point, and the type IIB theory. The full string theory is propagating on $\mathbb{R}^6 \times (\mathbb{R}^4/\mathbb{Z}_2)$, which arises from imposing a symmetry under the reflection $\mathbf{R} : (x^6, x^7, x^8, x^9) \rightarrow (-x^6, -x^7, -x^8, -x^9)$, which we used before in section 7.6. Now we can place a D1-brane in this plane at $x^2, \ldots, x^9 = 0$. Let's draw a little table to help keep track of where everything is.

	x^0	x^1	x^2	x^3	x^4	x^5	x^6	x^7	x^8	x^9
D1	—	—	•	•	•	•	•	•	•	•
ALE	—	—	—	—	—	—	•	•	•	•

(We have represented the $\mathbb{R}^4/\mathbb{Z}_2$ (ALE) space as a sort of five dimensional extended object in the table, since it only has structure in the directions x^6, x^7, x^8, x^9.)

The D1-brane can quite trivially sit at the origin and respect the symmetry $\mathbf{R}$, but if it moves off the fixed point, it will break the $\mathbb{Z}_2$ symmetry. In order for it to be able to move off the fixed point there also needs to be an image brane moving to the mirror image position. We therefore need two Chan–Paton indices: one for the D1-brane and the other for its $\mathbb{Z}_2$ image. So (to begin with) the gauge group carried by our D1-brane system living at the origin appears to be $U(2)$, but this will be modified by the following considerations. Since $\mathbf{R}$ exchanges the D1-brane with its image, it can be chosen to act on an open string state as the exchange $\gamma = \sigma^1$, and we shall use the Pauli matrices

$$\sigma^0 \equiv \begin{pmatrix} 1 & 0 \\ 0 & 1 \end{pmatrix}, \quad \sigma^1 \equiv \begin{pmatrix} 0 & 1 \\ 1 & 0 \end{pmatrix}, \quad \sigma^2 \equiv \begin{pmatrix} 0 & -i \\ i & 0 \end{pmatrix}, \quad \sigma^1 \equiv \begin{pmatrix} 1 & 0 \\ 0 & -1 \end{pmatrix}. \tag{13.1}$$

So we can write the representation of the action of $\mathbf{R}$ as:

$$\mathbf{R}|\psi, ij\rangle = \gamma_{ii'} |\mathbf{R}\psi, i'j'\rangle \gamma_{j'j}^{-1}, \quad \text{that is,}$$

$$\mathbf{R}|\psi, ij\rangle = \sigma_{ii'} |\mathbf{R}\psi, i'j'\rangle \sigma_{j'j}^{-1}. \tag{13.2}$$

So it acts on the oscillators in the usual way but also switches the Chan–Paton factors for the brane and its image. The idea[132] is that we must choose an action of the string theory orbifold symmetry on the Chan–Paton factors when there are branes present and make sure that the string theory is consistent in that sector too. Note that the action on the Chan–Paton factors is again chosen to respect the manner in which they appear in amplitudes, just as in section 2.5.

We can therefore compute what happens. In the NS sector, the massless $\mathbf{R}$-invariant states are, in terms of vertex operators:

$$\begin{aligned} &\partial_t X^\mu \sigma^{0,1}, & \mu &= 0, 1 \\ &\partial_n X^i \sigma^{0,1}, & i &= 2, 3, 4, 5 \\ &\partial_n X^m \sigma^{2,3}, & m &= 6, 7, 8, 9. \end{aligned} \tag{13.3}$$

The first row is the vertex operator describing a gauge field with $U(1) \times U(1)$ as the gauge symmetry. The next row constitutes four scalars in the adjoint of the gauge group, parametrising the position of the D1-brane within the six-plane $\mathbb{R}^6$, and the last row is four scalars in the 'bifundamental' charges $(\pm 1, \mp 1)$ of the gauge group the transverse position on x^6, x^7, x^8, x^9. Let us denote the corresponding D-string fields A^μ, X^i, X^m, all 2×2 matrices. We may draw a 'quiver diagram'[188] displaying this gauge and matter content (see figure 13.1).

Such diagrams have in general an integer m inside each node, representing a factor $U(m)$ in the gauge group. An arrowed edge of the diagram represents a hypermultiplet transforming as the fundamental (for the sharp end) and antifundamental (for the blunt end) of the two gauge groups corresponding to the connected nodes. The diagram is simply a decorated version of the extended Dynkin diagram associated to A_1. This will make even more sense shortly, since there is geometric meaning to this. Finally, note that one of the $U(1)$s, (the σ_0 one) is trivial: nothing transforms under it, and it simply represents the overall centre of mass of the brane system.

The bosonic action for the fields is the $D = 10$ $U(2)$ Yang–Mills action, dimensionally reduced and $\mathbb{Z}_2$-projected (which breaks the gauge symmetry to $U(1) \times U(1)$). This dimensional reduction is easy to do. There are kinetic terms:

$$T = -\frac{1}{4g_{\mathrm{YM}}^2} \left(F^{\mu\nu} F_{\mu\nu} + \sum_i \mathcal{D}_\mu X^i \mathcal{D}^\mu X^i + \sum_m \mathcal{D}_\mu X^m \mathcal{D}^\mu X^m \right), \quad (13.4)$$

and potential terms:

$$U = -\frac{1}{4g_{\mathrm{YM}}^2} \left(2 \sum_{i,m} \mathrm{Tr}\,[X^i, X^m]^2 + \sum_{m,n} \mathrm{Tr}\,[X^m, X^n]^2 \right), \quad (13.5)$$

where by using (8.13), we have $g_{\mathrm{YM}}^2 = (2\pi)^{-1} \alpha'^{-1/2} g_s$. (Another potentially non-trivial term disappears since the gauge group is Abelian.)

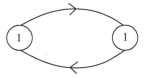

Fig. 13.1. A diagram showing the content of the probe gauge theory. The nodes give information about the gauge groups, while the links give the amount and charges of the mattter hypermultiplets.

The resulting theory has $\mathcal{N} = (4, 4)$ supersymmetry in $D = 2$, which has an $SU(2)$ R-symmetry, and can be thought of as the $SU(2)_R$ left over from parametrising the $\mathbb{Z}_2$ as an action in the $SU(2)_L$ of the natural $SO(4)$. See insert 7.4.

13.1.2 *The moduli space of vacua*

The important thing to realise is that there are large *families* of vacua (here, $U = 0$) of the theory. The space of such vacua is called the 'moduli space' of vacua, and they shall have an interesting interpretation. The moduli space has two branches.

On one, the 'Coulomb Branch', $X^m = 0$ and $X^i = u^i \sigma^0 + v^i \sigma^1$. This corresponds to two D-branes moving independently in the $\mathbb{R}^6$, with positions $u^i \pm v^i$, but staying at the origin of the $\mathbb{R}^4$. The gauge symmetry is unbroken, giving independent $U(1)$s on each D-brane.

On the other, the 'Higgs Branch', X^m is non-zero and $X^i = u^i \sigma^0$. The σ^1 gauge invariance is broken and so we can make the gauge choice $X^m = w^m \sigma^3$. This corresponds to the D1-brane moving off the fixed plane, the string and its image being at $(u^i, \pm w^m)$. We see that this branch has the geometry of the $\mathbb{R}^6 \times \mathbb{R}^4/\mathbb{Z}_2$ which we built in.

Now let us turn on twisted-sector fields which we uncovered in section 7.6, where we learned that they give the blow-up of the geometry. They will appear as parameters in our D-brane gauge theory. Define complex q^m by $X^m = \sigma^3 \mathrm{Re}(q^m) + \sigma^2 \mathrm{Im}(q^m)$, and define two doublets of the $SU(2)_R$:

$$\Phi_0 = \begin{pmatrix} q^6 + iq^7 \\ q^8 + iq^9 \end{pmatrix}, \qquad \Phi_1 = \begin{pmatrix} \bar{q}^6 + i\bar{q}^7 \\ \bar{q}^8 + i\bar{q}^9 \end{pmatrix}. \tag{13.6}$$

These have charges ± 1 respectively under the σ^1 $U(1)$. The three NS–NS moduli can be written as a vector $\boldsymbol{\xi}$ of the $SU(2)_R$, and the potential is proportional to

$$(\boldsymbol{\xi} - \boldsymbol{\mu})^2 \equiv (\Phi_0^\dagger \boldsymbol{\tau} \Phi_0 - \Phi_1^\dagger \boldsymbol{\tau} \Phi_1 + \boldsymbol{\xi})^2, \tag{13.7}$$

where the Pauli matrices are now denoted τ^I to emphasise that they act in a different space. They are assembled into a vector $\boldsymbol{\tau} = (\tau^1, \tau^2, \tau^3)$. (The vector $\boldsymbol{\mu}$ is called a 'moment map' in the mathematical understanding of this construction, which we shall discuss later.) Its form is determined by supersymmetry, and it should be checked that it reduces to the second term of the earlier potential (13.5) when $\boldsymbol{\xi} = 0$. The entire potential arises in supersymmetric constructions using superfields as a 'D-term', and its vanishing to find the vacua is the 'D-flatness condition'. The vector $\boldsymbol{\xi}$ enters as a 'Fayet–Iliopoulos' term[222] in the D-term, and is allowed whenever there is an Abelian factor in the gauge group. The $SU(2)_R$ symmetry requires that the FI term and the entire D-term come as a vector. These are all of the facts we will need about such supersymmetry techniques. Unfortunately, a fuller discussion of these matters will take us too far afield, and we refer to reader to the literature[223].

Notice that equation (13.7) implies a coupling between the open string sector and the twisted sector fields. This can be checked directly by a disc computation, where a twist field is in the interior of the disc and the open string fields are on the edge[184].

For $\boldsymbol{\xi} \neq 0$ the orbifold point is blown up. The moduli space of the gauge theory is simply the set of possible locations of the probe i.e., the blown up ALE space. (Note that the branch of the moduli space with $v^i \neq 0$ is no longer present.)

Let us count parameters and constants. The X^m contain eight scalar fields. Three of them are removed by the '$\boldsymbol{\xi}$-flatness' condition that the potential vanishes, and a fourth is a gauge degree of freedom, leaving the expected four moduli. In terms of supermultiplets, the system has the equivalent of $D = 6$ $N = 1$ supersymmetry. The D-string has two hypermultiplets and two vector multiplets, which are Higgsed down to one hypermultiplet and one vector multiplet.

13.1.3 ALE space as metric on moduli space

The idea[184] is that the metric on this moduli space, as seen in the kinetic term for the D-string fields, should be the smoothed ALE metric. Given the fact that we have eight supercharges, it should be a hyper-Kähler manifold[185], and the ALE space has this property. Let us explore this[187].

Three coordinates on our moduli space are conveniently defined as (there are dimensionful constants missing from this normalisation which we shall ignore for now):

$$\mathbf{y} = \Phi_0^\dagger \boldsymbol{\tau} \Phi_0. \tag{13.8}$$

The fourth coordinate, z, can be defined

$$z = 2 \arg(\Phi_{0,1}\Phi_{1,1}). \tag{13.9}$$

The $\boldsymbol{\xi}$-flatness condition implies that

$$\Phi_1^\dagger \boldsymbol{\tau} \Phi_1 = \mathbf{y} + \boldsymbol{\xi}, \tag{13.10}$$

and Φ_0 and Φ_1 are determined in terms of $\mathbf{y}$ and z, up to a gauge choice.

The original metric on the space of hypermultiplet vevs is just the flat metric $ds^2 = d\Phi_0^\dagger d\Phi_0 + d\Phi_1^\dagger d\Phi_1$. We must project this onto the space orthogonal to the $U(1)$ gauge transformation. This is performed (for example) by coupling the Φ_0, Φ_1 for two dimensional gauge fields according to their charges, and integrating out the gauge field. The result is

$$ds^2 = d\Phi_0^\dagger d\Phi_0 + d\Phi_1^\dagger d\Phi_1 - \frac{(\omega_0 + \omega_1)^2}{4(\Phi_0^\dagger \Phi_0 + \Phi_1^\dagger \Phi_1)}, \qquad (13.11)$$

with

$$\omega_i = i(\Phi_i^\dagger d\Phi_i - d\Phi_i^\dagger \Phi_i). \qquad (13.12)$$

We can express the metric in terms of $\mathbf{y}$ and t using the identity:

$$(\alpha^\dagger \tau^a \beta)(\gamma^\dagger \tau^a \delta) = 2(\alpha^\dagger \delta)(\gamma^\dagger \beta) - (\alpha^\dagger \beta)(\gamma^\dagger \delta)$$

for $SU(2)$ arbitrary doublets $\alpha, \beta, \gamma, \delta$. This gives:

$$\Phi_0^\dagger \Phi_0 = |\mathbf{y}|, \qquad \Phi_1^\dagger \Phi_1 = |\mathbf{y} + \boldsymbol{\xi}|,$$
$$d\mathbf{y} \cdot d\mathbf{y} = |\mathbf{y}|d\Phi_0^\dagger d\Phi_0 - \omega_0^2 = |\mathbf{y} + \boldsymbol{\xi}|d\Phi_1^\dagger d\Phi_1 - \omega_1^2, \quad (13.13)$$

and we find that our metric can be written as the $N = 2$ case of the Gibbons–Hawking metric:

$$ds^2 = V^{-1}(dz - \mathbf{A} \cdot d\mathbf{y})^2 + V d\mathbf{y} \cdot d\mathbf{y}$$
$$V = \sum_{i=0}^{N-1} \frac{1}{|\mathbf{y} - \mathbf{y}_i|}, \qquad \boldsymbol{\nabla} V = \boldsymbol{\nabla} \times \mathbf{A}, \qquad (13.14)$$

which is in fact a '*hyper-Kähler*' metric, as we shall see.

Up to an overall normalisation (which we will fix later), we have $\mathbf{y}_0 = 0$, $\mathbf{y}_1 = \boldsymbol{\xi}$, and the vector potential is

$$\mathbf{A}(\mathbf{y}) \cdot d\mathbf{y} = |\mathbf{y}|^{-1}\omega_0 + |\mathbf{y} + \boldsymbol{\xi}|^{-1}\omega_1 + dz, \qquad (13.15)$$

and the field strength is readily obtained by taking the exterior derivative and using the identity

$$\epsilon_{abc}(\alpha^\dagger \tau^b \beta)(\gamma^\dagger \tau^c \delta) = i(\alpha^\dagger \tau^a \delta)(\gamma^\dagger \beta) - i(\alpha^\dagger \delta)(\gamma^\dagger \tau^a \beta).$$

Under a change of variables[92], this metric (for $N = 2$) becomes the Eguchi–Hanson metric, (7.53) which we first identified as the blow-up of the orbifold point. The three parameters in the vector $\mathbf{y}_1 = \boldsymbol{\xi}$ are the NS–NS fields representing the size and orientation of the blown up $\mathbb{CP}^1$.

It is easy to carry out the generalisation to the full A_{N-1} series, and get the metric (13.14) on the moduli space for a D1-brane probing a $\mathbb{Z}_N$ orbifold. The gauge theory is just the obvious generalisation derived from the extended Dynkin diagram: $U(1)^N$, with $N + 1$ bifundamental hypermultiplets with charges $(1, -1)$ under the neighbouring $U(1)$s. (See figure 13.2.)

There will be $3(N - 1)$ NS–NS moduli which will become the $N - 1$ differences $\mathbf{y}_i - \mathbf{y}_0$ in the resulting Gibbons–Hawking metric (13.14). Geometrically, these correspond to the size and orientation of $N - 1$ separate $\mathbb{CP}^1$'s which can be blown up. In fact, we see that the there is another meaning to be ascribed to the Dynkin diagram: each node

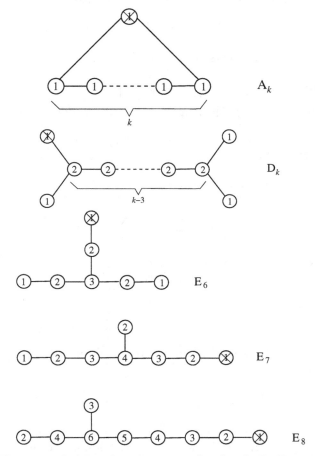

Fig. 13.2. The extended Dynkin diagrams for the A–D–E series. As quiver diagrams, they give the gauge and matter content for the probe gauge theories which compute the resolved geometry of an ALE space. At the same time they also denote the actual underlying geometry of the ALE space, as each node denotes a $\mathbb{CP}^1$, with the connecting edge representing a non-zero intersection.

(except the trivial one) represents a $\mathbb{CP}^1$ in the spacetime geometry that the probe sees on the Higgs branch. We shall expand upon this intriguing picture in section 13.2.

13.1.4 D-branes and the hyper-Kähler quotient

This entire construction which we have just described is a '*hyper-Kähler quotient*', a powerful technique[189] for describing hyper-Kähler metrics of various types, and which has been used to prove the existence of the full family of ALE metrics[190]. Hyper-Kähler spaces are complex $4k$-dimensional manifolds ($k \in \mathbb{Z}$) which admit not just one complex structure, but three, and they transform under an $SU(2)$ symmetry which, for us, will often become and $SU(2)$ R-symmetry of some system with eight supercharges. In fact, the complex structure becomes a quaternionic structure for such manifolds. Flat $\mathbb{R}^4$, presented in the manner done in insert 7.4, is a simple example, and the $SU(2)$ is either of the $SU(2)$ isometries manifest there. The ALE spaces are non-trivial examples, as is the K3 manifold. Two other important four dimensional examples we shall encounter later are the Taub-NUT space and the Atiyah-Hitchin space. Multi-instanton and BPS multi-monopole moduli spaces[218] are higher dimensional cases which we shall also meet.

The hyper-Kähler quotient technique is a powerful mathematical method for showing the existence of (and sometimes exhibiting explicitly) such spaces. It is remarkable that using D-branes we can encounter this construction, physically realised in terms of supersymmetric gauge theory variables-precisely the same variables which appear in the mathematical description of the construction. We shall see this connection arising a number of other times in these pages[191]. Just as we got a $U(1)^2$ gauge theory from the A_1 example, and $U(1)^N$ for the A_{N-1}, the rest of the A–D–E series gives a family of associated gauge theories on the brane as well. These families, and the correspondence to the A–D–E classification arises as follows[81] (see figure 13.2). We start with D-branes on $\mathbb{R}^4/\Gamma$, where Γ is any discrete subgroup of $SU(2)$ (the cover of the $SO(3)$ which acts as rotations at fixed radii). It turns out that the Γ are classified in an 'A–D–E classification', as shown by McKay[87]. The $\mathbb{Z}_N$ are the A_{N-1} series. For the D_N and $E_{6,7,8}$ series, we have the binary dihedral ($\mathcal{D}_{N-2}$), tetrahedral ($\mathcal{T}$), octahedral ($\mathcal{O}$) and icosahedral ($\mathcal{I}$) groups. Let us list them.

• The A_k series ($k \geq 1$). This is the set of cyclic groups of order $k+1$, denoted $\mathbb{Z}_{k+1}$. Their action on the z^i is generated by

$$g = \begin{pmatrix} e^{\frac{2i\pi}{k+1}} & 0 \\ 0 & e^{-\frac{2i\pi}{k+1}} \end{pmatrix}. \tag{13.16}$$

• The D_k series $(k \geq 4)$. This is the binary extension of the dihedral group, of order $4(k-2)$, denoted $\mathcal{D}_{k-2}$. Their action on the z^i is generated by

$$\mathcal{A} = \begin{pmatrix} e^{\frac{i\pi}{k-2}} & 0 \\ 0 & e^{-\frac{i\pi}{k-2}} \end{pmatrix} \quad \text{and} \quad \mathcal{B} = \begin{pmatrix} 0 & i \\ i & 0 \end{pmatrix}. \tag{13.17}$$

In this representation the central element is $\mathcal{Z} = -1 (= \mathcal{A}^2 = \mathcal{B}^2 = (\mathcal{AB})^2)$. Note that the generators $\mathcal{A}$ form a cyclic subgroup $\mathbb{Z}_{2k-4}$.

• The $E_{6,7,8}$ series. These are the binary tetrahedral $(\mathcal{T})$, octahedral $(\mathcal{O})$ and icosahedral $(\mathcal{I})$ groups of order 24, 48 and 120, respectively.

The group $\mathcal{T}$ is generated by taking the elements of $\mathcal{D}_2$ and combining them with

$$\frac{1}{\sqrt{2}} \begin{pmatrix} \varepsilon^7 & \varepsilon^7 \\ \varepsilon^5 & \varepsilon \end{pmatrix} \tag{13.18}$$

where ε is an eighth root of unity.

The group $\mathcal{O}$ is generated by taking the elements of $\mathcal{T}$ and combining them with

$$\begin{pmatrix} \varepsilon & 0 \\ 0 & \varepsilon^7 \end{pmatrix}. \tag{13.19}$$

Finally $\mathcal{I}$ is generated by

$$-\begin{pmatrix} \eta^3 & 0 \\ 0 & \eta^2 \end{pmatrix} \quad \text{and} \quad \frac{1}{\eta^2 - \eta^3} \begin{pmatrix} \eta + \eta^4 & 1 \\ 1 & -\eta - \eta^4 \end{pmatrix}, \tag{13.20}$$

where η is a fifth root of unity.

Given the action of these groups, in order to have the D-branes form a faithful representation on the covering space of the quotient, we need to start with a number equal to the order $|\Gamma|$ of the discrete group. This was two previously, and we started with $U(2)$. So we now start with a gauge group $U(|\Gamma|)$, and then project, as before.

After projecting $U(|\Gamma|)$, the gauge group turns out to be[191]:

$$F = \prod_i U(n_i),$$

where i labels the irreducible representations R_i of dimension n_i. Pictorially (see figure 13.2), the gauge group associated with a D-string on a ALE singularity is simply a product of unitary groups associated to the extended A–D–E Dynkin diagram, with a unitary group coming from each vertex[190]. In the Dynkin diagrams, each vertex represents an irreducible representation of Γ. The integer in the vertex denotes its dimension. The special vertex with the '×' sign is the trivial representation, the one dimensional conjugacy class containing only the identity. The specific connectivity of each graph encodes the information about the following decomposition:

$$Q \otimes R_i = \bigoplus_j a_{ij} R_j, \qquad (13.21)$$

where R_i is the ith irreducible representation and Q is the defining two dimensional representation. Here, the a_{ij} are the elements of the adjacency matrix A of the simply laced extended Dynkin diagrams.

Turning to the hypermultiplets, as stated before, we trivially have $\dim(F)$ hypermultiplets transforming in the adjoint of F. These come from the x^2, x^3, x^4, x^5 sector. They are simply the internal components of the six dimensional vectors after dimensional reduction.

More interestingly, we have hypermultiplets coming from the x^6, x^7, x^8, x^9 sector. These hypermultiplets transform in the fundamentals of the unitary groups, according to the representations

$$\bigoplus_i a_{ij}(\mathbf{n}_i, \bar{\mathbf{n}}_j). \qquad (13.22)$$

Pictorially, the hypermultiplets are simply the links of the extended Dynkin diagrams. These hypermultiplets together with the D–flatness conditions, etc., are precisely the variables and algebraic condition which appear in Kronheimer's constructive proof of the existence of the smooth ALE metrics[190, 191]. The hyper-Kähler quotient is a more general method for constructing manifolds, and this is a well-known example. Another is the construction of moduli spaces of instantons, and we shall see that D-branes capture that rather explicitly in chapter 15.

For example, the simplest model in the D-series is $\mathcal{D}_4$, which would require eight D1-branes on the covering space. The final probe gauge theory after projecting is $F = U(2) \times U(1)^4$, with four copies of a hypermultipet in the $(\mathbf{2}, \mathbf{1})$ of this group.

Unfortunately, it is a difficult and unsolved problem to obtain explicit metrics for the resolved spaces in the D and E cases. This is in a certain sense closely related to the problem of finding an explicit metric on K3, a long-standing goal which the D-brane technique described here implicitly gives a recipe to tackle. We leave it as an exercise to the reader to apply these methods, and suggest that they publish the result if successful.

13.2 Fractional D-branes and wrapped D-branes

13.2.1 Fractional branes

Let us pause to consider the following. In the previous section, we noted that in order for the probe brane to move off the fixed point, we needed to make sure that there were enough copies of it (on the covering space) to furnish a representation of the discrete symmetry Γ that we were going to orbifold by. After the orbifold, we saw that the Higgs branch corresponds to a *single* D-brane moving off the fixed point to non-zero position in x^6, x^7, x^8, x^9. It is made up of the $|\Gamma|$ D-branes we started with on the cover, which are now images of each other under Γ. We can blow up the fixed point to a smooth surface by setting the three NS–NS fields $\boldsymbol{\xi}$ non-zero.

When $\boldsymbol{\xi} = 0$, there is a Coulomb branch. There, the brane is at the fixed point $x^6, x^7, x^8, x^9 = 0$. The $|\Gamma|$ D-branes are free to move apart, independently, as they are no longer constrained by Γ projection. So in fact, we have (as many as *)) $|\Gamma|$ independent branes, which therefore have the interpretation as a fraction of the full brane. None of these individual fractional branes can move off. They have charges under the twisted sector R–R fields. Twisted sector strings have no zero mode, as we have seen, and so cannot propagate.

For an arbitrary number of these fractional branes (and there is no reason not to consider any number that we want) a full $|\Gamma|$ of them must come together to form a closed orbit of Γ, in order for them to move off onto the Higgs branch as one single brane. This fits with the pattern of hypermultiplets and subsequent Higgs-ing which can take place. There simply are not the hypermultiplets in the model corresponding to the movement of an individual fractional brane off the fixed point, and so they are 'frozen' there, while they can move within it[182], in the x^2, x^3, x^4, x^5 directions.

13.2.2 *Wrapped branes*

Notice that when the ALE space is blown-up, we don't see the fractional branes. The fancy language often used at this point is that the Coulomb branch is 'lifted', which is to say it is no longer a branch of degenerate vacua whose existence is protected by supersymmetry. While it is possible to blow-up the point with the separated fractional branes, it is not a supersymmetric operation. We shall see why presently. First, let us set up the geometry of this description.

As we have already mentioned, each node (except for the extended one) in a Dynkin diagram corresponds to a $\mathbb{CP}^1$ which can be blown-up in the smooth geometry. This is in fact a cycle on which a D3-brane can be wrapped in order to make a D1-brane on $\mathbb{R}^6$. For the A_{N-1}-series, where

---

* In the D and E cases, some of the branes are in clumps of size n (according to the nodes in figure 13.2) and carry non-Abelian $U(n)$.

things are simple, there are $N - 1$ such cycles, giving that many different species of D1-brane. This matches with the picture of the previous section, and extends to the whole A–D–E case, since Γ is the number of $\mathbb{CP}^1$s.

Where exactly is this $\mathbb{CP}^1$ in the metric (13.14)? Notice that the 4π periodic variable z, while actually a circle, has a radius that depends upon the prefactor V^{-1}, which varies with $\mathbf{y}$ in a way that is set by the parameters ('centres') $\mathbf{y}_i$. When $\mathbf{y} = \mathbf{y}_i$, the z-circle shrinks to zero size. There is a $\mathbb{CP}^1$ between successive $\mathbf{y}_i$s, which is the minimal surface made up of the locus of z-circles which start out at zero size, grow to some maximum value, and then shrink again to zero size, where a $\mathbb{CP}^1$ then begins again as the neighbouring cycle, having intersected with the previous one in a point. The straight line connecting this will give the smallest cycle, and so the area is $4\pi|\mathbf{y}_i - \mathbf{y}_j|$ for the $\mathbb{CP}^1$ connecting centres $\mathbf{y}_{i,j}$. See figure 13.3. This is just like the case of wrapping a closed string on a circle, as we saw in chapter 4. Winding number is conserved. We saw that even if the circle shrinks away to zero size, the string cannot be pulled off. We worked in T-dual variables and saw that the winding survives as a conserved momentum. Similarly, a closed brane wrapped on a cycle is stuck there, even if the cycle shrinks away. If we don't use some sort of dual description using a large cycle, we need to find a remnant of the wrapped brane after the cycle has shrunk away completely.

Perhaps this is responsible for the fractional brane description. Let us get it to work for a single cycle[187, 201] (crucially, we need to get rid of the

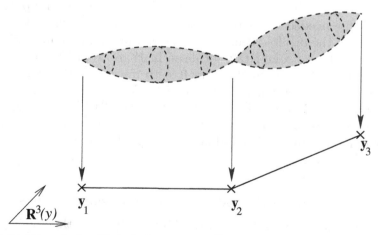

Fig. 13.3. The circles fibred above the $\mathbb{R}^3$ in which lies the centres of the ALE space metric. The collapsing of the circles above the centres results in a network of $\mathbb{CP}^1$ cycles. Their possible intersections are isomorphic to the A–D–E Dynkin diagrams.

total D3-brane charge), and the entire A–D–E series of ALE spaces will follow from what we've already said.

Imagine[202] a D3-brane with some non-zero amount of $B + 2\pi\alpha'F$ on its world-volume. Recall that this corresponds to some D1-brane dissolved into the worldvolume. We deduced this from T-duality in sections 5.2.1 and 9.1. (We did it with pure F, but we can always gauge in some B.) Since we need a total D3-brane charge of zero in our final solution, let us also consider a D3-brane with opposite charge, and with some non-zero $B + 2\pi\alpha'\widetilde{F}$ on its world-volume. We write $\widetilde{F}$ to distinguish it from the F on the other brane's worldvolume, but the Bs are the same, since this is a spacetime background field. So we have a worldvolume interaction:

$$\mu_3 \int C^{(2)} \wedge \left\{ (B + 2\pi\alpha'F) - (B + 2\pi\alpha'\widetilde{F}) \right\}, \qquad (13.23)$$

where we are keeping the terms separate for clarity. Our net D3-brane charge is zero. Now let us choose $2\pi\alpha'(\int_\Sigma F - \widetilde{F}) = \mu_1/\mu_3$, and $\Phi_B \equiv (\mu_3/\mu_1) \int_\Sigma B = 1/2$ for some two dimensional spatial subspace Σ of the three-volume. (Note that $\Phi_B \sim \Phi_B + 1$.) This gives a net D1-brane charge of $1/2 + 1/2 = 1$. The two halves shall be our fractional branes. Right now, they are totally delocalised in the world-volume of the D3-anti D3 system. We can make the D1s more localised by identifying Σ (the parts of the three-volume where B and F are non-zero) with the $\mathbb{CP}^1$ of the ALE space. The smaller the $\mathbb{CP}^1$ is, the more localised the D1s are. In the limit where it shrinks away we have the orbifold fixed point geometry. (Note that we still have $\Phi_B = 1/2$ on the shrunken cycle. Happily, this is just the value needed to be present for a sensible conformal field theory description of the orbifold sector[89], described for example in section 7.6).

Once the D1s are completely localised in x^6, x^7, x^8, x^9 from the shrinking away of the $\mathbb{CP}^1$, then they are free to move supersymmetrically in the x^2, x^3, x^4, x^5 directions. This should be familiar as the general facts we uncovered in chapter 11 about the Dp-D$(p+2)$ bound state system: if the D$(p+2)$ is extended, the Dp cannot move out of it and preserve supersymmetry. This is also T-dual to a single brane at an angle and we shall see this next.

13.3 Wrapped, fractional and stretched branes

There is yet another useful way of thinking of all the of the above physics, and even more aspects of it will become manifest here. It requires exploring a duality to another picture altogether. This duality is a T-duality, although since it is a non-trivial background that is involved, we should be careful. It is best trusted at low energy, as we cannot be sure that the string theories are completely dual at all mass levels without further analysis. We will study only the massless fields, so we should probably claim only that the backgrounds give the same low energy physics. Nevertheless, once we arrive at our dual, we can forget about where it came from and study it directly in its own right. Recent work, using extensions of the techniques of this chapter, has directly proven the duality [340].

13.3.1 *NS5-branes from ALE spaces*

Up to a change of variables, in the supergravity background (13.14), $\mathbf{y}$ can be taken to be the vector $\mathbf{y} = (x^7, x^8, x^9)$ while we will take x^6 to be our periodic coordinate z. (There are some dimensionful parameters which were left out of the derivation of (13.14), for clarity, and we shall put them in by hand, and try to fix the pure numbers with T-duality.)

Then, using the T-duality rules (5.4) we can arrive at another background (note that we have adjoined the flat transverse spacetime $\mathbb{R}^6$ to make a ten dimensional solution, and restored an α' for dimensions):

$$ds^2 = -dt^2 + \sum_{m=1}^{5} dx^m dx^m + V(y)(dx^6 dx^6 + d\mathbf{y} \cdot d\mathbf{y})$$

$$e^{2\Phi} = V(y) = \sum_{i=0}^{N-1} \frac{\sqrt{\alpha'}}{|\mathbf{y} - \mathbf{y}_i|}, \tag{13.24}$$

which is also a ten dimensional solution if taken with a non-trivial background field[203, 204] $H_{mns} = \epsilon_{mns}{}^r \partial_r \Phi$, which defines the potential B_{6i} ($i = 7, 8, 9$) as a vector A_i that satisfies $\nabla V = \nabla \times \mathbf{A}$. Non-zero B_{6i} arose because the T-dual solution had non-zero G_{6i}.

In fact, this is not quite the solution we are looking for. What we have arrived at is a solution which is independent of the x^6 direction. This is necessary if we are to use the operation (5.4). In fact, we expect that the full solution we seek has some structure in x^6, since translation invariance is certainly broken there. This is because the x^6-circle of the ALE space has N places where something special happens to the winding states, since the circle shrinks away there. So we expect that the same must be true for momentum in the dual situation[205]. A simple guess for a solution which is localised completely in the x^6, x^7, x^8, x^9 directions is to simply ask that it be harmonic there. We simply take $\mathbf{x} = (x^6, \mathbf{y})$ to mean a position in the full $\mathbb{R}^4$, and replace $V(y)$ by:

$$V(x) = 1 + \sum_{i=0}^{N-1} \frac{\alpha'}{(\mathbf{x} - \mathbf{x}_i)^2} \tag{13.25}$$

We have done a bit more than just delocalised. By adding the 1 we have endowed the solution with an asymptotically flat region. However, adding

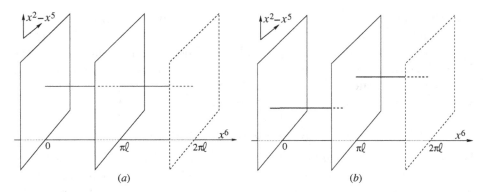

Fig. 13.4. (a) This configuration of two NS5-branes on a circle with D2-branes stretched between them is dual to a D1-brane probing an A_1 ALE space. (b) The Coulomb branch where the D2-brane splits into two 'fractional branes'.

the 1 is consistent with $V(\mathbf{x})$ being harmonic in x^6, x^7, x^8, x^9, and so it is still a solution.

The solution we have just uncovered is made up of a chain of N objects which are pointlike in $\mathbb{R}^4$ and magnetic sources of the NS–NS potential $B_{\mu\nu}$. They are in fact the 'NS5-branes' we discovered by various arguments in chapter 12, with the result (12.8). Here, the NS5-branes are arranged in a circle on x^6, and distributed on the rest of $\mathbb{R}^4$ according to the centres $\mathbf{x}_i$, $i = 0, \ldots, N - 1$.

13.3.2 Dual realisations of quivers

Recall that we had a D1-brane lying along the x^1 direction, probing the ALE space. By the rules of T-duality on a D-brane, it becomes a D2-brane probing the space, with the extra leg of the D2-brane extended along the compact x^6 direction. The D2-brane penetrates the two NS5-branes as it winds around once. The point at which it passes through an NS5-brane is given by four numbers $\mathbf{x}_i$ for the ith brane. The intersection point can be located anywhere within the fivebrane's worldvolume in the directions x^2, x^3, x^4, x^5. (See figure 13.4 (a).)

In the table below, we show the extension of the D2 in the x^6 direction as a $| - |$ to indicate that it may be of finite extent, if it were ending on an NS5-brane.

	x^0	x^1	x^2	x^3	x^4	x^5	x^6	x^7	x^8	x^9
D2	—	—	•	•	•	•	$\|-\|$	•	•	•
NS5	—	—	—	—	—	—	•	•	•	•

This arrangement, with the branes lying in the directions which we have described, preserves the same eight supercharges we discussed before. Starting with the 32 supercharges of the type IIA supersymmetry, the NS5-branes break a half, and the D2-brane breaks half again. The infinite part of the probe, an effective one-brane (string), has a $U(1)$ on its worldvolume, and its tension is $\mu = 2\pi\ell\mu_2$, where ℓ is the as yet unspecified length of the new x^6 direction. However, just as in the discussion in section 10.4, we may consider different values of ℓ if we allow ourselves to consider different densities of branes in the dual picture. Let us focus on $N = 2$. If the two fivebranes (with positions $\mathbf{x}_1, \mathbf{x}_2$; we can set $\mathbf{x}_0$ to zero) are located at the same $\mathbf{y} = (x^7, x^8, x^9)$ position, then the D2-brane can break into two segments, giving a $U(1) \times U(1)$ (one from each segment) on the one-brane part stretched in the infinite x^1 direction. The two segments can move independently within the NS5-brane worldvolume, while still remaining parallel, preserving supersymmetry.

N.B. It makes sense that the D2-brane can end on an NS5-brane, as already discussed in section 12.6.2. There is a 2-form potential in the world-volume for which the string-like end can act as an electric source.

This is the precise analogue of the Coulomb branch of the D1-brane probing the ALE space that we saw earlier. The hypermultiplets of the $U(1) \times U(1)$ theory are made here by stretching fundamental strings across the NS5-branes in x^6 to make a connection between the D-brane segments[206]. The three differences $\mathbf{y}_1 - \mathbf{y}_2$ are the T-dual of the NS–NS parameters representing the size and orientation of the ALE space's $\mathbb{CP}^1$. The x^6 separation of the NS5-branes is dual to the flux $2\pi\ell\Phi_B$. This is the length of one segment while $2\pi\ell(1 - \Phi_B)$ is the length of the other. (Note that the symmetry $\Phi_B \sim \Phi_B + 1$ is preserved, as it just swaps the segments.) Notice also that there is an interesting duality between the quiver diagram and the arrangement of branes in the dual picture. (See figure 13.5.)

The original setup had the lengths equal, but we can change them at will, and this is dual to changing Φ_B. There is the possibility of one of the lengths becoming zero. The NS-branes become coincident, and at the same time a fractional brane becomes a tensionless string, and we get an A_1 enhancement of the gauge symmetry carried by the two-form potential which lives on the type IIA NS5-brane[160]. If we had D1-branes stretched between NS5s in type IIB instead, we would get massless particles, and an enhanced $SU(2)$ gauge symmetry.

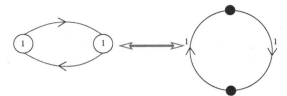

Fig. 13.5. There is a duality between the extended Dynkin diagram which gives the probe gauge theory and the diagram representing D-branes stretched between NS5-branes. The nodes in one are replaced by links in the other. In particular, the number inside the Dynkin nodes become the number of D-branes in the links in the dual diagram. The hypermultiplets associated with links in the Dynkin diagram arise from strings connecting the D-brane fragment ending on one side of an NS5-brane with the fragment on the other.

N.B. This fits nicely with the discovery we made in insert 7.5 that the type IIA string on K3 was dual to the heterotic string on T^4. There are indeed enhanced gauge symmetries on the type IIA side as well. They are not visible in the usual conformal field theory approach because there the flux Φ_B is non-zero and fixed. But now we see that if it is tuned to zero, we can then get the enhanced A–D–E symmetries[161], corresponding to wrapped D2-branes becoming shrunk to zero size with *no* remaining flux, or D1-brane segments shrinking to zero length.

If the segments are separated, and thus attached to the NS5-branes, then when we move the NS5-branes out to different x^{789} positions, the segments must tilt in order to remain stretched between the two branes. They will therefore be oriented differently from each other and will break supersymmetry. This is how the Coulomb branch is 'lifted' in this language. (See figure 13.6(c).) A segment at an orientation gives a contribution $\sqrt{(2\pi\ell\Phi_B)^2 + (\mathbf{y}_1 - \mathbf{y}_2)^2}$ to the D1-brane's tension. This formula should be familiar: it is of the form for the more general formula for a D1–D3 bound state (see section 11.2), to which this tilted D2-brane segment is T-dual.

For supersymmetric vacua to be recovered when the NS5-brane are moved to different positions (the dual of smoothing the ALE space) the branes segments must first rejoin with the other (Higgs-ing), giving the single D-brane. Then it need not move with the NS5-branes as they separate in $\mathbf{y}$, and can preserve supersymmetry by remaining stretched as a single component. (See figure 13.6(d).) Its **y** position and an x^6 Wilson line constitute the Higgs branch parameters. Evidently the metric on these Higgs branch parameters is that of an ALE space, since the 1+1 dimensional gauge theory is the same as the discussion in section 13.1, and hence the moduli spaces match. It is worth sharpening this into a field theory proof of the low energy validity of the T-duality, but we will not do that here.

It is worth noting here that once we have uncovered the existence of fractional D-branes with a modulus for their separation, there is no reason why we cannot separate them infinitely far from each other and consider them in their own right. We also have the right to take a limit where we focus on just one segment with a finite separation between two NS5-branes, but with a *non-compact* x^6 direction. This is achievable from what we started with here by sending $\Phi_B \to 0$, but changing to scaled variables in which there is still a finite separation, and hence a finite gauge coupling on the brane segment in question. (U-duality will then give us various species of branes ending on branes which we will discuss later.)

Fractional branes, and their duals the stretched brane segments, are useful objects since they are less mobile than a complete D-brane, in that

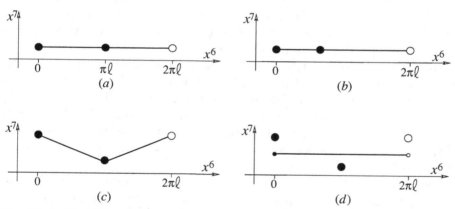

Fig. 13.6. Possible deformations of the brane arrangements, and their gauge theory interpretation. (a) The configuration dual to the standard orbifold limit with the traditional 'half unit' of B-flux. (b) Varying the distribution of B-flux between segments. Sending it to zero will make the NS5-brane coincide and give an enhanced gauge symmetry. (c) Switching on a deformation parameter (an FI term in gauge theory) 'lifts' the Coulomb branch: if there are separated D-brane fragments, supersymmetry cannot be retained. (d) First Higgs-ing to make a complete brane allows smooth movement onto the supersymmetric Higgs branch.

they cannot move in some directions. One use of this is the study of gauge theory on branes with a reduced number of supersymmetries and a reduced number of charged hypermultiplets *à la* Hanany–Witten,[206, 212]. This has many applications[213], some of which we will consider later.

13.4 D-branes as instantons

Consider a D0-brane and N coincident D4-branes. There is a $U(1)$ on the D0 and $U(N)$ on the D4's, which we shall take to be extended in the x^6, x^7, x^8, x^9 directions. The potential terms in the action are

$$\frac{\chi_i^\dagger \chi_i}{(2\pi\alpha')^2} \sum_{a=1}^{5} (X_a - Y_a)^2 + \frac{1}{4g_0^2} \sum_{I=1}^{3} (\chi_i^\dagger \tau^I \chi_i)^2. \qquad (13.26)$$

Here a runs over the dimensions transverse to the D4-brane, and X_a and Y_a are respectively the D0-brane and D4-brane positions, and for now we ignore the position of the D0-brane within the D4-branes' world-volume. This is the same action as in the earlier case (11.7), but here the D4-branes have infinite volume and so their D-term drops out relative to that of the D0-brane. We have also written the 0–4 hypermultiplet field χ with a D4-brane index i. (The $SU(2)_R$ index is suppressed.) The potential (13.26) is exact on grounds of $\mathcal{N} = 2$ supersymmetry. The first term is the $\mathcal{N} = 2$ coupling between the hypermultiplets χ and the vector multiplet scalars X, Y. The second is the $U(1)$ D-term.

For $N > 1$ there are two branches of moduli space, in direct analogy with the ALE case. The Coulomb branch is $(X \neq Y, \quad \chi = 0)$, which is simply the position of the D0-brane transverse to the D4-branes. There is a mass for χ and so its vev is zero. The Higgs branch $(X = Y, \quad \chi \neq 0)$ represents the physics of the D0-brane being stuck on the world-volume of the D4-branes. The non-zero vev of χ Higgses away the $U(1)$ and some of the $U(N)$.

Let us count the dimension of moduli space. There are $4N$ real degrees of freedom in χ. The vanishing of the $U(1)$ D-term imposes three constraints, and moding by the (broken) $U(1)$ removes another degree of freedom leaving $4N - 4$. There are four moduli for the position of the D0-brane inside the the D4-branes, giving a total of $4N$ moduli. This is in fact the correct dimension of moduli space for an $SU(N)$ instanton when we do not mod out also the $SU(N)$ identifications. For k instantons this dimension becomes $4Nk$.

Another clue that the Higgs branch describes the D0-brane as a D4-brane gauge theory instanton is the fact that the Ramond–Ramond couplings include a term $\mu_4 C_{(1)} \wedge \text{Tr}(F \wedge F)$. As shown in section 9.2, when

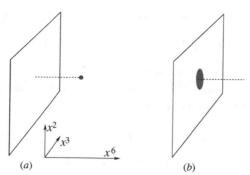

Fig. 13.7. Instantons and the Dp–D$(p + 4)$ system. (a) The Coulomb branch of the Dp-brane theory represents a pointlike brane away from the D$(p + 4)$-brane. (b) The Higgs branch corresponds to it being stuck inside the D$(p+4)$-brane as a finite sized instanton of the D$(p+4)$-brane's gauge theory.

there is an instanton on the D4-brane it carries D0-brane charge. The position of the instanton is given by the 0–0 fields, while the 0–4 fields should give the size and shape (see figure 13.7).

13.4.1 Seeing the instanton with a probe

Actually, we can really see the resulting instanton gauge fields by using a D-brane as a probe. We will use a D9–D5 system[130] instead of D4–D0, and so we won't have the Coulomb branch, since D9-branes fill all of spacetime, and so the D5-branes cannot move out of them. We will us a D1-brane to probe the D9–D5 system[184]. It breaks half of the supersymmetries left over from the 9–5 system, leaving four supercharges overall. The effective 1+1 dimensional theory is $(0, 4)$ supersymmetric and is made of 1–1 fields, which has two classes of hypermulitplets. One represents the motions of the probe transverse to the D5, and the other parallel. The 1–5 and 1–9 fields are also hypermultiplets, while the 9–5 and 5–5 fields are parameters in the model.

In what follows, we shall borrow a lot of the notation of the original papers on the subject[130, 184]. Let us place the D5-branes such that they are pointlike in the x^6, x^7, x^8, x^9 directions. The D1-brane probe will lie along the x^1 direction, as usual.

	x^0	x^1	x^2	x^3	x^4	x^5	x^6	x^7	x^8	x^9
D1	—	—	•	•	•	•	•	•	•	•
D5	—	—	—	—	—	—	•	•	•	•

This arrangement of branes breaks the Lorentz group up as follows:

$$SO(1,9) \supset SO(1,1)_{01} \times SO(4)_{2345} \times SO(4)_{6789}, \qquad (13.27)$$

where the subscripts denote the sub-spacetimes in which the surviving factors act. We may label[207, 184] the world-sheet fields according to how they transform under the covering group:

$$[SU(2)' \times \widetilde{SU(2)}']_{2345} \times [SU(2)_R \times SU(2)_L]_{6789}, \qquad (13.28)$$

with doublet indices $(A', \tilde{A}', A, Y)$, respectively.

The analysis that we did for the D1-brane probe in the type I string theory in section 12.2 still applies, but there are some new details. Now ξ_- is further decomposed into ξ_-^1 and ξ_-^2, where superscripts 1 and 2 denote the decomposition into the (2345) sector and the (6789) sector, respectively. So we have that the fermion ξ_-^1 (hereafter called $\psi_-^{A\tilde{A}'}$) is the right-moving superpartner of the four component scalar field $b^{A'\tilde{A}'}$, while ξ_-^2 (called $\psi_-^{A'Y}$) is the right-moving superpartner of b^{AY}. The supersymmetry transformations are:

$$\delta b^{A'\tilde{A}'} = i\epsilon_{AB}\eta_+^{A'A}\psi_-^{B\tilde{A}'}$$
$$\delta b^{AY} = i\epsilon_{A'B'}\eta_+^{AA'}\psi_-^{B'Y}. \qquad (13.29)$$

In the 1–5 sector, there are four DN coordinates, and four DD coordinates giving the NS sector a zero point energy of zero, with excitations coming from integer modes in the 2345 directions, giving a four component boson. The R sector also has zero point energy of zero, with excitations coming from the 6789 directions, giving a four component fermion χ.

The GSO projections in either sector reduce us to two bosonic states $\phi^{A'}$ and decomposes the spinor χ into left- and right-moving two component spinors, χ_-^A and χ_+^Y, respectively. We see that χ_-^A is the right-moving superpartner of $\phi^{A'}$. Taking into account the fact that there is a D5-brane index for these fields, we can display the components $(\phi^{A'm}, \chi^{Am})$ which are related by supersymmetry:

$$\delta\phi^{A'm} = i\epsilon_{AB}\eta_+^{A'A}\chi_-^{Bm}, \qquad (13.30)$$

and the $(0,4)$ supersymmetry parameter is denoted by $\eta_+^{A'A}$. Here, m is a D5-brane group theory index. Also, χ_+^Y has components χ_+^{Ym}.

The supersymmetry transformation relating them to the left-moving fields are:

$$\delta\lambda_+^M = \eta_+^{AA'}C_{AA'}^M$$
$$\delta\chi_+^{Ym} = \eta_+^{AA'}C_{AA'}^{Ym}, \qquad (13.31)$$

where $C_{AA'}^M$ and $C_{AA'}^{Ym}$ shall be determined shortly. They will be made of the bosonic 1–1 fields and other background couplings built out of the 5–5 and 5–9 fields.

The 5–5 and 5–9 couplings descend from the fields in the D9–D5 sector. There are some details of those fields which are peculiarities of the fact that we are in type I string theory. First, the gauge symmetry on the D9-branes is $SO(32)$. Also, for k coincident D5-branes, there is a gauge symmetry $USp(2k)^{130}$, since there is an extra -1 in the action of Ω on D5-brane fields, as explained already[132] in section 8.7. The 5–5 sector hypermultiplet scalars (fluctuations in the transverse $x^{6,7,8,9}$ directions) transform in the antisymmetric of $USp(2k)$, which we call X_{mn}^{AY}, matching the notation in the literature[184]. Meanwhile, the 5–9 sector produces a $(\mathbf{2k}, \mathbf{32})$, denoted h_M^{Am}, with m and M as in D5- and D9-brane labels.

Using the form of the transformations (13.31) allows us to write the non-trivial part of the $(0, 4)$ supersymmetric 1+1 dimensional Lagrangian containing the 'Yukawa' couplings and the potential of the $(0, 4)$ model:

$$\mathcal{L}_{\text{tot}} = \mathcal{L}_{\text{kinetic}} - \frac{i}{4} \int d^2\sigma \left[\lambda_+^M \left(\epsilon^{BD} \frac{\partial C_{BB'}^M}{\partial b^{DY}} \psi_-^{B'Y} + \epsilon^{B'D'} \frac{\partial C_{BB'}^M}{\partial \phi^{D'm}} \chi_-^{Bm} \right) \right.$$
$$+ \chi_+^{Ym} \left(\epsilon^{BD} \frac{\partial C_{BB'}^{Ym}}{\partial b^{DY}} \psi_-^{B'Y} + \epsilon^{B'D'} \frac{\partial C_{BB'}^{Ym}}{\partial \phi^{D'm}} \chi_-^{Bm} \right)$$
$$\left. + \frac{1}{2} \epsilon^{AB} \epsilon^{A'B'} \left(C_{AA'}^M C_{BB'}^M + C_{AA'}^{Ym} C_{BB'}^{Ym} \right) \right]. \tag{13.32}$$

This is the most general[207] $(0, 4)$ supersymmetric Lagrangian with these types of multiplets, providing that the C satisfy the D-flatness condition:

$$C_{AA'}^M C_{BB'}^M + C_{AA'}^{Ym} C_{BB'}^{Ym} + C_{BA'}^M C_{AB'}^M + C_{BA'}^{Ym} C_{AB'}^{Ym} = 0, \tag{13.33}$$

where $\mathcal{L}_{\text{kinetic}}$ contains the usual kinetic terms for all of the fields. Notice that the fields $b^{A'\bar{A}'}$ and $\psi_-^{AA'}$ are free.

Now equation (13.32) might appear somewhat daunting, but is in fact mostly notation. The trick is to note that general considerations can allow us to fix what sort of things can appear in the matrices $C_{AA'}$. The distance between the D1-brane and the D5-branes should set the mass of the 1–5 fields, $\phi^{A'm}$ and its fermionic partners χ_-^{Am}, χ_+^{Ym}. So there should be terms of the form:

$$\phi_{A'}^m \phi^{A'n} (X_{mn}^{AY} - b^{AY}\delta_{mn})^2, \quad \chi_-^{Am} \chi_+^{Yn} (X_{mn}^{AY} - b^{AY}\delta_{mn}), \tag{13.34}$$

where the term in brackets is the unique translation invariant combination of the appropriate 1–1 and 5–5 fields. There are also 1–5–9 couplings, which would be induced by couplings between 1–9, 1–5 and 5–9 fields, in the form $\lambda_+^M \chi_{m-}^A h_{AM}^m$.

In fact, the required Cs which satisfy the requirements (13.33) and give us the coupling which we expect are[184].

$$C_{AA'}^M = h_A^{Mm} \phi_{A'm}$$
$$C_{AA'}^{Ym} = \phi_{A'}^n (X_{An}^{Ym} - b_A^Y \delta_n^m). \tag{13.35}$$

The $(0,4)$ D-flatness conditions (13.33) translate directly into a series of equations for the D5-brane hypermultiplets to act as data specifying an instanton via the 'ADHM description'[208]. The crucial point is[207] that the vacua of the sigma model gives a space of solutions which is isomorphic to those of ADHM.

One can see that one has the right number of parameters as follows: The potential is of the form $V = \phi^2((X - b)^2 + h^2)$. So the term in brackets acts as a mass term for ϕ. The potential vanishes for $\phi = 0$, leaving this space of vacua to be parametrised by X and h, with b giving the position of the D1-brane in the four transverse directions. Let us write $\widehat{X}^{AY} = (X^{AY} - b^{AY})$ as the centre of mass field.

Notice that for these vacua ($\phi = 0$), the Yukawa couplings are of the form $\sum_a \lambda_{Am}^a B_{Am}^a \chi^{Am}$ where $B_{Am}^a = \partial C_{AB'}^a / \partial \phi_{B'm}$, and the index a is the set (M, Y, m). There are $4k$ fermions in χ_- and so this pairs with $4k$ fermions in the set $\lambda_+^a = (\chi_+^{Ym}, \lambda_+^M)$, leaving a subspace of 32 massless modes describing the non-trivial gauge bundle.

The idea is to write the low energy sigma model action for these massless fields. This is done as follows: a basis of massless components is given by v_i^a ($i = 1, \ldots, 32$) defined by $\sum_a v_v^a B_{Am}^a = 0$, and we choose it to be orthonormal: $\sum_a v_i^a v_j^a = \delta_{ij}$. The basis v_i^a depends on $\widehat{X}$. So substituting $\lambda_+^a = \sum_i v_i^a \lambda_+^i$ into the kinetic energy gives[207]:

$$\lambda_+^a \partial_- \lambda_+^a = \sum_{i,j} \left\{ \lambda_{+i} \left(\delta_{ij} \partial_- + \partial_- \widehat{X}^\mu A_{\mu,ij} \right) \lambda_{+j} \right\}, \tag{13.36}$$

where

$$A_{\mu,ij} \equiv A_{BY,ij} = \sum_a v_i^a \frac{\partial v_j^a}{\partial \widehat{X}^{BY}}; \tag{13.37}$$

we have used the x^6, x^7, x^8, x^9 spacetime index μ on our 1–1 field $\widehat{X}^{BY}$ instead of the indices (B, Y), for clarity.

So we see that the second term in (13.36) shows the sigma model couplings of the fermions to a background gauge field A_μ. Since we have generically

$$B_{Am}^a : \quad \left(\widehat{X}^{AY}, h_A^{Mm} \right), \tag{13.38}$$

the orthonormal basis v_i^a is

$$v^a : \quad \left(\frac{h_A^{Mm}}{\sqrt{\widehat{X}^2 + h^2}}, \frac{-\widehat{X}^{AY}}{\sqrt{\widehat{X}^2 + h^2}} \right), \tag{13.39}$$

and from (13.37), it is clear that the background gauge field is indeed of the form of an instanton: the 5–9 field h indeed sets the scale size of the instanton, and the 5–5 field X sets its position.

13.4.2 Small instantons

Notice that this model gives a meaning to the instanton when its size, set by h, drops to zero[130]. This limit of the instanton is simply singular in field theory. Here we see that the size is just the vev of a 9–5 field, for which zero is a perfectly fine value. Generically, in the Dp-D $(p+4)$ description, zero scale size is the place where the Higgs branch joins onto the Coulomb branch representing the Dp-brane becoming pointlike (getting an enhanced gauge symmetry on its world-volume), and moves out of the world-volume of the parent brane. (For $p = 5$ this branch is not present, but the connection is clear via T-duality.)

This supplies a method for making rather different sorts of gauge group for the heterotic or type I string theory, beyond the perturbative $SO(32)$ that we are used to[130]. The inclusion of k type I D5-branes allows for additional $SU(2)^k$ if they are all separated, or if m are coincident, a $USp(2m) \times SU(2)^{k-m}$, as we have seen in section 8.7. In a compact model, the number, k, of D5-branes is restricted by Gauss's law, and we shall see some examples of this in chapter 14. On the heterotic side, we see that this origin of the gauge group is not visible in any perturbative description, and so the description is best done at strong coupling, in terms of type I strings with D5-branes.

Recall from section 12.5 that the chain of dualities involving T-duality between the heterotic strings and type I/heterotic duality leads to the picture of the strongly coupled $E_8 \times E_8$ heterotic string as eleven dimensional supergravity on a line interval. The same reasoning leads to a picture of small instantons in that case too[341]. They are simply M5-branes in a special situation. An ordinary instanton would be embedded in one or other E_8, and this corresponds to the M5-brane being located at one or other end of the interval. In the intermediate picture denoted in figure 12.4, there is a D4-brane which lives inside the worldvolume of the eight D8-branes located at the O8-plane, at one end or the other. The 4–8 strings can take vevs and allow them to fatten up into fully fledged instantons of the $SO(16)$ which will be enhanced with the spinor representation become the E_8 as they become M5-branes when the extra dimension opens up. Setting the 4–8 strings' vev (vacuum expectation value) to zero allows them to give vevs to the 4–4 strings which can move them away from the ends of the interval into the interior. In the fully eleven dimensional picture, this is the M5-brane moving into the interior. The $E_8 \times E_8$ is restored, but there is something extra from the M5-brane, just as in the $SO(32)$ case there was something extra from the D5-branes. In this case, it is not an extra $\mathcal{N} = 2$ six dimensional vector multiplet, $(\mathbf{2}, \mathbf{2}) + 4(\mathbf{1}, \mathbf{1})$ giving extra gauge symmetry, but an extra $\mathcal{N} = 2$ *tensor* multiplet, $(\mathbf{3}, \mathbf{1}) + 5(\mathbf{1}, \mathbf{1})$. Even after the return to the weakly coupled $E_8 \times E_8$ string by shrinking the interval, this structure remains as the result of shrinking an E_8 instanton to zero size.

This is a rather nice result, for many reasons. One is that we see that the number of scalars in the multiplet reflect the fact that the brane (and hence the instanton) indeed has an eleven dimensional origin, representing its strongly coupled roots even after the return to the weakly coupled heterotic string. The other is that the intermediate picture in type IA allows us to use the result that upon dimensional reduction from six to five dimensions (which happens to the $SO(32)$ D5-brane on its way to becoming an M5-brane), a vector multiplet and a tensor multiplet both reduce to the same multiplet (a vector), and so it is possible to make transitions between these multiplets by making a dimension compact and then decompactifying one afterwards[341].

13.5 D-branes as monopoles

Consider the case of a pair of parallel D3-branes, extended in the directions x^1, x^2, x^3, and separated by a distance L in the x^6 direction. Let us now stretch a family of k parallel D1-branes along the x^6 direction, and have them end on the D3-branes. (This is U-dual to the case of D2-branes ending on NS5-branes, as stated earlier in section 13.3.) Let us call the x^6 direction s, and place the D3-branes symmetrically about the origin, choosing our units such that they are at $s = \pm 1$.

	x^0	x^1	x^2	x^3	x^4	x^5	x^6	x^7	x^8	x^9
D1	—	•	•	•	•	•	|−|	•	•	•
D3	—	—	—	—	•	•	•	•	•	•

This configuration preserves eight supercharges, as can be seen from our previous discussion of fractional branes. Also, a T_6-duality yields a pair of D4-branes (with a Wilson line) in x^1, x^2, x^3, x^6 with k (fractional) D0-branes. (We naively expect that this construction should be related to our previous discussion of instantons, but instead of on $\mathbb{R}^4$, they are on $\mathbb{R}^3 \times S^1$.) We can see it directly from the fact that the presence of the D3- and D1-branes world-volumes place the following constraints on the available supercharges:

$$\epsilon_{\text{L}} = \Gamma^0 \Gamma^1 \Gamma^2 \Gamma^3 \epsilon_{\text{R}}; \quad \epsilon_{\text{L}} = \Gamma^0 \Gamma^6 \epsilon_{\text{R}}, \tag{13.40}$$

which taken together give eight supercharges, satisfying the condition

$$\epsilon_{\text{L}} = \Gamma^1 \Gamma^2 \Gamma^3 \Gamma^6 \epsilon_{\text{L}}. \tag{13.41}$$

The 1–1 massless fields are simply the $(1+1)$-dimensional gauge field $A^\mu(t, s)$ and eight scalars $\Phi^m(t, s)$ in the adjoint of $U(k)$, the latter representing the transverse fluctuations of the branes. There are fluctuations in x^1, x^2, x^3 and others in x^4, x^5, x^7, x^8, x^9. We shall really only be interested in the motions of the D1-brane within the D3-brane's directions x^1, x^2, x^3, which is the Coulomb branch of the D1-brane moduli space. So of the Φ^m, we keep only the three for $m = 1, 2, 3$. There are additionally 1–3 fields transforming in the $(\pm 1, k)$. They form a complex doublet of $SU(2)_{\text{R}}$ and are $1 \times k$ matrices. Crucially, these flavour fields are massless only at $s = \pm 1$, the locations where the D1-branes touch the D3-branes. If we were to write a Lagrangian for the massless fields, there will be a delta function $\delta(s \mp 1)$ in front of terms containing those. The structure of the Lagrangian is very similar to the one written for the $p - (p + 4)$ system, with the additional features of $U(k)$ non-Abelian structure. Asking that the D-terms vanish, for a supersymmetric vacuum, we get:[209]

$$\frac{d\Phi^i}{ds} - [A_s, \Phi^i] + \frac{1}{2}\epsilon^{ijk}[\Phi^j, \Phi^k] = 0, \tag{13.42}$$

where we have ignored possible terms on the right hand side supported only at $s = \pm 1$. These would arise from the interactions induced by massless 1–3 fields there[210]. We shall derive those effects in another way by carefully considering the boundary conditions in a short while.

If we choose the gauge in which $A_s = 0$, our equation (13.42) can be recognised as the Nahm equations[216], known to construct the moduli space[218] of N $SU(2)$ monopoles, *via* an adaptation of the ADHM construction[208]. The covariant form $A_s \neq 0$, is useful for actually solving for the metric on the moduli space of monopole solutions and for the spacetime monopole fields themselves, as we shall show[211].

If our k D1-branes were reasonably well separated, we would imagine that the boundary condition at $s = \pm 1$ is clearly $2\pi\alpha'\Phi^i(s = 1)) =$ diag$\{x_1^i, x_2^i, \ldots, x_k^i\}$, where x_n^i, $i = 1, 2, 3$ are the three coordinates of the end of the nth D1-brane (similarly for the other end). In other words, the off-diagonal fields corresponding to the 1–1 strings stretching between the individual D1-branes are heavy, and therefore lie outside the description of the massless fields. However, this is not quite right. In fact, it is very badly wrong. To see this, note that the D1-branes have tension, and therefore must be pulling on the D3-brane, deforming its shape somewhat. In fact, the shape must be given, to a good approximation, by the following description. The function $s(\mathbf{x})$ describing the position of the D3-brane along the x^6 direction as a function of the three coordinates x^i should satisfy the equation $\nabla^2 s(x) = 0$, where ∇^2 is the three dimensional Laplacian. A solution to this is

$$s = 1 + \frac{c}{|\mathbf{x} - \mathbf{x}_0|}, \tag{13.43}$$

where 1 is the position along the s direction and c and $\mathbf{x}_0$ are constants. So, far away from $\mathbf{x}_0$, we see that the solution is $s = 1$, telling us that we have a description of a flat D3-brane. Nearer to $\mathbf{x}_0$, we see that s increases away from 0, and eventually blows up at $\mathbf{x}_0$.

We sketch this shape in figure 13.8(b). It is again our BIon-type solution, described before in section 5.7. The D3-brane smoothly interpolates between a pure D1-brane geometry far away and a spiked shape resembling D1-brane behaviour at the centre. A multi-centred solution is easy to construct as a superposition of harmonic solutions of the above type.

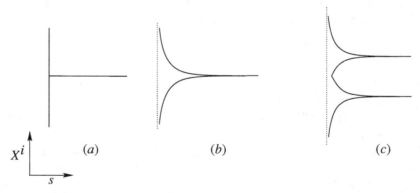

Fig. 13.8. (a) A D3-brane (vertical) with a D1-brane ending on it (horizontal) is actually pulled (b) into a smooth interpolating shape. (c) Finitely separated D1-branes can only be described with non-commutative coordinates (see text)

Considering two of them, we see that in fact for any finite separation of the D1-branes (as measured far enough along the s-direction), by time we get to $s = 1$, they will be arbitrarily close to each other (see figure 13.8(b)). We therefore cannot forget[213] about the off-diagonal parts of Φ^i corresponding to 1–1 strings stretching between the branes, and in fact we are forced to describe the geometry of the branes' endpoints on the D3-brane using non-Abelian Φ^i. This is another example of the 'natural' occurrence of a non-commutativity arising in what we would have naively interpreted as ordinary spacetime coordinates.

We can see precisely what the boundary conditions must be, since we are simply asking that there be a pole in $\Phi^i(s)$ as $s \rightarrow \pm 1$:

$$\Phi^i(s) \rightarrow \frac{\Sigma^i}{s \mp 1}, \tag{13.44}$$

and placing this into (13.42), we see that the $k \times k$ residues must satisfy

$$[\Sigma^i, \Sigma^j] = 2i\epsilon_{ijk}\Sigma^k. \tag{13.45}$$

In other words, they must form a k-dimensional representations of $SU(2)$ (in an unusual normalisation). This representation *must* be irreducible, as we have seen. Otherwise it necessarily captures only the physics of m infinitely separated clumps of D1-branes, for the case where the representation is reducible into m smaller irreducible representations.

13.5.1 Adjoint Higgs and monopoles

The problem we have constructed is that of k monopoles[214, 215] of $SU(2)$ spontaneously broken to $U(1)$ *via* an adjoint Higgs field[217] $H = H^a t^a$. Ignoring the centre of mass of the D3-brane pair, this $SU(2)$ is on their world-volume, and the separation is set by the vacuum expectation value (vev), $\sqrt{H^a H^a} = h$ of the Higgs field. In our problem, we have $(2\pi\alpha')h = L/2$, where L is the separation of our D3-branes, where we label the positions of the branes with $L\sigma_3/2$, with a factor of $2\pi\alpha'$ to convert the Higgs field to a length, as we have done before.

Generically, such a model is given by the following effective Lagrangian:

$$\mathcal{L} = -\frac{1}{4}\text{Tr}(F_{\mu\nu}F^{\mu\nu}) - \frac{1}{2}\text{Tr}(D_\mu H D^\mu H) - V(H), \tag{13.46}$$

$$F_{\mu\nu} = \partial_\mu A_\nu - \partial_\nu A_\mu + e[A_\mu, A_\nu], \qquad D_\mu H = \partial_\mu H + e[A_\mu, H],$$

where H is valued in the adjoint of the gauge group, and there is the usual gauge invariance ($g(\mathbf{x}) \in SU(2)$):

$$A_\mu \to g^{-1}A_\mu g + g^{-1}\partial_\mu g; \quad H \to g^{-1}Hg. \tag{13.47}$$

N.B. Note that we have put explicitly a coupling e into the model so that we can see how the physics depends on it. Our usual conventions do not have that coupling there, but this is for convenience. We will remove it at a later juncture.

The potential V is taken to be positive (but see later), and a typical choice is $V \sim \lambda(\text{Tr}(H \cdot H)/2 - h^2)$. We can imagine a non-zero vev for H which breaks the $SU(2)$ to $U(1)$, such as $H = h\sigma_3$, or, in an $SO(3)$ language, $\vec{H} = (0,0,h)$. The $U(1)$ left over is the σ_3 generator, or just a rotation about the x^3 axis. Notice that the family of values of H which minimise V form an S^2 of radius h, each point on the S^2 being equivalent, as they can be reached by an $SU(2)$ rotation from any other point. This S^2 is the coset space $SU(2)/U(1)$.

More interestingly, we shall seek static solutions with configurations of H and A which have non–trivial dependence on the spatial coordinate $\mathbf{x} = (x_1, x_2, x_3)$. Let us work in a gauge in which $A_0 = 0$, and seek static configurations of finite energy. The Lagrangian reduces to a potential energy density:

$$\mathcal{L} = -\mathcal{E} = -\frac{1}{4}F_{ij}^a F^{aij} - \frac{1}{2}D_i H^a D^i H^a - V(H^a H^a). \tag{13.48}$$

Each term, being positive define, must give a finite value as the result for integrating it over all space. In particular, V requires us to have that H approaches a constant value, h at infinity. We can think of the choice of $H(x)$ at infinity as a map from the sphere at infinity to the vacuum manifold. This map can in fact wind the S^2 of H-vacua around the S^2 at infinity k times, where k is an integer. (The fancy way of saying this is that $\pi_2(SU(2)/U(1)) = \mathbb{Z}$.) This will give a stable solution whose magnetic charge will turn out to be a fixed number times k.

A standard choice for $k = 1$ is that of 't Hooft and Polyakov[214]

$$\text{as} \quad r \longrightarrow \infty, \quad H_i \longrightarrow h\frac{x_i}{r^2} \quad \text{and} \quad A_{ai} \longrightarrow \epsilon_{aij}\frac{x^j}{r^2}, \tag{13.49}$$

where $r^2 = x_1^2 + x_2^2 + x_3^2$. To seek lowest energy configurations, a spherical ansatz

$$H = \hat{x}_i \sigma_i h F(r); \qquad A_{ai} = \epsilon_{aij}\frac{\hat{x}_j}{r}G(r), \tag{13.50}$$

where finite energy requires that $F(r)$ and $G(r)$ tend to unity as $r \to \infty$ and $F'(r) \to 0$ in the limit.

The gauge field strength can then be seen to go at infinity as

$$F_{aij} = \frac{\epsilon_{ijk} x_k x_a}{r^4}.$$

If we pick a gauge where the unbroken $U(1)$ is in the three-direction, we see that, since the magnetic field vector is

$$B_i = \frac{1}{2} \epsilon_{ijk} F^{jk},$$

the magnetic charge is given by

$$g = \frac{1}{4\pi} \int_{S^2} B_i dS_i = \frac{1}{e}, \tag{13.51}$$

which fixes the relation between the magnetic charge and the winding solution. It is possible to show that higher winding, which is a topological invariant, will simply give integers times $1/e$, but we will not do that here, and refer the reader instead to the literature.

A special class of solutions to this model are the *Bogomol'nyi–Prasad–Summerfeld* solutions[61, 62], which have the smallest energy that such a solution can posses, the lower bound being set by the magnetic charge and the potential V. In supersymmetric cases, V actually vanishes, and we recover the familiar situation which we have been seeing all through this book, which is a supersymmetric solution whose mass is essentially equal to its charge (in appropriate units). See insert 13.1 for a discussion.

13.5.2 BPS monopole solution from Nahm data

In fact, we can construct the Higgs field and gauge field of BPS monopole solutions of the 3+1 dimensional gauge theory directly from the Nahm data as follows. Given $k \times k$ Nahm data $(\Phi^1, \Phi^2, \Phi^3) = 2\pi\alpha'(T_1, T_2, T_3)$ solving equation (13.42), there is an associated differential equation for a $2k$ component vector $\mathbf{v}(s)$:

$$\left\{ \mathbf{1}_{2N} \frac{d}{ds} + \left(\frac{x^a}{2} \mathbf{1}_k + iT_a \right) \otimes \sigma^a \right\} \mathbf{v} = 0.$$

There is a unique solution normalisable with respect to the inner product

$$\langle \mathbf{v}_1, \mathbf{v}_2 \rangle = \int_{-1}^{1} \mathbf{v}_1^\dagger \mathbf{v}_2 ds.$$

Insert 13.1.   The prototype BPS object

Let us see why BPS monopoles are BPS in the sense that we have been using in many places in this book. This is in fact the *original* BPS solution. The energy density that we presented in equation (13.48) can be written in a suggestive way:

$$\mathcal{E} = \frac{1}{4}\left(F_{aij} \pm \epsilon_{ijk}D_k H_a\right)^2 \pm \epsilon_{ijk}F_{aij}D_k H_a + V(H),$$

as can be checked by direct reexpansion (we've just completed the square). The second term in this form can be written as

$$\pm\epsilon_{ijk}D_k(F_{aij}H_a) = \pm\frac{1}{2}\epsilon_{ijk}\partial_k(F_{aij}H_a),$$

where we have used the Bianchi identity for the electromagnetic field strength. Since we are interested in the total energy, observe that if we integrate the second term, we get:

$$\pm\frac{1}{2}\epsilon_{ijk}\int d^3x\,\partial_k(F_{aij}H_a) = \pm\frac{1}{2}\epsilon_{ijk}\int_{S^2} F_{aij}H_a\,dS_k,$$

but this just integrates the magnetic field at infinity, in the $U(1)$ picked out by H^a, (which we can choose, as before, to be in the three-generator), and so the integral gives $\pm 4\pi gh$. Since all of the other terms are manifestly positive, we have the bound on the total energy

$$E \geq 4\pi|g|h.$$

Well, it is easy to see how to saturate this bound. We can make the first term vanish with the *Bogomol'nyi condition*:

$$F_{aij} = \mp\epsilon_{ijk}D_k H_a,$$

and then choose to make V as small as we can, which means that $\lambda \ll e$. In fact, we know that in supersymmetric cases, we have that V vanishes for supersymmetry preserving vacua, and so we can saturate the bound precisely in such a situation, giving an energy (mass) for the monopole which is equal to the charge, in appropriate units. Actually, putting the condition above directly into the ansatz (13.50) gives a soluble first order differential equation, with solution written in equations (13.54) and (13.55). We will obtain this solution directly from Nahm data in the next section.

In fact, the space of normalisable solutions to the equation is four dimensional, or complex dimension two. Picking an orthonormal basis $\widehat{\mathbf{v}}_1, \widehat{\mathbf{v}}_2$, we construct the Higgs and gauge potential as:

$$\mathbf{H} = i \begin{bmatrix} \langle s\widehat{\mathbf{v}}_1, \widehat{\mathbf{v}}_1 \rangle & \langle s\widehat{\mathbf{v}}_1, \widehat{\mathbf{v}}_2 \rangle \\ \langle s\widehat{\mathbf{v}}_2, \widehat{\mathbf{v}}_1 \rangle & \langle s\widehat{\mathbf{v}}_2, \widehat{\mathbf{v}}_2 \rangle \end{bmatrix},$$

$$A_i = \begin{bmatrix} \langle \widehat{\mathbf{v}}_1, \partial_i \widehat{\mathbf{v}}_1 \rangle & \langle \widehat{\mathbf{v}}_1, \partial_i \widehat{\mathbf{v}}_2 \rangle \\ \langle \widehat{\mathbf{v}}_2, \partial_i \widehat{\mathbf{v}}_1 \rangle & \langle \widehat{\mathbf{v}}_2, \partial_i \widehat{\mathbf{v}}_2 \rangle \end{bmatrix}. \tag{13.52}$$

The reader may notice a similarity between this means of extracting the gauge and Higgs fields, and the extraction (13.36)(13.37) of the instanton gauge fields in the previous section. This is not an accident. The Nahm construction is in fact a hyper-Kähler quotient which modifies the ADHM procedure. The fact that this arrangement of branes is T-dual to that of the p-$(p+4)$ system is the physical realisation of this fact, showing that the basic families of hypermultiplet fields upon which the construction is based (in the brane context) are present here too.

It is worth studying the case $k = 1$, for orientation, and since we can get an exact solution for this value. In this case, the solutions T_i are simply real constants $(2\pi\alpha')\Phi_i = -ia_i/2$, having the meaning of the position of the monopole at $\mathbf{x} = (a_1, a_2, a_3)$. Let us place it at the origin. Furthermore, as this situation is spherically symmetric, we can write $\mathbf{x} = (0, 0, r)$. Writing components $\mathbf{v} = (w_1, w_2)$, we get a pair of simple differential equations with solution

$$w_1 = c_1 e^{-rs/2}, \quad w_2 = c_2 e^{rs/2}. \tag{13.53}$$

An orthonormal basis is given by

$$\widehat{\mathbf{v}}_1 : \left(c_1 = 0, c_2 = \sqrt{\frac{r}{e^{2r} - 1}} \right); \quad \widehat{\mathbf{v}}_2 : \left(c_2 = 0, c_1 = \sqrt{\frac{r}{1 - e^{-2r}}} \right)$$

and the Higgs field is simply:

$$\mathbf{H}(r) = \hat{x}_i \sigma_i \frac{\varphi(r)}{r}, \quad \text{with}$$

$$\varphi(r) = \frac{r}{(e^{2r} - 1)} \int_{-1}^{1} s e^{rs} ds = r \coth r - 1, \tag{13.54}$$

while the gauge field is:

$$A_i(r) = \epsilon_{ijk} \sigma_j \hat{x}_k \frac{\sinh r - r}{r^2 \sinh r}. \tag{13.55}$$

This is the standard one-monopole solution of Bogomol'nyi, Prasad and Sommerfield, the prototypical 'BPS monopole'[61, 62]. (See insert 13.1.)

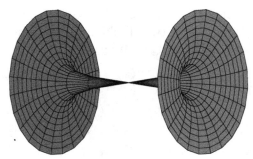

Fig. 13.9. A slice through part of two (horizontal) D3-branes with a (vertical) D1-brane acting as a single BPS monopole. This is made by plotting the Higgs field of the exact BPS solution.

We can insert the required dimensionful quantities:

$$\varphi(r) \to \varphi\left(Lr/4\pi\alpha'\right), \tag{13.56}$$

to get the Higgs field:

$$\mathbf{H} = \frac{\sigma_3}{r}\varphi\left(\frac{Lr}{4\pi\alpha'}\right) \longrightarrow \frac{L}{4\pi\alpha'}\sigma_3, \quad \text{as} \quad r \to \infty, \tag{13.57}$$

showing the asymptotic positions of the D3-branes to be $\pm L/2$, after multiplying by $2\pi\alpha'$ to convert the Higgs field (which has dimensions of a gauge field) to a distance in x^6. A picture of the resulting shape[59, 220] of the D3-brane is shown in figure 13.9.

There is also a simple generalisation of the purely magnetic solution which makes a 'dyon', a monopole with an additional n units of electric charge[221]. It interpolates between the magnetic monopole behaviour we see here and the spike electric solution we found in section 5.7. It is amusing to note[47] that an evaluation of the mass of the solution gives the correct formula for the bound state mass of a D1-string bound to n fundamental strings, as it should, since an electric point source is in fact the fundamental string.

13.6 The D-brane dielectric effect

13.6.1 Non-Abelian world-volume interactions

Consider the familiar non-Abelian term in the D-brane's world-volume action corresponding to the familiar scalar potential of the Yang–Mills theory. This of course appears in the Yang–Mills theory in the usual way, and can be thought of as resulting from the reduction of the ten dimensional Yang–Mills theory. It also arises as the leading part of the expansion of the $\det(Q^i_j)$ term in the non-Abelian Born–Infeld action, in the case when the brane is embedded in the trivial flat background $G_{\mu\nu} = \eta_{\mu\nu}$, as discussed in section 5.6:

$$V = \tau_p \operatorname{Tr}\sqrt{\det(Q^i_{\ j})} = N\tau_p + \frac{\tau_p(2\pi\alpha')^2}{4}\operatorname{Tr}([\Phi^i,\Phi^j][\Phi^j,\Phi^i]) + \cdots, \quad (13.58)$$

where $i = p+1,\ldots,9$. As we have discussed in a number of cases before, the simplest solution extremising V is that the Φ^i all commute, in which case we can write them as diagonal matrices $\Phi^i = (2\pi\alpha')^{-1}X^i$, where $X^i = \operatorname{diag}(x^i_1, x^i_2, \ldots, x^i_N)$. The interpretation is that x^i_n is the coordinate of the nth Dp-brane in the X^i direction; we have N parallel flat Dp-branes, identically oriented, at arbitrary positions in a flat background, $\mathbb{R}^{9-p}$. The centre of mass of the Dp-branes is at $x^i_0 = \operatorname{Tr}(X^i)/N$. The potential is $N\tau_p$, which is simply the sum of all of the rest energies of the branes. We shall discard it in much of what follows.

When we look for situations with non-zero commutators, things become more complicated in interesting ways, giving us the possibility of new interesting extrema of the potential in the presence of non-trivial backgrounds. This is because the commutators appear in many parts of the worldvolume action, and in particular appear in couplings to the R–R fields, as we have seen in section 9.7. Furthermore, the background fields themselves depend upon the transverse coordinates X^i even in the Abelian case, and so will expected to depend upon their non-Abelian generalisation.

In general, this is all rather complicated, but we shall focus on one of the simpler cases as an illustration of the rich set of physical phenomena waiting to be uncovered[51]. Imagine that we have N Dp-branes in a constant background R–R $(p+4)$-form field strength $G_{(p+4)} = dC_{(p+3)}$, with non-trivial components:

$$G_{01\ldots pijk} \equiv G_{tijk} = -2f\varepsilon_{ijk} \quad i,j,k \in \{1,2,3\}. \quad (13.59)$$

(We have suppressed the indices $1,\ldots,p$, as there is no structure there, and will continue to do so in what follows.) Let the Dp-brane be pointlike in the directions x^i ($i = 1,2,3$), and extended in p other directions. None of these Dp-branes in isolation is an electric source of this R–R field strength. Recall, however, that there is a coupling of the Dp-branes to the R–R $(p+3)$-form potential in the non-Abelian case, as shown in equation (9.34). We will assume a static configuration, choosing static gauge

$$\zeta^0 = t, \quad \zeta^\mu = X^\mu, \quad \text{for } \mu = 1,\ldots,p, \quad (13.60)$$

and get (see equation (9.34)):

$$(2\pi\alpha')\mu_p \int \mathrm{Tr}\, P\,[i_\Phi i_\Phi C]$$
$$= (2\pi\alpha')\mu_p \int dt \mathrm{Tr}\left[\Phi^j \Phi^i \left(C_{ijt}(\Phi,t) + (2\pi\alpha')C_{ijk}(\Phi,t)\, D_t \Phi^k\right)\right].$$

We can now do a 'non-Abelian Taylor expansion'[48, 253] of the background field about Φ^i. Generally, this is defined as:

$$F(\Phi^i) = \sum_{n=0}^{\infty} \frac{(2\pi\alpha')^n}{n!} \Phi^{i_1} \dots \Phi^{i_n} \partial_{x^{i_1}} \dots \partial_{x^{i_n}} F(x^i)|_{x=0}, \qquad (13.61)$$

and so:

$$C_{ijk}(\Phi,t) = C_{ijk}(t) + (2\pi\alpha')\Phi^k \partial_k C_{ijk}(t)$$
$$+ \frac{(2\pi\alpha')^2}{2} \Phi^l \Phi^k \partial_l \partial_k C_{ijk}(t) + \cdots. \qquad (13.62)$$

Now since $C_{ijt}(t)$ does not depend on Φ^i, the quadratic term containing it vanishes, since it is antisymmetric in (ij) and we are taking the trace. This leaves the cubic parts:

$$(2\pi\alpha')^2 \mu_p \int dt\, \mathrm{Tr}\left(\Phi^j \Phi^i \left[\Phi^k \partial_k C_{ijt}(t) + C_{ijk}(t)\, D_t \Phi^k\right]\right)$$
$$= \frac{1}{3}(2\pi\alpha')^2 \mu_p \int dt\, \mathrm{Tr}\left(\Phi^i \Phi^j \Phi^k\right) G_{tijk}(t), \qquad (13.63)$$

after an integration by parts. Note that the final expression only depends on the gauge invariant field strength, $G_{(p+4)}$. Since we have chosen it to be constant, this interaction (13.63) is the only term that need be considered, since of the higher order terms implicit in equation (13.62) will give rise to terms depending on derivatives of G.

13.6.2 Stable fuzzy spherical D-branes

Combining equation (13.63) with the part arising in the Dirac–Born-Infeld potential (13.58) yields our effective Lagrangian. This is a static configuration, so there are no kinetic terms and so $\mathcal{L} = -V(\Phi)$, with

$$V(\Phi) = -\frac{(2\pi\alpha')^2 \tau_p}{4} \mathrm{Tr}([\Phi^i, \Phi^j]^2) - \frac{1}{3}(2\pi\alpha')^2 \mu_p \mathrm{Tr}\left(\Phi^i \Phi^j \Phi^k\right) G_{tijk}(t). \qquad (13.64)$$

Let us substitute our choice of background field (13.59). The Euler–Lagrange equations $\delta V(\Phi)/\delta \Phi^i = 0$ yield

$$[[\Phi^i, \Phi^j], \Phi^j] + f\varepsilon_{ijk}[\Phi^j, \Phi^k] = 0. \qquad (13.65)$$

Now of course, the situation of N parallel static branes, $[\Phi^i, \Phi^j] = 0$ is still a solution, but there is a far more interesting one[51]. In fact, the non-zero commutator:

$$[\Phi^i, \Phi^j] = f \, \varepsilon_{ijk} \Phi^k, \tag{13.66}$$

is a solution. In other words we can choose

$$\Phi^i = -i \frac{f}{2} \Sigma^i \tag{13.67}$$

where Σ^i are any $N \times N$ matrix representation of the $SU(2)$ algebra (we have chosen a non-standard normalisation for convenience)

$$[\Sigma^i, \Sigma^j] = 2i \, \epsilon_{ijk} \Sigma^k. \tag{13.68}$$

The $N \times N$ irreducible representation of $SU(2)$ has

$$(\Sigma^i)^2 = \frac{1}{3}(N^2 - 1)\mathbf{I}_{N \times N} \quad \text{for } i = 1, 2, 3, \tag{13.69}$$

where $\mathbf{I}_{N \times N}$ is the identity. Now the value of the potential (13.64) for this solution is

$$V_N = -\frac{T_p (2\pi\alpha')^2 f^2}{6} \sum_{i=1}^{3} \text{Tr}[(\Phi^i)^2] = -\frac{(2\pi)^{-p+2} \alpha'^{\frac{3-p}{2}} f^4}{12g} N(N^2 - 1). \tag{13.70}$$

So our non-commutative solution solution has *lower energy* than the commuting solution, which has $V = 0$ (since we threw away the constant rest energy). This means that the configuration of separated Dp-branes is unstable to collapse to the new configuration.

What is the geometry of this new configuration? Well, the Φs are the transverse coordinates, and so we should try to understand their geometry, despite the fact that they do not commute. In fact, the choice (13.67) with the algebra (13.68) is that corresponding to the non-commutative or 'fuzzy' two-sphere[252]. The radius of this sphere is given by

$$R_N^2 = (2\pi\alpha')^2 \frac{1}{N} \sum_{i=1}^{3} \text{Tr}[(\Phi^i)^2] = \pi^2 \alpha'^2 f^2 (N^2 - 1), \tag{13.71}$$

and so at large N: $R_N \simeq \pi\alpha' f N$. The fuzzy sphere construction may be unfamiliar, and we refer the reader to the references for the details[252]. It suffices to say that as N gets large, the approximation to a smooth sphere improves.

Note that the *irreducible* $N \times N$ representation is not the only solution. A reducible $N \times N$ representation can be made by direct product of k smaller irreducible representations. Such a representation gives a $\text{Tr}[(\Sigma^i)^2]$ which is less than that for the irreducible representation (13.66), and therefore yields higher values for their corresponding potential. Therefore, these smaller representations representations, corresponding geometrically to smaller spheres, are unstable extrema of the potential which again would collapse into the single large sphere of radius R_N. It is amusing to note that we can adjust the solution representing an sphere of size n by

$$\Phi^i = -i\frac{f}{2}\,\Sigma_n^i + x^i\,\mathbf{I}_{n\times n}. \tag{13.72}$$

This has the interpretation of shifting the position of its centre of mass by x^i.

What we have constructed is a D$(p+2)$-brane with topology $\mathbb{R}^p \times S^2$. The $\mathbb{R}^p$ part is where the N Dp-branes are extended and the S^2 is the fuzzy sphere. There is no net D$(p+2)$-brane charge, as each infinitesimal element of the spherical brane which would act as a source of $C_{(p+3)}$ potential has an identical oppositely oriented (and hence oppositely charged) partner. There is therefore a 'dipole' coupling due to the separation of these oppositely oriented surface elements. This type of construction is useful in matrix theory (described in chapter 16), where one can construct for example, spherical D2-brane backgrounds in terms of N D0-branes variables[253, 254, 255].

13.6.3 Stable smooth spherical D-branes

One way[59, 51] to confirm that we have made a spherical brane at large N, is to *start* with a spherical D$(p+2)$-brane, (topology $\mathbb{R}^p \times S^2$) and bind N Dp-branes to it, aligned along an $\mathbb{R}^p$. We can then place it in the background R–R field we first thought of and see if the system will find a static configuration keeping the topology $\mathbb{R}^p \times S^2$, with radius R_N. Failure to find a non-zero radius as a solution of this probe problem would be a sign that we have not interpreted our physics correctly.

Let us write the ten dimensional flat space metric with spherical polar coordinates on the part where the sphere is to be located (x^1, x^2, x^3):

$$ds^2 = -dt^2 + dr^2 + r^2\left(d\theta^2 + \sin^2\theta\,d\phi^2\right) + \sum_{i=4}^{9}(dx^i)^2. \tag{13.73}$$

Our constant background fields in these coordinates is (again, suppressing the $1, \ldots, p$ indices):

$$G_{tr\theta\phi} = -2fr^2\sin\theta \quad \text{and so} \quad C_{t\theta\phi} = \frac{2}{3}fr^3\sin\theta. \tag{13.74}$$

As we have seen many times before, N bound Dp-branes in the D$(p+2)$-brane's world-volume correspond to a flux due to the coupling:

$$(2\pi\alpha')\mu_{p+2} \int_{\mathcal{M}^3} C_{(p+1)} \wedge F = \frac{\mu_p}{2\pi} \int dt\, C_{(p+1)} \wedge F, \qquad (13.75)$$

where $C_{(p+1)}$ is the R–R potential to which the Dp-branes couple, and is not to be confused with the $C_{(p+3)}$ we are using in our background, in (13.74). We need exactly N Dp-branes, so let us determine what F-flux we need to achieve this. If we work again in static gauge, with the D$(p+2)$-brane's world-volume coordinates in the interesting directions being:

$$\zeta^0 = t, \quad \zeta^1 = \theta, \quad \zeta^2 = \phi, \qquad (13.76)$$

then

$$F_{\theta\phi} = \frac{N}{2}\sin\theta, \qquad (13.77)$$

is correctly normalised magnetic field to give our desired flux.

We now have our background, and our N bound Dp-branes, so let us seek a static solution of the form

$$r = R \quad \text{and } x^i = 0, \text{ for } i = 4,\ldots,9. \qquad (13.78)$$

The world volume action for our D$(p+2)$-brane is:

$$S = -\tau_{p+2} \int dt\, d\theta\, d\phi\, e^{-\Phi} \det{}^{\frac{1}{2}}(-G_{ab} + 2\pi\alpha' F_{ab}) + \mu_{p+2} \int C_{(p+3)}. \quad (13.79)$$

Assuming that we have the static trial solution (13.78), inserting the fields (13.74), a trivial dilaton, and the metric from (13.73), the potential energy is:

$$\begin{aligned}
V(R) &= -\int d\theta\, d\phi\, \mathcal{L} \\
&= 4\pi\tau_{p+2}\left(\left[R^4 + \frac{(2\pi\alpha')^2 N^2}{4}\right]^{\frac{1}{2}} - \frac{2f}{3}R^3\right) \\
&= N\tau_p + \frac{2\tau_p}{(2\pi\alpha')^2 N}R^4 - \frac{4\tau_p}{3(2\pi\alpha')}fR^3 + \cdots. \qquad (13.80)
\end{aligned}$$

In the above we expanded the square root assuming that $2R^2/(2\pi\alpha' N) \leq 1$, and kept the first two terms in the expansion. As usual we have substituted $\tau_p = 4\pi^2\alpha'\tau_{p+2}$.

The constant term in the potential energy corresponds to the rest energy of N Dp-branes, and we discard that as before in order to make our comparison. The case $V = 0$ corresponds to $R = 0$, the solution representing flat Dp-branes. Happily, there is another extremum:

$$R = R_N = \pi \alpha' f N \quad \text{with} \quad V = -\frac{(2\pi)^{-p+2} \alpha'^{\frac{3-p}{2}} f^4}{12 g_{\mathrm{s}}} N^3.$$

To leading order in $1/N$, we see that we have recovered the radius (and potential energy) of the non-commutative sphere configuration which we found in equations (13.71) and (13.70). It is appropriate that it only matches at large N, since the fuzzy geometry only approximates a smooth one in this limit.

As noted before, this spherical D$(p + 2)$-brane configuration carries no net D$(p+2)$-brane charge, since each surface element of it has an antipodal part of opposite orientation and hence opposite charge. However, as the sphere is at a finite radius, there is a finite dipole coupling.

There is one major limitation of this whole discussion which is worth remarking upon. There is no such solution as a constant flux in flat space. So the analysis above need to be taken with a pinch of salt. Actually, this sort of brane expansion mechanism was anticipated in an earlier supergravity study before the identification of the precise world-volume mechanisms behind it[59] and so it is worthwhile revisiting the supergravity technology. A flux would create a gravitational back-reaction and so the flat metric that we've been using should really be replaced by some other metric. The prototype such solution of four dimensional Einstein–Maxwell gravity is the Melvin solution[224], which has an infinite magnetic flux threading a four dimensional universe.

This is the sort of solution which we need, with the flux identified with the R–R sector. There is a Kaluza–Klein version of the Melvin solution [225] and this fact has been used[226] to make a $C^{(1)}$ R–R flux solution of type IIA using a reduction from eleven dimensions, and other related solutions. Doing this with a twist allows one to include M5-branes, which upon reduction give a solution representing D4-branes expanding dielectrically into D6-branes via the dielectric mechanism in the magnetic $C^{(1)}$ flux.

All of this that we have described here is the D-brane analogue[51] of the dielectric effect in electromagnetism. If we place Dp-branes in a background R–R field under which the Dp-branes would normally be regarded as neutral, the external field 'polarises' the Dp-branes, making them puff out into a (higher dimensional) non-commutative world-volume geometry. Just as in electromagnetism, where an external field may induce a separation of charges in neutral materials, the D-branes respond through the production of electric dipole and possibly higher multipole couplings via the non-zero commutators of the world-volume scalars.

There is clearly a rich set of physical phenomena to be uncovered by considering non-commuting Φs. Already there have been applications of this mechanism to the understanding of a number of systems, such as large N gauge theory via the AdS/CFT correspondence and other gauge/gravity duals.[256]

The phenomena of branes being able to deform their shape and change their dimensionality, turning into other branes, etc., is a very important direction to explore further. This represents a quite mature physical arena, but currently we are limited to only a few solutions, and quite indirect description. It is possible that we need a whole new language to efficiently describe this physics, which may well be formulated directly in terms of non-commutative variables at the outset.

14

K3 orientifolds and compactification

In section 7.6, we constructed type IIA and IIB compactifications on K3 with pure geometrical orbifolds. In that same chapter, we met the prototype orientifold, which constructs the type I string theory from the type IIB string theory. Now that we have understood the behaviour of D-branes in the neighbourhood of orbifold fixed points, we are in a good position to revisit the orbifold construction of K3 compactification, by combining our ideas with those about the orientifold to make K3 compactifications of type I string theory. There are many models that can be made, and we will present only a small sample of them here[236], in order to illustrate the key ideas.

14.1 $\mathbb{Z}_N$ orientifolds and Chan–Paton factors

Let us consider constructing K3 in its orbifold limits using the spacetime symmetry group $\mathbb{Z}_N$, which acts as described in subsection 7.6.5, which the reader should consult. Recall that we denoted the generator of $\mathbb{Z}_N$ by α_N, the group elements being the powers α_N^k, for $k \in \{0, 1, 2, \ldots, N-1\}$. There are spacetime symmetries, acting on $\mathbb{R}^4$ with coordinates (x^6, x^7, x^8, x^9).

We now define what might be called the 'orientifold group', combining both spacetime and world-sheet symmetries, which contains the elements α_N^k and also $\Omega \cdot \alpha_N^k$ (which we shall sometimes denote Ω_k), where Ω is world-sheet parity. Gauging the action of α_N^k (i.e. projecting onto states invariant under it) will require the introduction of the familiar closed string twisted sectors for the orbifold part, while gauging Ω_k will result in unoriented world-sheets, as described in section 2.6.

At this level, we have a choice as to the elements we wish to consider in our orientifold group, the only constraint being closure. There are two distinct choices of $\mathbb{Z}_N$ orientifold group, which we can denote* as $\mathbb{Z}_N^A$ and $\mathbb{Z}_N^B$. One choice is to have

$$\mathbb{Z}_N^A = \{1, \Omega, \alpha_N^k, \Omega_j\}, \quad k, j = 1, 2, \dots, N-1. \tag{14.1}$$

A second choice (only available for N even) is

$$\mathbb{Z}_N^B = \{1, \alpha_N^{2k-2}, \Omega_{2j-1}\}, \quad k, j = 1, 2, \dots, \frac{N}{2}. \tag{14.2}$$

Both of these orientifold groups close under group multiplication since $\Omega^2 = 1$.

Let us consider the presence of D-branes in this situation, introducing Chan–Paton factors, λ. As in section 2.6, an open string state will be denoted $\lambda_{ij}|\psi, ij\rangle$, where ψ is the state of the world-sheet fields and i and j label the endpoint states. Note that there should be no non-zero elements of the Chan–Paton matrices, λ, which connect D-branes which are at different points in spacetime.

The action of an orientifold group element $g \in \mathbb{Z}_N^{A,B}$ on this complete set will be represented by the unitary matrices denoted γ_g. We have for example, generalising expressions in section 2.6:

$$\alpha_N^k : \quad |\psi, ij\rangle \quad \rightarrow \quad (\gamma_k)_{ii'}|\alpha_N^k \cdot \psi, i'j'\rangle(\gamma_k^{-1})_{j'j} \tag{14.3}$$

while for $\Omega \cdot \alpha_N^k \equiv \Omega_k$,

$$\Omega_k : \quad |\psi, ij\rangle \quad \rightarrow \quad (\gamma_{\Omega_k})_{ii'}|\Omega_k \cdot \psi, j'i'\rangle(\gamma_{\Omega_k}^{-1})_{j'j}. \tag{14.4}$$

As before, when the group element includes Ω, the ends of the string are transposed. Composing various actions of the group elements, we see that since $(\alpha_N^k)^N = 1$, then

$$(\alpha_N^k)^N : \quad |\psi, ij\rangle \quad \rightarrow \quad (\gamma_k^N)_{ii'}|\psi, i'j'\rangle(\gamma_k^{-N})_{j'j} \tag{14.5}$$

and so

$$\gamma_k^N = \pm 1.$$

Similarly, since $\Omega^2 = 1$

$$\Omega^2 : \quad |\psi, ij\rangle \quad \rightarrow \quad (\gamma_\Omega \, (\gamma_\Omega^T)^{-1})_{ii'}|\psi, i'j'\rangle(\gamma_\Omega^T \gamma_\Omega^{-1})_{j'j}, \tag{14.6}$$

resulting in

$$\gamma_\Omega = \pm \gamma_\Omega^T. \tag{14.7}$$

Other examples of such conditions will be put to explicit use later.

---

* Here A and B have nothing to do with the A and B of the type II strings.

14.2 Loops and tadpoles for ALE $\mathbb{Z}_M$ singularities

14.2.1 One-loop diagrams and tadpoles

In open and/or unoriented string theory, certain divergences arise at the one-loop level, which may be interpreted[31] as inconsistencies in the field equations for the R–R potentials in the theory. They manifest themselves as 'tadpoles': amplitudes for emission of quanta from the vacuum. We must ensure that these are absent from the theory, and the way to do this is to cancel them against each other, possibly form different sectors of the theory. We saw this in the prototype orientifold, constructing the $SO(32)$ type I theory from the type IIB superstring. Converting the earlier language of chapter 7 to the one we use here, the tadpoles are of two types, disc tadpoles and $\mathbb{RP}^2$ tadpoles. They are perhaps best visualised as the process of emitting an R–R closed string state from a Dp-brane (for the disc), or from an orientifold plane (for $\mathbb{RP}^2$). In that prototype case the disc and $\mathbb{RP}^2$ produce a divergence proportional to $(2n_9-32)^2$ for $SO(2n_9)$ Chan–Paton factors (i.e. there are n_9 D9-branes) and $(2n_9+32)^2$ for $USp(2n_9)$. Cancellation of the divergences therefore requires gauge group $SO(32)$ (i.e. 16 D9-branes). Here, the orientifold group is simply $\{1, \Omega\}$, as there are no spacetime symmetries to consider. The tadpole cancellation ensures consistency of the ten-form potential's field equation.

Just as in the example of computing the D-brane and orientifold tensions in chapter 7, the most efficient way of computing the divergent contribution of the tadpoles is to compute the one-loop diagrams (the Klein bottle (KB), Möbius strip (MS) and cylinder (C)) and then to take a limit which extracts the divergent pieces. The fact that these diagrams yield the disc and $\mathbb{RP}^2$ tadpoles in terms of the sums of three different products means that the requirement of factorisation of the final expression is a strong consistency check on the whole computation.

The three types of diagram which can be drawn, labelled by the possible elements of the orientifold group under consideration, are depicted in figure 14.1. In the figure, the crosscaps show the action of Ω_m as one goes half way around the open string channel (*around* the cylinder). (Recall that the the crosscap is a disc with the edges identified. See figure 2.13,

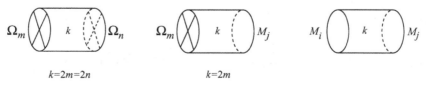

Fig. 14.1. One-loop diagrams from which we will extract the tadpoles.

page 56.) Going around all the way picks up another action of Ω_m, yielding the $\mathbb{Z}_N$ element $\Omega_m^2 = \alpha_N^{2m}$ which is the twist which propagates in the closed string channel (*along* the cylinder). If there is a crosscap with Ω_n at the other end, forming a Klein bottle, then $\Omega_n^2 = \alpha_N^{2n}$ should yield the same twist in the closed string channel, i.e. $2n = 2m$. For $\mathbb{Z}_N$ with N odd, there is only one solution to this: $m = n$ mod N. When N is even however, we can have also the solution $n = m + N/2$ mod N.

14.2.2 Computing the one-loop diagrams

We may parametrise the surfaces as cylinders with length $2\pi l$ and circumference 2π with either boundaries or crosscaps on their ends with boundary conditions on a generic field $\phi(\sigma^1, \sigma^2)$ (and its derivatives):

$$
\begin{aligned}
\text{KB}: \quad & \phi(0, \pi + \sigma^2) = \Omega_m \cdot \phi(0, \sigma^2) \\
& \phi(2\pi l, \pi + \sigma^2) = \Omega_n \cdot \phi(2\pi l, \sigma^2) \\
& \phi(\sigma^1, 2\pi + \sigma^2) = \alpha_N^k \cdot \phi(\sigma^1, \sigma^2); \quad k = 2m = 2n; \\
\text{MS}: \quad & \phi(2\pi l, \sigma^2) \in M_j \\
& \phi(0, \pi + \sigma^2) = \Omega_m \cdot \phi(0, \sigma^2) \\
& \phi(\sigma^1, 2\pi + \sigma^2) = \alpha_N^k \cdot \phi(\sigma^1, \sigma^2); \quad k = 2m; \\
\text{C}: \quad & \phi(0, \sigma^2) \in M_i \\
& \phi(2\pi l, \pi + \sigma^2) \in M_j \\
& \phi(\sigma^1, 2\pi + \sigma^2) = \alpha_N^k \cdot \phi(\sigma^1, \sigma^2).
\end{aligned}
\tag{14.8}
$$

In computing the traces to yield the one-loop expressions, it is convenient to parametrise the Klein bottle and Möbius strip in the region $0 \le \sigma^1 \le 4\pi l, 0 \le \sigma^2 \le \pi$ as follows:

$$
\begin{aligned}
\text{KB}: \quad & \phi(\sigma^1, \pi + \sigma^2) = \Omega_m \cdot \phi(4\pi l - \sigma^1, \sigma^2) \\
& \phi(4\pi l, \sigma^2) = \alpha_N^{m-n} \cdot \phi(0, \sigma^2); \\
\text{MS}: \quad & \phi(0, \sigma^2) \in M_j \\
& \phi(4\pi l, \sigma^2) \in M_j \\
& \phi(\sigma^1, \pi + \sigma^2) = \Omega_m \cdot \phi(4\pi l - \sigma^1, \sigma^2).
\end{aligned}
\tag{14.9}
$$

After the standard rescaling of the coordinates such that open strings are length π while closed strings are length 2π, the amplitudes are

$$
\begin{aligned}
\text{KB}: \quad & \text{Tr}_{c,k}\left(\Omega_m (-1)^{F + \tilde{F}} e^{\pi(L_0 + \bar{L}_0)/2l}\right), \\
\text{MS}: \quad & \text{Tr}_{o,jj}\left(\Omega_m (-1)^F e^{\pi L_0/4l}\right), \\
\text{C}: \quad & \text{Tr}_{o,ij}\left(\alpha_N^k (-1)^F e^{\pi L_0/l}\right).
\end{aligned}
\tag{14.10}
$$

(In the above, 'o' and 'c' mean 'open' and 'closed', respectively.)

The complete one-loop amplitude is

$$\int_0^\infty \frac{dt}{t} \left\{ \text{Tr}_c \left(\mathbf{P}(-1)^{\mathbf{F}} e^{-2\pi t(L_0 + \bar{L}_0)} \right) + \text{Tr}_o \left(\mathbf{P}(-1)^{\mathbf{F}} e^{-2\pi t L_0} \right) \right\}. \quad (14.11)$$

The projector $\mathbf{P}$ includes the GSO and group projections and $\mathbf{F}$ is the fermion number. The traces are over transverse oscillator states and include sums over spacetime momenta. After we evaluate the traces, the $t \to 0$ limit will yield the divergences. Note also that the loop modulus t is related to the cylinder length l as $t = 1/4l, 1/8l$ and $1/2l$ for the Klein bottle, Möbius strip and cylinder, respectively. See figure 6.2, page 148.

Note that the elements α_N^k act as follows on the bosons and in the Neveu–Schwarz (NS) sector:

$$\alpha_N^k : \begin{cases} z_1 = X^6 + iX^7 \to e^{\frac{2\pi i k}{N}} z_1, \\ z_2 = X^8 + iX^9 \to e^{-\frac{2\pi i k}{N}} z_2, \end{cases} \quad (14.12)$$

and it acts in the Ramond (R) sector as

$$\alpha_N^k = e^{\frac{2\pi i k}{N}(J_{67} - J_{89})}. \quad (14.13)$$

As a consequence of this latter convention, notice for example that α_N^k gives the result $4\cos^2 \frac{\pi k}{N}$ when evaluated on the R ground states while $(-)^F \alpha_N^k$ gives $4\sin^2 \frac{\pi k}{N}$.

For the closed string with integer of half-integer modes labelled by r, we may write the action of Ω:

$$\Omega \alpha_r \Omega^{-1} = \tilde{\alpha}_r, \quad \Omega \psi_r \Omega^{-1} = \tilde{\psi}_r, \quad \Omega \tilde{\psi}_r \Omega^{-1} = -\psi_r. \quad (14.14)$$

For open strings with mode expansion

$$X^\mu(\sigma, 0) = x^\mu + i\sqrt{\frac{\alpha'}{2}} \sum_{\substack{m=-\infty \\ m \neq 0}}^{\infty} \frac{\alpha_m}{m} (e^{im\sigma} \pm e^{-im\sigma}). \quad (14.15)$$

Being more explicit, we must compute the following amplitudes,

$$\text{KB}: \quad \text{Tr}_{\text{NS}-\text{NS}+\text{R}-\text{R}}^{U+T} \left\{ \frac{\Omega}{2} \sum_{k=0}^{N-1} \frac{\alpha_N^k}{N} \cdot \frac{1 + (-1)^F}{2} \cdot e^{-2\pi t(L_0 + \bar{L}_0)} \right\},$$

$$\text{MS}: \quad \text{Tr}_{\text{NS}-\text{R}}^{99+55} \left\{ \frac{\Omega}{2} \sum_{k=0}^{N-1} \frac{\alpha_N^k}{N} \cdot \frac{1 + (-1)^F}{2} \cdot e^{-2\pi t L_0} \right\},$$

$$\text{C}: \quad \text{Tr}_{\text{NS}-\text{R}}^{99+55+95+59} \left\{ \frac{1}{2} \sum_{k=0}^{N-1} \frac{\alpha_N^k}{N} \cdot \frac{1 + (-1)^F}{2} \cdot e^{-2\pi t L_0} \right\}, \quad (14.16)$$

Insert 14.1. Jacobi's ϑ-functions

There are key seventeenth-century special functions which play a role in organising one-loop amplitudes:

$$\vartheta_1(z|t) = 2q^{1/4} \sin \pi z \tag{14.17}$$
$$\times \prod_{n=1}^{\infty} (1 - q^{2n}) \prod_{n=1}^{\infty} (1 - q^{2n} e^{2\pi i z}) \prod_{n=1}^{\infty} (1 - q^{2n} e^{-2\pi i z}),$$

$$\vartheta_2(z|t) = 2q^{1/4} \cos \pi z$$
$$\times \prod_{n=1}^{\infty} (1 - q^{2n}) \prod_{n=1}^{\infty} (1 + q^{2n} e^{2\pi i z}) \prod_{n=1}^{\infty} (1 + q^{2n} e^{-2\pi i z}),$$

$$\vartheta_3(z|t) = \prod_{n=1}^{\infty} (1 - q^{2n}) \prod_{n=1}^{\infty} (1 + q^{2n-1} e^{2\pi i z}) \prod_{n=1}^{\infty} (1 + q^{2n-1} e^{-2\pi i z}),$$

$$\vartheta_4(z|t) = \prod_{n=1}^{\infty} (1 - q^{2n}) \prod_{n=1}^{\infty} (1 - q^{2n-1} e^{2\pi i z}) \prod_{n=1}^{\infty} (1 - q^{2n-1} e^{-2\pi i z}),$$

where $q = e^{-\pi t}$. We will need their asymptotics at $t \to 0$. Since the asymptotics as $t \to \infty$ are straightforward we can obtain the $t \to 0$ asymptotia from the modular transformations ($\tau = it$)

$$\vartheta_1(z|\tau) = \tau^{-1/2} e^{3i\pi/4} e^{-i\pi z^2/\tau} \vartheta_1\left(\frac{z}{\tau} \Big| -\frac{1}{\tau}\right),$$

$$\vartheta_3(z|\tau) = \tau^{-1/2} e^{i\pi/4} e^{-i\pi z^2/\tau} \vartheta_3\left(\frac{z}{\tau} \Big| -\frac{1}{\tau}\right),$$

$$\vartheta_2(z|\tau) = \tau^{-1/2} e^{i\pi/4} e^{-i\pi z^2/\tau} \vartheta_4\left(\frac{z}{\tau} \Big| -\frac{1}{\tau}\right),$$

$$\vartheta_4(z|\tau) = \tau^{-1/2} e^{i\pi/4} e^{-i\pi z^2/\tau} \vartheta_2\left(\frac{z}{\tau} \Big| -\frac{1}{\tau}\right). \tag{14.18}$$

---

where $U(T)$ refers to the untwisted (twisted) sector of the closed string. Since Ω forces the left- and right-moving sector to be identical, there is no need to include $\frac{1}{2}(1 + (-1)^F)$ in the trace in the Klein bottle. The open string traces include a sum over Chan–Paton factors.

The results can be written in terms of Jacobi's elliptic functions (see insert 14.1), generalising what we had before with a new twist. In the various loop channels, there is a twist by α_N^k, which introduces a $z = \frac{k}{N}$

Insert 14.2.   The abstruse indentity

The useful abstruce identity ('*aequatio identico satis abstrusa*') ensures the cancellation between the R–R and NS–NS sectors (see chapter 7):

$$f_3(q)^8 - f_4(q)^8 - f_2^8(q) = 0. \tag{14.19}$$

It follows from the more general identities

$$\vartheta_3^2(0|t)\vartheta_3^2(z|t) - \vartheta_4^2(0|t)\vartheta_4^2(z|t) - \vartheta_2^2(0|t)\vartheta_2^2(z|t) = 0$$
$$\vartheta_3^2(0|t)\vartheta_2^2(z|t) - \vartheta_4^2(0|t)\vartheta_1^2(z|t) - \vartheta_2^2(0|t)\vartheta_3^2(z|t) = 0$$
$$\vartheta_3^2(0|t)\vartheta_4^2(z|t) - \vartheta_4^2(0|t)\vartheta_3^2(z|t) - \vartheta_2^2(0|t)\vartheta_1^2(z|t) = 0, \tag{14.20}$$

for the full ϑ-functions. The twisting in the loop channels by α_N^k introduces $z = k/N$ into the oscillator sums, and hence we find that supersymmetry of the models are ensured by these more general identities representing the cancellations between twisted sectors.

---

into the oscillator sums since there is a shift in the energy of the twisted states.

In the case of the Klein bottle, there is also a twist in the closed string loop channel by α_N^{n-m}. Such a space twist will in general change the moding of the fermion and bosons. This should manifest itself as another type of twist of the ϑ-functions. To write this relationship, we use the notation

$$\vartheta_1 = \vartheta[{}^1_1], \quad \vartheta_2 = \vartheta[{}^0_1], \quad \vartheta_3 = \vartheta[{}^0_0], \quad \vartheta_4 = \vartheta[{}^1_0], \tag{14.21}$$

in which we can succinctly write[227]:

$$\vartheta[{}^\epsilon_{\epsilon'}](z - \zeta t|t) = e^{i\pi\{-\tau\zeta^2 + \zeta\epsilon' + 2\zeta z\}}\vartheta[{}^{\epsilon - 2\zeta}_{\epsilon'}](z|t), \tag{14.22}$$

where $(\epsilon, \epsilon') \in \{0, 1\}$ for the familiar ϑ-functions. In evaluating the Klein bottle amplitude, the identities (14.22) are used to rewrite twisted expressions in terms of ϑ-functions.

For the twisted 99 cylinders the one-loop amplitudes are ($z = k/N$):

$$\frac{V_6}{2^3 N} \sum_{k=1}^{N-1} \frac{(\mathrm{Tr}(\gamma_{k,9}))^2}{(4\sin^2 \pi z)^2} \int_0^\infty \frac{dt}{t} (8\pi^2\alpha' t)^{-3} \, 4\sin^2 \pi z \, f_1^{-6}(t)\vartheta_1^{-2}(z|t)$$
$$\times \left\{ \vartheta_3^2(0|t)\vartheta_3^2(z|t) - \vartheta_4^2(0|t)\vartheta_4^2(z|t) - \vartheta_2^2(0|t)\vartheta_2^2(z|t) \right\}, \tag{14.23}$$

while for the twisted 55 cylinders they are:

$$\frac{V_6}{2^3 N} \sum_{k=1}^{N-1} (\mathrm{Tr}(\gamma_{k,5}))^2 \int_0^\infty \frac{dt}{t} (8\pi^2 \alpha' t)^{-3} \; 4\sin^2 \pi z \; f_1^{-6}(t) \vartheta_1^{-2}(z|t)$$

$$\times \left\{ \vartheta_3^2(0|t)\vartheta_3^2(z|t) - \vartheta_4^2(0|t)\vartheta_4^2(z|t) - \vartheta_2^2(0|t)\vartheta_2^2(z|t) \right\}. \tag{14.24}$$

The 95 cylinders give:

$$2\frac{V_6}{2^3 N} \sum_{k=1}^{N-1} \mathrm{Tr}(\gamma_{k,9})\mathrm{Tr}(\gamma_{k,5}) \int_0^\infty \frac{dt}{t} (8\pi^2 \alpha' t)^{-3} \; f_1^{-6}(t) \vartheta_4^{-2}(z|t)$$

$$\times \left\{ \vartheta_3^2(0|t)\vartheta_2^2(z|t) - \vartheta_4^2(0|t)\vartheta_1^2(z|t) - \vartheta_2^2(0|t)\vartheta_3^2(z|t) \right\}. \tag{14.25}$$

The twisted Möbius strip amplitudes are, for the D5-branes ($z=m/N$):

$$-\frac{V_6}{2^3 N} \sum_{m=1}^{N-1} \mathrm{Tr}(\gamma_{\Omega_m,5}^{-1}\gamma_{\Omega_m,5}^T) \int_0^\infty \frac{dt}{t} (8\pi^2 \alpha' t)^{-3}$$

$$\times 4\cos^2 \pi z \; f_1^{-6}(iq) \vartheta_2^{-2}(iq, z)$$

$$\times \left\{ \vartheta_3^2(iq, 0)\vartheta_4^2(iq, z) - \vartheta_4^2(iq, 0)\vartheta_3^2(iq, z) - \vartheta_2^2(iq, 0)\vartheta_1^2(iq, z) \right\}, \tag{14.26}$$

and for the D9-branes:

$$-\frac{V_6}{2^3 N} \sum_{m=1}^{N-1} \frac{\mathrm{Tr}(\gamma_{\Omega_m,9}^{-1}\gamma_{\Omega_m,9}^T)}{(4\sin^2 \pi z)^2} \int_0^\infty \frac{dt}{t} (8\pi^2 \alpha' t)^{-3}$$

$$\times 4\sin^2 \pi z \; f_1^{-6}(iq) \vartheta_1^{-2}(iq, z)$$

$$\times \left\{ \vartheta_3^2(iq, 0)\vartheta_3^2(iq, z) - \vartheta_4^2(iq, 0)\vartheta_4^2(iq, z) - \vartheta_2^2(iq, 0)\vartheta_2^2(iq, z) \right\}. \tag{14.27}$$

Finally, the Klein bottle gives ($t^+ = t + \xi t$, $t^- = t - \xi t$):

$$\frac{V_6}{2^3 N} \sum_{m,n=1}^{N-1} \frac{1}{(4\sin^2 \pi z)^2} \int_0^\infty \frac{dt}{t} (4\pi^2 \alpha' t)^{-3} \; 4\sin^2 2\pi(z - \zeta t)$$

$$\times f_1^{-6}(2t) \vartheta_1^{-1}(z|2t^-) \vartheta_1^{-1}(z|2t^+)$$

$$\times \left\{ -\vartheta_4^2(0|2t)\vartheta_4(z|2t^-)\vartheta_4(z|2t^+) + \vartheta_3^2(0|2t)\vartheta_3(z|2t^-)\vartheta_3(z|2t^+) \right.$$

$$\left. - \vartheta_2^2(0|2t)\vartheta_2(z|2t^-)\vartheta_2(z|2t^+) \right\}. \tag{14.28}$$

In the Klein bottle amplitudes, we have the twist $\zeta = (m - n)/N$ in the closed string channel, resulting in a zero point energy shift for the bosons and fermions which contribute. V_6 is the regularised six dimensional space-time volume.

N.B. The f-functions we met in computations in chapters 4 and 7 are a special case of the functions (14.17), at $z = 0$, as shown in equation (4.44) (here, a prime denotes $\partial/\partial z$):

$$f_1(q) = q^{1/12} \prod_{n=1}^{\infty} (1 - q^{2n}) = (2\pi)^{-1/3} \vartheta_1'(0|t)^{1/3}$$

$$f_2(q) = \sqrt{2} q^{1/12} \prod_{n=1}^{\infty} (1 + q^{2n}) = (2\pi)^{1/6} \vartheta_2'(0|t)^{1/2} \vartheta_1'(0|t)^{-1/6}$$

$$f_3(q) = q^{-1/24} \prod_{n=1}^{\infty} (1 + q^{2n-1}) = (2\pi)^{1/6} \vartheta_3'(0|t)^{1/2} \vartheta_1'(0|t)^{-1/6}$$

$$f_4(q) = q^{-1/24} \prod_{n=1}^{\infty} (1 - q^{2n-1}) = (2\pi)^{1/6} \vartheta_4'(0|t)^{1/2} \vartheta_1'(0|t)^{-1/6}.$$

$$(14.29)$$

The factor of $(4 \sin^2 \pi z)^{-2}$ is a non-trivial contribution from evaluating the trace of the operator $\mathcal{O}$ in the z^1 and z^2 complex planes in the NN sector. The operator $\mathcal{O}$ is the rotation

$$\mathcal{O}: \quad z^{1,2} \to e^{\pm \frac{2\pi i k}{N}} z^{1,2}. \tag{14.30}$$

We have

$$\mathrm{Tr}[e^{\frac{2\pi i k}{N}}] = \int dz^1 dz^2 \langle z^1, z^2 | \mathcal{O} | z^{1\prime}, z^{2\prime} \rangle = \left(4 \sin^2 \frac{\pi k}{N} \right)^{-2}, \tag{14.31}$$

where we have used the basis

$$\langle z^1, z^2 | z^{1\prime}, z^{2\prime} \rangle = \frac{1}{V_{T^4}} \delta(z^1 - z^{1\prime}) \delta(z^2 - z^{2\prime}). \tag{14.32}$$

Supersymmetry is manifest here, as due to the identities (14.20) each of these amplitudes vanishes identically. However, we wish to extract the tadpoles for closed string massless NS–NS fields from these amplitudes, and we do so by identifying the contribution of this sector from each of these amplitudes.

14.2.3 *Extracting the tadpoles*

The next step is to extract the asymptotics as $t \to 0$ of the amplitudes, relating this limit to the $l \to \infty$ limit for each surface, (using the relation between l and t for each surface given earlier). Here, the asymptotic behaviour of the ϑ-functions given in equations (14.18) are used. This extracts the (divergent) contribution of the massless closed string R–R fields, which we shall list below. In what follows, we shall neglect the overall factors of $1/N$ and powers of two which accompany all of the amplitudes.

First, we list the tadpoles for the untwisted R–R potentials. For the tenform we have the following expression (proportional to $(1-1)v_6v_4\int_0^\infty dl$):

$$\text{Tr}(\gamma_{0,9})^2 - 64\text{Tr}(\gamma_{\Omega,9}^{-1}\gamma_{\Omega,9}^T) + 32^2, \tag{14.33}$$

corresponding to the diagrams in figure 14.2

Here, $v_D = V_D(4\pi^2\alpha')^{-D/2}$, where V_D is a D dimensional volume. The limit where we focus upon the neighbourhood of one ALE point is equivalent to taking the non-compact limit $v_4 \to \infty$ while keeping $v_{10} = v_6v_4$ finite.

For the six-form we have (proportional to $(1-1)\frac{v_6}{v_4}\int_0^\infty dl$):

$$\text{Tr}(\gamma_{0,5})^2 - 64\text{Tr}(\gamma_{\Omega\frac{N}{2},5}^{-1}\gamma_{\Omega\frac{N}{2},5}^T) + 32^2, \tag{14.34}$$

which arise from the diagrams in figure 14.3.

In the non-compact limit we are considering here, this last contribution does not survive, as it is proportional to v_6/v_4. The fact that it vanishes is consistent with the fact that if space is not compact, there is no restriction from charge conservation on the number of D5-branes which may be present: the analogue of Gauss's Law for the six-form potential's field strength does not apply, as the flux lines can stretch to infinity. In the compact case, they must begin and end all within the compact volume. So this equation will be relevant only when we return to the study of global six-form charge cancellation in the compact K3 examples.

Notice also in this case that the last two diagrams obviously vanish in the case when N is odd. An immediate consequence of this is that $\mathbb{Z}_3$ fixed points have no six-form charge.

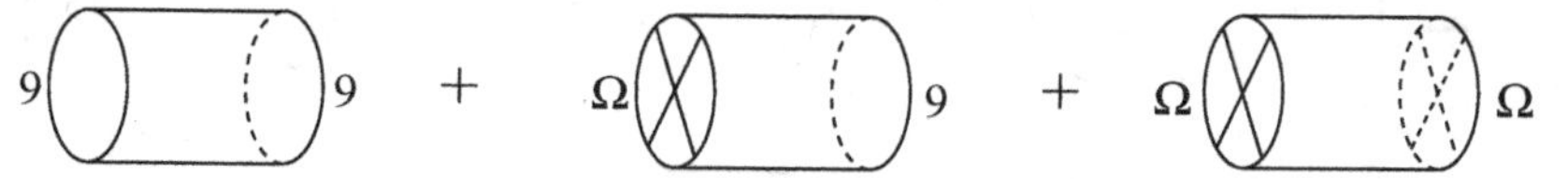

Fig. 14.2. The tadpoles for D9-branes.

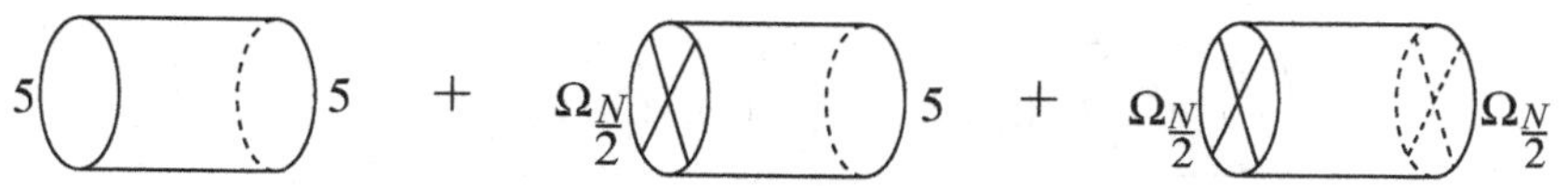

Fig. 14.3. The tadpoles for D5-branes.

The twisted sector tadpoles are (proportional to $(1-1)v_6 \int_0^\infty dl$):

$$\sum_{\substack{k=1 \\ k \neq \frac{N}{2}}}^{N-1} \left[\frac{1}{4 \sin^2 \frac{\pi k}{N}} \mathrm{Tr}(\gamma_{k,9})^2 - 2\mathrm{Tr}(\gamma_{k,9})\mathrm{Tr}(\gamma_{k,5}) + 4 \sin^2 \frac{\pi k}{N} \mathrm{Tr}(\gamma_{k,5})^2 \right]$$

$$- 16 \sum_{\substack{k=1 \\ k \neq \frac{N}{2}}}^{N-1} \left[4 \cos^2 \frac{\pi k}{N} \mathrm{Tr}(\gamma_{\Omega_k,5}^{-1} \gamma_{\Omega_k,5}^T) + \frac{1}{4 \sin^2 \frac{\pi k}{N}} \mathrm{Tr}(\gamma_{\Omega_k,9}^{-1} \gamma_{\Omega_k,9}^T) \right]$$

$$+ 64 \sum_{\substack{k=1 \\ k \neq \frac{N}{2}}}^{N-1} \left[\frac{\cos^2 \frac{\pi k}{N}}{\sin^2 \frac{\pi k}{N}} - \delta_{N \bmod 2,0} \right]. \tag{14.35}$$

These tadpoles correspond to the diagrams in figure 14.4.

Notice that since α_N^k and $\alpha_N^{k+N/2}$ both square to the same element, α_N^{2k}, we can make opposite phase choices in the composition algebra of the γ_{Ω_k} matrices:

$$\mathrm{Tr}[\gamma_{\Omega_k,9}^{-1} \gamma_{\Omega_k,9}^T] = \mathrm{Tr}[\gamma_{2k,9}]$$
$$\mathrm{Tr}[\gamma_{\Omega_{k+\frac{N}{2}},9}^{-1} \gamma_{\Omega_{k+\frac{N}{2}},9}^T] = -\mathrm{Tr}[\gamma_{2k,9}] \tag{14.36}$$

for D9-branes and

$$\mathrm{Tr}[\gamma_{\Omega_k,5}^{-1} \gamma_{\Omega_k,5}^T] = -\mathrm{Tr}[\gamma_{2k,5}]$$
$$\mathrm{Tr}[\gamma_{\Omega_{k+\frac{N}{2}},5}^{-1} \gamma_{\Omega_{k+\frac{N}{2}},5}^T] = \mathrm{Tr}[\gamma_{2k,5}] \tag{14.37}$$

for D5-branes. This is more than an aesthetic choice, as the first line of each of these conditions is simply the crucial result derived in section 8.7

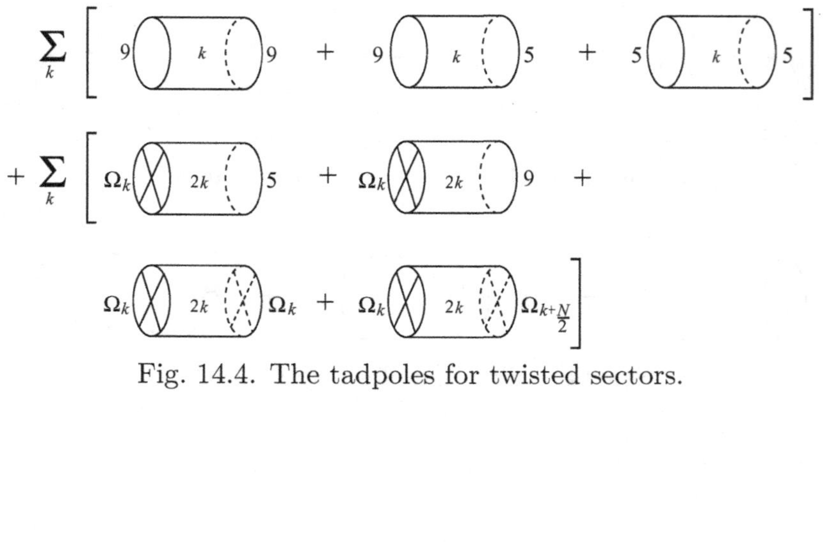

Fig. 14.4. The tadpoles for twisted sectors.

that $\Omega^2 = 1$ in the 99 sector, but -1 in the 55 sector[132]. The second line in each is the statement that $\gamma_{N/2}^2 = -1$ in each sector.

With (14.36) and (14.37), the expression (14.35) can be factorised, for even N:

$$\sum_{k=1}^{\frac{N}{2}} \frac{1}{4\sin^2\frac{(2k-1)\pi}{N}} \left[\mathrm{Tr}(\gamma_{2k-1,9}) - 4\sin^2\frac{(2k-1)\pi}{N} \, \mathrm{Tr}(\gamma_{2k-1,5}) \right]^2$$

$$\sum_{k=1}^{\frac{N}{2}} \frac{1}{4\sin^2\left(\frac{2\pi k}{N}\right)} \left[\mathrm{Tr}(\gamma_{2k,9}) - 4\sin^2\left(\frac{2\pi k}{N}\right)\mathrm{Tr}(\gamma_{2k,5}) - 32\cos\left(\frac{2\pi k}{N}\right) \right]^2$$

(14.38)

and for odd N:

$$\sum_{k=1}^{M-1} \frac{1}{4\sin^2\frac{\pi k}{N}} \left[\mathrm{Tr}(\gamma_{2k,9}) - 4\sin^2\left(\frac{2\pi k}{N}\right)\mathrm{Tr}(\gamma_{2k,5}) - 32\cos^2\frac{\pi k}{N} \right]^2 .$$

(14.39)

Having extracted the divergences and factorised them, revealing the tadpole equations (which may be also interpreted as charge cancellation equations, as discussed earlier) we are ready to find ways of solving these equations for the various orientifold groups.

14.3 Solving the tadpole equations

14.3.1 *T-duality relations*

Compact manifolds which can be constructed as $T^4/\mathbb{Z}_N$ (as described in section 7.6.5) exist only for $N = 2, 3, 4$ and 6. From the discussion in section 14.1, we can therefore construct orientifolds of type A for all these values of N, but of type B only for $N = 2, 4$ and 6. We list below explicitly the orientifold groups:

$$\mathbb{Z}_2^A = \{1, \alpha_2^1, \Omega, \Omega\alpha_2^1\}, \quad \mathbb{Z}_2^B = \{1, \Omega\alpha_2\},$$
$$\mathbb{Z}_3^A = \{1, \alpha_3^1, \alpha_3^2, \Omega, \Omega\alpha_3^1, \Omega\alpha_3^2\},$$
$$\mathbb{Z}_4^A = \{1, \alpha_4^1, \alpha_4^2, \alpha_4^3, \Omega, \Omega\alpha_4^1, \Omega\alpha_4^2, \Omega\alpha_4^3, \}, \quad \mathbb{Z}_4^B = \{1, \alpha_4^2, \Omega\alpha_4^1, \Omega\alpha_4^3, \},$$
$$\mathbb{Z}_6^A = \{1, \alpha_6^1, \ldots, \alpha_6^5, \Omega, \Omega\alpha_6^1, \ldots, \Omega\alpha_6^5\},$$
$$\mathbb{Z}_6^B = \{1, \alpha_6^2, \alpha_6^4, \Omega\alpha_6^1, \Omega\alpha_6^3, \Omega\alpha_6^5\},$$

(14.40)

where $\alpha_N^{\frac{N}{2}} \equiv R$. In equation (14.33) for the untwisted ten-form potential, $\mathrm{Tr}(\gamma_{0,9}) = 2n_9$, where n_9 is the number of D9-branes. All of the orientifold groups of type A contain the element Ω, and therefore there will be an equation of the form (14.33), telling us that there are 16 D9-branes.

Similarly, all type A models except $\mathbb{Z}_3^A$ will contain 16 D5-branes also, as the presence of an element ΩR means that there will be an equation of the form (14.34).

In contrast, the models of type B all lack the element Ω and therefore have only the first term of equation (14.33). The only solution is that the number of D9-branes in these models is zero. All type B models except $\mathbb{Z}_4^B$ have the element ΩR, and therefore have 16 D5-branes. So $\mathbb{Z}_4^B$ has the distinction of having no open string sectors at all: it is a consistent unoriented closed string theory

Now T-duality in the (x^6, x^7, x^8, x^9) directions exchanges the elements Ω and ΩR. This also exchanges D9-branes with D5-branes. So models $\mathbb{Z}_2^A$, $\mathbb{Z}_4^A$, $\mathbb{Z}_6^A$ and $\mathbb{Z}_4^B$ are self T_{6789}-dual. Meanwhile $\mathbb{Z}_3^A$, which has only D9-branes, is dual to Z_6^B which has only D5-branes. $\mathbb{Z}_2^B$, which has only ΩR as a non-trivial element of its orientifold group, is dual to ordinary type I string theory (which we may denote as $\mathbb{Z}_1^A$), whose orientifold group has only Ω as its non-trivial element. This is summarised in figure 14.5.

14.3.2 Explicit solutions

Let us write out the tadpole equations explicitly in each case. For the $\mathbb{Z}_2^A$ case, for which there is one twisted tadpole equation (recall $\alpha_2^1 \equiv R$):

$$\text{Tr}[\gamma_{1,9}] - 4\text{Tr}[\gamma_{1,5}] = 0. \tag{14.41}$$

A solution is

$$\gamma_{\Omega,9} = \gamma_{\Omega R,5} = I_{32}$$
$$\gamma_{R,9} = \gamma_{\Omega,5} = \begin{pmatrix} 0 & I_{16} \\ -I_{16} & 0 \end{pmatrix}.$$

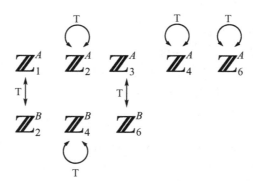

Fig. 14.5. T-duality relations between the models.

For the $\mathbb{Z}_3^A$ case we have

$$\mathrm{Tr}[\gamma_{1,9}] - 3\mathrm{Tr}[\gamma_{1,5}] = 8$$
$$\mathrm{Tr}[\gamma_{2,9}] - 3\mathrm{Tr}[\gamma_{2,5}] = 8, \tag{14.42}$$

and since we have already learned that the number of D5-branes is zero in this case, we have solution $\gamma_5 = 0$ for all orientifold elements in the D5-brane sector, and we can write

$$\gamma_{\Omega,9} = \begin{pmatrix} 0 & 1 & 0 & 0 \\ 1 & 0 & 0 & 0 \\ 0 & 0 & 1 & 0 \\ 0 & 0 & 0 & 1 \end{pmatrix} \otimes I_8,$$

$$\gamma_{1,9} = \mathrm{diag}\{e^{\frac{2\pi i}{3}} \,(8\text{ times}), e^{-\frac{2\pi i}{3}} \,(8\text{ times}), 1\,(16\text{ times})\},$$

from which it is trivially verified that (14.42) is satisfied.

Notice that γ_Ω acts by exchanging the roots and their complex conjugates: $e^{\frac{2\pi i}{3}} \leftrightarrow e^{-\frac{2\pi i}{3}}$. This will be the case in all of the later models, and so we will no longer list it explicitly in the later solutions. Note also that in the other type A orientifolds, $\gamma_{\Omega,9} = \gamma_{\Omega R,5}$, and $\gamma_{\Omega R,9} = \gamma_{\Omega,5}$. We can always choose a phase such that we can always write $\gamma_{1,9} = e^{\frac{2\pi i m}{N}} \gamma_{1,5}$, for m any odd integer. This simple relationship between the γ matrices in the D5- and D9-brane sectors is a manifestation of T_{6789} duality.

For the $\mathbb{Z}_4^A$ case we have:

$$\mathrm{Tr}[\gamma_{1,9}] - 2\mathrm{Tr}[\gamma_{1,5}] = 0$$
$$\mathrm{Tr}[\gamma_{2,9}] - 4\mathrm{Tr}[\gamma_{2,5}] = 0$$
$$\mathrm{Tr}[\gamma_{3,9}] - 2\mathrm{Tr}[\gamma_{3,5}] = 0. \tag{14.43}$$

Note that the middle case correctly reproduces the $\mathbb{Z}_2^A$ example, and therefore the $\mathbb{Z}_2^A$ example appears as a substructure. This will be true for $\mathbb{Z}_6^A$ also.

The solution is ($\alpha_4^2 \equiv R$):

$$\gamma_{1,9} = \mathrm{diag}\{e^{\frac{\pi i}{4}}\,(8\text{ times}), e^{-\frac{\pi i}{4}}\,(8\text{ times}), e^{\frac{3\pi i}{4}}\,(8\text{ times}), e^{-\frac{3\pi i}{4}}\,(8\text{ times})\}.$$

For the $\mathbb{Z}_6^A$ case we have:

$$\mathrm{Tr}[\gamma_{1,9}] - \mathrm{Tr}[\gamma_{1,5}] = 0$$
$$\mathrm{Tr}[\gamma_{2,9}] - 3\mathrm{Tr}[\gamma_{2,5}] = 16$$
$$\mathrm{Tr}[\gamma_{3,9}] - 4\mathrm{Tr}[\gamma_{3,5}] = 0$$
$$\mathrm{Tr}[\gamma_{4,9}] - 3\mathrm{Tr}[\gamma_{4,5}] = -16$$
$$\mathrm{Tr}[\gamma_{5,9}] - \mathrm{Tr}[\gamma_{5,5}] = 0, \tag{14.44}$$

for which we have ($\alpha_4^2 \equiv R$):

$$\text{Tr}[\gamma_{1,9}] = \text{Tr}[\gamma_{1,5}] = 0, \;\; \text{Tr}[\gamma_{3,9}] = \text{Tr}[\gamma_{3,5}] = 0, \;\; \text{Tr}[\gamma_{5,9}] = \text{Tr}[\gamma_{5,5}] = 0,$$

$$\text{Tr}[\gamma_{2,9}] = \text{Tr}[\gamma_{2,5}] = -8, \;\; \text{Tr}[\gamma_{4,9}] = \text{Tr}[\gamma_{4,5}] = 8,$$

$$\gamma_{1,9} = \text{diag}\{e^{\frac{\pi i}{6}} \,(4 \text{ times}), e^{-\frac{\pi i}{6}} \,(4 \text{ times}),$$

$$e^{\frac{5\pi i}{6}} \,(4 \text{ times}), e^{-\frac{5\pi i}{6}} \,(4 \text{ times}), -i \,(8 \text{ times}), i \,(8 \text{ times})\}.$$

Note here that:

$$\gamma_{1,9} \equiv \text{diag}\{e^{\frac{2\pi i}{3}} \,(4 \text{ times}), e^{-\frac{2\pi i}{3}} \,(4 \text{ times}), 1 \,(8 \text{ times})\} \otimes \text{diag}\{-i, i\},$$

which shows a $\mathbb{Z}_3 \times \mathbb{Z}_2$ structure, using the solutions previously obtained for the $\mathbb{Z}_2^A$ and $\mathbb{Z}_3^A$ models.

Also notice that in all cases above, the coefficient of the $\gamma_{k,5}$ trace is the square root of the number of fixed points invariant under α_N^k. Also interesting is that (generalising the $\mathbb{Z}_2$ case,) the same choice made for D9-branes can be made for D5-branes, up to a phase.

The tadpoles for the case $\mathbb{Z}_6^B$ will turn out to be isomorphic to those listed above for $\mathbb{Z}_3^A$, while there are no tadpoles to list for the $\mathbb{Z}_4^B$ model, as there are no D-branes required.

Let us now turn to the closed string spectra.

14.4 Closed string spectra

Just as we did in section 7.6, we should construct the closed string spectrum for each model, which we will combine with the open string spectrum later. The procedure is much the same as we did for the K3 orbifold, except that we will apply the orientifold projection, which will throw out more states.

The right-moving untwisted sector has the massless states given below.

Sector	State	α_N^k	$SO(4)$ charge
NS	$\psi^\mu_{-1/2}\|0\rangle$	1	**(2,2)**
	$\psi^{1\pm}_{-1/2}\|0\rangle$	$e^{\pm\frac{2\pi ik}{N}}$	$2(\mathbf{1,1})$
	$\psi^{2\pm}_{-1/2}\|0\rangle$	$e^{\mp\frac{2\pi ik}{N}}$	$2(\mathbf{1,1})$
R	$\|s_1 s_2 s_3 s_4\rangle$		
	$s_1 = +s_2, \, s_3 = +s_4$	1	$2(\mathbf{2,1})$
	$s_1 = -s_2, \, s_3 = -s_4$	$e^{\pm\frac{2\pi ik}{N}}$	$2(\mathbf{1,2})$

Meanwhile, the right-moving sector twisted by $\frac{m}{N} \neq \frac{1}{2}$ has the following states.

Sector	State	α_N^k	$SO(4)$ charge
NS	$\psi_{-1/2+m/N}^{1+}\|0\rangle$	$e^{\frac{2\pi i k}{N}(1-2\frac{m}{N})}$	$(\mathbf{1},\mathbf{1})$
	$\psi_{-1/2+m/N}^{2+}\|0\rangle$	$e^{\frac{2\pi i k}{N}(1-2\frac{m}{N})}$	$(\mathbf{1},\mathbf{1})$
R	$\|s_1 s_2\rangle,\ s_1 = -s_2$	$e^{\frac{2\pi i k}{N}(1-2\frac{m}{N})}$	$(\mathbf{1},\mathbf{2})$

The exception to the situation depicted there is when we have an $\frac{m}{N}=\frac{1}{2}$ twist.

Sector	State	α_N^k	$SO(4)$ charge
NS	$\|s_3 s_4\rangle,\ s_3 = +s_4$	1	$2(\mathbf{1},\mathbf{1})$
R	$\|s_1 s_2\rangle,\ s_1 = -s_2$	1	$(\mathbf{1},\mathbf{2})$

We have imposed the GSO projection, and decomposed the little group of the spacetime Lorentz group as $SO(4) = SU(2) \times SU(2)$, just as in section 7.6. We form the spectrum for orientifold group of type A by taking products of states from the left and right sectors (to give states invariant under α_N^k), symmetrised by the Ω projection in the NS–NS sector, while antisymmetrising in the R–R sector, since we have fermions there.

In this way, we have from the untwisted closed string sector of the type A $\mathbb{Z}_N$ orientifold ($N \neq 2$).

Sector	$SO(4)$ charge
NS–NS	$(\mathbf{3},\mathbf{3})+5(\mathbf{1},\mathbf{1})$
R–R	$(\mathbf{3},\mathbf{1})+(\mathbf{1},\mathbf{3})+4(\mathbf{1},\mathbf{1})$

This is the content of the $\mathcal{N}=1$ supergravity multiplet in six dimensions, accompanied by one tensor multiplet and two hypermultiplets (see table 14.3, p. 341, for a list of the types of multiplet). In the case of $\mathbb{Z}_2$ it is as follows[132].

Sector	$SO(4)$ charge
NS–NS	$(\mathbf{3},\mathbf{3})+11(\mathbf{1},\mathbf{1})$
R–R	$(\mathbf{3},\mathbf{1})+(\mathbf{1},\mathbf{3})+6(\mathbf{1},\mathbf{1})$

This is the $D=6$, $\mathcal{N}=1$ supergravity multiplet in six dimensions, accompanied by one tensor multiplet and four hypermultiplets

The twisted sectors will produce additional multiplets. The bosonic content of a hypermultiplet is four scalars $4(\mathbf{1},\mathbf{1})$, while that of a tensor multiplet is $(\mathbf{3},\mathbf{1})+(\mathbf{1},\mathbf{1})$. By combining on the left and right the sectors twisted by $\frac{m}{N}$ and $(1-\frac{m}{N})$, we find that the NS–NS sector produces one hypermultiplet while the R–R sector produces a tensor multiplet. A sector twisted by $\frac{1}{2}$ simply produces one hypermultiplet: one quarter coming from the R–R sector and three quarters from the NS–NS sector, just as in section 7.6.

To evaluate the number of hypermultiplets (and, as we shall see, tensor multiplets) coming from the twisted sectors of a K3 orbifold, we need to recall the structure of the fixed points and their transformation properties, done in section 7.6.5. Using the data above, we see that in the case of $\mathbb{Z}_2^A$, we simply multiply by the number of $\mathbb{Z}_2$ fixed points, and we find that there are 16 hypermultiplets from the twisted sectors, giving a total of 20 hypermultiplets when combined with the four from the untwisted sector. For $\mathbb{Z}_3^A$ there are nine fixed points, each supplying a hypermultiplet and a tensor multiplet (for twists by $(\frac{1}{3}, \frac{2}{3})$), giving a total of 11 hypermultiplets and 10 tensor multiplets when added to those arising in the untwisted sector.

For $\mathbb{Z}_4^A$ the four $\mathbb{Z}_4$ invariant fixed points give four hypermultiplets and four tensor multiplets. They are also $\mathbb{Z}_2$ fixed points and so supply an additional four hypermultiplets. The other 12 $\mathbb{Z}_2$ fixed points form six $\mathbb{Z}_4$ invariant pairs, from which arise six hypermultiplets. This gives a total of 16 hypermultiplets and five tensor multiplets for the complete model.

Finally, for the model $\mathbb{Z}_6^A$, the $\mathbb{Z}_6$ fixed point gives two hypermultiplets and two tensor multiplets from $(\frac{1}{6}, \frac{5}{6})$ and $(\frac{1}{3}, \frac{2}{3})$ twists. It also gives six hypermultiplets from the $\frac{1}{2}$ twisted sector. The four pairs of $\mathbb{Z}_3$ points give four tensor multiplets and four hypermultiplets for $(\frac{1}{3}, \frac{2}{3})$ twists, while the five $\mathbb{Z}_2$ triplets of fixed points supply five hypermultiplets. This gives 14 hypermultiplets and seven tensor multiplets in all.

For $\mathbb{Z}_N$ orientifolds of type B the situation is as follows. For closed string states, prior to making the theory unorientable, the relevant orbifold states to consider are those for the group formed by the remaining pure $\mathbb{Z}_N$ elements in the orientifold group, which is therefore $\mathbb{Z}_{\frac{N}{2}}$. The possible left and right states are evaluated as before, and then they are projected to the unoriented theory invariant under $\Omega \cdot \alpha_N^1$.

It is thus easy to see that the closed string spectra for $\mathbb{Z}_6^B$ and $\mathbb{Z}_3^A$ are isomorphic, as are those of $\mathbb{Z}_2^B$ and $\mathbb{Z}_1^A$, (the latter being simply ten dimensional type I string theory: there is no orbifold to perform for $\mathbb{Z}_2^B$).

There remains only the spectrum of $\mathbb{Z}_4^B$ to compute, which is self T-dual. The pure orbifold states to consider are those of $\mathbb{Z}_2$. Tensoring left and right to form the Ω invariant spectrum, we obtain 12 hypermultiplets and nine tensor multiplets in total.

In summary, we have (in addition to the usual gravity and tensor multiplet) a spectrum of hypermultiplets and tensor multiplets from the closed string sector for each model, given in table 14.1.

Table 14.1. *The spectrum of hypermultiplets and tensor multiplets in the various orientifold models*

Model	Neutral hypermultiplets	Extra tensor multiplets
$\mathbb{Z}_2^A$	20	0
$\mathbb{Z}_3^A$	11	9
$\mathbb{Z}_4^A$	16	4
$\mathbb{Z}_6^A$	14	6
$\mathbb{Z}_4^B$	12	8
$\mathbb{Z}_6^B$	11	9

Notice that in the $\mathbb{Z}_2^A$ case we have a total of 20 hypermultiplets. Looking at table 14.3 we see that this is a total of 80 scalar fields: the 80 moduli of the K3 surface[228]. However, in the other other orientifold examples, some of the potential scalars have instead combine into tensor multiplets, leaving us with fewer hypermultiplets in the final model. This is a reduction in the dimension of the moduli space of K3 deformations available to these models: some of the K3 moduli are frozen.

To complete the story, let us turn to the open string sector, computing what the allowed gauge groups are.

14.5 Open string spectra

Let us study first the 99 open string sector. The massless bosonic spectrum arises as follows.

For the 55 states at a fixed point we have the following.

state	$\alpha_N^k = +$	$\Omega = +$	$SO(4)$ charge
$\psi_{-1/2}^{\mu}\lvert 0, ij\rangle\lambda_{ij}$	$\lambda = \gamma_{k,9}\lambda\gamma_{k,9}^{-1}$	$\lambda = -\gamma_{\Omega,9}\lambda^T\gamma_{\Omega,9}^{-1}$	$(\mathbf{2,2})$
$\psi_{-1/2}^{1\pm}\lvert 0, ij\rangle\lambda_{ij}$	$\lambda = e^{\pm\frac{2\pi k}{N}}\gamma_{k,9}\lambda\gamma_{k,9}^{-1}$	$\lambda = -\gamma_{\Omega,9}\lambda^T\gamma_{\Omega,9}^{-1}$	$2(\mathbf{1,1})$
$\psi_{-1/2}^{2\pm}\lvert 0, ij\rangle\lambda_{ij}$	$\lambda = e^{\mp\frac{2\pi k}{N}}\gamma_{k,9}\lambda\gamma_{k,9}^{-1}$	$\lambda = -\gamma_{\Omega,9}\lambda^T\gamma_{\Omega,9}^{-1}$	$2(\mathbf{1,1})$

state	$\alpha_N^k = +$	$\Omega = +$	$SO(4)$ charge
$\psi_{-1/2}^{\mu}\lvert 0, ij\rangle\lambda_{ij}$	$\lambda = \gamma_{k,5}\lambda\gamma_{k,5}^{-1}$	$\lambda = -\gamma_{\Omega,5}\lambda^T\gamma_{\Omega,5}^{-1}$	$(\mathbf{2,2})$
$\psi_{-1/2}^{1\pm}\lvert 0, ij\rangle\lambda_{ij}$	$\lambda = e^{\pm\frac{2\pi k}{N}}\gamma_{k,5}\lambda\gamma_{k,5}^{-1}$	$\lambda = \gamma_{\Omega,5}\lambda^T\gamma_{\Omega,5}^{-1}$	$2(\mathbf{1,1})$
$\psi_{-1/2}^{2\pm}\lvert 0, ij\rangle\lambda_{ij}$	$\lambda = e^{\mp\frac{2\pi k}{N}}\gamma_{k,5}\lambda\gamma_{k,5}^{-1}$	$\lambda = \gamma_{\Omega,5}\lambda^T\gamma_{\Omega,5}^{-1}$	$2(\mathbf{1,1})$

For the 55 states away from a fixed point we have the following.

state	$\Omega = +$	$SO(4)$ charge
$\psi^\mu_{-1/2}\|0,ij\rangle\lambda_{ij}$	$\lambda = -\gamma_{\Omega,5}\lambda^T\gamma_{\Omega,5}^{-1}$	**(2,2)**
$\psi^{1\pm}_{-1/2}\|0,ij\rangle\lambda_{ij}$	$\lambda = \gamma_{\Omega,5}\lambda^T\gamma_{\Omega,5}^{-1}$	2(1,1)
$\psi^{2\pm}_{-1/2}\|0,ij\rangle\lambda_{ij}$	$\lambda = \gamma_{\Omega,5}\lambda^T\gamma_{\Omega,5}^{-1}$	2(1,1)

For the 59 states we have the following at a fixed point.

state	$\alpha^k_N = +$	$SO(4)$ charge
$\|s_3 s_4, ij\rangle\lambda_{ij}, s_3 = s_4$	$\lambda = \gamma_{k,5}\lambda\gamma_{k,9}^{-1}$	**2(1,1)**

Away from a fixed point we have the following.

state	$SO(4)$ charge
$\|s_3 s_4, ij\rangle\lambda_{ij},\ s_3 = s_4$	2(1,1)

Using the solution presented in section 14.4 for the γ matrices, the solutions for the open string spectra of the models are given in table 14.2.

Table 14.2.   *The gauge groups for the various orientifold models*

Model	Sector	Gauge group	Charged hypermultiplets
$\mathbb{Z}_2^A$	99	$U(16)$	$2 \times \mathbf{120}$
	55	$U(16)$	$2 \times \mathbf{120}$
	59		$(\mathbf{16,16})$
$\mathbb{Z}_3^A$	99	$U(8) \times SO(16)$	$(\mathbf{28,1}),\ (\mathbf{8,16})$
$\mathbb{Z}_4^A$	99	$U(8) \times U(8)$	$(\mathbf{28,1}),\ (\mathbf{1,28}),\ (\mathbf{8,8})$
	55	$U(8) \times U(8)$	$(\mathbf{28,1}),\ (\mathbf{1,28}),\ (\mathbf{8,8})$
	59		$(\mathbf{8,1;8,1}),\ (\mathbf{1,8;1,8})$
$\mathbb{Z}_6^A$	99	$U(4) \times U(4) \times U(8)$	$(\mathbf{6,1,1}),\ (\mathbf{1,6,1})$ $(\mathbf{4,1,8}),\ (\mathbf{1,4,8})$
	55	$U(4) \times U(4) \times U(8)$	$(\mathbf{6,1,1}),\ (\mathbf{1,6,1})$ $(\mathbf{4,1,8}),\ (\mathbf{1,4,8})$
	59		$(\mathbf{4,1,1;4,1,1})$ $(\mathbf{1,4,1;1,4,1})$ $(\mathbf{1,1,8;1,1,8})$
$\mathbb{Z}_4^B$		—	—
$\mathbb{Z}_6^B$	55	$U(8) \times SO(16)$	$(\mathbf{28,1}),\ (\mathbf{8,16})$

14.6 Anomalies for $\mathcal{N} = 1$ in six dimensions

As chiral models in six dimensions, these orientifold vacua that we have constructed have a chance of being afflicted by anomalies. There are anomaly polynomials which we can write for each sector, just as was done in ten dimensions in chapter 7, and in six dimensions in section 7.6.2.

Of course, we can be confident that if we have done everything properly at the string level, checking the anomaly is nothing more than a formality, but it is instructive anyway, as we have seen before. Before we proceed, we must pause to note the structure of the multiplets. In fact, they arise naturally from splitting the $\mathcal{N} = 2$ multiplets which we have seen in table 7.1. Both the vector and tensor multiplets become smaller by yielding up a hypermultiplet, whose bosonic part is four scalars, as we have already seen. Table 14.3 lists the multiplets.

So again we see that the orientifolding has thrown away many pieces of the pure K3 spectrum, and so the marvellous cancellation in equation (7.52) will not happen. As in the prototype orientifold, we have additional pieces as well, coming from the open string sectors, which may give some new interesting structures.

The first thing to check is that the irreducible parts of the anomaly cancel. For the gravitational part, we must look at the coefficient of $\mathrm{tr}R^4$. In fact, for $\mathcal{N} = 1$ models in $D = 6$ it is easy to see that the vanishing of this coefficient is equivalent to:

$$n_{\mathrm{H}} - n_{\mathrm{V}} = 244 - 29n_{\mathrm{T}}. \tag{14.45}$$

The reader should verify this. Here, $n_{\mathrm{H}}, n_{\mathrm{V}}$ and $n_{\mathrm{T}} + 1$ are respectively the numbers of hypermultiplets, vector multiplets and tensor multiplets in the six dimensional supergravity model. This follows from direct use of the anomaly polynomials in insert 7.2, remembering to divide the polynomials for the complex fermions listed there by two to match the real fermions we

Table 14.3. *The structure of the $\mathcal{N} = 1$ multiplets in $D = 6$*

Multiplet	Bosons	Fermions
vector	**(2,2)**	2**(1,2)**
hyper	4**(1,1)**	2**(2,1)**
SD tensor	**(1,3)**+**(1,1)**	2**(2,1)**
ASD tensor	**(3,1)**+**(1,1)**	2**(2,1)**
supergravity	$(\mathbf{3},\mathbf{3}) + (\mathbf{3},\mathbf{1}) + (\mathbf{1},\mathbf{3}) + (\mathbf{1},\mathbf{1})$	$2(\mathbf{3},\mathbf{2}) + 2(\mathbf{2},\mathbf{1})$ or $2(\mathbf{2},\mathbf{3}) + 2(\mathbf{1},\mathbf{2})$

are counting with here. It is natural that the vectors and hypers contribute equal and oppositely, since they are components of the non-chiral $\mathcal{N} = 2$ vectors multiplet, as is clear from tables 7.1 and 14.3.

In fact, in the spirit of the miraculous cancellation for pure type IIB string theory in ten dimensions, and for the spectrum of type IIB on K3 to six dimensions, shown in sections 7.1.2 and 7.6.2, there is another such purely closed string example, that of the $\mathbb{Z}_4^B$ model. Indeed, equation (14.45) is satisfied, but the coefficients of the $(\text{tr} R^2)^2$ terms vanish also, giving:

$$12\hat{I}_8^{(2,1)} + 8\hat{I}_8^{(1,3)} + 8\hat{I}_8^{(2,1)} + \hat{I}_8^{(3,2)} + \hat{I}_8^{(3,1)} + \hat{I}_8^{(1,3)} + \hat{I}_8^{(2,1)} = 0. \quad (14.46)$$

This is, again, another remarkable purely closed string solution of the anomaly equations, supplying an $\mathcal{N} = 1$ supersymmetric solution of orientifolded type IIB strings on K3.

Let us turn to the models which need the addition of D-branes, and hence have gauge contributions to the anomaly. For the irreducible $\text{Tr} F^4$ terms, everything again cancels. Again, it is recommended that the reader who is interested should verify this. This is done with the aid of the following information, which should be set alongside that given in equations (7.39) and (7.40). For $SU(n)$ we have:

$$\begin{aligned}
\text{Tr}_{\text{adj}}(t^2) &= 2n\text{Tr}_{\text{f}}(t^2), \\
\text{Tr}_{\text{adj}}(t^4) &= 2n\text{Tr}_{\text{f}}(t^4) + 6\text{Tr}_{\text{f}}(t^2)\text{Tr}_{\text{f}}(t^2), \quad (14.47)
\end{aligned}$$

and for completeness, we also list the result for $Sp(n) \equiv USp(2n)$:

$$\begin{aligned}
\text{Tr}_{\text{adj}}(t^2) &= (n+2)\text{Tr}_{\text{f}}(t^2), \\
\text{Tr}_{\text{adj}}(t^4) &= (n+8)\text{Tr}_{\text{f}}(t^4) + 3\text{Tr}_{\text{f}}(t^2)\text{Tr}_{\text{f}}(t^2). \quad (14.48)
\end{aligned}$$

Crucially, note that for symmetric tensor representations of any of the groups mentioned, we can use the $Sp(n)$ relations just mentioned in equation (14.48), while for antisymmetric tensor representations we can use the $SO(n)$ relations given in equation (7.39).

Now let us see how the irreducible $\text{Tr} F^4$ terms cancel for one example, that of $\mathbb{Z}_3^A$. Table 14.2 shows that the gauge group is $U(8) \times SO(16)$ with hypermultiplets charged as $(\mathbf{28}, \mathbf{1})$ and $(\mathbf{8}, \mathbf{16})$. We must do each separate gauge group independently. Let us first do $U(8)$, and write everything in terms of the fundamental representation. Doing so, we see that we get $16\text{Tr}_{\text{f}} F^4$, ignoring (and for the rest of the paragraph) the purely numerical denominator in equation (7.2), of course. The first hypermultiplet, $(\mathbf{28}, \mathbf{1})$, which in fact in the antisymmetric of $U(8)$, and so using the second line in equation (7.39), we see that its coefficient of $\text{Tr}_{\text{f}} F^4$ is in fact zero. This

---

Insert 14.3.   **Another string–string duality in $D = 6$**

Very interesting is the case of the $\mathbb{Z}_2^A$ model. The $U(1)$ factors can be shown to be absent non-perturbatively[196], leaving gauge group $SU(16) \times SU(16)$. Meanwhile the remaining anomaly factorises as:

$$I_8 \sim (\mathrm{tr}\,R^2 - 2\mathrm{Tr}_\mathrm{f}F^2)(\mathrm{tr}\,R^2 - 2\mathrm{Tr}_\mathrm{f}F^2),$$

for both the D9-brane and D5-brane sectors. This similar factorisation between the two sectors is of course a reflection of the fact that there is a T_{6789}-duality exchanging the two types of brane, but there is more. There is a signal of another duality, now between this formulation and that of a strong/weak coupling dual model also with strings in six dimensions. Looking at the anomaly two-forms in the factors (say, on the right hand side), these dual strings have a similar structure to those we associate with the two-forms on the left hand side. This is of course very special to six dimensions, where we have a chance of such a string–string duality, and the reader can probably guess what the dual string theory might be. It is in fact a K3 compactification of the *heterotic* string, of a very special sort[229, 196]. One way to make it is as follows: Recall that from equation (7.48), the field equation of $\tilde{H}^{(3)}$ is:

$$d\tilde{H}^{(3)} = -\frac{\alpha'}{4}\left[\mathrm{Tr}\,F^2 - \mathrm{tr}\,R^2\right],$$

for which, if we were to integrate this over K3, would get a contribution of 24 (up to an overall factor) which can be cancelled by precisely 24 instantons. So an interesting compactification is achieved by taking the $E_8 \times E_8$ heterotic string on K3, with 12 instantons in one E_8 and 12 in the other. The details are interesting to uncover, but we shall have to leave it to the reader to study the literature, since it will take us too far afield[229]. Note also that we can see the dual strings. They are perturbatively manifest on the orientifold side as solitons. One is a D1-brane which one can place in the six non-compact directions and the other is its T_{6789} dual, a D5-brane with four of its dimensions wrapped on the compact directions. On the heterotic side, these map over to a pair of strong/weak dual heterotic strings. One is the heterotic string itself, and the other is a K3-wrapped NS5-brane.

leaves us with the hypermultiplet $(\mathbf{8}, \mathbf{16})$, which can be treated as sixteen copies of the fundamental of the $\mathbf{8}$, and therefore contributes $-16\mathrm{Tr}_{\mathrm{f}}F^4$, giving a cancellation.

For the mixed anomalies, the anomalies all factorise in a way which allows for their cancellation by a generalisation[193] of the Green–Schwarz mechanism. In the models with multiple tensors, some subset of them can be given a classically anomalous gauge transformation to produce the required cancellations. In some cases, the factorisation gives a sign of interesting physics, since there is a stringy interpretation of both four-form factors, suggesting new dualities (see insert 14.3).

15

D-branes and geometry II

In a number of the previous chapters, we probed various systems while remaining largely in the limit where D-branes are pointlike in their transverse directions. However, we learned in chapter 10 that D-branes have an intrinsic geometry of their own, which can be seen when we place a lot of them together to produce a large back-reaction on the spacetime geometry, or if we were to turn up the string coupling (for fixed string tension) such that Newton's constant is strong. Both sorts of situation can and will be forced upon us later, so it is worthwhile trying to understand what we can learn by probing the supergravity geometry with different types of branes (we have already probed extremal p-branes with Dp-branes in section 10.3). If we choose things such that there is some supersymmetry preserved, we can use it to help us learn many useful things.

15.1 Probing p with D$(p-4)$

Let us probe the geometry of the extremal p-branes with a D$(p-4)$-brane. From our analysis of chapter 11, we know that this system is supersymmetric. Therefore, we expect that there should still be a trivial potential for the result of the probe computation, but there is not enough super-symmetry to force the metric to be flat. There are actually two sectors within which the probe brane can move transversely. Let us choose static gauge again, with the probe aligned so that its $p - 4$ spatial directions $\xi^1 - \xi^{p-4}$ are aligned with the directions $x^1 - x^{p-4}$. Then there are four transverse directions *within* the p-brane background, labelled $x^{p-3} - x^p$, and which we can call $x_{\parallel}^i$ for short. There are $9 - p$ remaining transverse directions which are transverse to the p-brane as well, labelled $x^{p+1} - x^9$, which we'll abbreviate to $x_{\perp}^m$. The 6–2 case is tabulated as a visual guide below.

	x^0	x^1	x^2	x^3	x^4	x^5	x^6	x^7	x^8	x^9
D2-brane	—	—	—	•	•	•	•	•	•	•
6-brane	—	—	—	—	—	—	—	•	•	•

The extremal p-brane supergravity solution is given in equation (10.38). As in section 10.3, we can probe this solution with D-branes, using the world-volume actions described in chapters 5 and 9. Following the same lines of reasoning as used in section 10.3, the determinant which shall go into our Dirac–Born–Infeld Lagrangian is:

$$\det[-G_{ab}] = Z_p^{-\frac{(p-3)}{2}} \left(1 - v_\parallel^2 - Z_p v_\perp^2\right), \tag{15.1}$$

where the velocities come from the time (ξ^0) derivatives of $x_\parallel$ and $x_\perp$. This is nice, since in forming the action by multiplying by the exponentiated dilaton factor and expanding in small velocities, we get the Lagrangian

$$\mathcal{L} = \frac{1}{2} m_{p-4} \left(v_\parallel^2 + Z_p v_\perp^2 - 2\right), \tag{15.2}$$

which again has a constant potential which we can discard, leaving pure kinetic terms. We see that there is a purely flat metric on the moduli space for the motion inside the four dimensions of the p-brane geometry, while there is a metric

$$ds^2 = Z_p(r)\delta_{mn}dx^m dx^n, \tag{15.3}$$

for the transverse motion. This is the Coulomb branch, in gauge theory terms, and the flat metric was on the Higgs branch. (In fact, the Higgs result does not display all of the richness of this system that we have seen. In addition to the flat metric geometry inside the brane that we see here, there is additional geometry describing the Dp–$D(p-4)$ fields corresponding to the full instanton geometry. This comes from the fact that the $D(p-4)$-brane behaves as an instanton of the non-Abelian gauge theory on the world-volume of the coincident Dp-branes. See section 13.4.)

Notice that for the fields we have studied, we obtained a trivial potential for free without having to appeal to a cancellation due to the coupling of the charge μ_{p-4} of the probe. This is good, since there is no electric source of this in the background for it to couple to. Instead, the form of the solution for the background makes it force-free automatically.

15.2 Probing six-branes: Kaluza–Klein monopoles and M-theory

Actually, when $p \geq 5$, something interesting happens. The electric source of $C_{(p+1)}$ potential in the background produces a *magnetic* source of

$C_{(7-p)}$. The rank of this is low enough for there to be a chance for the D$(p-4)$-probe brane to couple to it even in the Abelian theory. For example, for $p = 5$ there is a magnetic source of C_2 to which the D1-brane probe can couple. Meanwhile for $p = 6$, there is a magnetic source of C_1. The D2-brane probes see this in an interesting way. Let us linger here to study this case a bit more closely. Since there is always a trivial $U(1)$ gauge field on the world volume of a D2-brane probe, corresponding to the centre of mass of the brane, we should include the coupling of the world-volume gauge potential A_a (with field strength F_{ab}) to any of the fields coming from the background geometry.

In fact, as we saw before in section 9.2, there is a coupling

$$2\pi\alpha'\mu_2 \int_M C_1 \wedge F, \tag{15.4}$$

where $C_1 = C_\phi d\phi$ is the magnetic potential produced by the six-brane background geometry, which is easily computed to be: $C_\phi = -(r_6/g_s) \cos\theta$, where $r_6 = g_s N \alpha'^{1/2}/2$.

The gauge field on the world volume is equivalent to one scalar, since we may exchange A_a for a scalar s by Hodge duality in the $(2+1)$-dimensional world-volume. (This is of course a feature specific to the $p=2$ case.) To get the coupling for this extra scalar correct, we should augment the probe computation. As we have seen, the Dirac–Born–Infeld action is modified by an extra term in the determinant:

$$-\det g_{ab} \rightarrow -\det(g_{ab} + 2\pi\alpha' F_{ab}). \tag{15.5}$$

We can[143, 171] introduce an auxiliary vector field v_a, replacing $2\pi\alpha' F_{ab}$ by the combination $e^{2\phi}\mu_2^{-2} v_a v_b$ in the Dirac action, and adding the term

$$2\pi\alpha' \int_M F \wedge v$$

overall. Treating v_a as a Lagrange multiplier, the path integral over v_a will give the action involving F as before. Alternatively, we may treat F_{ab} as a Lagrange multiplier, and integrating it out enforces

$$\epsilon^{abc} \partial_b(-\mu_2 \hat{C}_c + v_c) = 0. \tag{15.6}$$

Here, $\hat{C}_c$ are the components of the pull-back of C_1 to the probe's world-volume. The solution to the constraint above is

$$-\mu_2 \hat{C}_a + v_a = \partial_a s, \tag{15.7}$$

where s is our dual scalar. We may now replace v_a by $\partial_a s + \mu_2 \hat{C}_a$ in the action.

The static gauge computation picks out only $\dot{s} + \mu_2 C_\phi \dot{\phi}$, and recomputing the determinant gives

$$\det = Z_6^{-\frac{3}{2}} \left(1 - v_\parallel^2 - Z_6 v_\perp^2 - \frac{Z_6^{\frac{1}{2}} e^{2\Phi}}{\mu_2^2} \left[\dot{s} + \mu_2 C_\phi \dot{\phi} \right]^2 \right). \qquad (15.8)$$

Again, in the full Dirac–Born–Infeld action, the dilaton factor cancels the prefactor exactly, and including the factor of $-\mu_2$ and the trivial integral over the worldvolume directions to give a factor V_2, the resulting Lagrangian is

$$\mathcal{L} = \frac{1}{2} m_2 (v_\parallel^2 - 2) + \frac{1}{2} V_2 \left(\frac{\mu_2 Z_6}{g} v_\perp^2 + \frac{g_s}{\mu_2 Z_6} \left(\dot{s} + \mu_2 C_\phi \dot{\phi} \right)^2 \right), \qquad (15.9)$$

which is (after ignoring the constant potential) again a purely kinetic Lagrangian for motion in *eight* directions. There is a non-trivial metric in the part transverse to both branes:

$$ds^2 = V(r) \left(dr^2 + r^2 d\Omega^2 \right) + V(r)^{-1} \left(ds + A_\phi d\phi \right)^2,$$

$$\text{with} \quad V(r) = \frac{\mu_2 Z_6}{g_s} \quad \text{and} \quad A = \frac{\mu_2 r_6}{g_s} \cos\theta\, d\phi, \qquad (15.10)$$

where $d\Omega^2 = d\theta^2 + \sin^2\theta\, d\phi^2$. There is a number of fascinating interpretations of this result. In pure geometry, the most striking feature is that there are now *eleven* dimensions for our spacetime geometry. The D2-brane probe computation has uncovered, in a very natural way, an extra transverse dimension. This extra dimension is compact, since s is periodic, which is inherited from the gauge invariance of the dual worldvolume gauge field. The radius of the extra dimension is proportional to the string coupling, which is also interesting. This eleventh dimension is of course the M-direction we saw in section 12.4. The D2-brane has revealed that the six-brane is a Kaluza–Klein monopole[168] of eleven dimensional supergravity on a circle[152], which is constructed out of a Taub–NUT geometry* in equation (15.10). This fits very well with the fact that the D6 is the Hodge dual of the D0-brane, which we already saw is a Kaluza–Klein electric particle.

15.3 The moduli space of 3D supersymmetric gauge theory

As before, the result also has a field theory interpretation. The $(2+1)$-dimensional $U(1)$ gauge theory (with eight supercharges) on the

---

* It is a very useful exercise for the reader to take the Taub–NUT metric, times seven flat directions, and use the reduction formula given in insert 12.1 (p. 274) to reproduce the six-brane metric of equation (10.38) directly.

world-volume of the D2-brane has $N_{\rm f} = N$ extra hypermultiplets coming from light strings connecting it to the $N_{\rm f} = N$ D6-branes. The $SU(N_{\rm f})$ symmetry on the worldvolume of the D6-branes is a global 'flavour' symmetry of the $U(1)$ gauge theory on the D2-brane. A hypermultiplet Ψ has four components Ψ_i corresponding to the four scalar degrees of freedom given by the four positions $\Psi^i \equiv (2\pi\alpha')^{-1}x^i_\parallel$. The vector multiplet contains the vector A_a and three scalars $\Phi^m \equiv (2\pi\alpha')^{-1}x^m_\perp$. The Yang–Mills coupling is $g^2_{\rm YM} = g_{\rm s}\alpha'^{-1/2}$.

The branch of vacua of the theory with $\Psi \neq 0$ is called the 'Higgs' branch of vacua while that with $\Phi \neq 0$ constitutes the 'Coulomb' branch, since there is generically a $U(1)$ left unbroken. There is a non-trivial four dimensional metric on the Coulomb branch. This is made of the three Φ^m, and the dual scalar of the $U(1)$s photon. Let us focus on the quantities which survive in the low energy limit or 'decoupling limit' $\alpha' \to 0$, holding fixed any sensible gauge theory quantities which appear in our expressions. The surviving parts of the metric (15.10) are:

$$ds^2 = V(U)(dU^2 + U^2 d\Omega^2_2) + V(U)^{-1}(d\sigma + A_\phi d\phi)^2$$

$$\text{where} \quad V(U) = \frac{1}{4\pi^2 g^2_{\rm YM}}\left(1 + \frac{g^2_{\rm YM}N_{\rm f}}{2U}\right); \qquad A_\phi = \frac{N_{\rm f}}{8\pi^2}\cos\theta, \quad (15.11)$$

where $U = r/\alpha'$ has the dimensions of an energy scale in the gauge theory. Also, $\sigma = \alpha's$, and we will fix its period shortly.

In fact, the naive tree level metric on the moduli space is that on $\mathbb{R}^3 \times S^1$, of form $ds^2 = g^{-2}_{\rm YM}dx^2_\perp + g^2_{\rm YM}d\sigma^2$. Here, we have the tree level and one loop result: $V(U)$ has the interpretation as the sum of the tree level and one-loop correction to the gauge coupling of the 2+1 dimensional gauge theory[237]. Note the factor $N_{\rm f}$ in the one loop correction. This multiplicity comes from the number of hypermultiplets which can run around the loop. Similarly, the cross term from the second part of the metric has the interpretation as a one-loop correction to the naive four dimensional topology, changing it to the (Hopf) fibred structure above.

Actually, the moduli space's dimension had to be a multiple of four, as it generally has to be hyper-Kähler for $D=2+1$ supersymmetry with eight supercharges[185]. Our metric is indeed hyper-Kähler since it is the Taub–NUT metric: the hyper-Kähler condition on the metric in the form it is written is the by-now familiar equation: $\nabla \times \mathbf{A} = \nabla V$, which is satisfied.

In fact, we are not quite done yet. With some more care we can establish some important facts quite neatly. We have not been careful about the period of σ, the dual to the gauge field, which is not surprising given all of the factors of 2, π and α'. To get it right is an important task, which will yield interesting physics. We can work it out in a number of ways, but the following is quite instructive. If we perform the rescaling $U = \rho/4g_{\text{YM}}^2$ and $\psi = 8\pi^2\sigma/N_{\text{f}}$, our metric is:

$$ds^2 = \frac{g_{\text{YM}}^2}{64\pi^2} \, ds_{\text{TN}}^2, \qquad \text{where}$$

$$ds_{\text{TN}}^2 = \left(1 + \frac{2N_{\text{f}}}{\rho}\right)(d\rho^2 + \rho^2 d\Omega_2^2)$$

$$+ 4N_{\text{f}}^2 \left(1 + \frac{2N_{\text{f}}}{\rho}\right)^{-1}(d\psi + \cos\theta d\phi)^2, \qquad (15.12)$$

which is a standard form for the Taub–NUT metric, with mass N_{f}, equal to the 'nut parameter' for this self-dual case[186].

This metric is apparently singular at $\rho = 0$, and in fact, for the correct choice of periodicity for ψ, this pointlike structure, called a 'nut', is removable, just like the case of the bolt singularity encountered for the Eguchi–Hanson space. (See insert 7.6, p. 188.) Just for fun, insert 15.1 carries out the analysis and finds that ψ should have period 4π, and so in fact the full $SU(2)$ isometry of the metric is preserved.

What does this all have to do with gauge theory? Let us consider the case of $N_{\text{f}} = 1$, which means one six-brane. This is 2+1 dimensional $U(1)$ gauge theory with one hypermultiplet, a rather simple theory. We

---

Insert 15.1.   The 'nut' of Taub–NUT

The metric (15.12) will be singular at at the point $\rho = 0$, for arbitrary periodicity of ψ. This will be a pointlike singularity which is called a 'nut'[83, 82], in contrast to the 'bolt' we encountered for the Eguchi–Hanson space in insert 7.6 (p. 188), which was an S^2. In this case, near $\rho = 0$, if we make the space look like the *origin* of $\mathbb{R}^4$, we can make this pointlike structure into nothing but a coordinate singularity. Near $\rho = 0$, we have, for $R = 2\rho^2$ (see also insert 7.4, p. 180):

$$ds_{\text{TN}}^2 = 2N_{\text{f}}(dR^2 + R^2 d\Omega_3^2),$$

which is just the right metric for $\mathbb{R}^4$ if $\Delta\psi = 4\pi$, the standard choice for the Euler coordinate. (This may have seemed somewhat heavy-handed for a result one would perhaps have guessed anyway, but it is worthwhile seeing it, in preparation for more complicated examples.)

see that after restoring the physical scales to our parameters, our original field σ has period $1/2\pi$, and so we see that the dual to the photon is more sensibly defined as $\tilde{\sigma} = 4\pi^2\sigma$, which would have period 2π, which is a more reasonable choice for a scalar dual to a photon. We shall use this choice later. With this choice, the metric on the Coulomb branch of moduli space is completely non-singular, as should be expected for such a simple theory.

Let us now return to arbitrary N_f. This means that we have N_f hypermultiplets, but still a $U(1)$ 2+1 dimensional gauge theory with a global 'flavour' symmetry of $SU(N_f)$ coming from the six-branes. There is no reason for the addition of hypermultiplets to change the periodicity of our dual scalar and so we keep it fixed and accept the consequences when we return to physical coordinates $(U, \tilde{\sigma})$: *the metric on the Coulomb branch is singular at $U = 0$!* This is so because insert 15.1 told us to give $\tilde{\sigma}$ a periodicity of $2\pi N_f$ for freedom from singularities, but we are keeping it as 2π. So our metric in physical units has $\tilde{\sigma}$ with period 2π appearing in the combination $(2d\tilde{\sigma} + N_f \cos\theta d\phi)^2$. This means that the metric is no longer has an $SU(2)$ acting, since the round S^3 has been deformed into a 'squashed' S^3, where the squashing is controlled by N_f. In fact, there is a deficit angle at the origin corresponding to an A_{N_f-1} singularity.

How are we to make sense of this singularity? Well, happily, this all fits rather nicely with the fact that for $N_f > 1$ there is an $SU(N_f)$ gauge theory on the six-branes, and so there is a Higgs branch, corresponding to the D2-brane becoming an $SU(N_f)$ instanton! The singularity of the Coulomb branch is indeed a signal that we are at the origin of the Higgs branch, and it also fits that there is no singularity for $N_f = 1$.

It is worthwhile carrying out this computation for the case of N_f D6-branes in the presence of a negative orientifold six-plane oriented in the same way. In that case we deduce from facts we learned before that the presence of the O6-plane gives global flavour group $SO(2N_f)$ for N_f D6-branes. The D2-brane, however, carries an $SU(2)$ gauge group. This is T-dual to the earlier statement made in section 13.4 about D9-branes in type I string theory carrying $SO(N_f)$ groups while D5s carry $USp(2M)$ groups as we learned in section 8.7: the orientifold forces a pair of D2-branes to travel as one, with a $USp(2) = SU(2)$ group.

So the story now involves 2+1 dimensional $SU(2)$ gauge theory with N_f hypermultiplets. The Coulomb branch for $N_f = 0$ must be completely non-singular, since again there is no Higgs branch to join to. This fits with the fact that there are no D6-branes; just the O6-plane. The result for the metric on moduli space can be deduced from a study of the gauge theory (with the intuition gained from this stringy situation), and has been proven to be the Atiyah–Hitchin manifold[231]. Some of this will be discussed in more detail in subsection 15.6. For the case of $N_f = 1$, the result is also non-singular (there is again no Higgs branch for one D6-brane) and the result is a certain cover of the Atiyah–Hitchin manifold[232, 248]. The case of general N_f gives certain generalisations of the Atiyah–Hitchin manifold[248, 250]. The manifolds have D_{N_f} singularities, consistent with the fact that there is a Higgs branch to connect to. Note also that a stringy interpretation of this result is that the strong coupling limit of these O6-planes is in fact M-theory on the Atiyah–Hitchin manifold, just like it is Taub–NUT for the D6-brane.

N.B. It is amusing to note – and the reader may have already spotted it – that the story above seems to be describing local pieces of K3, which has ADE singularities of just the right type, with the associated $SU(N)$ and $SO(2N)$ enhanced gauge symmetries appearing also (global flavour groups for the 2+1 dimensional theory here). (The existence of three new exceptional theories, for E_6, E_7, E_8, is then immediate[237].) What we are actually recovering is the fact[153] that there is a strong/weak coupling duality between type I (or $SO(32)$ heterotic) string theory on T^3 and M-theory on K3. We'll recover this fact again via another route in section 16.2.2.

15.4 Wrapped branes and the enhançon mechanism

Despite the successes we have achieved in the previous section with interpretation of supergravity solutions in terms of constituent D-branes, we should be careful, even in the case when we have supersymmetry to steer us away from potential pathologies. It is not always the case that if someone presents us with a solution of supergravity with R–R charges that we should believe that it has an interpretation as being 'made of D-branes'.

Consider again the case of eight supercharges. We studied brane configurations with this amount of supersymmetry by probing the geometry of N (large) Dp-branes with a single D$(p-4)$-brane. As described in previous sections, another simple way to achieve a geometry with eight supercharges from D-branes is to simply wrap branes on a manifold which already breaks half of the supersymmetry[117]. The example mentioned was the four dimensional case of K3. In this case, we learned that if we wrap a D$(p+4)$-brane (say) on K3, we induce precisely one unit of negative Dp-brane charge[115] supported on the unwrapped part of the world-volume (see equation (9.36)). At large N therefore, we might expect[239] that there is a simple supergravity geometry which might be obtained by taking the known solution for the D$(p+4)$-Dp system, and modifying the asymptotic charges to suit this situation. The resulting geometry naively should have the interpretation as that due to a large number N of wrapped D$(p+4)$ branes ($p = 1, 2, 3$):

$$ds^2 = Z_p^{-1/2} Z_{p+4}^{-1/2} \eta_{\mu\nu} dx^\mu dx^\nu + Z_p^{1/2} Z_{p+4}^{1/2} dx^i dx^i$$
$$+ V^{1/2} Z_p^{1/2} Z_{p+4}^{-1/2} ds^2_{K3},$$
$$e^{2\Phi} = g_s^2 Z_p^{(3-p)/2} Z_{p+4}^{-(p+1)/2},$$
$$C_{(p+1)} = (Z_p^{-1} - 1) g_s^{-1} dx^0 \wedge dx^1 \wedge \cdots \wedge dx^{p+1}$$
$$C_{(p+5)} = (Z_{p+4}^{-1} - 1) g_s^{-1} dx^0 \wedge dx^1 \wedge \cdots \wedge dx^{p+5}. \tag{15.13}$$

Here, μ, ν run over the $x^0 - x^{p+1}$ directions, which are tangent to all the branes. Also i runs over the directions transverse to all branes, $x^{p+2} - x^5$, and in the remaining directions, transverse to the induced brane but inside the large brane, ds^2_{K3} is the metric of a K3 surface of unit volume. V is the volume of the K3 as measured at infinity, but the supergravity solution adjusts itself such that $V(r) = V Z_p / Z_{p+4}$ is the measured volume of the K3 at radius r.

In the above,

$$Z_{p+4} = 1 + \frac{r_{p+1}^{3-p}}{r^{3-p}}, \quad \text{while} \quad Z_p = 1 - \frac{(2\pi\sqrt{\alpha'})^4 r_{p+1}^{3-p}}{V r^{3-p}}, \tag{15.14}$$

where the normalisations are related to those in section 10.2. We have precisely N units of D$(p + 4)$-brane charge and $-N$ units of Dp-brane charge. Note that the smaller brane is delocalised in the K3 directions, as it should be, since the same is true of K3's curvature.

15.4.1 Wrapping D6-branes

Let us focus on the case $p = 2$, where we wrap D6-branes, getting induced D2-branes.[†] The orientations are given as follows.

	x^0	x^1	x^2	x^3	x^4	x^5	x^6	x^7	x^8	x^9
D2	—	—	—	•	•	•	•	•	•	•
D6	—	—	—	•	•	•	—	—	—	—
K3	—	—	—	—	—	—	•	•	•	•

[†] This will also teach us a lot about the pure $SU(N)$ gauge theory on the remaining 2+1 dimensional world-volume. Wrapping D7-branes ($p = 3$) teaches us[239] about pure $SU(N)$ gauge theory in 3+1 dimensions, where we should make a connection to Seiberg–Witten theory[240, 241] at large N.

The harmonic functions are

$$Z_2 = 1 + \frac{\hat{r}_2}{r}, \qquad \hat{r}_2 = -\frac{(2\pi)^4 g_s N \alpha'^{5/2}}{2V},$$

$$Z_6 = 1 + \frac{r_6}{r}, \qquad r_6 = \frac{g_s N \alpha'^{1/2}}{2}, \tag{15.15}$$

normalised such that the D2- and D6-brane charges are $Q_2 = -Q_6 = -N$.

We worked out the spectrum of type IIA supergravity theory compactified to six dimensions on K3 in section 7.6. Let us remind ourselves of some of the salient features. The six dimensional supergravity theory has an additional 24 $U(1)$s in the R–R sector. Of these, 22 come from wrapping the ten dimensional three-form on the 19+3 two-cycles of K3. The remaining two are special $U(1)$s for our purposes: one of them arises from wrapping IIAs five-form entirely on K3, while the final one is simply the plain one-form already present in the uncompactified theory.

15.4.2   The repulson geometry

It is easy to see that there is something wrong with the geometry which we have just written down, representing the wrapping of the D6-branes on the K3. There is a naked singularity at $r = |\hat{r}_2|$, known as the 'repulson', since[‡] it represents a repulsive gravitational potential for small enough r. The curvature diverges there, which is related to the fact that the volume of the K3 goes to zero, and the geometry stops making sense (see figure 15.1).

To characterise the repulsive nature of the geometry we can consider it as a background for particle motion and study geodesics. There is the usual obvious pair of Killing vectors, $\boldsymbol{\xi} = \partial_t$ and $\boldsymbol{\eta} = \partial_\phi$, and so a probe with ten-velocity $\mathbf{u}$ has conserved quantities

$$e = -\xi \cdot u = -G_{tt} u^t$$

and

$$\ell = \eta \cdot u = G_{\phi\phi} u^\phi,$$

where e and ℓ are the total energy and angular momentum per unit mass, respectively. Since the particle is massive, we have $-1 = u \cdot u$. In other words, picking

$$\mathbf{u} = \left(\frac{dt}{d\tau}, \frac{dr}{d\tau}, \frac{d\theta}{d\tau}, \frac{d\phi}{d\tau}, \vec{0} \right),$$

[‡] This is because it is dual[239] to solutions which had earlier become known by that name[257].

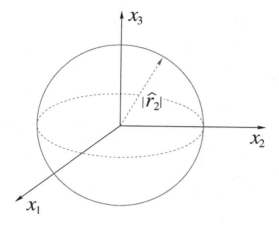

Fig. 15.1. The repulson locus of points; an unphysical naked singularity.

we have, working in the equatorial plane $\theta = \pi/2$,

$$-1 = -G_{tt}\left(\frac{dt}{d\tau}\right)^2 + G_{rr}\left(\frac{dr}{d\tau}\right)^2 + G_{\phi\phi}\left(\frac{d\phi}{d\tau}\right)^2,$$

and so

$$-1 = -\frac{e^2}{G_{tt}} + G_{rr}\left(\frac{dr}{d\tau}\right)^2 + \frac{\ell^2}{G_{\phi\phi}},$$

which we can rewrite as

$$E = \frac{1}{2}\left(\frac{dr}{d\tau}\right)^2 + V_{\text{eff}},$$

where

$$\frac{dr}{d\tau} = \pm\sqrt{E - V_{\text{eff}}(r)}, \qquad E = \frac{e^2 - 1}{2},$$

$$V_{\text{eff}}(r) = \frac{1}{2}\left[\frac{1}{G_{rr}}\left(1 + \frac{\ell^2}{G_{\phi\phi}}\right) - 1\right], \qquad (15.16)$$

and the metric components in the above are in string frame, and we have used that $-G_{tt} = 1/G_{rr}$. For what we wish to analyse, we can consider only purely radial motion, and hence set to zero the angular momentum ℓ which would correspond to a non-zero impact parameter. We sketch the resulting effective potential in figure 15.2.

For large enough r, the effective potential is attractive, and so we need only seek a vanishing first derivative of $V_{\text{eff}}(r)$ to see where it becomes

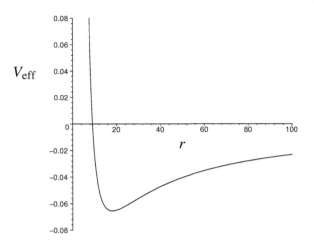

Fig. 15.2. The effective potential for massive particle motion in the geometry. The minimum is at $r = r_{\mathrm{e}}$.

negative. This gives the condition:

$$G_{rr}^{-1} = -G_{tt}' = 0, \tag{15.17}$$

which we can write in a number of interesting ways as

$$(Z_2 Z_6)' = Z_6 Z_6' \left(\frac{Z_2}{Z_6} + \frac{Z_2'}{Z_6'} \right) = Z_2 Z_6 \left(\frac{Z_2'}{Z_2} + \frac{Z_6'}{Z_6} \right) = 0, \tag{15.18}$$

and the particle begins to be repelled at radii smaller than this. Particles with non-zero angular momentum will of course experience additional centrifugal repulsion, but $r = r_{\mathrm{e}}$ is the boundary of the region where there is an intrinsic repulsion in the geometry.

15.4.3 Probing with a wrapped D6-brane

Let us look carefully to see if this is really the geometry produced by the wrapped branes. The object we have made should be a BPS membrane made of N identical objects. These objects feel no force due to each other's presence, and therefore the BPS formula for the total mass is simply (see equation (9.37))

$$\tau_N = \frac{N}{g_{\mathrm{s}}} (\mu_6 V - \mu_2) \tag{15.19}$$

with $\mu_6 = (2\pi)^{-6} \alpha'^{-7/2}$ and $\mu_2 = (2\pi)^{-2} \alpha'^{-3/2}$. In fact, the BPS membrane is actually a monopole of one of the six dimensional $U(1)$s. It is obvious which $U(1)$ this is; the diagonal combination of the two special ones we mentioned above. The D6-brane component is already a monopole of the IIA R–R one-form, and the D2 is a monopole of the five-form, which gets wrapped.

N.B. As we shall see, the final combination is a non-singular BPS monopole, having been appropriately dressed by the Higgs field associated to the volume of K3. Also, it maps[165] (under the strong/weak coupling duality of the type IIA string on K3 to the heterotic string on T^4) to a bound state of a Kaluza–Klein monopole[168] and an H-monopole[242], made by wrapping the heterotic NS5-brane.[239, 243, 244]

If we are to interpret our geometry as having been made by bringing together N copies of our membrane, we ought to be able to carry out the procedure we described in the previous sections. We should see that the geometry as seen by the probe is potential-free and well-behaved, allowing us the interpretation of being able to bring the BPS probe in slowly from infinity.

The effective action for a D6-brane probe (wrapped on the K3) is:

$$S = -\int_M d^3\xi\, e^{-\Phi(r)}(\mu_6 V(r) - \mu_2)(-\det g_{ab})^{1/2} + \mu_6 \int_{M \times K3} C_7 - \mu_2 \int_M C_3.$$
(15.20)

Here M is the part of the world-volume in the three non-compact dimensions. As discussed previously (see equation (9.39) and surrounding discussion), the first term is the Dirac–Born–Infeld action with the position dependence of the tension (15.19) taken into account; in particular, $V(r) = V Z_2(r)/Z_6(r)$. The second and third terms are the couplings of the probe charges $(\mu_6, -\mu_2)$ to the background R–R potentials, following from equation (9.36), and surrounding discussion.

Having derived the action, the calculation proceeds very much as we outlined in previous sections, with the result:

$$\mathcal{L} = -\frac{\mu_6 V Z_2 - \mu_2 Z_6}{Z_6 Z_2 g_s} + \frac{\mu_6 V}{g_s}(Z_6^{-1} - 1) - \frac{\mu_2}{g_s}(Z_2^{-1} - 1)$$
$$+ \frac{1}{2g_s}(\mu_6 V Z_2 - \mu_2 Z_6)v^2 + O(v^4).$$
(15.21)

The position-dependent potential terms cancel as expected for a super-symmetric system, leaving the constant potential $(\mu_6 V - \mu_2)/g$ and a nontrivial metric on moduli space (the $O(v^2)$ part) as expected with eight supersymmetries. The metric is proportional to

$$ds^2 = \frac{1}{g_s} \left(\mu_6 V Z_2 - \mu_2 Z_6 \right) dx_\perp^2 = \frac{\mu_6 Z_6}{g_s} \left(\frac{Z_2}{Z_6} - \frac{V_*}{V} \right) (dr^2 + r^2 d\Omega_2^2),$$

(15.22)

where we have used $\mu_2/\mu_6 = V_*$. We assume that $V > V_* \equiv (2\pi)^4 \alpha'^2$, so that the metric at infinity (and the membrane tension) is positive. However, as r decreases the metric eventually becomes negative, and this occurs at a radius

$$r = \frac{2V}{V - V_*} |\hat{r}_2| \equiv r_{\mathrm{e}},$$

(15.23)

which is greater than the radius $r_{\mathrm{r}} = |\hat{r}_2|$ of the repulson singularity. Furthermore, it is precisely the radius at which the geometry becomes repulsive, since $Z_2'/Z_6' = -V_*/V$, and that radius is determined by equation (15.18).

In fact, our BPS monopole is becoming massless as we approach the special radius. This should mean that the $U(1)$ under which it is charged is becoming enhanced to a non-Abelian group. This is the case. There is a purely stringy phenomenon which lies outside supergravity which we have not included thus far. The W-bosons are wrapped D4-branes, which enhance the $U(1)$ to an $SU(2)$. The masses of wrapped D4-branes is computed just like that of the membrane, and so becomes zero when the K3's volume reaches the value $V_* \equiv (2\pi\sqrt{\alpha'})^4$.

The point is that the repulson geometry represents supergravity's best attempt to construct a solution with the correct asymptotic charges. In the solution, the volume of the K3 decreases from its asymptotic value V as one approaches the core of the configuration. At the centre, the K3 radius is zero, and this is the singularity. This ignores rather interesting physics, however. At a finite distance from the putative singularity (where $V_{\mathrm{K3}} = 0$), the volume of the K3 gets to $V = V_*$, so the stringy phenomena – including new massless fields – giving the enhanced $SU(2)$, should have played a role[§]. So the aspects of the supergravity solution near and inside the special radius, called the 'enhançon radius', should not be taken seriously at all, since it ignored this stringy physics.

The supergravity solution should only be taken as physical down to the neighbourhood of the enhançon radius r_{e}. That locus of points, a two-sphere S^2, is itself called[239] an 'enhançon' (see figure 15.3).

[§] Actually, this enhancement of $SU(2)$ is even less mysterious in the heterotic-on-T^4 dual picture mentioned two pages ago. It is just the $SU(2)$ of a self-dual circle in this picture, which we studied extensively in section 4.3.

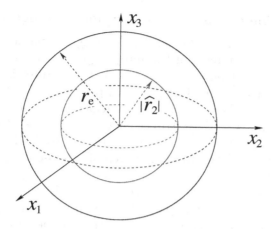

Fig. 15.3. The enhançon locus at which new physics beyond supergravity appears. This happens before the singular repulson locus, signalling that the original geometry inside the enhançon radius was unphysical.

Recall also (see section 13.5.1) that the *size* of the monopole is inverse to the mass of the W-bosons (or the Higgs vev), and so in fact by time our probe gets to the enhançon radius, it has smeared out considerably, and in fact merges into the geometry, forming a 'shell' with the other monopoles at that radius. Since by this argument we cannot place sharp sources inside the enhançon radius, and so the geometry on the inside must be very different from that of the repulson. In fact, to a first approximation, it must simply be flat, forming a junction with the outside geometry at $r = r_e$.

In general, the same sort of reasoning applies for all p. The enhançon locus results from wrapping a D$(p + 4)$-brane on K3 is $S^{4-p} \times \mathbb{R}^{p+1}$, whose interior is $(5 + 1)$-dimensional. This must work since the ratio $\mu_p/\mu_{p+4} = V_*$ and so there will always be wrapped branes becoming massless at the same loci in the geometry, giving physics which goes beyond supergravity. For even p the theory in the interior has an $SU(2)$ gauge symmetry, while for odd p there is the A_1 two-form gauge theory, carried by tensionless strings. The details of the smoothing will be very case dependent, and it should be interesting to work out those details.

One can also study $SO(2N)$, $SO(2N+1)$ and $USp(2N)$ gauge theories with eight supercharges in various dimensions using similar techniques, placing an orientifold O6-plane into the system parallel to the D6-branes. The enhançon then becomes[245] an $\mathbb{RP}^2$.

15.5 The consistency of excision in supergravity

We can actually use classic General Relativity techniques[259, 260] to carry out the procedure of removing the interior geometry and replacing it by flat space. We should be able to see if this procedure is consistent and makes some physical sense. The standard procedure for this is as follows. If we join two solutions of Einstein's equations across some surface, there will be a discontinuity in the extrinsic curvature at the surface. A rewriting of the equations of motion can be done to show that this discontinuity can be interpreted as a δ-function source of stress-energy located at the surface.

Let's carry this out here[264], performing an incision at arbitrary radius $r = r_{\rm i}$, and then gluing in flat space. The computation must be performed in Einstein frame to enable an interpretation of the discontinuity in the extrinsic curvature as a stress-energy. So we work with the ten dimensional Einstein metric $ds_{\rm E}^2 = e^{-\Phi/2} ds^2$ denoting the generic metric components as G_{AB}:

$$
\begin{aligned}
g_{\rm s}^{1/2}\, ds^2 &= Z_2^{-5/8} Z_6^{-1/8} \eta_{\mu\nu} dx^\mu dx^\nu + Z_2^{3/8} Z_6^{7/8} dx^i dx^i \\
&\quad + V^{1/2} Z_2^{3/8} Z_6^{-1/8} ds_{K3}^2 \\
&= G_{\mu\nu} dx^\mu dx^\nu + G_{ij} dx^i dx^j + G_{ab} dx^a dx^b,
\end{aligned}
\tag{15.24}
$$

where Z_2 and Z_6 are given by (15.15).

Since we make a radial slice, we can define unit normal vectors (see insert 10.2):

$$
n_\pm^A = \mp \frac{1}{\sqrt{G_{rr}}} \left(\frac{\partial}{\partial r} \right)^A,
\tag{15.25}
$$

where n_+ (n_-) is the outward pointing normal for the spacetime region $r > r_{\rm i}$ $(r < r_{\rm i})$. Referring to insert 10.2, we see that the extrinsic curvature of the junction surface for each region is

$$
K_{AB}^\pm = \frac{1}{2} n_\pm^C \partial_C G_{AB} = \mp \frac{1}{2\sqrt{G_{rr}}} \frac{\partial G_{AB}}{\partial r}.
\tag{15.26}
$$

We next define the discontinuity in the extrinsic curvature across the junction as $\gamma_{AB} = K_{AB}^+ + K_{AB}^-$. The stress-energy tensor supported at the junction is defined in terms of these as:

$$
S_{AB} = \frac{1}{\kappa^2} \left(\gamma_{AB} - G_{AB}\, \gamma^C{}_C \right),
\tag{15.27}
$$

where κ is the gravitational coupling defined in (7.44).

In choosing the metric for flat space, we should ensure that all fields are continuous through the incision by writing the interior solution in appropriate coordinates and gauge:

$$
\begin{aligned}
g_s^{1/2} ds^2 &= Z_2(r_i)^{-5/8} Z_6(r_i)^{-1/8} \eta_{\mu\nu} dx^\mu dx^\nu + Z_2(r_i)^{3/8} Z_6(r_i)^{7/8} dx^i dx^i \\
&\quad + V^{1/2} Z_2(r_i)^{3/8} Z_6(r_i)^{-1/8} ds_{K3}^2, \\
e^{2\Phi} &= g_s^2 Z_2^{1/2}(r_i) Z_6^{-3/2}(r_i), \\
C_{(3)} &= (Z_2(r_i) g_s)^{-1} dx^0 \wedge dx^1 \wedge dx^2, \\
C_{(7)} &= (Z_6(r_i) g_s)^{-1} dx^0 \wedge dx^1 \wedge dx^2 \wedge V \varepsilon_{K3}.
\end{aligned}
\tag{15.28}
$$

It is straightforward to derive the following results for the discontinuity tensor, and the reader should check the result:

$$
\begin{aligned}
\gamma_{\mu\nu} &= \frac{1}{16} \frac{1}{\sqrt{G_{rr}}} \left(5 \frac{Z_2'}{Z_2} + \frac{Z_6'}{Z_6} \right) G_{\mu\nu}, \\
\gamma_{ij} &= -\frac{1}{16} \frac{1}{\sqrt{G_{rr}}} \left(3 \frac{Z_2'}{Z_2} + 7 \frac{Z_6'}{Z_6} \right) G_{ij}, \\
\gamma_{ab} &= -\frac{1}{16} \frac{1}{\sqrt{G_{rr}}} \left(3 \frac{Z_2'}{Z_2} - \frac{Z_6'}{Z_6} \right) G_{ab},
\end{aligned}
\tag{15.29}
$$

where a prime denotes ∂_r and all quantities are evaluated at the incision surface $r = r_i$. The trace is:

$$
\gamma^C{}_C = -\frac{1}{16} \frac{1}{\sqrt{G_{rr}}} \left(3 \frac{Z_2'}{Z_2} + 7 \frac{Z_6'}{Z_6} \right),
\tag{15.30}
$$

and the $\mu, \nu = 0, 1, 2$ index directions along the brane, i, j index the two angular directions (θ, ϕ) transverse to the brane, and a, b index the four K3 directions.

So finally we have the stress-energy tensor of the discontinuity:

$$
\begin{aligned}
S_{\mu\nu} &= \frac{1}{2\kappa^2 \sqrt{G_{rr}}} \left(\frac{Z_2'}{Z_2} + \frac{Z_6'}{Z_6} \right) G_{\mu\nu}, \\
S_{ij} &= 0, \\
S_{ab} &= \frac{1}{2\kappa^2 \sqrt{G_{rr}}} \left(\frac{Z_6'}{Z_6} \right) G_{ab}.
\end{aligned}
\tag{15.31}
$$

Let us consider the physical properties of this object[264]. The last line gives the components of the stress-energy along the K3 direction. It involves only the harmonic function for the pure D6-brane part which is consistent with the fact that there are only D6-branes wrapped there. The middle line shows that there is no stress in the directions transverse to the branes, which dovetails nicely with the fact that the constituent branes are BPS with no interaction forces needed to support the shell in the transverse space.

As a first check of this interpretation, we can expand the results in equation (15.31) for large r_i. Up to an overall sign, the coefficient of the metric components gives an effective tension in the various directions. The leading contributions are simply:

$$\tau(r_i) = \frac{1}{2\kappa^2} \frac{r_6}{r_i^2} \left(1 - \frac{V_*}{V}\right) \tag{15.32}$$

$$= \frac{N}{(2\pi)^6 (\alpha')^{7/2} g_s} (V - V_*) \frac{1}{4\pi r_i^2 V} = N(\tau_6 V - \tau_2) \left(\frac{1}{4\pi r_i^2 V}\right),$$

$$\tau_{K3}(r_i) = \frac{1}{2\kappa^2} \frac{r_6}{r_i^2} = \frac{N}{(2\pi)^6 (\alpha')^{7/2} g_s} \left(\frac{1}{4\pi r_i^2}\right) = N\tau_6 \left(\frac{1}{4\pi r_i^2}\right), \tag{15.33}$$

which is in precise accord with expectations. In the K3 directions, the effective tension matches precisely that of N fundamental D6-branes, with an additional averaging factor $(1/4\pi r_i^2)$ coming from smearing the branes over the transverse space. In the x^0, x^1, x^2 directions, we have an effective membrane tension which, up to the appropriate smearing factor, again matches that for N D6-branes including the subtraction of N units of D2-brane tension as a result of wrapping on the K3 manifold[128] .

Notice that the result for the stress-energy in the unwrapped part of the brane is proportional to $(Z_2 Z_6)'$. As we have already observed in equations (15.18) and (15.23), this vanishes at precisely $r = r_e$, where the probe starts to become unphysical, and where the supergravity starts to become repulsive. So, for incision at the enhançon radius, there is a shell of branes of zero tension, as the probe computation showed.

For $r < r_e$ we would get a negative tension from the stress-energy tensor, which is problematic even in supergravity. Notice, however, that nothing in our computation shows that we cannot make an incision at any radius of our choosing for $r \geq r_e$, and place a shell of branes of the appropriate tension (as in the calculation of the effective tensions at large r_i above). This corresponds physically to the fact that constituent branes experience no potential, so they can consistently be placed at any arbitrary position outside the enhançon.

15.6 The moduli space of pure glue in 3D

Note that the Lagrangian (15.21) depends only on three moduli space coordinates, (x^3, x^4, x^5), or (r, θ, ϕ) in polar coordinates. As mentioned before, a (2+1) dimensional theory with eight supercharges, should have a moduli space metric which is hyper-Kähler[185]. So we need at least one extra modulus, s. A similar procedure to that used in section 15.2 can be used to introduce the gauge field's correct couplings and dualise to introduce the scalar s. A crucial difference is that one must replace $2\pi\alpha' F_{ab}$ by $e^{2\phi}(\mu_6 V(r) - \mu_2)^{-2} v_a v_b$ in the Dirac–Born–Infeld action, the extra complication being due to the r dependent nature of the tension. The static gauge computation gives for the kinetic term:

$$\mathcal{L} = F(r)\left(\dot{r}^2 + r^2\dot{\Omega}^2\right) + F(r)^{-1}\left(\dot{s}/2 - \mu_2 C_\phi \dot{\phi}/2\right)^2, \qquad (15.34)$$

where

$$F(r) = \frac{Z_6}{2g_s}(\mu_6 V(r) - \mu_2) = \frac{\alpha'^{-3/2}}{(2\pi)^2 g_s}\left(\frac{V}{V_*} - 1 - \frac{g_s N \alpha'^{1/2}}{r}\right), \qquad (15.35)$$

and $\dot{\Omega}^2 = \dot{\theta}^2 + \sin^2\theta\,\dot{\phi}^2$.

Again, there is gauge theory information to be extracted here. We have pure gauge $SU(N)$ theory with no hypermultiplets, and eight supercharges. We should be able to cleanly separate the gauge theory data from everything else by taking the decoupling limit $\alpha' \to 0$ while holding the gauge theory coupling $g_{YM}^2 = g_{YM,p}^2 V^{-1} = (2\pi)^4 g_s \alpha'^{3/2} V^{-1}$ and the energy scale $U = r/\alpha'$ (proportional to M_W) fixed. In doing this, we get the metric:

$$ds^2 = f(U)\left(\dot{U}^2 + U^2 d\Omega^2\right) + f(U)^{-1}\left(d\sigma - \frac{N}{4\pi^2}A_\phi d\phi\right)^2,$$

$$\text{where} \quad f(U) = \frac{1}{4\pi^2 g_{YM}^2}\left(1 - \frac{g_{YM}^2 N}{U}\right), \qquad (15.36)$$

the $U(1)$ monopole potential is $A_\phi = \pm 1 - \cos\theta$, and $\sigma = s\alpha'$, and the metric is meaningful only for $U > U_e = \lambda = g_{YM}^2 N$, the "'t Hooft coupling', a natural gauge theory quantity to hold fixed in the limit of large N, where we make contact with supergravity. This metric, which should be contrasted with equation (15.11), is the hyper-Kähler Taub–NUT metric, but this time with a negative mass. It is singular. For $N = 2$, the full metric, obtained by instanton corrections to this one-loop result, is smooth, as we will discuss. For large N, the instantons are suppressed. We shall discuss this some more in the next section.

15.6.1 Multi-monopole moduli space

Recall that the membrane resulting from wrapping the six-brane is s BPS monopole. Therefore the moduli space of the entire wrapped system should be related to the moduli space of N BPS monopoles. In fact, since the low energy dynamics of the branes is $SU(N)$ gauge theory, we learn that BPS monopole moduli space is to be identified with the Coulomb branch of the gauge theory as well[231]. The part of the moduli space corresponding to the motion of a single sub-brane (the probe discussed above) is evidently a submanifold of the full $4N-4$ dimensional metric on the *relative* moduli space[218] of N BPS monopoles which is smooth[219].

This should remind the reader of our study in section 15.3. Recalling that this is also a study of $SU(N)$ gauge theory with no hypermultiplets, we know the result for $N = 2$: the metric on the moduli space must be smooth, as there is no Higgs branch to connect to via the singularity. This is true for all $SU(N)$, and matches the monopole result. For $N = 2$, we stated that the metric on the moduli space[247] is actually the Atiyah–Hitchin manifold[232]. The metric may be written in the following manifestly $SO(3)$ invariant manner[232, 251]:

$$ds^2_{\text{AH}} = f^2 d\rho^2 + a^2 \sigma_1^2 + b^2 \sigma_2^2 + c^2 \sigma_3^2;$$
$$\frac{2bc}{f} \frac{da}{d\rho} = (b-c)^2 - a^2, \text{ and cyclic perms.}; \quad \rho = 2K\left(\sin\frac{\beta}{2}\right),$$

$$(15.37)$$

where the choice $f = -b/\rho$ can be made, the σ_i are defined in (7.4), and $K(k)$ is the elliptic integral of the first kind:

$$K(k) = \int_0^{\frac{\pi}{2}} (1 - k^2 \sin^2 \tau)^{\frac{1}{2}} d\tau. \qquad (15.38)$$

Also, $k = \sin(\beta/2)$, the 'modulus', runs from 0 to 1, so $\pi \le \rho \le \infty$. In fact, the solution for a, b, c can be written out in terms of elliptic functions, but we shall not do that here. All of the functions entering the metric can be expanded in large ρ, and the result is:

$$ds^2_{\text{TN}-} = \left(1 - \frac{2}{\rho}\right)\left(d\rho^2 + \rho^2 d\Omega^2\right) + 4\left(1 - \frac{2}{\rho}\right)^{-1}(d\psi + \cos\theta d\phi)^2. \qquad (15.39)$$

Comparing to equation (15.12), we see that this is the Taub–NUT metric, but with a negative mass parameter, i.e. $N_{\text{f}} = -1$. Now, as already stated, Taub–NUT has an $SU(2)$ isometry, and the full Atiyah–Hitchin metric has an $SO(3)$. Furthermore, the metric we have here is singular at $\rho = 2$, whereas the full metric is smooth everywhere. Therefore there is a lot missing from this approximate metric. In fact, these key differences are invisible at any order in the large ρ expansion, being exponentially small in ρ, of the form $e^{-\rho}$. These exponential corrections for smaller ρ remove the singularity: $\rho = 2$ is just an artifact of the large ρ metric in the above form (15.39). As for the isometry, the fact that it is really $SO(3)$ follows from the fact that ψ started out with periodicity 2π and not 4π in the full metric, as required by the requirement that there is no bolt spherical singularity at finite ρ. Expanding in large ρ will not change that periodicity of course, but if one was just presented with the expanded result one would not know of the non-perturbative no-bolt condition. So in this case of two monopoles, there is an $SO(3) = SU(2)/\mathbb{Z}_2$ isometry in the problem, and not the naive $SU(2)$ of the Taub–NUT space, since ψ has period 2π and not 4π. The $SO(3)$ isometry, smoothness, and the condition of hyper-Kählerity actually picks out *uniquely* the Atiyah–Hitchin manifold as the completion of the negative mass Taub–NUT.

Actually, we have described a trivial cover of the true Atiyah–Hitchin space. The two monopole problem has an obvious $\mathbb{Z}_2$ symmetry coming from the fact that the monopoles are identical. Some field configurations described by the manifold as described up to now are overcounted, and so we must divide by this $\mathbb{Z}_2$, resulting in an $\mathbb{RP}^2$ for the bolt instead of an S^2.

What is the relation to our probe result? To see it[258], change variables in our probe metric (15.36) by absorbing a factor of $\lambda/2 = g_{\text{YM}}^2 N/2$ into the radial variable U, defining $\rho = 2U/\lambda$. Further absorb $\psi = \sigma 8\pi^2/N$ and gauge transform to $A_\phi = -\cos\theta$. Then we get:

$$ds^2 = \frac{g_{\text{YM}}^2 N^2}{32\pi^2} ds_{\text{TN}-}^2, \qquad (15.40)$$

showing that we have precisely the form of the Taub–NUT metric that one gets by expanding the Atiyah–Hitchin metric in large ρ and neglecting exponential corrections.

Now for the same reasons as in section 15.3, the periodicity of σ is $1/2\pi$, and we will use $\tilde{\sigma} = 4\pi^2\sigma$ as our 2π periodic scalar dual to the photon on the probe's world-volume. Looking at the choices we made above, this implies that for the $SU(2)$ case, the coordinate ψ has period 2π, which fits what we stated about the Atiyah–Hitchin manifold above.

The exponential corrections have the expected interpretation in the gauge theory as the instanton corrections which maintain positivity of the metric and the gauge coupling[249]. Translating back to physical variables, we see that these corrections go as $\exp(-U/g_{\text{YM}}^2)$, which has the correct form of action for a gauge theory instanton. (We have just described a *cover* of the Atiyah–Hitchin manifold needed for the $SU(2)$ case. There is an additional identification to be discussed below.) This completes the story for the $SU(2)$ gauge theory moduli space problem[248].

Can we learn anything from this for our case of general N, especially for large N, to teach us about the enhançon geometry? Notice[258] that the instanton corrections are suppressed at large N if we hold the 't Hooft coupling λ (which sets the Taub–NUT mass) fixed, since there is a bare N in the exponential: $\exp\left(-NU/\lambda\right)$. So the smoothing is suppressed at large N, and we recover the macroscopic sharp (relatively) enhançon locus at large N in the supergravity geometry. Notice that if we've fixed our period of $\tilde{\sigma}$ to be 2π as before, for general N the resulting period of ψ in the scaled variables is $\Delta\psi = 4\pi/N$. Therefore our isometry is not $SO(3)$ but is only $SU(2)/\mathbb{Z}_N$, which is not an isometry at all.

16

Towards M- and F-theory

As we saw in chapter 12, there is an extremely tantalising picture of the fate of string theory at strong coupling, obtained using certain 'duality' transformations. In fact, D-branes were rather useful, as they allowed for an explicit constructive method for finding evidence of the products of duality, for example exhibiting stable states which must exist – with special properties – on both sides of the duality.

One major task is to try to understand how to write better formulations of the physics of strong coupling. There are two main goals to be achieved by this. The first is simply to find better ways of finding new and interesting backgrounds (vacua) for string theory, with techniques which allow for better handing of strongly coupled regions of the solution. The second is to attempt to find the 'correct' manner in which to describe the complete M-theory from which all string theories are supposed to arise as weakly coupled limits.

Both '*Matrix theory*'[157] and '*F-theory*'[199] are ideas in these directions, putting together the strongly coupled brane and string data in ways which allow for new geometric ways of describing and connecting string vacua, and giving insights into the next generation of formulations of the physics. In this chapter we shall uncover aspects of both, while learning much more about the properties of various branes.

16.1 The type IIB string and F-theory

One of the remarkable dualities which we observed in chapter 12 was the 'self-duality' of the type IIB superstring theory. Its fullest expression is in terms of a rich family of transformations which generate the group $SL(2, \mathbb{Z})$. The consequences of this duality group are profound, and we shall uncover some of them in this chapter.

16.1.1 $SL(2, \mathbb{Z})$ duality

Recall that we saw that the coupling inverted under the 'duality transformation': $g_s \to 1/g_s$, or $\Phi \to -\Phi$, since $g_s = e^\Phi$. The fundamental string was exchanged with the D1-brane, from which it follows that the NS–NS two-form potential and the R–R two-form potential (to which those strings couple electrically, respectively), are also exchanged.

In fact, as has been discussed earlier as well, this is all part of a larger duality, whose complete transformation group is $SL(2, \mathbb{Z})$, which is parametrised by matrices of the form:

$$\Lambda = \begin{pmatrix} a & b \\ c & d \end{pmatrix}, \quad ab - cd = 1; \quad a, b, c, d \in \mathbb{Z}. \tag{16.1}$$

Combining the R–R scalar $C_{(0)}$ and the dilaton into a complex coupling $\tau = C_{(0)} + ie^{-\Phi}$, the duality group acts on it as:

$$\tau \longrightarrow \frac{a\tau + b}{c\tau + d}, \tag{16.2}$$

and acts on the two-form potentials as

$$\begin{pmatrix} B_{(2)} \\ C_{(2)} \end{pmatrix} \longrightarrow (\Lambda^T)^{-1} \begin{pmatrix} B_{(2)} \\ C_{(2)} \end{pmatrix} \longrightarrow \begin{pmatrix} d & -c \\ -b & a \end{pmatrix} \begin{pmatrix} B_{(2)} \\ C_{(2)} \end{pmatrix}. \tag{16.3}$$

So the basic strong weak coupling duality we discovered first is the case

$$\Lambda = S = \begin{pmatrix} 0 & 1 \\ -1 & 0 \end{pmatrix}, \tag{16.4}$$

for which we get $\tau \to -1/\tau$, $B_{(2)} \to -C_{(2)}$, $C_{(2)} \to B_{(2)}$. While all of this is taking place, the R–R four-form $C_{(4)}$ is invariant, which has remarkable consequences for the D3-branes which couple to it, as we shall see later in this and other chapters.

In fact, at low energy and tree level, the $SL(2, \mathbb{Z})$ symmetry is only $SL(2, \mathbb{R})$, as the integer restriction to the former case is only visible beyond tree level. The quantisation of the charges of the D-instanton (and by supersymmetry, their action) which couple electrically to $C_{(0)}$ arises in the quantum theory, as we saw in chapter 8. It is very instructive to rewrite the low energy supergravity action (7.42) in a manifestly $SL(2, \mathbb{R})$ invariant way, with the understanding that at this level we can restrict to integers by hand. We work in Einstein frame metric, defined by $G_{\mu\nu}^E = e^{-\Phi/2} G_{\mu\nu}^s$, and find that it is useful to define a field strength doublet $\widehat{G}_{(3)} = (H_{(3)}, G_{(3)})$ and a matrix

$$\mathcal{M} = \frac{1}{\tau_2} \begin{pmatrix} |\tau|^2 & -\mathrm{Re}\tau \\ -\mathrm{Re}\tau & 1 \end{pmatrix} = e^\Phi \begin{pmatrix} |\tau|^2 & -C_{(0)} \\ -C_{(0)} & 1 \end{pmatrix}, \quad \tau = \tau_1 + i\tau_2, \tag{16.5}$$

and the action is:

$$S_{\text{IIB}} = \frac{1}{2\kappa^2} \int d^{10}x \sqrt{-G} \left(R + \frac{1}{4} \text{Tr}[\partial_\mu \mathcal{M} \partial^\mu \mathcal{M}] - \frac{1}{12} \widetilde{G}_{(3)}^T \mathcal{M} \widetilde{G}_{(3)} \right.$$

$$\left. - \frac{1}{480} G_{(5)}^2 \right) - \frac{\epsilon_{ij}}{4\kappa^2} \int C_{(4)} \wedge G_{(3)}^{[i]} \wedge G_{(3)}^{[j]}, \quad (16.6)$$

(where ϵ_{ij} is antisymmetric with $\epsilon_{12} = 1$) and the $SL(2, \mathbb{R})$ invariance is under:

$$\mathcal{M} \to \Lambda \mathcal{M} \Lambda^T; \qquad \widetilde{G}_{(3)} \to (\Lambda^T)^{-1} \widetilde{G}_{(3)}. \quad (16.7)$$

In fact, $\mathcal{M}$ parametrises the coset $SL(2, \mathbb{R})/SO(2)$, (the dimension of the coset, $3 - 1 = 2$ corresponds correctly to the number of scalars) and the kinetic term for the scalars can also be written as

$$-\frac{\partial_\mu \bar{\tau} \partial^\mu \tau}{2(\text{Im}\tau)^2},$$

showing that the metric on the coset space is essentially[*][†] $(\text{Im}\tau)^{-2}$.

16.1.2 The (p, q) strings

We saw in chapter 11 that we can construct a family of strings as bound states of fundamental strings (denoted $(1, 0)$) and D1-branes or 'D-strings' (denoted $(0, 1)$). It is instructive to construct the supergravity solutions corresponding to these bound states[133]. The metric resembles the Einstein frame version of the D-string metric which we wrote in chapter 10, reproduced here (lying along x_1):

$$ds^2 = H_1^{-3/4}(-dt^2 + dx_1^2) + H_1^{1/4} \sum_{i=2}^{9} dx_i^2,$$

$$e^\Phi = g_s H_1^{-1/2}, \qquad C_{(2)} = g_s^{-1} H_1^{-1} dt \wedge dx_1,$$

$$H_1 = 1 + \left(\frac{r_1}{r} \right)^6, \quad (16.8)$$

---

[*] This form should also be familiar from chapter 2 when we discovered how to write modular invariant partition functions.

[†] As an aside, it is worth noting that this is the simplest non-trivial example of a supergravity model for which we find that the scalars are valued on a coset G/H for some non-compact G and compact H. This example will be embedded in more complicated examples later. For example, we have already seen a five dimensional example at the end of chapter 12, arising from compactifying on T^5 to five dimensions. There the scalars live on the coset $E_{6(6)}/USp(8)$, and there are $78 - 36 = 42$ of them.

where r_1^6 is given in equation (10.36) (where we choose $N = 1$ for a single brane), which is normalised so that the $C_{(2)}$ charge of the D1-brane is $\mu_1 = (2\pi\alpha')^{-1}$. It is possible to use the $SL(2,\mathbb{R})$ transformations to write a more general solution[133], which has an asymptotic value of $C_{(0)}$ which is non-zero as well, which we shall call $c_0 = \theta/2\pi$, giving us an asymptotic coupling $\tau_0 = c_0 + i/g_s$. Such a solution is to be interpreted as being in a different vacuum from the usual case where we just have the string coupling switched on.

Defining the asymptotic value of $\mathcal{M}$ to be $\mathcal{M}_0$, (made out of τ_0 in the obvious way, in view of equation (16.5) we define for the (p,q) case:

$$\Delta_{p,q} = (p \;\; q)\mathcal{M}_0^{-1} \begin{pmatrix} p \\ q \end{pmatrix} = g_s(p - qc_0)^2 + g_s^{-1}q^2, \qquad (16.9)$$

and we get the same form for the metric above, but with

$$H_1 = 1 + \Delta_{p,q} \left(\frac{r_1}{r}\right)^6,$$

$$C_{(2)}^{[i]} = \frac{(\mathcal{M}_0^{-1})_{ij}q^{[j]}}{\Delta_{p,q}^{1/2}}(g_s H_1)^{-1},$$

$$\tau = \frac{pc_0 - q|\tau_0|^2 + ipH_1^{1/2}g_s^{-1}}{p - qc_0 + iqH_1^{1/2}g_s^{-1}}, \qquad (16.10)$$

where $q^{[1]} = p$, $q^{[2]} = q$, $C_{(2)}^{[1]} = B_{(2)}$ and $C_{(2)}^{[2]} = C_{(2)}$. The special case $(1,0)$ is the solution for the fields around the fundamental string[163]. We see from the first line in the above that the tension of the string solution is in fact

$$\tau_{p,q}^1 = \frac{1}{2\pi\alpha'}\sqrt{(p - qc_0)^2 + g_s^{-2}q^2} = \sqrt{([p - qc_0]\tau_{1,0})^2 + (q\tau_{0,1})^2}$$

$$= \frac{1}{2\pi\alpha'}|p - q\tau_0|. \qquad (16.11)$$

Notice that we have reproduced the formula (11.12), but generalised to include non-zero asymptotic $C_{(0)}$, denoted c_0. This is a generalisation to a different vacuum than the previous case. In fact, it is interesting to notice that various values of c_0, g_s give interesting patterns for the lightest string, which determines what we would call the perturbative string spectrum!

In the case $c_0 = 0$, the fundamental string $(1,0)$ is indeed the lightest, for small g_s, as is familiar. Generically, one can always find one such string which is the lightest, for a given value of c_0. This is the dominant string at weak coupling. However, at special values, we can obtain degeneracies. For example, notice that if $|\tau_0| = 1$, we get $\tau_{p,q} = \tau_{q,p}$. Meanwhile $\tau_{p,q} = \tau_{p,p-q}$ if $c_0 = 1/2$ and $g_s^{-2} = 3/4$. Amusingly, at $\tau_0 = e^{\pi i/3}$, all three of the 'simplest' strings are degenerate: $\tau_{1,0} = \tau_{0,1} = \tau_{1,1}$. Also, for $\tau_0 = e^{2\pi i/3}$, which differs from the previous τ_0 by one, we have $\tau_{-1,0} = \tau_{0,-1} = \tau_{-1,-1}$, which are the strings we encountered before in reverse orientation. Geometrically, $\tau = e^{\pi i/3}, e^{2\pi i/3}$ are the special 'orbifold' points of the fundamental region of the $SL(2, \mathbb{Z})$ shown in figure 3.3. This fits rather well with what we already discussed in chapter 11, where we saw that we could form a three string junction, by balancing the tensions of the three types of string. At this point of the moduli space of (p, q) string theories the junction diagram is $\mathbb{Z}_3$ symmetric.

16.1.3 String networks

Recalling the three string junction[135, 137, 140] that we encountered in section 11.4, it must have already occurred to the reader that there is an amusing construction that follows. We can make a network[138] of such string junctions, preserving some supersymmetry. Let us see how this junction must work.

First, note that when three strings meet, with charges (p_i, q_i) for the ith string, the sum of the charges must vanish:

$$\sum_{i=1}^{3} p_i = 0 = \sum_{i=1}^{3} q_i. \tag{16.12}$$

In addition, we must balance the forces exerted by each string, so as to achieve a stable configuration. Let the ith string by oriented along a unit vector $\hat{n}_i$. Then, given that it has tension τ_{p_i,q_i}, the balance condition is:

$$\sum_{i=1}^{3} \tau_{p_i,q_i} \hat{n}_i = 0. \tag{16.13}$$

Now recall that our tension formula is simply

$$\tau_{p,q} = |p + q\tau|.$$

Consider the complex number $p + q\tau$. Its modulus is the tension given above, while its argument shall be denoted $\phi(p, q, \tau)$:

$$p + q\tau = |p + q\tau|e^{i\phi(p,q,\tau)} = \tau_{p,q}e^{i\phi(p,q,\tau)}.$$

Let us now rewrite our force and charge balancing conditions in terms of this. First, the charge conditions (16.12) tell us that

$$\sum_{i=1}^{3} (p_i + iq_i) = 0,$$

and therefore:

$$\sum_{i=1}^{3} \tau_{p_i,q_i} e^{i\phi(p_i,q_i,\tau)} = 0.$$

This is two equations, a real and imaginary part, which we can get to agree with the force balance equation (16.13) if we simply set

$$\hat{n}_i = \begin{pmatrix} \cos\phi(p_i, q_i, \tau) \\ \sin\phi(p_i, q_i, \tau) \end{pmatrix}.$$

What does this mean? Well, our result tells us that we can achieve a completely balanced string network of (p,q) strings if any string with charges (p,q) is oriented at angle $\phi(p_i, q_i, \tau)$ in the plane, i.e. pointing in the direction given by $p + q\tau$. Note that this result does not depend on the location of any string within the network, just its orientation. So we can build a string network of arbitrary size out of (p,q) strings (see figure 16.1).

This solution, and the fact that it preserves eight supercharges, is very interesting, and perhaps suggestive of something remarkable, like a new non-perturbative building block of the type IIB string theory. It is particularly suggestive because it reminds one of a number of diagrams that occur elsewhere in theoretical physics, such as planar diagrams for large N gauge theory, dual triangulations of string world sheets, etc. Speculations of this sort based on pictures alone are of course easy to do, and so it would be interesting to see if there are connections with firmer foundations which might be exploited fruitfully.

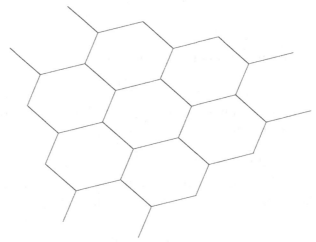

Fig. 16.1. A string network.

16.1.4 The self-duality of D3-branes

As has been remarked upon previously, the four-form potential $C_{(4)}$ is invariant under the $SL(2, \mathbb{Z})$ duality transformation. This must mean something quite remarkable for the D3-brane which couples to it, since the world-volume action of the D3-brane couples to all of the background fields we have been discussing so far that *do* have non-trivial $SL(2, \mathbb{Z})$ duality transformation properties. In Einstein frame, the action is:

$$S = -\tau_3 \int_{\mathcal{M}_4} d^4\xi \, \det^{1/2}[G_{ab} + e^{-\Phi/2}\mathcal{F}_{ab}]$$

$$+ \mu_3 \int_{\mathcal{M}_4} \left(C_{(4)} + C_{(2)} \wedge \mathcal{F} + \frac{1}{2}C_{(0)}\mathcal{F} \wedge \mathcal{F} \right), \tag{16.14}$$

where $\mathcal{F}_{ab} = B_{ab} + 2\pi\alpha' F_{ab}$, and $\mathcal{M}_4$ is the world-volume of the D3-brane, with coordinates $\xi^0, \ldots, \xi^3$. As usual, the parameters μ_3 and τ_3 are the basic R–R charge and tension of the D3-brane:

$$\mu_3 = \tau_3 g_s = (2\pi)^{-3}(\alpha')^{-2}. \tag{16.15}$$

Also, G_{ab} and B_{ab} are the pulls-back of the ten dimensional metric (in Einstein frame) and the NS–NS two-form potential, respectively.

Before we do anything else, let us stop to think about what is going on at low energy, in flat space. Let us also switch off the the background antisymmetric tensor fields. The theory then becomes gauge theory, in fact, the $\mathcal{N} = 4$ supersymmetric four dimensional $SU(N)$ gauge theory (if we have N branes and neglected the overall centre of mass). This theory has a number of special properties. It is supposed to be conformally invariant in the full quantum theory. That it is classically scale invariant is of course trivial. For a start, all of the fields are massless. Furthermore a quick dimensional analysis shows that the coupling g_{YM} has to be dimensionless, and indeed, our formula for it in terms of the closed string coupling sets it to be $g_{YM}^2 = 2\pi g_s$. The theory's θ-angle is set by the R–R scalar $C_{(0)}$. The statement that it is quantum mechanically conformally invariant is highly non-trivial. This means that the β-function vanishes, or that the trace of the full energy-momentum tensor vanishes, etc. This is more involved, and we shall see that this does follow from the properties of D3-branes, in chapter 18, in remarkably interesting ways.

Another property that this theory is supposed to have is exact $SL(2, \mathbb{Z})$ 'S-duality', generalising the following electromagnetic duality which one would expect for the Abelian case:

$$S = -\tau_3 \int d^4x \mathcal{L}$$

$$\mathcal{L} = -\frac{1}{4}e^{-\Phi}F_{\mu\nu}F^{\mu\nu} + \frac{1}{4}C_{(0)}F_{\mu\nu}{}^*F^{\mu\nu}. \tag{16.16}$$

We have the electromagnetic field $\vec{E}$ and the magnetic induction $\vec{B}$ arising from $F_{\mu\nu}$ as $E_i = F_{i0}$ and $B_i = \frac{1}{2}\epsilon_{ijk}F_{jk}$. $F_{\mu\nu}$ satisfies a Bianchi identity $\partial_{[\lambda}F_{\mu\nu]} = 0$. The (source-free) field equations are given in terms of another antisymmetric tensor $\widetilde{F}_{\mu\nu}$ as $\partial_{[\lambda}{}^*\widetilde{F}_{\mu\nu]} = 0$. In the absence of $C_{(0)}$, the theta-angle, this would simply be the $F_{\mu\nu}$ we first thought of, but more generally it is[276]:

$$\widetilde{F}_{\mu\nu} \equiv -2\frac{\delta S}{\delta F_{\mu\nu}} \tag{16.17}$$

and from it we get the electric induction $\vec{D}$ from $D_i = \widetilde{F}_{i0}$ and the magnetic field $\vec{H}$ as $H_i = \frac{1}{2}\epsilon_{ijk}\widetilde{F}_{jk}$. These are related to the previous fields as:

$$\vec{D} = \frac{\partial \mathcal{L}}{\partial \vec{E}} = e^{-\Phi}\vec{E} + C_{(0)}\vec{B}$$

$$\vec{H} = \frac{\partial \mathcal{L}}{\partial \vec{B}} = e^{-\Phi}\vec{B} - C_{(0)}\vec{E}. \tag{16.18}$$

In components, the Bianchi identities and field equations are the familiar ones:

$$\nabla \cdot \vec{B} = 0, \qquad \nabla \times \vec{E} = -\frac{\partial \vec{B}}{\partial t}$$

$$\nabla \cdot \vec{D} = 0, \qquad \nabla \times \vec{H} = \frac{\partial \vec{D}}{\partial t}. \tag{16.19}$$

These 'constituitive relations' may be written in terms of our earlier defined matrix $\mathcal{M}$:

$$\begin{pmatrix} \vec{H} \\ \vec{E} \end{pmatrix} = \mathcal{M} \begin{pmatrix} \vec{B} \\ \vec{D} \end{pmatrix}. \tag{16.20}$$

The $SL(2,\mathbb{Z})$ duality transformations are then easily written as:

$$\begin{pmatrix} \vec{H} \\ \vec{E} \end{pmatrix} \to (\Lambda^T)^{-1}\begin{pmatrix} \vec{H} \\ \vec{E} \end{pmatrix}, \qquad \begin{pmatrix} \vec{B} \\ \vec{D} \end{pmatrix} \to \Lambda \begin{pmatrix} \vec{B} \\ \vec{D} \end{pmatrix}, \tag{16.21}$$

which leave the relations (16.20) invariant, in view of the transformation of $\mathcal{M}$ given in equation (16.7).

Going to the full Born–Infeld Lagrangian, it has been shown (we will not do it here) that the duality still holds. Furthermore, inclusion of the coupling to the two-form potential preserves the $SL(2,\mathbb{Z})$ duality, provided that they transform according to equation (16.3).

Considering two D3-branes gives an $SU(2)$ gauge group, (neglecting the overall $U(1)$) and the S-duality is still supposed to hold, but with the dual theory having the dual $SO(3)$ gauge group. More generally, in this

'Montonen–Olive duality'[277], gauge group G is replaced by a gauge group G^* whose weight lattice is dual to that of G. This is not a subject we shall go into here, although it is a beautiful one[277].

Note, however, that we can translate the expected spectrum of BPS monopoles and dyons in the gauge theory to the case here. Recall from section 13.5 that if the branes separate by some distance L, these are made by stretching the (p, q) strings between them, and ending on the D3-branes' surface, the $SU(2)$ having been broken to a $U(1)$, and the Higgs vev is set by L. Observe that we can surround a string with an S^7. This means that the point at the end of the string can be surrounded by an S^8. Meanwhile, we can locate the D3-brane world-volume as a point in $\mathbb{R}^6$, and so it can be surrounded by an S^5. Finally, to specify the location of the endpoint inside the worldvolume, we can surround it by an S^2. So the source equation for the string in ten dimensions is supplemented by a contribution from the D3-brane action[276]:

$$d^* \widetilde{G}_{(3)}^{[i]} = \mu_1^{[i]} \delta^8(\mathbf{x}) + \sum_\alpha \frac{\delta S^\alpha}{\delta C_{(2)}^{[i]}} \wedge \delta^6(\mathbf{x}), \qquad (16.22)$$

where $C_{(2)}^{[1]} = B_{(2)}$ and $C_{(2)}^{[2]} = C_{(2)}$, the NS–NS and R–R form potentials respectively, and $\mu_1^{[i]}$ are the charge per unit length of the fundamental string and D-string. Also α labels each D3-brane. Here, the Hodge dual is performed in ten dimensions, and so on both sides we have something which can be integrated over S^8 in order to measure the charge. Performing the integral, and observing how the R–R and NS–NS potentials couple in the action (16.14), we have explicitly:

$$0 = \mu_1^{[1]} + \int_{S^2} \mathcal{F}, \qquad 0 = \mu_1^{[2]} + \int_{S^2} {}^* \widetilde{F}. \qquad (16.23)$$

This shows that the charges of the string endpoints are correlated with the spacetime charges of the strings, allowing them to furnish the complete set of (p, q) dyons in the field theory, and the $SL(2, \mathbb{Z})$ strong/weak coupling duality descends correctly to these states as well, and they have masses $m_{p,q} = \tau_{p,q} L$.

16.1.5 (p, q) Fivebranes

In a very similar way to the construction of the supergravity solution for the (p, q) strings, a family of $(p.q)$ fivebranes may be constructed, filling out the expectation that such objects ought to exist in view of ten dimensional string/fivebrane duality, hence sourcing the doublet of two form potentials magnetically. The solution may be written in Einstein frame as:

$$ds^2 = H_5^{1/4}\left(-dt^2 + \sum_{i=1}^{5} dx_i^2\right) + H_5^{-3/4} \sum_{i=6}^{9} dx_i^2,$$

$$H_5 = 1 + \Delta_{p,q} \left(\frac{r_5}{r}\right)^2, \tag{16.24}$$

with $\Delta_{p,q}$ given in equation (16.9), and expressions for $C_{(6)}^{[i]}$ and τ similar to the ones written for the (p, q)-strings in equation (16.10). The of these solutions therefore comes out to be:

$$\tau_{p,q}^5 = \sqrt{([p - qc_{(0)}]\tau_{1,0}^5)^2 + (q\tau_{0,1}^5)^2}, \tag{16.25}$$

the expected analogous equation to the (p, q) string tension (16.11).

16.1.6 SL(2, ℤ) and D7-branes

Let us consider the case of the action (16.6) with all of the higher rank potentials switched off. Furthermore, let us worry only about non-trivial structure in the x_8 and x_9 directions, leaving the directions $t, x_1, \ldots, x_8$ untouched. Let us write a complex coordinate $z = x_8 + ix_9$, in terms of which the action and equations of motion from varying it with respect to $\bar{\tau}$ are:

$$S = \frac{1}{2\kappa^2} \int d^{10}x \sqrt{-G} \left(R - \frac{\partial\tau\bar{\partial}\bar{\tau}}{2(\mathrm{Im}\tau)^2}\right),$$

$$\partial\bar{\partial}\tau + \frac{2\partial\tau\bar{\partial}\tau}{\bar{\tau} - \tau} = 0. \tag{16.26}$$

A simple trial solution to this which preserves half the supersymmetries is to ask that τ is in fact holomorphic: $\partial\tau(z, \bar{z}) = 0$. Now recall that a D7-brane carries the magnetic charge of $C_{(0)}$. Notice further that we have its $C_{(8)}$ charge is $\mu_7 = (2\pi)^{-7}(\alpha')^{-4}$, which happens to match the normalisation of our action, $1/(2\kappa^2)$, and so in circling a single D7-brane once, $C_{(0)}$ should change by precisely 1 in order to register the correct amount of D7-brane charge (recall that we integrate $^*dC_{(8)}$ around the S^1 to measure a D7-brane's charge).

Using this information, a suitable choice for a D7-brane located at $z = 0$ would seem to be:

$$\tau(z) = \frac{1}{2\pi i} \log(z), \tag{16.27}$$

since circling the origin will produce a jump $\tau \to \tau + 1$. This is a good description of the object for a range of distances, but there are problems.

Near the origin, Imτ is becoming large and negative, which cannot make sense, since it should be positive, given that it is the inverse string coupling. So there the solution breaks down, but this is perhaps fine, since we can simply use open string perturbation theory there, in the spirit of previous brane solutions which break down near the origin. We can generalise this trivially to many branes located at points z_i by writing:

$$\tau(z) = \frac{1}{2\pi i} \log(z - z_i). \tag{16.28}$$

Unfortunately, at large z, the solution is not very good either. If there was a four dimensional problem (i.e. with only one other spatial direction) this solution would be a 'cosmic string', and as such, the energy per unit length diverges for this solution, and so we cannot also solve the gravity equations.

Recall however that τ is allowed to jump by an $SL(2,\mathbb{Z})$ transformation. This can be exploited[279], since now τ is not just any number. The inequivalent values of it are restricted to lie in the fundamental domain $\mathcal{F}$ in figure 3.3. So the energy density is now controlled by:

$$\frac{1}{2\kappa^2} \int d^2z \left(\frac{1}{2} \frac{\partial\tau\bar{\partial}\bar{\tau}}{(\tau - \bar{\tau})^2} \right) = \frac{1}{2\kappa^2} \int d^2z \left(\frac{1}{2} \partial\bar{\partial} \log(\tau - \bar{\tau}) \right), \tag{16.29}$$

but we can convert this to an integral over the fundamental domain in the τ plane via:

$$d\tau d\bar{\tau} = dz d\bar{z} \frac{\partial\tau}{\partial z} \frac{\partial\bar{\tau}}{\partial\bar{z}}$$

to give:

$$\frac{1}{2\kappa^2} \int_{\mathcal{F}} d^2\tau \left(\frac{1}{2} \partial\bar{\partial} \log(\tau - \bar{\tau}) \right), \tag{16.30}$$

and we can integrate by parts to perform a boundary integral over the edge of the domain to give $2\pi/12$ for the integral, which is the mass density in units of $1/2\kappa^2$. Actually, we have assumed that we have flat space for the solution. This is not correct, really, since the energy density in the τ field ought to have a non-trivial back reaction on the geometry. Let us attempt to find a solution which looks like the following (inspired by the structure of the case $p = 7$ in equation (10.38)):

$$ds^2 = -dt^2 + \sum_{i=1}^{7} dx_i^2 + H_7(z, \bar{z}) dz d\bar{z}. \tag{16.31}$$

In fact, the equations of motion for the τ field are not modified by this ansatz, since they would have included contributions from the combination $(-G)^{1/2} G^{z\bar{z}}$, which remains unchanged with the above ansatz. The only non-trivial equation which results from this is

$$R_{00} - \frac{1}{2}G_{00}R = -\frac{1}{H_7}\frac{1}{8\tau_2^2}G_{00}\partial\tau h\bar{\partial}\bar{\tau}. \tag{16.32}$$

In fact, this can be written as:

$$\partial\bar{\partial}\log H_7 = \frac{\partial\tau\bar{\partial}\bar{\tau}}{\tau_2^2} = \partial\bar{\partial}\log\tau_2. \tag{16.33}$$

This is just Poisson's equation in two dimensions. The source is $\partial\bar{\partial}\log\tau_2$, and its energy density of $2\pi/12$ is the total charge in the problem. An obvious long distance solution is:

$$\log H_7 = -\frac{1}{12}\log|z|.$$

Looking back at the metric, we see that the z-plane has metric $ds^2 \sim |z^{-1/12}dz|^2$. We can change variables to $\tilde{z} = z^{1-1/12}$, and see that the metric is flat $ds^2 \sim |d\tilde{z}|^2$, but there is a deficit angle of $2\pi/12$, since as we do a complete circle in z, $\tilde{z}$ only goes around part of the way.

It is straightforward to see that if there are N copies of this sort of solution, the result is $\log H_7 = -\frac{N}{12}\log|z|$ and so the metric is $ds^2 \sim |z^{-N/12}dz|^2$. There is a deficit angle of $2\pi N/12$. Let us consider the case of $N = 24$. Well, by a change of variables similar to what we did previously, $\tilde{z} = z^{1-N/12}$, for $N = 24$, we get $\tilde{z} = 1/z$, and then the metric is $ds^2 \sim |d\tilde{z}|^2$, but *the periodicity of z and $\tilde{z}$ are the same*. So there is no conical singularity. We have just built a familiar space, $\mathbb{CP}^1$, or in more familiar terms, S^2, which of course has 'deficit' angle 4π. This is highly suggestive, as we shall see.

Let us try to make an exact solution of the equations of motion (16.33). Actually, to be careful, we should construct a solution to which is manifestly modular invariant. A guess at a solution is obviously $\log H_7 = \tau_2$, but this fails because τ_2 is not modular invariant. Because the operator $\partial\bar{\partial}$ acts, we are free to add anything which is annihilated by this to our guess, in other words, the real part of any holomorphic function. Well, this is where our experience with modular invariance from one-loop string theory in chapter 3 suddenly becomes useful. A nice candidate is in fact to replace τ_2 with $\tau_2\eta^2\bar{\eta}^2$, where η is Dedekind's function, which we met in equation (3.58), since that combination is modular invariant, being a one-loop string partition function. Recall that $q = e^{2\pi i\tau}$. A final requirement is that we must not let the metric function H_7 go to zero. With our present prescription, it goes to zero at a generic point z_i where a seven-brane is located. This is because near there, we have the behaviour given by equation (16.27) and so $q \sim z - z_i$ with the result $H_7 \sim |(z - z_i)^{1/12}|^2$. So, multiplying in the inverse of such a factor for each of the N points, we have finally[279]:

$$H_7 = \tau_2 \eta^2 \bar{\eta}^2 \left| \prod_{i=1}^{N} (z - z_i)^{-1/12} \right|^2 . \tag{16.34}$$

16.1.7 Some algebraic geometry

Let us step back and see what we are doing. We actually are studying a background in which $\tau(z)$ and hence the string coupling varies as we move around the plane transverse to the sevenbrane. We can solve the full equations of motion if we have 24 of the branes present, and the transverse space curls up into an S^2, or $\mathbb{CP}^1$. The function τ varies over the $\mathbb{CP}^1$ and is acted on by $SL(2, \mathbb{Z})$, the physically distinct values being given by the fundamental region $\mathcal{F}$ given in figure 3.3. We can visualise this geometry by thinking of an auxiliary torus T^2 which is fibred over the $\mathbb{CP}^1$, since τ can always be thought of as the modulus of the torus. The torus can change as $\tau \to \tau + 1$ as it circles a sevenbrane. However, as we shrink that circle to a point, maintaining this condition is rather singular, and the result is that a cycle of the torus must degenerate over the point. We have the idea that as we encircle the point, there is a 'monodromy', meaning that everything that can transform under $SL(2, \mathbb{Z})$ gets multiplied by the matrix

$$T = \begin{pmatrix} 1 & 1 \\ 0 & 1 \end{pmatrix}.$$

This happens generically in 24 places, and the physics of it will become much clearer later.

We can describe this all in a rather amusing (and powerful) way, using a small amount of algebraic geometry. Consider three complex coordinates x, y, w. We will identify points as follows: $(x, y, w) \sim (\lambda x, \lambda y, \lambda w)$, for some complex number λ. The resulting four dimensional space is $\mathbb{CP}^2$. This is a generalisation of the more familiar $\mathbb{CP}^1$ which is simply the sphere, as described in insert 16.1.

Starting with our $\mathbb{CP}^2$ coordinates (x, y, w), consider the following homogeneous equation of degree three, giving the 'Weierstrass' form:

$$W(x, y, w) = y^2 w - x^3 - fxw^2 - gw^3 = 0, \tag{16.35}$$

where f and g are constants. Here, homogeneous of degree three means that $W(\lambda x, \lambda y, \lambda w) = \lambda^3 W(x, y, w)$. This equation will give us some one complex dimensional object as a subspace of $\mathbb{CP}^2$. In fact, it is a torus

---

Insert 16.1. S^2 or $\mathbb{CP}^1$ from affine coordinates

As a simple example of the use of affine coordinates to define something familiar, let us look at the sphere, S^2, which in this language is better called $\mathbb{CP}^1$. We start with two complex coordinates, (x, y). Our space of interest has one complex dimension made by identifying $(x, y) \sim (\lambda x, \lambda y)$. Now lets find the space we want. If $y \neq 0$, then we can set y to one by an appropriate choice of λ. Then we have one complex coordinate x, giving a plane. A plane differs from S^2 or $\mathbb{CP}^1$ by the addition of the point at infinity. Indeed, we have this point in the description. It is the case $y = 0$, for which we can set $x = 1$ by the scaling, giving our final point. In other words, we can recover the standard North and South pole preferred projections of the S^2 to a plane seen in elementary geometry: one is the x plane with $y = 1$, and the other is the y plane with $x = 1$.

T^2. This is true for any such cubic in $\mathbb{CP}^2$, and we can see it as follows. A single complex equation in $\mathbb{CP}^2$ gives a one complex dimensional (or Riemann) surface Σ, and so all we need to do is determine its genus, or Euler number, which completely classifies it, as stated in chapter 2. After a change of variables, we can write our equation as $w^3 = x^3 + y^3$. Let us first assume that $x^3 + y^3$ does not vanish. Then our equation yields three generically distinct values of w for each (x, y), which on their own each form a $\mathbb{CP}^1$. So naively, the equation has the Euler number of three $\mathbb{CP}^1$s, which is $3 \times 2 = 6$. But there are three roots of $x^3 + y^3 = 0$, and so the equation requires that $w = 0$ in that situation. These make three points, each of which are represented three times, once on each $\mathbb{CP}^1$. Let us remove the three points from the $\mathbb{CP}^1$s, and hence the Euler number of Σ-{points} is $3(2 - 3) = -3$ and then we must add back in the missing three points, giving a total of zero, the Euler number of a torus.

We can see a torus more directly as follows. Let us first assume that $w \neq 0$, and so we can set it to unity. Then we have $y^2 = x^3 + fx + g$. The solutions for y are double valued, giving two copies of the $\mathbb{CP}^1$ given by x. (We have added the point at infinity in x.) However, there are three places where the cubic vanishes, giving us a place where y is single valued. Together with the point at infinity, this allows us to draw two branch cuts through which to join the two 'branches' of y. We connect the two $\mathbb{CP}^1$s through two separate cuts forming tubes which construct for us a torus. See figure 16.2.

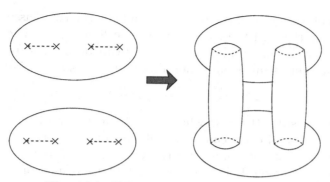

Fig. 16.2. Why a cubic gives a torus.

Let stay with the $w = 1$ or 'affine' form for a while. This form keeps us in the picture where x forms a plane over which y takes its values; y is double valued everywhere except where the cubic $x^3 + fx + g = 0$ has roots. It is an elementary fact that the nature of the roots of this cubic is determined by the discriminant which is proportional to $\Delta = 4f^3 + 27g^2$. We have three situations,

- $\Delta > 0$ There is one real root and a pair of complex ones.

- $\Delta = 0$ All of the roots are real, and at least two are equal.

- $\Delta < 0$ There are three distinct real roots.

We sketch these cases in various ways in figure 16.3 for (y, x) real.

In the case where the roots are distinct ($\Delta \neq 0$), we can make a torus as described above and depicted in figure 16.2. We can see how the generic

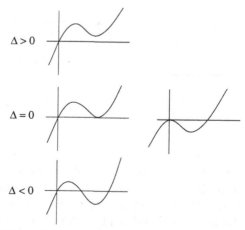

Fig. 16.3. Real cubics and their roots.

two classes of one-cycle of the torus are made by the two classes of journey one can make through the cuts, as shown in figure 16.4. There, we have also noted that a shift and a scaling on x can be used to put a root at zero and another at unity, and then the final root is at λ, giving the form

$$y^2 = x(x-1)(x-\lambda).$$

However, consider the case when $\Delta = 0$ and two roots coincide. Then one or other class of cycle can pinch off, causing the torus to degenerate. One may ask what the complex structure τ of a torus presented in the form (16.35) might be. It is given by the famous j-function:

$$j(\tau) \equiv \frac{\left(\theta_2^8(\tau) + \theta_3^8(\tau) + \theta_3^8(\tau)\right)^3}{\eta^{24}(\tau)} = \frac{4(24f)^3}{4f^3 + 27g^2}. \tag{16.36}$$

The function $j(z)$ is a very special one. It is a modular invariant complex number, and is in fact a one-to-one map of the fundamental region $\mathcal{F}$ to the complex plane. Since the denominator is the discriminant, we see that when the torus degenerates ($\Delta = 0$), $j(\tau)$ diverges.

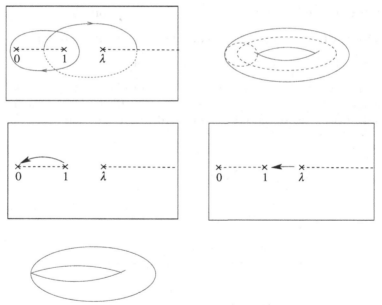

Fig. 16.4. The top sketches show one sheet of the cut complex x plane and the generic torus made from it, including the two classes of one-cycle (cf. figure 16.2). (Note that the dotted half of one of the curves is in fact on the other sheet.) The bottom sketches show how the torus can degenerate if roots collide, giving $\Delta = 0$ (cf. figure 16.3).

16.1.8 F-theory, and a dual heterotic description

Let us return to our problem of describing seven branes. The degeneration of a torus is exactly what happens when we are located at a seven brane, since if encircling one of charge 1 produces the jump $\tau \to \tau + 1$, then shrinking the loop to a point shows that the torus associated to that point must be degenerate.

We saw that we had a sensible solution of the equations of motion if we have generically 24 sevenbranes located on a sphere. The coupling τ can be allowed to vary as we move around the sphere, with coordinate z, between the sevenbranes. We can then associate a torus with every value of $\tau(z)$, thus making a *fibred structure*[199] of T^2 over $\mathbb{CP}^1$. At the location of a sevenbrane, we must have the torus degenerate, which is a statement that our fibration has 24 places where the torus fibre degenerate (see figure 16.5). We can describe this using the language above by allowing the numbers f, g become functions $f(z), g(z)$. Then we have that $\Delta(z) = 4f^3(z) + 27g^2(z)$ must vanish in 24 places. We can achieve this by making $f(z)$ an eighth order polynomial in z and $g(z)$ a twelfth order polynomial, and so we have:

$$W(x, y, z) = y^2 - x^3 - f(z)x - g(z) = 0. \tag{16.37}$$

Now observe that there are nine coefficients to specify $f(z)$ and thirteen for g. Four of these are parameters are redundant, however. For the first, scale $f \to \lambda^2 f, g \to \lambda^3 g$ which gives no change of the torus, as is evident from equation (16.36). For the other three, recall from chapter 3 that there is an $SL(2, \mathbb{C})$ action on the $\mathbb{CP}^1$ of z that allows up to three points to be placed at positions of one's choice (typically $z = 0, 1, \infty$). So there are 18 complex parameters which go into this solution.

Mathematically, this all fits the fact that the moduli space of K3 manifolds which can be written as an *'elliptic'* (i.e. torus) fibration is 18 complex dimensional, with a local description as:

$$\mathcal{M}_{\text{K3elliptic}} = \frac{O(18, 2)}{O(18) \times O(2)}. \tag{16.38}$$

Our fibration of T^2 over $\mathbb{CP}^1$ builds our friend the K3 manifold for us (see figure 16.5).

Furthermore, the reader might recognise this local structure from section 7.4. It is the local description of the moduli space of the heterotic string compactified on T^2. Let's check the counting. We get two complex parameters from the internal components of the graviton and the antisymmetric tensor: G_{ij} is symmetric and B_{ij} is antisymmetric, and $i, j = 8, 9$. Also, the rank 16 gauge group ($SO(32)$ or $E_8 \times E_8$) can have 16 Wilson

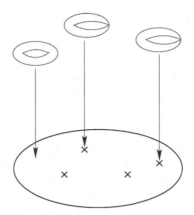

Fig. 16.5. The F-theory description of 24 sevenbranes (located at the crosses) as a torus fibration of T^2 over $\mathbb{CP}^1$. In fact, this is a description of K3 as an elliptic fibration!

lines on each direction of the torus. This gives 18 complex moduli in total, with generic gauge group $U(1)^{18} \times U(1)^2$. The extra $U(1)^2 \times U(1)^2$ supplementing the generic $U(1)^{16}$ gauge group from the current algebra sector, comes from the internal components $G_{\mu i}$, $B_{\mu i}$.

There is one more important parameter we ought to consider, the heterotic string coupling. This is identified with the size of the $\mathbb{CP}^1$ base of the fibration, which we are free to specify in making the elliptic fibration. We shall see this explicitly later. The other parameters we have naively available to us on the IIB side are not accessible. We cannot switch on either of the two-form fields since they transform under the $SL(2,\mathbb{Z})$. Furthermore, the torus fibres only have complex structure parameters; we should not think of them as tori whose Kähler structure (i.e. their size) can vary. By construction, only τ has physical meaning, at least in this type IIB picture.

The fact that the size of the $\mathbb{CP}^1$ is essentially the heterotic string coupling fits nicely with the expectation that the limit where we have a very small sphere over which the IIB coupling is varying greatly (due to the presence of 24 branes) would benefit from a weakly coupled dual string theory description.

16.1.9 (p, q) Sevenbranes

So far this duality is motivated by plausibility arguments. It would be nice to demonstrate this duality more in detail, and happily we have the tools to do it. The first thing to note is that we have 24 seven branes, but the duality to the heterotic string suggests that we only have $U(1)^{18} \times U(1)^2$ as the generic gauge group. Now a $U(1)^2$ of this (on this type IIB side) comes from internal components of the metric, $G_{\mu i}$, $(i = 8, 9)$ leaving a prediction that somehow, as many as six sevenbranes are not able to contribute. We can resolve this as follows. The description of $U(1)$s on the world-volume of D-branes is in terms of fundamental strings, or, more specifically, $(1, 0)$ strings, using the description of section 16.1.2. Correspondingly, since $\tau \to \tau + 1$ as we encircle one, the monodromy matrix about the sevenbrane is

$$T = \begin{pmatrix} 1 & 1 \\ 0 & 1 \end{pmatrix},$$

which leaves these string charges invariant. Clearly, we have the useful idea of a (p, q) sevenbrane[199, 200], which is a sevenbrane on which a (p, q) string can end. What is the monodromy about such a brane? Well, let us imagine that we transform from $(1, 0)$ string to a (p, q) string using an $SL(2, \mathbb{Z})$ matrix $M_{(p,q)}$:

$$M \begin{pmatrix} 1 \\ 0 \end{pmatrix} = \begin{pmatrix} p \\ q \end{pmatrix}.$$

Then the monodromy is derived by simply conjugating the problem, as follows:

$$T \begin{pmatrix} 1 \\ 0 \end{pmatrix} = \begin{pmatrix} 1 \\ 0 \end{pmatrix} \longrightarrow TM^{-1}M \begin{pmatrix} 1 \\ 0 \end{pmatrix} = \begin{pmatrix} 1 \\ 0 \end{pmatrix}$$

$$TM^{-1} \begin{pmatrix} p \\ q \end{pmatrix} = M^{-1} \begin{pmatrix} p \\ q \end{pmatrix} \longrightarrow MTM^{-1} \begin{pmatrix} p \\ q \end{pmatrix} = \begin{pmatrix} p \\ q \end{pmatrix}$$

$$\Rightarrow M_{(p,q)} = MTM^{-1} = \begin{pmatrix} 1 - pq & p^2 \\ -q^2 & 1 + pq \end{pmatrix}. \tag{16.39}$$

This is illustrated in figure 16.6. Now the condition that two sevenbranes can both be treated in perturbation theory at the same time is if their monodromy matrices commute. In other words, if they are (p_1, q_1) and

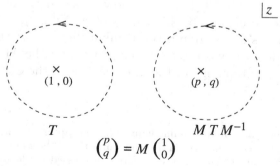

Fig. 16.6. The monodromy around a (p, q) sevenbrane, on which a (p, q) string can end, and its relation, by conjugation, to the $(1, 0)$ case.

(p_2, q_2), then $p_1 q_2 - p_2 q_1 = 0$. Branes which satisfy this condition are said to be 'mutually local'. Furthermore, the total monodromy around all of the points in the $\mathbb{CP}^1$ must be the identity $\prod_{i=1}^{24} M_{(p_i, q_i)} = 1$. This means that all of the seven-branes definitely cannot be of type $(1, 0)$, since $T^{24} \neq 1$. The slightly weaker locality condition allows a maximum of 18 mutually local branes, and hence $U(1)^{18}$ as the generic gauge group from the sevenbranes.

16.1.10 Enhanced gauge symmetry and singularities of K3

There is even more structure to the theory than that which we have already uncovered, since as we might expect from previous examples, there are enhanced gauge symmetries. The $U(1)^{18}$ can be enhanced to any of an A–D–E family of gauge groups of the same rank, of which the A-series is most obvious. We can tune parameters such that n of the branes are coincident, giving $U(n)$ as the gauge group. Actually, it is prudent to cast this into the language of the K3 geometry. Asking that n branes coincide is equivalent to asking that n of the basic singularities that can occur in the fibre coincide. What really happens is that the singularity becomes of a stronger type, measured by n.

In fact, we already know the description from chapter 13. We should think of the whole of the K3 as developing a singularity, and not just the fibre. We have already encountered the A–D–E singularities of K3 before, and it is instructive to observe how they are to be found in this elliptic description. In the purely brane description, an enhanced gauge symmetry arises because a fundamental string stretched between the branes becomes of zero length and hence there are extra massless sectors. The origin of this string in the F-theory description is as a the base of a $\mathbb{CP}^1$ fibred over the line which is the string. This $\mathbb{CP}^1$ shrinks to zero size when the seven branes coincide. See figure 16.7.

This is precisely the same description of the ALE singularity which we encountered in chapter 13. It is easy now to see how the other A–D–E singularities are described. It is in terms of n $\mathbb{CP}^1$s, c_i, with a set of intersection numbers $c_i \cdot c_j$ giving the Dynkin diagram of the appropriate group. The reader may wish to turn to insert 4.3 for the ADE Dynkin diagrams, showing the topology of the intersections of the $\mathbb{CP}^1$s (represented by the circles).[‡]

[‡] Alternatively, the reader may examine figure 13.2 in chapter 13, where we established the connection between the Dynkin diagrams and the $\mathbb{CP}^1$s underlying an ALE singularity, but they must remember to delete the crossed circle to get the Dynkin diagrams.

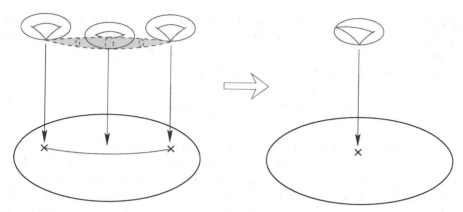

Fig. 16.7. When branes collide: A fundamental string stretching between them goes to zero length when they become coincident. This lifts to a $\mathbb{CP}^1$ which stretches between the locations of the two sevenbranes, which shrinks to zero size when the sevenbranes coincide. The resulting fibre is more singular.

Let us now turn to a few special points in the moduli space of this K3 description, where we will uncover some of this in detail in a more familiar setting.

16.1.11 F-theory at constant coupling

The main facility of the F-theory description is that it provides an economical geometrical way of describing the physics of type IIB vacua with sevenbranes together with varying coupling $\tau = C_{(0)} + ie^{-\Phi}$. This goes beyond our powerful but still only perturbative description of sevenbrane vacua. When we have multiple sevenbranes in the perturbative description, we must cancel the sevenbrane charge locally (using orientifolds) so as not to source any varying coupling away from the branes which would take us outside of perturbation theory.

Nevertheless, in understanding the statements of the previous few subsections better (especially the appearance of the heterotic string!), we ought to try to make contact with the perturbative type II description. What we need to do is find a limit where the torus fibration has all of its structure trapped a few points, between which τ is a constant[197]. There are a number of ways of doing this, as can be seen by looking at the expression (16.36) for the j-function. There, we see that we have two obvious choices, either $g(z) = 0$ or $f(z) = 0$. In the first case, we have only five moduli left to describe this possibility, and we see that $j = 24^3 = 13\ 824$ for which $\tau = i$. This is one of the very special points in the moduli space of tori, as we have already seen. The second choice gives us the other special point. There we have only nine moduli, and we see that $j = 0$, which is indeed $\tau = e^{2\pi i/3}$, the orbifold point of $\mathcal{F}$.

Returning to the first case, our K3 is given by

$$y^2 = x^3 + f(z)x, \qquad f(z) = \prod_{i=1}^{8}(z - z_i), \qquad (16.40)$$

and we have generically eight zeros, z_i, giving the discriminant $\Delta = 4\prod_{i=1}^{8}(z - z_i)^3$. The 24 branes must have split into eight groups of three sevenbranes. Recalling that a basic sevenbrane in this description has deficit angle $\pi/6$, we uncover that there is a deficit of $\pi/2$ at each of the eight points[197].

There is a way of splitting the eight points up differently. We can use up all of our remaining five moduli to have singularities at only three points, two of order three and one of order one

$$\Delta = 4(z - z_1)^6(z - z_2)^9(z - z_3)^9.$$

This gives deficit angles $3\pi/2$, $3\pi/2$ and π. These values for the deficit angles can be described as orbifold fixed points, since a $\mathbb{Z}_N$ orbifold has deficit angle $2\pi(N - 1)/N$. The first two points are therefore $\mathbb{Z}_4$ fixed points, while the last is fixed under a $\mathbb{Z}_2$. We have seen this description before in chapter 7. We really have $T^2/\mathbb{Z}_4$. Let us see what has happened to the constant fibre, by studying the monodromy around its base points. We have $f = (z - z_1)^2(z - z_2)^3(z - z_3)^3$. Looking at a $\mathbb{Z}_4$ fixed point (at $z = z_2$ or z_3) as we encircle it once $z \to e^{2\pi i}z$, we see that $f \to e^{6\pi i}f$. Looking at the form of the defining cubic in equation (16.40), we see that the K3 remains invariant if we also send

$$x \to e^{3\pi i}x = -x \quad y \to e^{\frac{9\pi i}{2}}y = iy.$$

So we see that the fibre has a $\mathbb{Z}_4$ orbifold action on it as well, and is therefore $T^2/\mathbb{Z}_4$. In fact there is another simple description of this same fact. The case $\tau = i$ is the unique point which is invariant under $S : \tau \to -1/\tau$, where

$$S = \begin{pmatrix} 0 & -1 \\ 1 & 0 \end{pmatrix},$$

is the standard $SL(2, \mathbb{Z})$ representation. The element S is of order four, and so the case $\tau = i$ is the situation of a square torus with a $\mathbb{Z}_4$ symmetry. Looking above the $\mathbb{Z}_2$ point (at z_1), we have

$$f \to e^{4\pi i}f, \quad x \to e^{2\pi i}x = x, \quad y \to e^{3\pi i}y = -y,$$

and so we have a $\mathbb{Z}_2$ symmetry, generated by $S^2 = -1$.

We observe that our torus fibre is also a $T^2/\mathbb{Z}_4$, with the correct correlation of its order four and order two points with the order four and order two points in the base, and so we discover that our K3 is in fact $T^4/\mathbb{Z}_4$, an orbifold description which we encountered previously in section 7.6.5. It should now be easy to anticipate what happens for branch two, using the knowledge we developed about K3's orbifold limits in section 7.6.5, or about the other special point of $\mathcal{F}$, the moduli space of the torus T^2 in insert 3.3. With $f = 0$, let us write

$$y^2 = x^3 + g(z), \quad g(z) = \prod_{i=1}^{12}(z - z_i), \tag{16.41}$$

and so we have we have generically twelve zeros, z_i, with $\Delta = 27\prod_{i=1}^{12}(z - z_i)^2$. The 24 branes are grouped into 12 pairs, with deficit angle $\pi/3$. Again, we cannot write this as an orbifold in general, but if we use up all of our moduli we can place them at three points z_1, z_2, z_3 in two distinct ways:

$$\Delta = 27(z - z_1)^6(z - z_2)^8(z - z_3)^{10},$$

or

$$\Delta = 27(z - z_1)^8(z - z_2)^8(z - z_3)^8.$$

The first way has gives a $\mathbb{Z}_2$ fixed point again, accompanied by a $\mathbb{Z}_3$ and a $\mathbb{Z}_6$. These are of course the fixed points of $T^2/\mathbb{Z}_6$. The second grouping has three $\mathbb{Z}_3$ points, which are the fixed points of $T^2/\mathbb{Z}_3$. The monodromy around a $\mathbb{Z}_6$ point in the first case gives a K3 invariant under

$$g \to e^{10\pi i}g, \quad x \to e^{\frac{10\pi i}{3}}x = e^{\frac{4\pi i}{3}}x, \quad y \to e^{5\pi i}y = -y,$$

which is again a $\mathbb{Z}_6$ action. Once again, we can also deduce this from that fact that the torus $\tau = e^{2\pi i/3}$ is the special point invariant under

$$ST = \begin{pmatrix} 0 & -1 \\ 1 & 0 \end{pmatrix}\begin{pmatrix} 1 & 1 \\ 0 & 1 \end{pmatrix} = \begin{pmatrix} 0 & -1 \\ 1 & 1 \end{pmatrix}, \tag{16.42}$$

which is of order six, $(ST)^6 = 1$. Above the $\mathbb{Z}_3$ point we get

$$g \to e^{8\pi i}g, \quad x \to e^{\frac{8\pi i}{3}}x = e^{\frac{2\pi i}{3}}x, \quad y \to e^{4\pi i}y = y, \tag{16.43}$$

which is a $\mathbb{Z}_3$ action, generated by $(ST)^2$. Lastly, over the $\mathbb{Z}_2$ point, we have

$$g \to e^{6\pi i}g, \quad x \to e^{\frac{6\pi i}{3}}x = x, \quad y \to e^{3\pi i}y = -y, \tag{16.44}$$

which is a $\mathbb{Z}_2$ action, generated by $(ST)^3 = S^2 = -1$. All of this information is simply the expression of the fact that K3 is now in its $T^4/\mathbb{Z}_6$ orbifold limit.

For the other grouping, things are even simpler, as all of the points are the same[198]. The monodromy around any of them gives that which we saw in equation (16.43), a $\mathbb{Z}_3$ action, showing that this limit represents K3 in its $T^4/\mathbb{Z}_3$ orbifold limit.

The missing orbifold is of course $T^4/\mathbb{Z}_2$. This is achieved by the symmetric choice of placing equal groups of branes at each of four orbifold points in the base, giving $T^4/\mathbb{Z}_2$ since in that case each singularity has deficit angle π. Slightly more generically, this can be achieved by asking that $f^3 = \alpha g^2$, for some parameter α. This does not fix τ's constant value, as should be clear from the j-function in equation (16.36). This is extremely useful, since we are then free to take the type IIB string coupling all the way to zero to achieve our goals of making contact with weakly coupled descriptions. This gives us:

$$\Delta = (4\alpha^3 + 27) \prod_{i=1}^{4} (z - z_i)^6.$$

The monodromy around one of these points is $\mathbb{Z}_2$, which is generated by $S^2 = -1$, as is clear from

$$g \to e^{6\pi i} g, \quad f \to e^{4\pi i} f, \quad x \to e^{\frac{6\pi i}{3}} x = x, \quad y \to e^{3\pi i} y = -y. \quad (16.45)$$

The next matter to consider is the precise way of identifying the A–D–E singularity which a fibre can develop over a point. This is a matter requiring some mathematical care and sophistication, and so as not to stray too far afield, we will not embark on such a discussion. We will simply note that this has been classified by Kodaira[183] in terms of the order, as polynomials in z, of the quantities $(f(z), g(z), \Delta(z))$ that we have been working with. Table 16.1 lists all of the types of singularity and the enhanced gauge symmetry they give[200].

Looking at table 16.1, we immediately see that the gauge groups associated to the special orbifold limits we have studied are given in table 16.1. There are a number of interesting general features of this result. The most obvious is the fact that we get exceptional gauge groups in the latter three cases. We have encountered no way of achieving this using perturbative D-branes up to now, and this remains the case. As we have already noted, although the coupling is constant in the last three models, it is not weak, and so the branes are not perturbative D-branes.

In the $\mathbb{Z}_2$ case however, we have something different[197]. We can achieve the required gauge group at weak coupling, and happily we have the freedom (by choice of α) to make the constant string coupling any value we

Table 16.1. *Kodaira's classification of the A–D–E singularities of K3 that can occur in the Weierstrass parametrisation given in equation (16.37)*

order(f)	order(g)	order(Δ)	fibre type	singularity
≥ 0	≥ 0	0	smooth	none
0	0	n	I_n	A_{n-1}
≥ 1	1	2	II	none
1	≥ 2	3	III	A_1
≥ 2	2	4	IV	A_2
2	≥ 3	$n+6$	I_n^*	D_{n+4}
≥ 2	3	$n+6$	I_n^*	D_{n+4}
≥ 3	4	8	IV^*	E_6
3	≥ 5	9	III^*	E_7
≥ 4	5	10	II^*	E_8

like. Choosing that the string coupling is zero (i.e. $\tau \to i\infty$) implies that we have completely cancelled the sevenbrane charge locally at each of the four points. In a perturbative description, this is achieved by using an O7-plane in the neighbourhood of an appropriate amount of D7-branes. Looking back to our computations of chapter 7, we see that the O7-plane charge is -4 in units where the D7-brane charge is 1. So we need to have four D7-branes and one O7-plane for charge cancellation. Actually, we also know precisely what gauge group this would give. It is in fact $SO(8)$. This is remarkably similar to have we have in the first line of table 16.1. There are four groups of six coincident sevenbranes. If we associate four of them with ordinary D7-branes, then two of them correspond to the orientifold sitting at the $\mathbb{Z}_2$ orbifold fixed point. We have arrived at the $T^4/\mathbb{Z}_2$ orientifold of type IIB, where a $(-1)^{F_L}\Omega$ also acts internally. From our experience with T-duality of simple orientifolds (see, for example, chapter 8), we see that this is simply T-dual to the $SO(32)$ type I string theory compactified on T^2. Accordingly, the orientifold (O9-plane) of charge -16 (in D9-brane units) splits into $2^2 = 4$ O7-planes of

Table 16.2. *The results for the gauge groups in the various constant coupling F-theory K3 orbifold limits*

K3 orbifold	Gauge group
$T^4/\mathbb{Z}_2$	$SO(8)^4$
$T^4/\mathbb{Z}_3$	E_6^3
$T^4/\mathbb{Z}_4$	$E_7^2 \times SO(8)$
$T^4/\mathbb{Z}_6$	$E_8 \times E_6 \times SO(8)$

charge -4. In order to achieve local charge cancellation, the 16 D7-branes are moved into four groups of four to sit at the O7-planes.

So we have obtained the weakly coupled description we sought. Furthermore, using the result of chapter 12 that the $SO(32)$ type I string is strong/weak coupling dual to the heterotic string, we also have the bonus of proving that we have a duality to the heterotic string on T^2. Deforming away from this special point using the moduli establishes the duality at all points on the moduli space.

Incidentally, in the spirit of the discussions in chapter 12, we can even see what the 'dual' heterotic string is in this picture. In ten dimensional type I, it would have been the D1-brane. We have T-dualised on a T^2, however, and so we see that the dual string becomes a D3-brane wrapped on the T^2. It is a useful exercise to check that the resulting heterotic string's coupling is set by the area of the torus. Tuning moduli to return to the general non-orbifold situation, we see that the dual heterotic string is a D3-brane wrapped on the $\mathbb{CP}^1$. The seven dimensional heterotic string coupling is set by the size of the $\mathbb{CP}^1$ in general.

16.1.12 The moduli space of $\mathcal{N} = 2$ $SU(N)$ with $N_f = 4$

Let us continue to focus on one of the four singular points for a while longer, placing everything at the origin $z = 0$. At weak coupling, we have seen that the branes carry an $SO(8)$ gauge symmetry and that the perturbative description is as four D7-branes and an orientifold O7-plane. Let us place a D3-brane probe into this background, oriented so that it is living in, say, the x_1, x_2, x_3 directions. This breaks half of the supersymmetries, leaving a total of eight supercharges. Observe further that when the D3-brane is located at the orientifold, the gauge theory on its world-volume is in fact $SU(2)$, since this situation is T_{89}-dual to a D5-brane in type I string theory, as we have seen. Because we have T-dualised, however, the D3-brane can move off the orientifold, and then the gauge group is $U(1)$. We can move the D7-branes to positions (z_1, z_2, z_3, z_4), which breaks the $SO(8)$ to $U(1)^4$ generically. There can be enhanced symmetry points to $U(n)$ if n of the D7-branes come together away from the O7-plane, and $SO(2n)$ if the coincide at the O7-plane.

What we have arrived at is the weakly coupled description of the Coulomb branch of the moduli space of $\mathcal{N} = 2$ four dimensional $SU(2)$ gauge theory with four flavours of quark in the fundamental. The latter come from the strings stretching between the D7-branes and the D3-branes. Their classical masses are given by the positions z_i. Moving the D3-brane from the origin is the process of giving a vacuum expectation value (vev) to the complex adjoint scalar in the $\mathcal{N} = 2$ vector multiplet, and the z-plane is the space of gauge inequivalent values of this vev. The origin remains as the naive classical $SU(2)$ gauge symmetry restoration, and the gauge groups associated to the D7-branes are *global* flavour symmetries in the D3-brane world-volume.

It is amusing that we have obtained this rich and beautiful theory as a piece of the F-theory background seen by probing with the D3-brane, and we can learn much about each from this. The first thing we can learn (assuming we did not now it before) is the gauge theory's β-function, encoded in the one-loop running of the gauge coupling. We can read this out from the weak coupling behaviour of the gauge coupling. Placing the orientifold at the origin, and the four D7-branes at positions we have:

$$\tau(z) = \tau_0 + \frac{1}{2\pi i} \left[\sum_{i=1}^{4} \ln(z - z_i) - 4 \ln z \right],$$

and use the fact that $\tau(z) = C_{(0)} + ie^{-\Phi}$. Remember also that $g_s(z) = e^{\Phi}(z)$ and that the Yang–Mills coupling and θ-angle are related to the string theory parameters by $g_{YM}^2 = 2\pi g_s$ and $\theta = 2\pi C_{(0)}$. The β-function for the pure glue is negative with respect to the contribution from the quarks. The quark masses are set by the positions z_i, since those positions set the length of the 3–7 strings. Notice that when all the $z_i = 0$, and we are at the $SU(2)$ point at the origin of moduli space, then we get no running of the coupling and $\tau = \tau_0$, the tree level value. This fits with the fact that the case of $N_f = 2N_c$ has vanishing β-function, and is in fact conformally invariant. We can also take the opposite limit, and send some of the z_i to infinity, thus reducing the number of quarks, all the way down to the case of pure glue, if we wish.

As we have seen before, we cannot trust the above one-loop expression near $z = \{0, z_i\}$, since the logarithm takes the expression large and negative, which is not acceptable behaviour for the gauge coupling. Of course, this is because we have neglected the instanton contribution, which produce non-perturbative effects which remove this singular behaviour. The beautiful results[240] of Seiberg and Witten address precisely this point, with the result that there is a complete solution of the problem in terms of the geometry of an auxiliary torus. The torus encodes the physics of the Coulomb branch, including the spectrum of masses of (p, q) dyons. The torus is singular over six points, four of them (the z_i) are the places where the quarks becomes effectively massless. The other two points originate from the single $SU(2)$ point at the origin: it has split (since instanton effects switch on to maintain positivity of the gauge coupling or, equivalently, the moduli space metric[240] and they are separated by a distance of order $e^{i\tau_0\pi}$, and they represent the places where $(0,1)$ monopoles and $(1,-1)$ dyons become massless.

From the point of view of the D-brane picture, it is extremely natural that an auxiliary torus appears in the description of the non-perturbative physics, as this is the torus of the underlying F-theory description. So what we learn is that the orientifold O7-plane splits into two seven-branes, of type $(0,1)$ and $(1,-1)$, beyond weak coupling, physics which is isomorphic to the removal of the gauge theory $SU(2)$ point by instanton effects[240]. We have seen that the full F-theory description, which allows the $SL(2,\mathbb{Z})$ behaviour of τ to come into play and keep it manifestly positive, maps to the same solution of the problem for the coupling in the gauge theory.

16.2 M-theory origins of F-theory

It is natural to wonder whether the appearance of the torus of F-theory is a sign of hidden twelve dimensional dynamics for which we should seek, in the spirit of the search for M-theory based on eleven dimensional dynamics seen by all of the branes of type IIA. A more conservative point of view is that the torus is merely a powerful bookkeeping device, and the type IIB theory is no more or less ten dimensional than it was before the advent of F-theory. This is perhaps supported in part by the fact that the only information about the torus which has physical meaning is its complex structure modulus τ. The Kähler structure, containing information about its size, is nowhere to be seen in the formulation. So the putative twelve dimensional dynamics would at best be purely (loosely speaking) topological, it would appear.

The spirit of string theory's history of advances is that one must keep one's mind and eyes open for new directions and often unexpected and fruitful changes of point of view. This is probably because we do not really know yet what the theory really is. So as long as a firm unambiguous computational advantage is obtained in exchange, most practitioners simply do not seem to care what explanatory words or terminology arises to decorate the new tools once they are found. It may well be that a formulation using dynamics in twelve dimensions does arise one day, and if it describes key pieces of physics in a manner more economical than current techniques, then it deserves a place alongside other important pieces of the puzzle of describing fundamental physics.

So no firm declaration is to be found in these pages concerning the twelve dimensional dynamical origins of F-theory. Instead, it is worth noting that there are also signs that many of the key pieces of F-theory – particularly the origin of the torus – can be seen directly to have more humble origins: It is simply a limit of the eleven dimensional picture of M-theory[133, 134].

Let us return to the duality between eleven dimensional supergravity on a circle of radius R_{10} and type IIA string theory. The type IIA string coupling is related to the circle radius by: $R_{10} = (g_s^A)^{2/3}\ell_p = g_s^A \ell_s$, since $\ell_p = (g_s^A)^{1/3}\ell_s$, recalling formulae from chapter 12. Once the circle is small enough, we are able to work with weakly coupled ten dimensional physics of the type IIA string to a good approximation. As we have described before, the D4-brane, the D2-brane, and the NS5-brane of type IIA arise from the M-branes reduced or wrapped on the circle, the D0-brane is a Kaluza–Klein momentum, and the D6-brane is a Kaluza–Klein monopole.

We can continue to compactify on another circle, this time of radius R_9, and shrink that one away as well. We know that this has a dual description in terms of the type IIB string theory, where now the theory is compactified on a circle of radius $R_9' = \ell_s^2/R_9$, and, crucially from equation (5.1), the type IIB string coupling is $g_s^B = g_s^A \ell_s/R_9$. We can go ahead and shrink away the second circle entirely as well, and use the ten dimensional type IIB description, which has no direct reference to the two circles we started with. However, we see that the type IIB string coupling can be expressed entirely in terms of the size of the two circles:

$$g_s^B = \frac{R_{10}}{R_9}. \tag{16.46}$$

So in fact, given the existence of M-theory, the type IIB string coupling can be interpreted entirely in terms of the ratio of the radii of two circles. These two circles make a torus, since they define a lattice upon which we can make an identification. Since equation (16.46) only refers to the ratio of the radii of the circles, we can rescale and write the lattice as of unit length in one direction (associated to x_{10}), and of length $1/g_s^B$ in the other (associated to x_9). See figure 16.8. Before making the identification on the lattice however, we are free to make a shift in the x_{10} direction before identifying to construct the torus. Different non-integer shifts give non-equivalent tori, while a shift by an integer gives the same torus. See figure 16.9. This shift is to be identified with the R–R periodic scalar $C_{(0)}$, a natural identification since it is correlated, by tracing backwards, with a familiar structure in the tenth direction. It is T-dual to the type IIA R–R potential $C_{(1)}$, which in turn is conjugate to momentum in the periodic direction x_{10} and so is directly related to a periodic shift.

What we have just described is our F-theory torus of the previous subsections, with complex structure $\tau = C_{(0)} + ie^{-\Phi}$. Notice that the fact that it seems to have no physical size is natural from this description. We arrived at it by sending the each circle to zero size, and so only the

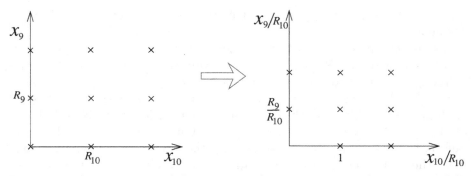

Fig. 16.8. The geometry of the compactification torus used to get type IIB string theory from M-theory.

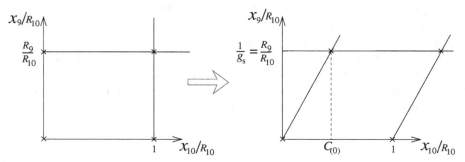

Fig. 16.9. Generalising the compactification lattice by including a shift. This is how the F-theory or type IIB theory torus arises from M-theory.

ratio of the circles has physical meaning in the resulting type IIB theory. Moreover, it is clear that the type IIB theory obtains its $SL(2,\mathbb{Z})$ structure in this way, and that it is truly and manifestly non-perturbative, given the construction.

So we see that at least locally, we can attribute the F-theory torus to the result of shrinking a physical torus in M-theory [133, 134]. Consequently, we should be able to make sense of, directly in M-theory, more complicated structures with varying type IIB couplings, like various branes, and even complete F-theory vacua.

16.2.1 M-branes and odd D-branes

The route of the previous subsection is just what we need to show the M-theory origin of type IIB's odd Dp-branes and NS5-brane. Of course, it is directly deducible from T-duality to the type IIA branes, but it is useful to recast things in terms of the M-theory reduction on the torus, following the steps above.

Imagine that we started in M-theory with an M2-brane, with one direction extended in x_{10}. It has tension $\tau_2^M = (2\pi)^{-2}\ell_p^{-3}$. Upon reduction, this becomes the type IIA string, with the correct tension $\tau_F \equiv \tau_{1,0} = \tau_2^M 2\pi R_{10} = (2\pi)^{-1}\ell_s^{-2}$, which becomes the type IIB string under the T_9-duality. We have used the fact that $\ell_p = (g_s^A)^{1/3}\ell_s$. Alternatively, the M2-brane could have been transverse to x_{10}, with one direction lying in x_9 instead. Then it would have become a D2-brane, with tension $\tau_2 = (2\pi)^{-2}\ell_s^{-3}(g_s^A)^{-1}$ and by T_9-duality a D1-brane in type IIB, with tension

$$\tau_1 \equiv \tau_{0,1} = \tau_2 2\pi R_9 = (2\pi)^{-1}\ell_s^{-2}(g_s^A)^{-1}R_9/\ell_s = (2\pi)^{-1}\ell_s^{-2}(g_s^B)^{-1},$$

where again we have used the fact that the type IIB string coupling is $g_s^B = g_s^A \ell_s/R_9$.

The two situations are related by a flip of the x_9 and x_{10} directions. This in turn is the S-transformation of the type IIB torus, and so we have correctly arrived at the S-duality action on the type IIB strings. It should be clear now how to get all of the (p,q) strings: we need to wrap the M2-brane p times on the x_{10} cycle and q times on the x_9 cycle. Let us check that we get the right tension formula. Wrapping as stated above, looking at figure 16.8 reveals that the length $2\pi R_{p,q}$ that the M2-brane is stretched is simply given by Pythagoras: $2\pi R_{p,q} = 2\pi\sqrt{(pR_{10})^2 + (qR_9)^2}$, and hence the resulting tension written in type IIB terms is:

$$\tau_{p,q} = \tau_2^M 2\pi R_{p,q} = (2\pi)^{-2}\ell_p^{-3} \cdot 2\pi R_{p,q} \tag{16.47}$$

$$= (2\pi)^{-1}\ell_s^{-3}(g_s^A)^{-1}\sqrt{(pR_{10})^2 + (qR_9)^2} \tag{16.48}$$

$$= (2\pi)^{-1}\ell_s^{-2}(g_s^B)^{-1}\sqrt{\frac{p^2 R_{10}^2}{R_9^2} + q^2} \tag{16.49}$$

$$= \sqrt{(p\tau_{1,0})^2 + (q\tau_{0,1})^2}, \tag{16.50}$$

where we have used the T-duality formula for the relation of the string couplings, the relation between ℓ_s and ℓ_p, etc. and we have recovered our earlier bound state formula (11.16). We can even go further and derive the more general formula for the case in which there is a background value, $c_{(0)}$, of the R–R scalar $C_{(0)}$ present. Recall that it is a shift in the x_{10} direction shown in figure 16.9. So in computing the length of the wrapped membrane, we ought to take into account this shift: looking at the diagram, it is elementary to see that every time ones goes around the x_9-cycle, one picks up a reduction of $c_{(0)} R_{10}$ in the total length stretched in the x_{10} direction. Therefore we should have, in this case, the more general expression $2\pi R_{p,q} = 2\pi\sqrt{([p - qc_{(0)}]R_{10})^2 + (qR_9)^2}$. Similar manipulations to the above give:

$$\tau_{p,q} = \sqrt{([p - qc_{(0)}]\tau_{1,0})^2 + (q\tau_{0,1})^2},\tag{16.51}$$

which is a rewriting of equation (16.11).

Turning to D3-branes, it is immediately clear from this picture what its origins must be. We can take an M5-brane and wrap two of its directions on the torus as its shrinks away. Following the type IIA route, it becomes first a D4-brane from shrinking x_{10}, and then a D3-brane after shrinking x_9 and T-dualising. We can check that we get the right tension directly:

$$\begin{aligned}\tau_3 &= \tau_5^M \cdot 2\pi R_{10} \cdot 2\pi R_9 = (2\pi)^{-5}\ell_p^{-6}(2\pi)^2 R_{10}R_9 \\ &= (2\pi)^{-3}\ell_s^{-6}(g_s^A)^{-2}\ell_s g_s^A R_9 = (2\pi)^{-3}\ell_s^{-4}(g_s^B)^{-1}.\end{aligned}\tag{16.52}$$

It is also clear that the D3-brane is invariant under $SL(2, \mathbb{Z})$ since it is wrapped entirely on both cycles of the torus.

For fivebranes, the story is similar to the case of the strings. There is a whole (p, q) family of them because there are two ways of getting a five dimensional extended object from the M5-brane: one either wraps it on the x_{10} cycle, in which case it becomes a D5-brane (which we ought to call $(1, 0)$), or we wrap it on the x_9 cycle and so it becomes an NS5-brane $(0, 1)$. It should be easy to see that the resulting tension of the (p, q) fivebrane made by wrapping the appropriate number of times on each cycle is (including the background $C_{(0)}$ field, and using the Pythagorean relation for $R_{p,q}$ above):

$$\tau_{p,q}^5 = \sqrt{([p - qc_{(0)}]\tau_{1,0}^5)^2 + (q\tau_{0,1}^5)^2},\tag{16.53}$$

which indeed gives the supergravity formula (16.25) given earlier. (In the computation, the above comes multiplied by $2\pi R_9'$, since that is what the resulting fivebrane is wrapped around on arrival in the type IIB theory.)

Finally, let us turn to the sevenbranes. In the stringy picture, these come from T-dualising transverse to D6-branes, but it is illuminating to think of it in the picture of reduction from M-theory. Recall that a D6-brane in M-theory comes from a clever twist of the geometry, making a Kaluza–Klein monopole. The metric is ($\mathbf{x} = (x_7, x_8, x_9)$):

$$ds_{11}^2 = -dt^2 + \sum_{i=1}^{6} dx_i^2 + V(r)(d\mathbf{x} \cdot d\mathbf{x}) + V(r)^{-1}(dx_{10} + \mathbf{A} \cdot d\mathbf{x})^2$$

$$V(r) = 1 + \frac{r_6}{r}, \qquad r^2 = \mathbf{x} \cdot \mathbf{x}, \qquad \nabla \times \mathbf{A} = \nabla V(r),\tag{16.54}$$

for a single brane located at $r = 0$ in the (x_7, x_8, x_9) plane. The key point is that the x_{10} circle shrinks to zero at the location of the D6-brane, since the metric vanishes there. So we see that shrinking the x_9 circle as well to go to the type IIB theory (after T-dualising), gives us the x_9, x_{10} torus, which we discover has a cycle which degenerates over the (x_7, x_8) plane. This is just how we describe a D7-brane in F-theory language, and so we have recovered yet another key F-theory phenomenon as a limit of M-theory. To do better, and get (p, q) sevenbranes, we may consider placing x_9 on a circle (on the M-theory side), giving a physical torus after identification (with a shift to include $c_{(0)}$). We may then consider more general S^1 fibration geometries than those in equation (16.54). The analysis of monodromies in the non-compact directions is then identical to the F-theory one.

A key phenomenon which we discovered was a description of the enhancement of symmetry when two seven-branes coincide, described as the collision of singularities in the F-torus. Since this is described by fundamental strings going to zero length in the type IIB picture, we drew this suggestively as an S^1 fibration over the string making a $\mathbb{CP}^1$ cycle, as depicted in figure 16.7, and then identified the appearance of extra massless fields with the shrinking of the cycle. Since the F-torus has no dynamics associated with it, in the way it was described, that suggestion could not be honestly taken as anything more than a strongly plausible description. Now we see in the M-theory origins of the torus that this is exactly the correct description: on the M-theory side, an M2-brane can wrap both of its directions on the cycle stretching between two lifted D6-brane fibrations of the type in equation (16.54). We have already learned that a fundamental string comes from such a wrapped M2-brane, and after shrinking the torus, we recover precisely figure 16.7. So the sevenbrane enhanced gauge symmetries in F-theory come from wrapped M2-branes on collapsing cycles in M-theory.

In summary we now see how to connect type IIB theory, and indeed the F-theory description, to M-theory. We can do the reverse now, and take various F-theory vacua and turn them into M-theory vacua. Here is a simple rule: Place the theory on any circle. Shrink the circle away, and in the limit the F-theory torus acquires a physical size, returning us to eleven dimensional M-theory.

16.2.2 *M-theory on K3 and heterotic on* T^3

We now have enough information to construct the M-theory versions of some of the data which we obtained in F-theory in earlier sections. In particular, we discovered that F-theory on K3 is in fact dual to the heterotic string on T^2.

Starting with the F-theory configuration described in earlier sections, let us now compactify a harmless direction (any of $x_1, \ldots, x_7$) on a circle, and shrink it away. The result is M-theory on K3. Actually, on the dual side, we are simply placing the heterotic string theory on an additional circle, and so derive the non-trivial result that M-theory on a K3 is dual to the heterotic string on T^3. The fundamental heterotic string is that string which originated as a D3-brane wrapped on the $\mathbb{CP}^1$ base of the elliptic K3. We now see that this string is now an M5-brane wrapped on the entire K3 in the M-theory picture. The pattern of enhanced gauge symmetries is enlarged somewhat on both sides, and the moduli space is now locally:

$$\mathcal{M} = \frac{O(19,3)}{O(19) \times O(3)}. \tag{16.55}$$

16.2.3   Type IIA on K3 and heterotic on T^4

Finally, we can in fact compactify another of the harmless circles on the M-theory side, and the result is type IIA string theory on K3. Since we have done nothing non-trivial to the heterotic side either, we discover as a result that there is a duality between type IIA on K3 and the heterotic string on T^4. We have already mentioned this duality previously in insert 7.5 (p. 186) and in chapter 12. The F-theory moduli space is now locally:

$$\mathcal{M} = \frac{O(20,4)}{O(20) \times O(4)}. \tag{16.56}$$

16.3 Matrix theory

One of the most striking features of string duality is the discovery that eleven dimensions is dynamically relevant to string theory. It had always been thought of as a useful bookkeeping device to start with eleven dimensional supergravity and derive the structure of type IIA supergravity by dimensional reduction, but it was thought of as nothing more than that. However, once one takes the loop-protected BPS spectrum of D0-branes seriously, one is forced to try to interpret the tower of light states they supply at large string coupling, and a Kaluza–Klein story appears inevitable[149].

Further study showed that the dynamics of D0-branes implied that they clearly were sensitive to shorter scales[106] than just ℓ_s. In fact, now we know (see the discussion surrounding equation (12.15)) that the physics they were sensitive to was the scale $\ell_p = g_s^{1/3}\ell_s$, which at weak coupling a lot is shorter than the supposed minimum distance ℓ_s perturbative strings know about.

This might lead one to attempt to capture some of the eleven dimensional physics in terms of that of D0-branes, hoping that it might lead to an understanding of the formulation of M-theory in its own right. This is not really a fully accurate picture of the thought processes that led to the presentation of Matrix theory[157], but then this is not an attempt at a history[158]. It suffices for us here to uncover a little of what we can with the above motivating remarks, and leave the matter of the history of it to be explored in the literature or elsewhere.

16.3.1 Another look at D0-branes

For reasons that will be stated shortly, let us focus on the low energy effective Lagrangian for N D0-branes. This is simply a $0+1$ dimensional theory (a quantum mechanics) involving the nine spatial transverse coordinates X^i, $i = 1, \ldots, 9$, and their superpartners. We start by considering the branes to be all in the same place, and so we have a $U(N)$ invariant system. We must remember to keep commutator terms which would normally vanish in the Abelian case.

The most efficient way of writing this action is in fact to start with ten dimensional maximally symmetric Yang–Mills theory and dimensionally reduce it all the way to $0+1$ dimensions. After a rescaling, the result is:

$$\mathcal{L} = \mathrm{Tr}\left[\frac{D_t X^i D_t X^i}{2g_s \ell_s} + \frac{[X^i, X^j]^2}{4g_s \ell_s (2\pi \ell_s^2)^2} - \frac{i}{2} \Theta D_t \Theta + \frac{1}{4\pi \ell_s^2} \Theta \Gamma^0 [X^i, \Gamma^i \Theta] \right].$$

We have indeed thrown away any terms with higher powers of velocity than quadratic, the trace is over $U(N)$. The X^is all come from internal components of the gauge field, and so there is the usual factor of $2\pi \ell_s^2$ to convert a gauge field to a coordinate. There are no remaining appearances of gauge fields except for A_0, which is inside the covariant derivative only, having no kinetic term. It may therefore be thought of as simply a constraint field, enforcing $U(N)$ gauge invariance. Also, Θ is a rescaled version of the $SO(9)$ sixteen component fermion which would have appeared in ten dimensions.

From the Lagrangian above, we can write a Hamiltonian. The details are left as an exercise to the reader, and the result is remarkably simple:

$$\mathcal{H} = \mathrm{Tr}\left[\frac{g_s \ell_s}{2} p_i p_i - \frac{[X^i, X^j]^2}{4g_s \ell_s (2\pi \ell_s^2)^2} - \frac{1}{4\pi \ell_s^2} \Theta \Gamma^0 [X^i, \Gamma^i \Theta] \right]$$

$$= R\mathrm{Tr}\left[\frac{1}{2} p_i p_i - \frac{[X^i, X^j]^2}{16\pi^2 \ell_p^6} - \frac{1}{4\pi \ell_p^3} \Theta \Gamma^0 [X^i, \Gamma^i \Theta] \right]. \tag{16.57}$$

Possibly the most immediately striking thing about this Hamiltonian is the fact that everything naturally assembles itself into eleven dimensional quantities, as shown in the second line above. We have pulled out an overall factor of the inverse mass of the D0-brane, $(g_s \ell_s)^{-1}$, which is the inverse of the radius of the eleventh direction, which we have called R.

16.3.2 The infinite momentum frame

There is a striking proposal for an interpretation of the physics of the above Hamiltonian[157]. The idea is that the system captures the physics of states with momentum $p_{10} = N/R$ in the limit that N and R go to infinity. This is the 'infinite momentum frame' (IMF), essentially a light cone frame. It uses the fact that D0-brane charge is momentum in the eleventh direction, and is quantised in units of $1/R$ if the direction is on a circle. We then take the limit in which the circle is large and the momentum in that direction is large, keeping the fraction fixed. This allows us to consider the decompactified limit where we are allowed to discuss a fully eleven dimensional choice like picking a boost direction.

To see that we have not neglected anything relevant in picking the original Lagrangian, notice that, if we separate momentum up into ten dimensional component, $\mathbf{p}$ and the eleven dimensional component $p_{10} = N/R$, we have:

$$E^2 = \frac{N^2}{R^2} + p^2 + m^2,$$

where m is the mass of the particle. In the limit that the eleven dimensional momentum is extremely large, we see that the dominant energy contribution is from states who have a finite fraction of the eleven dimensional momentum in the limit. In other words, since

$$E = \frac{N}{R} + \frac{1}{2} \frac{R}{N} (p^2 + m^2) + O\left(\frac{R}{N}\right)^2,$$

in the limit of $N/R \to \infty$, the energy a state with contribution mostly from the second terms will not be significant, and so it will not play a role in the dynamics.

In fact, this justifies our dropping of higher order terms in the basic Lagrangian, since those corrections (subleading in *ten dimensional* momentum) will not have a chance to contribute to the limit. Actually, the only sector which has a chance of contributing (from the ten dimensional perspective) are the D0-branes, together with the *lightest open strings* connecting them. These are precisely the sectors which appear in the Hamiltonian in equation (16.57).

The Hamiltonian above may therefore be studied in the light of this proposal in purely eleven dimensional terms. Apparently, we are to somehow recover all of the physics of M-theory this way, since eleven dimensional Lorentz invariance (*assumed* to be preserved) would suggest that we can always boost any situation into this frame. Of course, we can only do this is we can understand how to extract the physics appropriate to questions we might ask. Now we see, for example, why the bound state questions of chapter 11 were pertinent. A graviton of momentum n is in fact a bound state of n D0-branes, and so we must establish that a normalisable wavefunction for such a system exists. This is not a solved problem for arbitrary n, as already stated in chapter 11.

The scattering of gravitons with no exchange of longitudinal (eleventh direction) momentum is nicely described in terms of matrices in this language. The αth graviton of momentum $p_\alpha = n_\alpha/R$ is represented by a $n_\alpha \times n_\alpha$ block of the X^i (each X^i representing matrix position in the ith transverse coordinate). The trace of the $n_\alpha \times n_\alpha$ block of the matrix is the centre-of-mass position of the graviton. Interaction between the block diagonal parts can be determined by integrating out off-diagonal degrees of freedom, which correspond to integrating out the massive open strings stretching between the widely separated clumps and and determining the effective interactions between the clumps in that way. It has been shown that this reproduces rather nicely the expected results for graviton-graviton scattering.

In fact, a lot more can be done along those lines, including recovering the basic lightcone world-volume M2-brane description by a change of variables, making contact with the much earlier work[255] on the M2-brane Lagrangian done back when it was thought to be a viable fundamental object[157].

Another striking feature of the description is that there is a natural statement about the importance of the onset of non-commutativity of the description of spacetime at high energy. Recall that the X^i are supposed to be related to spacetime coordinates as well. They are naturally (and essentially) presented as $N \times N$ matrices here. It is only when the X^i are large that we recover the usual picture of them as commuting spacetime coordinates, for only in that limit is is favourable for the Hamiltonian to select sectors for which $[X^i, X^j]$ vanishes. Then, the X^i can all be simultaneously diagonalised into their eigenvalues x^i, the nine transverse spacetime positions[26, 157].

Note that an interpretation of the model at *finite* N has also been proposed[280]. It is simply a *discrete light cone quantisation* (DLCQ) of the theory. In other words, at finite N, the fact that the theory is on a circle of radius R is taken seriously. The theory is taken as being in the light cone frame, with a compact null direction. Such techniques have been used successfully elsewhere in order to supply the non-perturbative definition of field theories such as QCD.[281]

Note also that there is another matrix model proposal for capturing important degrees of freedom. It is based on structures in the type IIB string and D-instantons in particular[342].

16.3.3 Matrix string theory

Of course, one thing which we ought to be able to recover is the fact that we get the type IIA superstring upon compactification of a dimension on a circle. In fact, we should be able to do this on *any* spatial circle. How are we to see this here?

What we would like to do is compactify one of the directions X^i. There are a number of ways of working out just what that means for our model, but there is a particularly simple way[282], given all that we have studied so far: by T-duality, working with D0-branes in the presence of one of the X^i compact is equivalent to working with D1-branes extended in that compact direction. It must be that the model we need is a large N model built from D1-branes wound on a circle. As the size of the circle shrinks to smaller and smaller size, this picture is increasingly the more useful one to use. In fact, an extremely important sector to include is the family of light strings stretching between D0-branes after winding around the circle some number of times.

We know how to write the just the model that we want. It is 1+1 dimensional Yang–Mills on a circle. We can write it down by starting from the beginning again, or we can simply obtain it from the present matrix model. To do so, if X^9 is to be our compact direction, of radius R_9, we need only replace X^9 by $R_9 \mathcal{D}_\sigma$, where $0 \le \sigma \le 2\pi$ and $\mathcal{D}_\sigma$ is the covariant derivative.

The model which results is:

$$
\begin{aligned}
\mathcal{H} = R \int_0^{2\pi} d\sigma \, \mathrm{Tr} \Bigg[&\frac{1}{2} p_i p_i - \frac{[X^i, X^j]^2}{16\pi^2 \ell_p^6} - \frac{1}{4\pi \ell_p^3} \Theta \Gamma^0 [X^i, \Gamma^i \Theta] \\
&- \frac{E^2}{16\pi^2 \ell_p^6} - \frac{R_9^2 (\mathcal{D}_\sigma X^j)^2}{16\pi^2 \ell_p^6} - \frac{R_9}{4\pi \ell_p^3} \Theta \Gamma^0 \mathcal{D}_\sigma \Theta \Bigg] \\
= R \int_0^{2\pi} d\sigma \, \mathrm{Tr} \Bigg[&\frac{1}{2} p_i p_i - \frac{[X^i, X^j]^2}{16\pi^2 g_s^2 \ell_s^6} - \frac{1}{4\pi g_s \ell_s^3} \Theta \Gamma^0 [X^i, \Gamma^i \Theta] \\
&- \frac{F^2}{16\pi^2 g_s^2 \ell_s^6} - \frac{(\mathcal{D}_\sigma X^j)^2}{16\pi^2 \ell_s^3} - \frac{1}{4\pi \ell_s^2} \Theta \Gamma^0 \mathcal{D}_\sigma \Theta \Bigg].
\end{aligned} \tag{16.58}
$$

Notice that at the end, we made the substitution

$$R_9 = g_s \ell_s, \quad \text{and} \quad \ell_p = g_s^{1/3} \ell_s,$$

as appropriate to the case of the type IIA model we expect to arrive at in the limit. Indeed, we see that the model naturally cleans itself up into the string variables. The electric field $F_{01} = \dot{A}_\sigma$ is the non-trivial gauge field strength of the model, an electric flux, in fact. The 16 component field Θ has naturally split into an $\mathbf{8}_c \oplus \mathbf{8}_s$ under the natural $SO(8)$ which acts here. One is left moving on the string and the other is right moving. The X^i transform as the $\mathbf{8}_v$, of course. This model therefore has the manifest supersymmetry we expect for the type IIA model and is in 'Green–Schwarz' form[108]. It is also the model we arrived at (but for a single D1-brane) in section 12.1 within the type IIB string theory. There, it represented the type IIB soliton string and the opposite chiralites of the left and right movers was appropriate to the expected zero modes on the soliton.

N.B. It is amusing to note that to describe compactification of space-time dimensions, one has to work with a *higher dimensional* matrix model. This exchanges the role of dimensional reduction and the inverse procedure, dimensional 'oxidation'.

Now this model, with $U(N)$ gauge symmetry, is to be interpreted not as a soliton, but as a matrix definition of the type IIA string theory[283, 284]. The limits we are taking to get the free string are two-fold: we must take $R \to \infty$ and $N \to \infty$, as before, and we must also take $g_s \to 0$, which is of course the same as $R_9 \to 0$.

To study the model, let us consider the supersymmetric vacua, i.e. the moduli space $[X^i, X^j] = 0$. The $X^i(\sigma)$ can be chosen as diagonal matrices:

$$X^i(\sigma) = \begin{pmatrix} x_1^i(\sigma) & 0 & 0 & \cdots & \cdots \\ 0 & x_2^i(\sigma) & 0 & \cdots & \cdots \\ 0 & 0 & x_3^i(\sigma) & \cdots & \cdots \\ \vdots & \vdots & \ddots & \ddots & \cdots \\ \cdots & \cdots & \cdots & \cdots & x_N^i(\sigma) \end{pmatrix}. \tag{16.59}$$

Naively the moduli space is just the space of eigenvalues, $(\mathbb{R}^8)^N$. Notice, however, that a discrete subgroup of the gauge symmetry still acts. It is the permutations of the eigenvalues, which we shall denote as $\mathcal{S}_N$. Since we must divide by this, the vacuum moduli space is therefore the orbifold $(\mathbb{R}^8)^N / \mathcal{S}_N$.

The strings we've defined are lying in the direction parametrised by σ, but we must study this a bit more carefully. A configuration representing strings which are of the same length of the σ circle satisfies

$$X^i(\sigma + 2\pi) = X^i(\sigma).$$

One way to think of this configuration is as representing N closed strings. The x_n^i may be thought of as the x^i coordinate of the nth string, parameterised by σ. The $x^i(\sigma)$ are otherwise arbitrary functions (subject to the equations of motion) of τ and σ, and so can truly represent arbitrary strings in various shapes. (See figure 16.10.) In fact, one of these strings has energy of order $1/N$ that required to contribute to the physics in the limit, since it is T-dual to a single D0-brane among the very large N of the whole model. What we need is a method of making a string with a larger share of the longitudinal momentum.

The matrix model naturally contains such strings too. First, note that there is a natural symmetry group which we shall denote $\mathcal{S}_N$, which acts on the strings by permuting the N eigenvalues of the matrices. The strings are all identical, and so this is a very natural model. We can use this permutation symmetry to make long strings, by making configurations which satisfy:

$$X^i(\sigma + 2\pi) = s_2 X^i(\sigma),$$

where s_n is the element of $\mathcal{S}_N$ representing the permutation of n objects. The following configuration is an example:

$$s_2 = \begin{pmatrix} 1 & 0 & 0 & \cdots & \cdots \\ 0 & 0 & 1 & \cdots & \cdots \\ 0 & 1 & 0 & \cdots & \cdots \\ \vdots & \vdots & \ddots & \ddots & \cdots \\ \cdots & \cdots & \cdots & \cdots & 1 \end{pmatrix}.$$

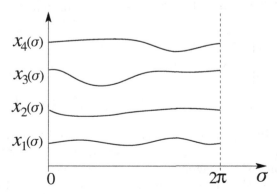

Fig. 16.10. Four minimum length strings in the matrix.

This should remind the reader of a twisted sector from our orbifold techniques in various previous chapters, such as in section 4.8. This matrix implements a permutation of the two eigenvalues x_2 and x_3 one goes around the σ circle. So, in fact, since $s_2^2 = 1$, in order to make a closed string with eigenvalues in the 2 and 3 position, one must go around the σ circle twice. So we have made a configuration representing a string of twice the length of the basic strings. See figure 16.11. In this way, we see that the model contains closed strings which possess a large enough fraction of their energy in momentum in the eleventh direction in order to survive the limit.

To see that we get the right sort of theory, note that the limit $g_\mathrm{s} \to 0$ actually defines a flow of the 1+1 dimensional Yang–Mills theory to the IR. There, the theory is expected to become a fully conformally invariant fixed point, representing the free type IIA matrix string. Notice that this is in fact a new way of constructing a *string field theory* of the strings, in the infinite momentum or light cone frame. It is a field theory in the sense that there are fields which create and destroy complete string configurations, the matrices $X^i(\tau, \sigma)$ themselves. The interactions between strings can be studied as well, and the splitting/joining operation is implemented by the addition of a special 'irrelevant' operator to the conformal field theory, deforming it away from the fixed point[284] towards the UV (see insert 3.1, p. 84).

It should be noted that the matrix string model at finite N has also been given an interpretation in its own right as a DLCQ definition of the theory. Also, matrix string theories (either DLCQ or IMF) for all of the other ten dimensional can be defined by similar methods. In fact, the technique has been used to supply a definition of theories (such as the special six dimensional non-gravitational string theories and their low

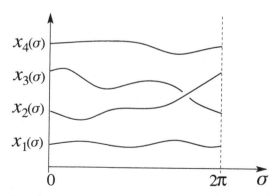

Fig. 16.11. A twisted sector representing two minimum length strings and one of twice the length.

energy field theory limits mentioned at the end of section 12.3.2) which the usual Lagrangian techniques seem to fail to define even perturbatively[285].

As already noted, describing further compactified spacetime dimensions leads us to study higher dimensional Yang–Mills field theories in various limits, implicitly related to world-volume theories of D-branes. Unfortunately, once one gets to the study of six uncompactified directions, progress seems to stop. This is because the matrix theory is now a $5 + 1$ dimensional Yang–Mills field theory, which in the required matrix theory limit does not seem to make sense[158].

For this and other reasons, it seems at the time of writing that Matrix theory, while apparently a tantalising glimpse into the correct direction which will lead to a definition of M-theory, is incomplete. In retrospect, this is perhaps not surprising, since it is still rather closely wedded to D-brane techniques, being largely a reinterpretation of the physics of open strings and D-branes in various limits, albeit a very instructive and useful one. The search for a definition of M-theory must continue.

17

D-branes and black holes

We've seen now many examples of the ways in which D-branes can be used as probes of the non-perturbative structure of string theory, with remarkable insights, including the one that string theory is not really a theory of strings beyond perturbation theory. It should not be forgotten that strings also have the intriguing feature that they insist on describing (at least) a perturbative quantum gravity. It is considerably significant that we can get insight into string theory's non-perturbative treatment of certain questions in quantum gravity, again using D-branes to probe and model the physics of black holes. This chapter will lay the foundations for how this works.

17.1 Black hole thermodynamics

17.1.1 The path integral and the Euclidean calculus

In an attempt to construct a path integral definition of quantum gravity, one might envision the following:

$$Z = \int \mathcal{D}[g, \varphi] e^{iI[g,\varphi]}, \tag{17.1}$$

for some appropriate choice of integration measure $\mathcal{D}[g, \varphi]$ over the metric g and matter fields φ. In the early days of studying the path integral for gravity, it was noticed that the gravity action for some region $\mathcal{M}$ should be supplemented by a term evaluated on its boundary $\partial\mathcal{M}$ which allows the contribution of variations which include configurations which vanish on $\partial\mathcal{M}$, but which might have non-vanishing normal derivatives on it. The result is (in units where $G_N = 1$):

$$I = \frac{1}{16\pi} \int_{\mathcal{M}} \sqrt{-g} \, R \, d^D x + \frac{1}{8\pi} \int_{\partial\mathcal{M}} \sqrt{-h} \, K \, d^{D-1} x, \tag{17.2}$$

where $h_{\mu\nu}$ is the induced metric on the boundary, and K is the trace of the extrinsic curvature of the boundary. (We learned how to compute these quantities in insert 10.2.) This term is required so that upon variation with metric fixed at the boundary, the action yields the Einstein equations.

Since I is real, there is the problem that the path integral has convergence problems, since the integral is in principle oscillatory. One way this is made sense of to 'Wick rotate' the time axis by 90° by the substitution $t \to -it$, and so the path integral becomes:

$$Z = \int \mathcal{D}[g, \varphi] e^{-I^{\mathrm{E}}[g,\varphi]}, \qquad (17.3)$$

where $I^{\mathrm{E}} = -iI$ is the Euclidean action, which is real for real fields, and now the integrand is seen to be a damped exponential, which improves convergence. The metric has gone from signature $(-++\cdots+)$ to signature $(+++\cdots+)$. In principle, we can evaluate our path integral on the Euclidean section and then rotate back to Lorentzian signature.

The Euclidean technology allows for the definition of the canonical thermodynamical ensemble as well. Let us see how this works. The amplitude to go from a configuration (g_1, φ_1) at time t_1 to a configuration (g_2, φ_2) at time t_2 is:

$$\langle (g_2, \varphi_2), t_2 | (g_1, \varphi_1), t_1 \rangle = \int \mathcal{D}[g, \varphi] e^{iI[g,\varphi]}.$$

This quantity has another representation, in the Schrödinger picture:

$$\langle (g_2, \varphi_2) | e^{-iHt_2} e^{iHt_1} (g_1, \varphi_1) \rangle = \langle (g_2, \varphi_2) | e^{-iH(t_2-t_1)} (g_1, \varphi_1) \rangle.$$

Let us study the situation that $(g_1, \varphi_1) = (g_2, \varphi_2)$. Writing $t_2 - t_1 = -i\beta$, and summing over a complete set of eigenstates (ψ_n, E_n) of the Hamiltonian, we get the partition function:

$$Z = \sum_n e^{-\beta E_n}. \qquad (17.4)$$

The system is at temperature $T = \beta^{-1}$, and we have the standard expression for the probability, p_n, of being in the nth state:

$$p_n = \frac{1}{Z} e^{-\beta E_n}.$$

The familiar representation given in equation (17.4) represents the same system represented by the Euclidean path integral given in equation (17.3), where the fields (g, φ) are periodic in τ with period β. We shall see how to extract other physical quantities from here a little later.

17.1.2 The semiclassical approximation

The evaluation of the entire path integral will not concern us here, since as string theorists, we take a rather different approach to the problem of quantum gravity. However, we expect from the reasoning that we have used many times already that we will arrive at a low energy action of the sort we studied above, regardless of the underlying microscopic model. So in fact, when we come to examine the macroscopic predictions of the microscopic details of our particular approach to fundamental physics (string and M-theory) – or any other approach, for that matter – they should make contact with the semiclassical results to be derived from the action above.

The expectation is that the configurations with the most dominant contribution to the path integral will be those which are near an extremum of the action, i.e. solutions to the equations of motion. This of course fits with our intuition about how the classical limit arises from the path integral approach.

In this limit, the path integral becomes

$$Z = e^{-I^{\mathrm{E}}} \equiv e^{-\beta W},$$

defining the thermodynamic (effective) potential W, which is

$$W = E - TS, \tag{17.5}$$

where T is the temperature and S is the entropy of the system. We can easily extract useful information in this limit. For example, the average energy of the system would be quite reasonably defined as the normalised quantity

$$\langle E \rangle = \frac{1}{Z} \sum_n E_n e^{-\beta E_n} = -\frac{1}{Z} \frac{\partial Z}{\partial \beta} = -\frac{\partial \log Z}{\partial \beta} = \frac{\partial I^{\mathrm{E}}}{\partial \beta}. \tag{17.6}$$

Another example of some importance is the entropy. This is defined in terms of the occupation probability p_n as:

$$
\begin{aligned}
S &= -\sum_n p_n \log p_n = -\frac{1}{Z} \sum_n e^{-\beta E_n} \log\left(\frac{e^{-\beta E_n}}{Z}\right) \\
&= -\frac{1}{Z} \sum_n e^{-\beta E_n} \left(-\beta E_n - \log Z\right) \\
&= \beta \frac{\partial I^{\mathrm{E}}}{\partial \beta} + \log Z = \beta \langle E \rangle - I^{\mathrm{E}}.
\end{aligned} \tag{17.7}
$$

The approximation will allow us to extract a number of key features of the physics. For example, the contribution of the fields φ to the effective action of quantum fields on various curved spacetime backgrounds will be sensitive to various features of the background and the properties of the fields themselves. Meanwhile, in the purely gravitational sector, we will find that there are dramatic effects which arise in our computations due to the non-trivial interplay of topology of the Euclidean section with the path integral[288]. An example of an immediate consequence of this is the result that black holes have an intrinsic temperature. Let us compute this for the Schwarzschild and Reissner–Nordström solutions to see how it works, since the computation in this framework is surprisingly straightforward.

17.1.3 The temperature of black holes

We begin with the Schwarzschild and Reissner–Nordström solutions which we met in given in chapter 10, and as we were instructed in the previous section, we continue the solution to Euclidean signature via $t \to -i\tau$, with period β for τ:

$$ds^2 = V d\tau^2 + V^{-1} dr^2 + r^2 d\theta^2 + r^2 \sin^2\theta d\phi^2, \qquad (17.8)$$

with

$$V = \left(1 - \frac{2M}{r} + \frac{Q^2}{r^2}\right).$$

This solution is taken as making sense in the range $r_+ \leq r \leq \infty$, where $r_+ = M + \sqrt{M^2 - Q^2}$. Now the neighbourhood of $r = r_+$ (what was the horizon) is trying to look like $\mathbb{R}^2 \times S^2$, but sadly, there is a conical singularity there, because the coordinates (r, τ), trying to look like polar coordinates in the plane, have the wrong periodicity for τ for arbitrary β.

In fact, the problem of computing the temperature reduces to the matter of removing this 'bolt singularity'[83, 82], ensuring the 'regularity of the Euclidean section'. This is quite easy to do: one has to make sure that the infinitesimal ratio of the circumference (going around in τ) to the radius (moving in r), is in fact 2π as one approaches the origin of $\mathbb{R}^2$, which is $r = r_+ = 2M$. This boils down to:

$$2\pi = \lim_{r \to r_+} \frac{\Delta\tau}{V^{-1/2}} \frac{d(V^{1/2})}{dr} \qquad \Longrightarrow \qquad \frac{4\pi}{\beta} = V'|_{r=r_+},$$

where $\Delta\tau = \beta = 1/T$. We then add a point (equivalent to a whole S^2) to repair $r = r_+$. From this we get:

$$\frac{1}{\beta} = T = \frac{Mr_+ - Q^2}{2\pi r_+^3} = \frac{\sqrt{M^2 - Q^2}}{4\pi M(M + \sqrt{M^2 - Q^2}) - 2\pi Q^2}, \qquad (17.9)$$

and for the case of $Q = 0$ (Schwarzschild), we have

$$T = \frac{1}{8\pi M},$$

which shows that large black holes are actually quite cold, and it is small black holes which are hot. This is actually a good thing for consistency with what we have already observed, since it means that astrophysical black holes (especially the really big ones apparently indirectly detected out there at the cores of galaxies, but even stellar-sized ones) have negligible mass loss due to this sort of radiation*. In fact, this means that asymptotically flat black holes (i.e. the sort we've been studying so far) have negative specific heat, since reducing the energy of the system (mass) increases the rate at which it is lost.

Notice furthermore that for the charged black hole, the temperature vanishes at extremality, since there $r_+ = Q = M$. This fits rather well with what we have learned previously: the extremal solution is supersymmetric and in fact a BPS state, and so zero temperature is consistent with its stability. In addition, we see that the thermodynamics protects the censorship idea, since it cannot radiate further mass away, making a sub-extremal object with a naked singularity.

In fact, the temperature can be related to a purely geometrical quantity known as the *surface gravity*, κ, of a black hole, which is a purely geometric quantity that exists at the horizon, and (crucially) is constant all over it[292]. If we had a test particle in the geometry connected to an observer at infinity by a long (light) string, the surface gravity is in fact the acceleration needed to hold the particle stationary at the horizon. It can be defined in terms of a Killing vector χ normal to the horizon:

$$\kappa^2 = -\tfrac{1}{2}(\nabla^\mu \chi^\nu)(\nabla_\mu \chi_\nu)\,|_{r=r_+}, \qquad (17.10)$$

where we perform the evaluation at the horizon.

For our solution, we have that $\chi^\mu = \xi^\mu = \delta_t^\mu$, and from the list of the non-vanishing components of the affine connection given in equation (10.5), we can compute the only non-zero component of the covariant derivative:

$$\nabla_r \chi_t = \partial_r \chi_t - \Gamma_{rt}^t \chi_t = \frac{-2Mr + 2Q^2}{r^3} + \frac{Mr - Q^2}{r^3} = \frac{-Mr + Q^2}{r^3},$$

which gives

$$\kappa = \frac{Mr_+ - Q^2}{r_+^3},$$

---

* The reader can multiply by $\hbar c^3 / G k_{\rm B}$ in order to restore the physical units.

and so we have:

$$T = \frac{\kappa}{2\pi}. \tag{17.11}$$

17.2 The Euclidean action calculus

The action is usually evaluated by computing with what is called the 'Euclidean section' of the spacetime, which arose in the previous sections. Since this removes the singularities from the integrand, it makes the integration procedure sensible[288, 290]. Furthermore, for asymptotically (locally) flat spacetimes, the action is interpreted as computed with reference to an appropriate background in order to give a finite answer. Later, we will see a different prescription in the context of asymptotically anti-de Sitter spacetimes, which allows for a computation of the action which does not require reference to another spacetime. Let us compute an example with the present methods to get used to how they work.

17.2.1 The action for Schwarzschild

The Schwarzschild spacetime is asymptotically flat, and so we can compute the action by using flat spacetime as a reference background. For both spacetimes, the Ricci scalar $R = 0$ and so the second part of the action is where we must concentrate our efforts.

We must evaluate the extrinsic curvature for both spacetimes. Let us pick for our boundary the spherical shell at $r = R$. The unit outward normal to this is (see insert 10.2, p. 229):

$$n^\mu = \frac{\delta_r^\mu}{\sqrt{G_{rr}}} = \frac{1}{\sqrt{G_{rr}}} \left(\frac{\partial}{\partial r}\right)^\mu.$$

The extrinsic curvature is

$$K_{\mu\nu} = \frac{1}{2} n_\alpha G^{\alpha\beta} \partial_\beta G_{\mu\nu},$$

which gives non-zero components

$$K_{tt} = -\frac{1}{2} \frac{1}{\left(1 - \frac{2M}{r}\right)^{1/2}} \left(1 - \frac{2M}{r}\right) \frac{2M}{r^2};$$

$$K_{\theta\theta} = \frac{1}{2} \frac{1}{\left(1 - \frac{2M}{r}\right)^{1/2}} \left(1 - \frac{2M}{r}\right) 2r;$$

$$K_{\theta\theta} = \frac{1}{2} \frac{1}{\left(1 - \frac{2M}{r}\right)^{1/2}} \left(1 - \frac{2M}{r}\right) 2r \sin^2 \theta;$$

$$K = G^{\mu\nu} K_{\mu\nu} = \frac{1}{\left(1 - \frac{2M}{r}\right)^{1/2} r^2} \left[2r - 3M\right], \qquad (17.12)$$

and by setting $M = 0$ we get the result $K = 2/r$ for Minkowski space. The measure for integration on the boundary is

$$\sqrt{h} = r^2 \left(1 - \frac{2M}{r}\right)^{1/2} \sin \theta$$

and recall that the period of the imaginary time is $\Delta \tau = \beta$. So for Schwarzschild we have

$$\int \sqrt{h}\, K d^3 x = \beta 4\pi (2r - 3M).$$

For Minkowski, we must be more careful. The measure is $\sqrt{h} = r^2 \sin \theta$, and $K = 2/r$, but we must choose our temperature carefully. Since Minkowski is regular for any period of τ, the temperature is arbitrary, and so we must fix it to match the Schwarzschild temperature. At radius r, the temperature is not β, but it is red shifted to $\beta \left(1 - 2M/r\right)^{1/2}$, so that is what we should use for the result of integrating over the compact time, with the result:

$$\int \sqrt{h}\, K d^3 x = \beta 4\pi 2r \left(1 - \frac{2M}{r}\right)^{1/2},$$

and so the action difference in the limit $R \to \infty$ is

$$I^{\mathrm{E}} = \beta \frac{M}{2}. \qquad (17.13)$$

Let us see that we can in fact extract useful information from this result. First, we note that M is a function of β ($M = \beta/8\pi$) and so we should be careful when differentiating with respect to β. A computation of the energy, using the formula (17.6) gives:

$$\langle E \rangle = \frac{M}{2} + \frac{\beta}{2} \frac{\partial M}{\partial \beta} = M,$$

which is an extremely intuitive result. We have seen that the system has a temperature, and so we should expect to compute a non-zero 'Bekenstein–Hawking'[262, 261] entropy, using equation (17.7):

$$S = \beta M - \frac{\beta M}{2} = \frac{8\pi M^2}{2} = \frac{\mathcal{A}}{4},$$

where $\mathcal{A}$ is the area of the black hole's horizon. So we see that these results combine nicely to confirm the expression for the thermodynamic potential

$$W \equiv \frac{I}{\beta} = \frac{M}{2} = M - TS.$$

17.2.2   The action for Reissner–Nordström

A similar computation of the gravity action can be done for the charged black hole, and in fact, the result is the same as in equation (17.13), with β now from the expression given in (17.9), which should be obvious to the reader who followed the computations. The term in the metric containing Q is subleading in a $1/r$ expansion. Now the action needs to be supplemented by a contribution from the Maxwell term, which can be manipulated into a boundary term, assuming that the equations of motion are obeyed:

$$I_{\mathrm{M}} = -\frac{1}{16\pi} \int_{\mathcal{M}} \sqrt{g} F_{\mu\nu} F^{\mu\nu} \, d^D x = -\frac{1}{8\pi} \int_{\partial\mathcal{M}} A_\mu F^{\mu\nu} dS_\nu, \qquad (17.14)$$

where the latter is a boundary integral, and we have used the on-shell condition that $\nabla_\nu F^{\mu\nu} = 0$.

Notice that the usual expression for the gauge potential, written as a one-form $A = A_t dt$, where $A_t = Q/r$, is singular, since the interval dt is infinite at the horizon. We can repair this problem by defining a value for the potential at the horizon, $\Phi = Q/r_+$, and then redefining the potential by a gauge transformation:

$$A = Q \left(\frac{1}{r} - \frac{1}{r_+} \right) dt.$$

Now since the non-zero components of $F_{\mu\nu}$ are just $F_{rt} = -Q/r^2$, the boundary integral for the action is easy to compute, giving, in the limit $R \to \infty$ the result:

$$I_{\mathrm{M}} = -\frac{\beta}{2} Q\Phi.$$

So the total action turns out to be

$$I^{\mathrm{E}} = \frac{\beta}{2} [M - Q\Phi].$$

Again, in the semiclassical limit we can equate this to βW, where the thermodynamic or Gibbs potential is

$$W = M - TS - Q\Phi,$$

since in the thermodynamic analogy, Φ is like a chemical potential for Q, the analogue of particle number. Now we can use the standard thermodynamic relations to compute:

$$E = \left(\frac{\partial I}{\partial \beta}\right)_\Phi - \frac{\Phi}{\beta}\left(\frac{\partial I}{\partial \Phi}\right)_\beta = M;$$

$$S = \beta\left(\frac{\partial I}{\partial \beta}\right)_\Phi - I = \frac{\mathcal{A}}{4};$$

$$Q = -\frac{1}{\beta}\left(\frac{\partial I}{\partial \Phi}\right)_\beta = Q, \tag{17.15}$$

where $\mathcal{A}$ is the area of the black hole's horizon. These canonical ensemble computations are best performed by working in terms of r_+ as much as possible, converting in the end to, for example, $\partial/\partial\beta = (\partial r_+/\partial\beta)\partial/\partial r_+$, etc.

17.2.3 The laws of thermodynamics

The reality of the thermodynamic behaviour of black holes begun to emerge from considering (among other things) the observation that was made by relativists that an isolated black hole's horizon area, $\mathcal{A}$, cannot be decreased by any physical process[292, 289]. This is, of course, reminiscent of the analogous law for entropy, S, in thermodynamics, where it is called the *Second Law* of thermodynamics.

Combining this with the result that there is in fact a temperature to be associated with black holes, because they are radiating their mass away quantum mechanically leads to the 'Bekenstein–Hawking' relation of the entropy to the area[262, 261], which we computed in two cases above:

$$S = \frac{\mathcal{A}}{4}. \tag{17.16}$$

In fact, a *First Law* can be formulated for black holes as well,

$$dE = TdS + pdV \quad\Longleftrightarrow\quad dM = \frac{1}{8\pi}\kappa d\mathcal{A} + \Omega_\text{H} dJ + Qd\Phi,$$

where on the left hand side are the usual quantities from the first law, and on the right are the analogous black hole quantities, the electric charge and potential at the horizon, and the angular velocity at the horizon Ω_H and angular momentum J such as could be computed for a rotating black hole (the Kerr solution).

Additionally, a *Third Law* can be stated[292]. For the Reissner–Nordström black hole, we saw that the extremal case has $T = 0$. However, to achieve such a case starting from finite temperature is intuitively physically impossible since approaching the extremal case would mean opening up the infinite volume spacetime which in chapter 10 was shown to live at the horizon of the extremal black hole.

17.3 $D = 5$ Reissner–Nordström black holes

It is a remarkable and profound fact that black holes obey the laws of thermodynamics, saying that gravity has some underlying structure which has yet to be fully understood. What one needs to find is (as for ordinary thermodynamics) an underlying microscopic description from which these laws arise. This is a big problem with quantum gravity. A universal microscopic description of the required degrees of freedom is not known.

Happily, the modern era has seen remarkable progress. String theory contains a theory of quantum gravity within it which is understood well enough to make progress in at least some of these questions. So far, we have only seen signs of gravity perturbatively, but black holes are firmly in the non-perturbative sector. Now, there are powerful arguments about the behaviour of strings at high energy density which can be followed to strong coupling to achieve a sharp, but qualitative understanding of the quantum behaviour of black holes as described by strings via a 'correspondence principle'[263]. There is marked qualitative agreement with the properties we have uncovered above[7].

However, by the study of a specific but large class of black holes in string theory, it is possible to find a microscopic description of them using D-branes which firmly establishes the precise (including all crucial universal numerical factors) thermodynamic relations we discussed semiclassically above. This is remarkable progress is a good sign that string theory (and M-theory) does indeed show mature signs of having a description of non-perturbative gravity. Let us begin to uncover some aspects of this description.

We shall work with five dimensions, for the simplest example. A five dimensional analogue of the charged black hole solution (10.4) that we already studied somewhat in chapter 10 is:

$$ds^2 = -\left(1 - \frac{2m}{R^2} + \frac{q^2}{R^4}\right) dt^2 + \left(1 - \frac{2m}{R^2} + \frac{q^2}{R^4}\right)^{-1} dR^2 + R^2 d\Omega_3^2,$$

$$A_t = \frac{q}{R^2}, \tag{17.17}$$

where

$$d\Omega_3^2 = d\theta^2 + \sin^2\theta(d\phi^2 + \sin^2\phi \, d\chi^2) \tag{17.18}$$

is the metric on a round three sphere, and $(t, R, \theta, \phi, \chi)$ constitute polar coordinates in the directions $(x^0, x^1, x^2, x^3, x^4)$.

As before, there is an outer horizon at the largest root of $G^{rr} = 0$:

$$R_\pm^2 = m \pm \sqrt{m^2 - q^2},$$

and a singularity at $R = 0$. From our previous discussion, we know that there is a Hawking temperature and Bekenstein–Hawking entropy set by the horizon. We would like to make a link to a microscopic description of the underlying structure of the black hole.

The challenge is therefore to attempt to embed this black hole into string theory in a manner which allows us to use some of the tricks we learned about D-branes to help us study its properties. It is useful to rewrite the hole in isotropic coordinates for this study, since we are going to build the black holes out of branes, and we have presented the supergravity solutions for them in chapter 10 in terms of such coordinates. To do this, let us write $R^2 = r^2 + R_-^2$ for some new radial coordinate r. Since we can write $-G_{tt} = G^{rr}$ as

$$\frac{1}{R^4}(R^2 - R_+^2)(R^2 - R_-^2) = \frac{1}{R^4}(r^2 - (R_+^2 - R_-^2))r^2 = \frac{r^4}{R^4}\left(1 - \frac{r_H^2}{r^2}\right),$$

where $r_H^2 = R_+^2 - R_-^2 = 2\sqrt{m^2 - q^2}$, we find the following pleasingly simple form:

$$ds^2 = -\mathcal{H}^{-2}f dt^2 + \mathcal{H}\left(f^{-1}dr^2 + r^2 d\Omega_3^2\right),$$

$$\text{where} \quad f \equiv 1 - \frac{r_H^2}{r^2}, \qquad \mathcal{H} \equiv \frac{R^2}{r^2} = 1 + \frac{R_-^2}{r^2}$$

$$A_t = \frac{q}{r^2 + R_-^2},$$

$$= \frac{q}{R_-^2}\left(1 - \mathcal{H}^{-1}\right), \tag{17.19}$$

where the horizon is at $r = r_H$. It has area $\mathcal{A} = 2\pi^2(r_H^2 + R_-^2)^{3/2} = 2\pi^2 R_+^3$. The interior region of the black hole containing the singularity is not covered by these coordinates. In the extremal limit where the horizon is degenerate $(m = q)$, we get $R_+^2 = R_-^2 = q$ and the solution in the original coordinates is:

$$ds^2 = -\left(1 - \frac{q}{R^2}\right)^2 dt^2 + \left(1 - \frac{q}{R^2}\right)^{-2} dR^2 + R^2 d\Omega_{3}^2,$$

$$A_t = \frac{q}{R^2}, \tag{17.20}$$

where the horizon is at $R^2 = q$. It has area $\mathcal{A} = 2\pi^2 q^{3/2}$. In isotropic coordinates we get simply:

$$ds^2 = -\mathcal{H}_e^{-2}dt^2 + \mathcal{H}_e\left(dr^2 + r^2 d\Omega_3^2\right),$$
$$A_t = \mathcal{H}_e^{-1}$$
$$\text{and}\quad \mathcal{H}_e \equiv 1 + \frac{Q^2}{r^2}, \tag{17.21}$$

where we write $Q^2 = q$ for later notational convenience. The horizon is now at $r = 0$.

Now comes the fun part. We have to see whether any of the structure of the solution is familiar to us from what we have learned so far. It is encouraging that we get something that looks like the correct type of harmonic function that we would like to come from a brane solution, but we have to achieve a constant dilaton, and see the gauge field arise from either pure metric geometry and/or the R–R sector, if we are to connect it entirely to D-branes.

17.3.1 Making the black hole

The most obvious thing to try would have been the D5-brane solution, wrapped on T^5, which would have given (ignoring the T^5 directions):

$$ds^2 = -H^{-1/4}dt^2 + H^{3/4}\left(dr^2 + r^2 d\Omega_3^2\right),$$
$$C_t^{(5)} = H^{-1}; \qquad e^{-\frac{\Phi}{2}} = H^{1/4}$$
$$\text{where}\qquad H \equiv 1 + \frac{Q^2}{r^2}. \tag{17.22}$$

Compare this to the solution (17.21). This comes close in the gauge field, but fails for a number of reasons. The first is that the powers of the function $H = 1 + Q^2/r^2$ are wrong in the parallel and transverse parts of the metric, and the second is that the dilaton is not a constant.

Looking at the transverse part to see what is missing, we observe that we really need an additional $H^{1/4}$. Perhaps we can combine this solution with something which has this behaviour. This behaviour is what we would get if were to attempt to make instead a hole by dimensionally reducing the D1-brane solution (delocalised in four of its transverse directions on a $T^4 \subset T^5$, so that we use r^{-2} and not r^{-6} in H):

$$ds^2 = -H^{-3/4}dt^2 + H^{1/4}\left(dr^2 + r^2 d\Omega_3^2\right),$$
$$C_t^{(1)} = H^{-1}; \qquad e^{-\frac{\Phi}{2}} = H^{-1/4}$$
$$\text{where}\qquad H \equiv 1 + \frac{Q^2}{r^2}. \tag{17.23}$$

Again, this solution on its own would have shortcomings. Notice that the dilaton goes inversely with that of the D5-brane solution, but that the reduced R–R field is again just what we want.

In fact, we can make a solution by combining these two in a manner analogous to that which we saw before in section 15.4, using the harmonic function sum rule to get a solution which has eight supercharges. The harmonic functions in the three sectors (i.e. directions transverse to both, transverse to the smaller, or parallel to both) combine by product. Ignoring the five directions of the T^5 this gives us:

$$ds^2 = -H^{-1}dt^2 + H\left(dr^2 + r^2 d\Omega_3^2\right),$$
$$C_t^{(1)} = H^{-1} = C_t^{(5)}; \qquad e^{-\frac{\Phi}{2}} = 1. \tag{17.24}$$

We could take the diagonal combination of the charge sector as our gauge field (thereby averaging $C^{(1)}$ and $C^{(5)}$ and so summing the charges) and things would be perfect there. So overall, this is very nearly what we want, but it sadly it fails because the power of H in G_{tt} is not correct.

Undaunted, we must search for some new component to the solution which does not modify what we have already got correct for the transverse directions and the dilaton and charge sector, but fixes the problematic power of H in G_{tt}. Switching off the contributions from the branes temporarily, we see that we must have a constant dilaton, and a metric:

$$ds^2 = -H^{-1}dt^2 + dr^2 + r^2 d\Omega_3^2,$$

and we can still possibly allow an electric potential $A_t = H^{-1}$, since we can take a linear combination of it with the other gauge sectors.

One recourse is to appeal to pure geometry. We have only so far been considering a direct reduction on the T^5 by simply ignoring it. We can be considerably more subtle and reduce on it (or part of it) with a Kaluza–Klein twist. This could achieve our modification of the metric without modifying the dilaton, since it would come the pure geometry of the reduction. Recall that we learned from earlier Kaluza–Klein studies in chapter 4 (see also insert 12.1) that we can modify a metric component which is G_{yy}^{D+1} in $D + 1$ dimensions by twisting the y direction with, say the x^5 direction along which we do the Kaluza–Klein reduction. The metric component G_{yy}^D in the D-dimensional metric is in fact $G_{yy}^{D+1} - G_{55}A_y^2$, and the gauge field $A_y = G_{5y}/G_{55}$. In the present case, our gauge field must be of the form (up to a gauge choice) $A_t = H^{-1}$, and so this fixes for us what we can achieve in the reduction.

N.B. Since the gauge field is electric, it must come from a metric component resulting from a twist of time t with a spatial component and so this is in fact equivalent to giving the entire solution some momentum in the internal direction x^5.

To see that this Kaluza–Klein will give the modification we need to get the five dimensional black hole metric, choose a six dimensional Kaluza–Klein ansatz (still with the D1- and D5-brane components switched off):

$$ds^2 = -\frac{1}{H}dt^2 + dr^2 + r^2 d\Omega_3^2 + H\left[dx_5 + \left(\frac{1}{H} - 1\right)dt\right]^2$$
$$= -dt^2 + dx_5^2 + \frac{Q^2}{r^2}(dt - dx_5)^2 + dr^2 + r^2 d\Omega_3^2, \qquad (17.25)$$

where we have shifted the gauge potential $A_t = H^{-1}$ by unity (this is just a gauge choice), and labelled the Kaluza–Klein dimension as x_5.

We see that the solution looks very simple as a six dimensional metric, but when written in the Kaluza–Klein ansatz, with the appropriate gauge field, we can achieve the desired modification of the coefficient of dt^2 which will appear in the reduced metric. When we introduce the D1 and D5 harmonic functions into the full solution, they will be multiplied back in according to the manner we have seen above, not modifying this structure at all.

Before writing the full solution, note that we can introduce orthogonal coordinates $\sqrt{2}u = x_5 - t$ and $\sqrt{2}v = x_5 + t$ and write the solution as

$$ds^2 = 2dudv + \frac{2Q^2}{r^2}du^2 + dr^2 + r^2 d\Omega_3^2.$$

There is a null vector with components $l_\mu = \partial_\mu u$, which is in fact co-variantly conserved. This shows that the solution (H is independent of the u, v directions and can have a variety of dependences on the transverse ones) is in fact a 'plane-fronted' wave, which has parallel wave fronts. It is often called a 'pp-wave' for this reason. (See insert 17.1 for a discussion.)

So we have in fact succeeded in our goal. By superposing these three components according to the sum rules, we can construct the five dimensional extremal black hole. To recapitulate, it corresponds to a D5-brane wrapped on a T^4 (in directions x_6, x_7, x_8, x_9) to make a string lying in the x_5-direction. This string is combined with a D1-brane also lying in x_5. We know from previous chapters that this is supersymmetric. Finally, we

Insert 17.1. pp-Waves as boosted Schwarzschild

Observe that we can write the pp-wave given in equation (17.25) in a manner in which is clearly a limit of a non-extremal form:

$$ds^2 = -dt^2 + dx_5^2 + \frac{r_H^2}{r^2}(\cosh\beta dt - \sinh\beta dx_5)^2 + \left(1 - \frac{r_H^2}{r^2}\right)^{-1} dr^2$$

$$+ r^2 d\Omega_3^2 + \sum_{i=6}^{9} dx_i^2. \tag{17.26}$$

This is written as a sort of boost, rather like we did for the p-brane solutions in subsection 10.2.2. It is actually a Lorentz boost of a familiar solution in the (t, x_5) plane. The supersymmetric solution we wrote previously is the limit of infinite boost, $\beta \to \infty$, with $r_H \to 0$ holding the combination $r_P^2 = r_H^2 e^{2\beta}/4$ held fixed, just like the infinite boost gives the supersymmetric extremal p-branes. The infinite boost gives a special supersymmetric solution with a null Killing vector $\partial/\partial u$, where $\sqrt{2}\,u = (x_5 - t)$. This is a momentum in the x_5 direction, as discussed in the main text. The correctly normalised value of r_P is

$$r_P^2 = g_s^2 \alpha' \frac{V_*}{V} \frac{\alpha'}{R^2} Q_P,$$

where Q_P is an integer, R is the radius of x_5, and V is the volume of the T^4. We were able to compute the momentum in this direction by Kaluza–Klein reduction to be $P = Q/R = RV/(g_s^2 \alpha'^4)$, where R is the length of x_5 and V is the volume of the T^4 on which we could put $x_6, \ldots, x_9$. More generally, we have now

$$P = \frac{Q_P^L - Q_P^R}{R} = \frac{RV r_H^2}{g_s^2 \alpha'^4} \frac{\sinh 2\beta}{2} = \frac{RV r_H^2}{g_s^2 \alpha'^4}\left(\frac{e^{2\beta} - e^{-2\beta}}{4}\right).$$

So we see that the supersymmetric limit is to have only a left-moving momentum excited. The general solution has both left and right momentum excited. What was it we boosted? Well, taking $\beta \to 0$:

$$ds^2 = -\left(1 - \frac{r_H^2}{r^2}\right)dt^2 + dx_5^2 + \left(1 - \frac{r_H^2}{r^2}\right)^{-1} dr^2 + r^2 d\Omega_3^2 + \sum_{i=6}^{9} dx_i^2,$$

simply the five dimensional Schwarzschild solution, times a T^5.

combine this with a third element, a wave in the x_5 direction. Compactifying on x_5 to five dimensions, we get a pointlike object, the extremal Reissner–Nordström black hole, where the $U(1)$ charge is in fact a diagonal combination of the $U(1)$s from the two R–R sector charges and the Kaluza–Klein charge of the momentum!

We can now be a bit more general. There is no reason why we cannot consider having different amounts of the various charges from the three independent sectors, since it is only their orientations which matter for the amount of preserved supersymmetry. So we can have Q_5, Q_1 and Q_P as three independent integers representing the number of D5-branes, D1-branes, and momentum in the compact x_5, respectively. Let us introduce the correctly normalised harmonic functions and write the solution representing this. The metric is (in Einstein frame)

$$g_s^{1/2}ds^2 = H_1^{-3/4}H_5^{-1/4}\left(-dt^2 + dx_5^2 + H_P\left(dt - dx_5\right)^2\right)$$
$$+ H_1^{1/4}H_5^{3/4}\left(dr^2 + r^2\,d\Omega_3^2\right) + V^{1/2}H_1^{1/4}H_5^{-1/4}ds_{T^4}^2,$$
$$(17.27)$$

where $ds_{T^4}^2$, in the (x^6, x^7, x^8, x^9) directions, is the metric on a T^4 with unit volume. Notice that given the orientations of the constituent branes, we can replace the T^4 by a K3 and preserve the same amount of supersymmetry. The results for the entropy count will turn out to be the same, but we will do it more carefully in a later section, since wrapping branes on K3 produces interesting subtleties, due to the enhançon mechanism which we discussed in chapter 15. The x_5 direction is compact with period $2\pi R$. The dilaton and Ramond–Ramond (R–R) fields are given by

$$e^{2\Phi} = g_s^2\frac{H_1}{H_5}, \qquad F_{rtz}^{(3)} = \partial_r H_1^{-1}, \qquad F_{\theta\phi\chi}^{(3)} = 2\,r_5^2\,\sin^2\theta\sin\phi. \quad (17.28)$$

The harmonic functions are given by

$$H_1 = 1 + \frac{r_1^2}{r^2}, \qquad H_5 = 1 + \frac{r_5^2}{r^2}, \qquad H_P = \frac{r_P^2}{r^2}, \qquad (17.29)$$

where the various scales are set by

$$r_5^2 = g_s\alpha'\,Q_5, \qquad r_1^2 = g_s\alpha'\,\frac{V_*}{V}\,Q_1, \qquad r_P^2 = g_s^2\alpha'\,\frac{V_*}{V}\,\frac{\alpha'}{R^2}\,Q_P, \quad (17.30)$$

where $V_* = (2\pi\sqrt{\alpha'})^4$. The properties of the event horizon at $r = 0$ can be computed (which the reader should do), yielding a vanishing surface gravity (and hence Hawking temperature) and a non-vanishing area and hence Bekenstein–Hawking entropy:

$$\mathcal{A}_H = 4\pi^3 V R_z\,r_1 r_5 r_p, \qquad S = 2\pi\sqrt{Q_1\,Q_5\,Q_P}. \quad (17.31)$$

Our goal is to find a microscopic description of this, and we do this next. Notice that the mass of the black hole is computed to be:

$$M = \frac{Q_P}{R} + \frac{Q_1 R}{g_s \alpha'} + \frac{Q_5 R V}{g_s \alpha'^3}, \qquad (17.32)$$

which is just the sum of the Kaluza–Klein mass and the constituent brane charges normalised by the appropriate volume factors arising from where they are wrapped. That there is no interaction energy is consistent with the fact that we are constructing this black hole out of BPS constituents.

Notice that, inevitably, there is an explicit dependence of the mass on the embedding parameters. This is in contrast to the entropy, which is independent of the embedding parameters and so appears to be much more universal. We shall see a reason for this much later.

17.3.2 Microscopic entropy and a 2D field theory

Now we can follow the logic which we used in chapter 10. This geometry is entirely constructed with R–R charged objects, with some momentum. We have established that D-branes are the smallest possible objects carrying those charges, and so we must be able to make the black hole out of D-branes, with some momentum[7].

The case which we consider here is a compactification in which Q_5 D5-branes wrap a T^4, appearing as strings in six dimensions, forming a composite with Q_1 D1-branes. The D1 can only move within the D5-brane world-volume, and so this configuration should remind us of the D1–D5 bound state, which preserves 1/4 of the spacetime supersymmetries. Adding BPS momentum (i.e. purely right-moving) to such a configuration breaks a further 1/2 of the supersymmetries, and so we have a total of four supercharges.

Let us consider the case of $g_s Q \ll 1$, where Q is any of the charges in the solution. Then from the form of the harmonic functions (17.30), it is clear that in this limit we are studying the weakly coupled system of D-branes in flat space. We shall perform the study of the system in this limit initially. The case of $g_s Q > 1$ is where we have a macroscopic black hole, and as we shall see, our results for the counting of the entropy will apply to this case as well. This will appear to be simply due to the fact that we are counting BPS states, but later we shall see that things are more robust than that.

The configuration yields the following decomposition of the spacetime Lorentz group:

$$SO(1,9) \supset SO(1,1) \otimes SO(4) \otimes SO(4), \qquad (17.33)$$

where the first factor acts along the D-string world sheet (t, x^5), the third acts in the rest of the D5-brane world-volume (x^6, x^7, x^8, x^9) and the second in the rest of spacetime (x^1, x^2, x^3, x^4). From the point of view of the D5-brane gauge theory, the D1-branes are bound states in the 'Higgs branch', in which the D1-branes are instantons inside the D5-branes (see section 13.4). This branch is parametrised by the vacuum expectation values (vevs) of 1–5 open strings, which give $4Q_1Q_5$ bosonic and fermionic states, simply the dimension of instanton moduli space. The 'Coulomb branch' of the gauge theory is the situation where the D1-branes become pointlike instantons and then leave the D5-branes[130], ceasing to be bound states. This branch is parameterised by the vevs of 1–1 and 5–5 strings, which ultimately separate the individual D-branes from each other. This takes us away from the black hole, the state of most degeneracy. So we study the 1–5 and 5–1 open string sector, i.e. oriented strings stretching between the D1- and D5-branes. From the counting in section 13.4, we know that we have $4Q_1Q_5$ boson-fermion ground states[7].

N.B. Another way of thinking of this theory is as follows. At strong coupling, it will flow to the infra-red and become a non-trivial conformal field theory (see insert 3.1). It turns out (this is essentially a property of the superconformal algebra) that the number of boson–fermion ground states is directly related to the central charge of the conformal field theory, which in turn is equal to the difference in the number of hypermultiplets and vector multiplets $n_{\mathrm{H}} - n_{\mathrm{V}}$. In this case (things will be different in the case of K3 wrapping later in section 17.5) the number of 1–1 and 5–5 hypermultiplets exactly cancel the number of 1–1 and 5–5 vector multiplets, leaving Q_1Q_5.

Our configuration must be made to carry momentum Q_P in the x^5 direction around which the D-string is wrapped. What we really have is an effective 2D field theory in the (t, x_5) directions on the world-volume of the effective string. The Hamiltonian is $H = Q_P/R$. We are trying to distribute this total momentum amongst the $4Q_1Q_5$ bosons and fermions. This should remind the reader of earlier studies in chapters 3 and 4. It is just like being at level n and trying to distribute the energy among the bosons and fermions in the two dimensional conformal field theories we discussed in chapter (see insert 3.4, p. 92). Here, we have a supersymmetric string moving in the $4Q_1Q_5$ dimensions of the moduli space.

The number of ways, $d(Q_P)$, of distributing a total momentum Q_P amongst the 1–5 and 5–1 strings is given by the partition function:

$$\sum d(Q_P)q^{Q_P} = \left(\prod_{n=1}^{\infty} \frac{1+q^n}{1-q^n}\right)^{4Q_1Q_5}. \tag{17.34}$$

For large Q_P, this gives $d(Q_P) \sim \exp(2\pi\sqrt{Q_1Q_5Q_P})$, and Boltzmann's relation $S = \ln d(Q_P)$ yields precisely the entropy (17.31) we computed for our black hole using the Bekenstein–Hawking area law, in the previous section[7].

Let us pause to admire this result. We have actually counted the degeneracy of BPS states in the limit $g_sQ \ll 1$ where we have D-branes in flat space. When we go to $g_sQ > 1$ and the geometry of the branes will take over, making the black hole with geometry given in (17.27), we can be assured that the degeneracy will be precisely the *same*, because this is not renormalised by any quantum effect. So we have actually found a microscopic description of the black holes, at least for the purposes of counting the entropy. This works for black holes in four dimensions too[268], and with other properties like spin, etc. There are excellent reviews of this in the literature[278]. In fact, as we shall see, it is not really supersymmetry that is protecting us from an awful mismatch between the strong and weak coupling limits, but an important universal structure which will be uncovered later in chapter 18. A sign of this is to perform the counting successfully for a non-extremal black hole[269], which we shall do next.

17.3.3 Non-extremality and a 2D dilute gas limit

A non-extremal generalisation[269] of the solution can be written by exploiting the boost forms of the various components which we noted in subsection 10.2.2 (see equation (17.26)) and insert 17.1, with the following result:

$$g_s^{1/2}ds^2 = Z_1^{-3/4}Z_5^{-1/4}\left(-dt^2 + dx_5^2 + \frac{r_H^2}{r^2}(\cosh\beta dt - \sinh\beta dx_5)^2\right)$$

$$+ Z_1^{1/4}Z_5^{3/4}\left(\left(1 - \frac{r_H^2}{r^2}\right)^{-1}dr^2 + r^2 d\Omega_3^2\right) + V^{1/2}Z_1^{1/4}Z_5^{-1/4}ds_{T^4}^2,$$

$$e^{2\Phi} = g_s^2 Z_1/Z_5, \tag{17.35}$$

where*

$$Z_1 = 1 + \sinh^2\beta_1 \frac{r_H^2}{r^2}; \quad Z_5 = 1 + \sinh^2\beta_5 \frac{r_H^2}{r^2}.$$

---

* The reader might find it worth checking that in the case that all of the R–R charges and the momentum are the same, a reduction to five dimensions gives the isotropic form of the five dimensional Reissner–Nordström black hole given in equation (17.19).

The R–R charges of this solution are as before, while as we learned in insert 17.1, there is now both left- and right-moving momentum in the x_5 direction, creating the non-extremality.

$$P = \frac{Q_P^L - Q_P^R}{R} = \frac{RVr_H^2}{g_s^2\alpha'^4}\frac{\sinh 2\beta}{2} = \frac{RVr_H^2}{g_s^2\alpha'^4}\left(\frac{e^{2\beta} - e^{-2\beta}}{4}\right).$$

The mass of this solution is

$$\widehat{M} = \frac{RVr_H^2}{g_s^2\alpha'^4}\left(\frac{\cosh 2\beta_1}{2} + \frac{\cosh 2\beta_5}{2} + \frac{\cosh 2\beta}{2}\right).$$

Now we can compute the entropy of the solution by computing the area of the horizon at $r = r_H$:

$$S = \frac{2\pi RVr_H^3}{g_s^2\alpha'^4}\left(\cosh\beta_1\cosh\beta_5\cosh\beta\right).$$

Now we study an interesting limit. We take the R–R charge densities to be greater than the momentum densities which in turn is larger than the string scale:

$$r_1^2, r_5^2 \gg r_P^2 \gg \alpha', \tag{17.36}$$

which has the effect of keeping the D-brane component close to extremality but allowing both left and right momenta to survive. We can check this by seeing that the energy above the amount at extremality, computed in equation (17.32), becomes:

$$\widehat{M} - M \simeq \frac{RVr_H^2}{g_s^2\alpha'^4}\frac{e^{2\beta}}{4} = \frac{Q_P^L}{R},$$

and so we see that the extra energy coming from the left-moving sector is simply additive, as though the left- and right-moving components of the system are non-interacting, despite the fact that we are non-extremal. This is called the 'dilute gas' limit, since in the 1+1 dimensional model, a 'gas' of $4Q_1Q_5$ boson-fermion pairs, there is no interaction between the left- and right-moving parts.

A little algebra shows that in this limit we get for the entropy

$$S = 2\pi\left(\sqrt{Q_1Q_2Q_P^L} + \sqrt{Q_1Q_5Q_P^R}\right). \tag{17.37}$$

The microscopic computation for the statistical entropy is just like the one we had before, but with both left- and right-moving sectors. In this dilute limit, since they are decoupled the result is just the sum of the entropies of the two sectors, as we have seen coming from the supergravity.

So again, we have exactly verified with a microscopic computation the entropy of a black hole, now even without the help of supersymmetry.

17.4 Near horizon geometry

Recall that in our earliest examination of extremal black holes in chapter 10, we found that the geometry of the horizon was an interesting place, since the geometry was highly symmetric. The extremal horizon size was controlled entirely by the asymptotic charge at infinity, and not by the details of the embedding of the solution into the supergravity. In fact, there are other special properties of the black hole apparent when the system is embedded in the supergravity.

Just as we saw in the case of the solution for the D6-brane wrapped on K3 in section 15.4, the parameters of the compact solution are just the asymptotic values of fields – the moduli – in the full supergravity. There, we studied a solution where the volume of K3. Here, the radius R of the x_5 circle, and the volume V of the T^4, are the asymptotic values of scalars. In fact, these scalars approach fixed universal values at the black hole horizon, due to what is called the 'attractor mechanism'[267]. The values are fixed by the underlying U-duality algebraic structure of the supergravity. In particular, the area of the horizon itself is fixed in terms of the $E_{6,(6)}$ U-duality invariant, and the parameters which make it up are determined only by the charges measured at infinity and not the details of the geometry or the embedding. In particular, the entropy itself is an $E_{6,(6)}$ U-duality invariant.

We won't study this general issue in any detail here, but refer the reader to the literature[267]. Let us instead directly examine the near-horizon geometry of the black hole that we constructed in the previous sections. Consider the non-extremal black hole solution given in equation (17.35), but in string frame:

$$ds^2 = Z_1^{-1/2} Z_5^{-1/2} \left(-dt^2 + dx_5^2 + \frac{r_H^2}{r^2} \left(\cosh\beta dt - \sinh\beta dx_5 \right)^2 \right)$$

$$+ Z_1^{1/2} Z_5^{1/2} \left(\left(1 - \frac{r_H^2}{r^2} \right)^{-1} dr^2 + r^2 \, d\Omega_3^2 \right) + V^{1/2} Z_1^{1/2} Z_5^{-1/2} ds_{T^4}^2,$$

$$e^{2\Phi} = g_s^2 Z_1 / Z_5. \tag{17.38}$$

The limit we will take is that $g_s Q_1, g_s Q_5$ are large, but Q_P are arbitrary. This means that $r^2 < r_1^2$ and r_5^2, and so we can neglect the 1 in the harmonic functions in which they appear. So we see that the volume of the T^4 has become fixed to $V r_1^2 / r_5^2$, and the dilaton has gone to $e^\Phi = g_s r_1 / r_5$.

In the limit, we get:

$$ds^2 = \frac{r^2}{r_1 r_5}\left(-dt^2 + dx_5^2 + \frac{r_H^2}{r^2}(\cosh\beta dt - \sinh\beta dx_5)^2\right)$$

$$+ \frac{r_1 r_5}{r^2}\left(\left(1 - \frac{r_H^2}{r^2}\right)^{-1}dr^2 + r^2 d\Omega_3^2\right) + \frac{r_1}{r_5}ds_{T^4}^2. \qquad (17.39)$$

It is useful to define

$$\rho_+^2 \equiv r_H^2\cosh^2\beta, \quad \rho_-^2 \equiv r_H^2\sinh^2\beta, \qquad (17.40)$$

and we get

$$ds^2 = \frac{r^2}{r_1 r_5}\left[-\left(1 - \frac{\rho_+^2}{r^2}\right)dt^2 + \left(1 + \frac{\rho_-^2}{r^2}\right)dx_5^2 + 2\frac{\rho_+\rho_-}{r^2}dtdx_5\right]$$

$$+ \frac{r_1 r_5}{r^2}\left(1 - \frac{(\rho_+^2 - \rho_-^2)}{r^2}\right)^{-1}dr^2 + r_1 r_5 d\Omega_3^2 + \frac{r_1}{r_5}ds_{T^4}^2. \qquad (17.41)$$

Finally, after a change of coordinates to $\rho^2 = r^2 + \rho_-^2$, the metric is:

$$ds^2 = -\frac{(\rho^2 - \rho_+^2)(\rho^2 - \rho_-^2)}{r_1 r_5 \rho^2}dt^2 + \frac{r_1 r_5 \rho^2}{(\rho^2 - \rho_+^2)(\rho^2 - \rho_-^2)}d\rho^2$$

$$+ \rho^2\left(dx_5 + \frac{\rho_+\rho_-}{\rho^2}dt\right)^2 + r_1 r_5 d\Omega_3^2 + \frac{r_1}{r_5}ds_{T^4}^2, \qquad (17.42)$$

which can be recognised[294, 295] as a three dimensional black hole solution called the 'BTZ black hole'[296] multiplied by an S^3 and T^4. In fact, the black hole solution can be seen to be asymptotically AdS$_3$, with a length scale ℓ set by $\ell^2 = r_1 r_5$. See insert 17.2. The case $\rho_+ = \rho_-$, gives the extremal 5D black hole, and the near-horizon metric becomes locally AdS$_3 \times S^3 \times T^4$, with an identification on the x_5 circle. This is a situation that we have seen before, where the extreme black hole has a simple, highly symmetric spacetime in the near-horizon limit, with the size of the solution controlled by the asymptotic charges.

The fact that the near-horizon geometry of the black hole is actually AdS$_3$, (times fixed compact spaces) with a black hole in it is interesting. As we shall see in the next chapter, there is remarkable duality proposed which – if correct – ensures that the physics of the 1+1 dimensional theory which was controlling our entropy count is captured entirely by the AdS$_3$ physics. Especially in the case of AdS$_3$, the aspects of the duality relevant to our problem are quite well understood. It is this AdS/CFT duality which seems to ensure that the entropy count was correct, even away from extremality. See insert 17.2.

Insert 17.2.   The BTZ black hole

Consider the action for (2+1)-dimensional gravity with negative cosmological constant $\Lambda = -1/\ell^2$:

$$S = \frac{1}{16\pi G_3} \int d^3x \sqrt{-g}\,(R - 2\Lambda). \qquad (17.43)$$

There is an interesting solution, whose metric is:

$$ds^2_{\mathrm{BTZ}} = -V(\rho)dt^2 + V(\rho)^{-1}d\rho^2 + \rho^2 \left(d\varphi + \frac{4G_3 J}{\rho^2}dt\right)^2,$$

$$V(\rho) = \left(-8G_3 M + \frac{\rho^2}{\ell^2} + \frac{16 G_3^2 J^2}{\rho^2}\right), \qquad (17.44)$$

where φ is periodic, with period 2π. This is the 'BTZ black hole' solution[296], and there are two event horizons are at $\rho = \rho_\pm$, in terms of which we arrived at the solution (17.42), and $\ell^2 = r_1 r_5$ there. The mass and angular momentum of this solution are given by

$$M = \frac{\rho_+^2 + \rho_-^2}{8\ell^2 G_3}, \qquad J = \frac{\rho_+ \rho_-}{4\ell G_3}.$$

Notice that the case $M = -1/8G_3, J = 0$ gives us AdS$_3$ in global coordinates, as given in equation (10.29). The case $M = 0, J = 0$ is also AdS$_3$, but now in local coordinates. In fact, the BTZ spacetime is locally AdS$_3$ everywhere. Since φ is compact, there is a global difference which makes it a non-trivial solution for arbitrary M and J.

Using the techniques presented at the beginning of this chapter, the entropy and temperature may be computed to be

$$S = \frac{2\pi\rho_+}{4G_3} = \frac{\mathcal{A}}{4G_3}, \qquad T = \frac{\rho_+^2 - \rho_-^2}{2\pi\ell^2\rho_+}.$$

The AdS/CFT correspondence, to be discussed in the next chapter, associates a dual (1+1)–dimensional CFT to the physics of AdS$_3 \times S^3$, with[297] $c = 3\ell/2G_3$. In fact, the $M = 0$ and $M = -1/8G_3$ cases can be identified[298] with the NS–NS and R–R ground states of the theory, with energy $E = 0$ or $E = -\ell/8G_3$, where the fermions are either periodic or antiperiodic around φ. (The factor of ℓ results from a conformal rescaling, see section 18.1.3.) Computations we know how to do from chapters 2 and 4 show that the zero point energy difference is $c/12$, which is the result one would get from converting $\ell/8G_3$.

17.5 Replacing T^4 with K3

An important variation on the constructions above is to replace the T^4 in the $x_6, \ldots, x_9$ directions by the K3 manifold instead. In fact, this does not break any more supersymmetries than the D5–D1 orientation, and so in principle, everything should go through trivially. However, as we know from chapters 9 and 15, the wrapping of the D5-branes on the K3 should change things considerably, since the enhançon mechanism ought to modify the geometry significantly in the limit of large charges where the black hole becomes manifest. In fact the original reference considered K3 first[7], and did not take into account the subtleties introduces by K3 in the macroscopic geometry. Our goal in this section is to examine this physics carefully[299]. Their answer for the entropy was not wrong, however, for reasons we shall see. Our careful analysis will produce a new result, however, since it will become clear that the enhançon mechanism works in precise conjunction with the second law of thermodynamics.

17.5.1 The geometry

The Einstein frame metric is:

$$ds^2 = H_1^{-3/4} H_5^{-1/4} \left(-dt^2 + dx_5^2 + H_P(dt - dx_5)^2 \right)$$
$$+ H_1^{1/4} H_5^{3/4} \left(dr^2 + r^2 \, d\Omega_3^2 \right) + V^{1/2} H_1^{1/4} H_5^{-1/4} ds_{\mathrm{K3}}^2, \quad (17.45)$$

where ds_{K3}^2 is the metric on a K3 manifold with unit volume. The other fields and harmonic functions are the same as those listed in equations (17.30).

Of course, the integers Q_1, Q_5 and Q_P appearing in the harmonic functions measure the asymptotic charges associated with the electric and magnetic R–R fluxes and the internal x_5-momentum, respectively. We must, however, introduce another set of integers, N_1 and N_5 to denote the number of D1-branes and D5-branes, respectively, in the system. Clearly we have $N_5 = Q_5$. However, as discussed in chapter 9 and in detail in section 15.4, wrapping the D5-branes on K3 induces a negative D1-brane charge and so we have $Q_1 = N_1 - N_5$ or alternatively $N_1 = Q_1 + Q_5$.

Just like in section 15.4, the volume of the K3 manifold (measured by the string frame metric) is:

$$V(r) = \frac{H_1}{H_5} V, \quad (17.46)$$

where V is the asymptotic volume of the K3. At the horizon, it is:

$$V_{\mathrm{H}} \equiv V(r = 0) = \frac{r_1^2}{r_5^2} V = \frac{Q_1}{Q_5} V_* = \frac{N_1 - N_5}{N_5} V_*, \quad (17.47)$$

and so if $r_1 < r_5$, then $V_H < V$. So we see that as long as $r_1 < r_5$, that the volume K3 is shrinks as we move in from $r \to \infty$. When we reach $V(r) = V_*$ at some radius, new physics will come into play, and this is the 'enhançon' locus we discovered in section 15.4. This radius is easily computed:

$$\hat{r}_e^2 = \frac{g_s \alpha' V_*}{(V - V_*)}(2N_5 - N_1), \qquad \begin{cases} > 0 & \text{for} \quad 2N_5 > N_1 \\ < 0 & \text{for} \quad 2N_5 < N_1 \end{cases}, \qquad (17.48)$$

where $\hat{r}_e^2 < 0$ simply indicates that the K3 volume reaches V_* inside the event horizon. Therefore we see that we can have the enhançon appearing either above or below the horizon, depending upon how we choose the parameters.

Let us consider the case of $\hat{r}_e^2 > 0$. Now when the K3 volume reaches V_*, at the enhançon radius, $\hat{r}_e$, the wrapped D5-branes will be unable to proceed supersymmetrically into smaller radius, due to the fact that their effective tensions are going through zero there. They are therefore forced to form an enhançon sphere at radius $\hat{r}_e$. By contrast, D1-branes and momentum modes can movie inside of $r = \hat{r}_e$: they are not wrapped on K3 and therefore do not care that it is approaching a special radius there. However, notice that the geometry can be made of D1–D5-bound states. The corrections of $-\tau_1$ to the effective tension of the wrapped D5-brane is precisely compensated by the $+\tau_1$ coming from the marginally bound D1-brane. Therefore we can make the above geometry in equations (17.45–17.29) by binding N_5 D1-branes to N_5 the D5-branes we wish to include in the geometry, and bring the resulting N_5 D1–D5 bound states in from infinity, together with Q_1 extra D1-branes.

17.5.2 The microscopic entropy

In the microscopic model we have some modifications to the T^4 situation. We have an effective 1+1 dimensional gauge theory on the effective D-string formed by wrapping the D5-branes and binding it with D1-branes. At strong coupling the theory will flow to a conformal field theory in the infra-red (see insert 3.1, p. 84). The important feature of the conformal field theory is its central charge, which can be computed from the gauge theory as proportional to $n_H - n_V$, the difference between the numbers of hypermultiplets and the number of vector multiplets. Counting the bosonic parts, the D1-branes contribute N_1^2 vectors and N_1^2 hypers, the latter coming from (x^6, x^7, x^8, x^9) fluctuations. The D5-branes contribute N_5^2 vectors, but there are no massless modes coming from oscillator excitations in the (x^6, x^7, x^8, x^9) (K3) directions. There are, in addition, 1–5 strings which give $N_1 N_5$ hypermultiplets. Evaluating the difference gives: $N_1 N_5 - N_5^2 = Q_1 Q_5$ hypermultiplets. Hence in total, there are $4Q_1 Q_5$ bosonic excitations and an equal number of fermions, since a hypermultiplet contains four scalars and their superpartners.

In another language all that we have done is evaluated the dimension the Higgs branch of the D5-brane moduli space of vacua, where the N_1 D1-branes can become instanton strings of the $U(N_5)$ gauge theory on the world-volume of the D5-branes. The vacuum expectation values of the 1–5 strings is precisely what constitutes this branch. In this language, the absence of hypers coming from the 5–5 sector corresponds to the absence of Wilson lines on the K3 surface (there are no non-trivial one-cycles). The entropy count then goes precisely along the lines of section 17.3.2.

Let's close this discussion by observing that we have a mild apparent conflict with the microscopic description. For $N_1 < 2N_5$, we know from the analysis of the previous section that, at any given value of the momentum, the entropy can be maximised by using only $N_1/2$ of the D5-branes in the problem. So, from the field theory point of view it appears to be favourable to Higgs the $U(N_5)$ gauge theory leaving massless only a $U(N_1/2)$ subgroup. But since all of these supersymmetric vacua are degenerate, all black holes appear to be on the same footing.

This is really an artifact of the thermodynamically peculiar situation that we are at zero temperature while having a finite entropy, so the entropy strictly has a meaning as a degeneracy of ground states. Processes which maximise the entropy require dynamics, and so we must take the system away from extremality in order that it can explore configuration space, and find the maximal entropy black hole.

17.5.3 Probing the black hole with branes

Let us illustrate the above statements with some probe computations[299]. Both D1- and D5-branes are natural probes of the geometry[266], since they preserve the same supersymmetries. Consider a composite probe brane consisting of n_5 D5-branes and n_1 D1-branes. It is important for the physics of the following that this composite probe is in the D5-branes' Higgs phase. That is, this composite probe is *not* simply a collection of individual D5-branes and D1-branes moving together, but rather the D1-branes have been absorbed as instanton strings lying along the z-direction in the D5-brane world-volume. These instantons are maximally smeared over the K3 directions and that we have chosen the orientation of the vevs of the hypermultiplets arising from 1–5 strings such that the instantons are of maximal rank in the $U(n_5)$ gauge theory. In this phase, the composite probe brane is then a true bound state, i.e. the fields describing the relative separation of the branes in the Coulomb phase are all massive.

The effective action for the composite brane probe regarded as an effective string is

$$S = - \int_\Sigma d^2\xi \; e^{-\Phi(r)} (n_5\tau_5 V(r) + (n_1 - n_5)\tau_1)(- \det g_{ab})^{1/2}$$

$$+ n_5\tau_5 \int_{\Sigma \times \text{K3}} C^{(6)} + (n_1 - n_5)\tau_1 \int_\Sigma C^{(2)}, \qquad (17.49)$$

where Σ is the unwrapped part of the brane's world-volume, with coordinates $\xi^{0,1}$. Remember in the above action that the wrapping of the D5-branes on the K3 introduces negative contributions to both the tension and two-form R–R charge terms. Recall that g_{ab} is the pull-back of the string-frame spacetime metric. The background fields in which the probe moves are those of the black hole solution given in equation (17.28). The corresponding R–R potentials may be written as

$$C^{(6)} = g_s^{-1} H_5^{-1} dx^0 \wedge dx^5 \wedge \varepsilon_{\text{K3}}, \quad C^{(2)} = g_s^{-1} H_1^{-1} dx^0 \wedge dx^5, \qquad (17.50)$$

where ε_{K3} denotes the volume four-form on the K3 space with unit volume. These R–R potentials do *not* vanish asymptotically because we choose a gauge which eliminates a constant contribution to the energy which would otherwise appear.

We will now choose static gauge, aligning the coordinates of the effective probe string with the x^5 direction and letting it move in the directions transverse to K3 while freezing and smearing the degrees of freedom on the K3:

$$\xi^0 = x^0 \equiv t, \quad x^i = x^i(t), \quad i = 1, 2, 3, 4. \qquad (17.51)$$

The result can be written as an effective Lagrangian $\mathcal{L}$ for a particle moving in the (x^1, x^2, x^3, x^4) directions:

$$\mathcal{L} = \frac{1}{2} (n_5\tau_5 V H_1 + (n_1 - n_5)\tau_1 H_5)(1 + H_P) \left[\dot{r}^2 + r^2 \dot{\Omega}_3^2 \right], \qquad (17.52)$$

where, as usual, a dot is used to denote $\partial/\partial t$, and

$$\dot{\Omega}_3^2 = \dot{\theta}^2 + \sin^2\theta(\dot{\phi}^2 + \sin^2\phi\, \dot{\chi}^2).$$

As should be expected by now, here is no non-trivial potential, since supersymmetry cancelled the mass against the R–R charge as in previous computations of this type.

The effective tension of the probe is given by the prefactor in equation (17.52). We can already see that there is the possibility that the tension will go negative when $n_5 > n_1$.

Putting in the definitions of the harmonic functions given in equations (17.29) and (17.30), we see that the tension remains positive as long as

$$(n_5 \tau_5 V H_1 + (n_1 - n_5) \tau_1 H_5) > 0 \qquad (17.53)$$

which translates into

$$r^2 > \tilde{r}_e^2 = g_s l_s^2 \, V_* \frac{(2N_5 - N_1)n_5 - N_5 n_1}{(V - V_*)n_5 + V_* n_1}. \qquad (17.54)$$

It is worth considering some special cases of this result. If we remove all of the D5-branes, the result for pure D1-brane probes is quite simple, as setting n_5 to zero in the above result gives:

$$\mathcal{L}_{D1} = \frac{1}{2} n_1 \tau_1 H_5 (1 + H_P) \left[\dot{r}^2 + r^2 \dot{\Omega}_3^2 \right], \qquad (17.55)$$

since the D1-brane is not wrapped on the K3 and so its tension is positive everywhere. It simply floats past the enhançon radius on its way to the origin without seeing anything particularly interesting there.

Note that the result (17.55) is the same as would have been obtained in the case of probing for a T^4 compactification, considering only motion in the directions transverse to the torus. Similarly in the case that $n_1 = n_5$, we get:

$$\mathcal{L} = \frac{1}{2} n_5 \tau_5 H_1 (1 + H_P) \left[\dot{r}^2 + r^2 \dot{\Omega}_3^2 \right], \qquad (17.56)$$

which is the same as the result for pure D5-brane probes in the case where they are wrapped on T^4. The cancellation of the induced tensions from K3 wrapping and non-trivial instanton number in constructing the bound state probe provided a simple result: the wrapped D5-branes, when appropriately dressed with instantons, can indeed pass through the enhançon shell.

If we instead remove all of the D1-branes, we just get the familiar result of section 15.4 that the probe, made of pure D5-branes, hangs up at the enhançon radius $\hat{r}_e$ given by equation (17.48). Now we discover that our earlier enhançon result is just a special case of a more general result: whenever there are more D5-branes than D1-branes making up the probe (i.e. $n_5 > n_1$), there is a generalisation of the enhançon radius, $\tilde{r}_e^2$, where the composite probe will become tensionless and must stop. Notice that this happens in a 'substringy' regime where $V_{K3} < V_*$.

17.5.4 *The enhançon and the second law*

The entropy and area of the black holes which we construct are given by the formula

$$S = \frac{\mathcal{A}}{4G} = 2\pi\sqrt{Q_1 Q_5 Q_P} = 2\pi\sqrt{(N_1 - N_5)N_5 Q_P}, \qquad (17.57)$$

where in the second equality we have written it in terms of the number of physical branes of each type. Let us consider the dependence of the entropy on the number of D5-branes. Fixing N_1 and Q_P, we see that it gives a half an ellipse, as depicted in figure 17.1. We see that there are *maximal entropy* black holes that we can make, (corresponding to the apside of the ellipse) which are those for which $N_5 = N_1/2$, or in other words $Q_1 = Q_5$.

If we wish to consider the maximum entropy that can be achieved for a given set of parameters, N_1, N_5 and Q_P, we observe that the behaviour of this entropy changes at precisely $N_1 = 2N_5$. In figure 17.2 is a plot of the (square of the) maximal entropy as a function of Q_1 for fixed N_5 and Q_P. For a 'large' number of D1-branes ($N_1 > 2N_5$), the maximal area squared is simply proportional to Q_1, as expected from equation (17.57). However, for a 'small' number of D1-branes ($N_1 < 2N_5$), the entropy is maximised if only $N_5' = N_1/2$ of the available D5-branes participate in the formation of the black hole. In this regime, we have

$$\mathcal{A}_{\text{max}}^2 \propto N_1^2 = (Q_1 + Q_5)^2 \qquad (17.58)$$

and so the curve becomes a parabola which only reaches zero at $Q_1 = -Q_5$.

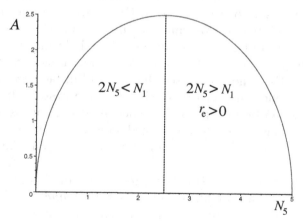

Fig. 17.1. The horizon area as a function of N_5, for fixed $Q_P \, (= 1)$ and $N_1 \, (= 5)$, which forms half of an ellipse. As the number of D5-branes increases past $N_1/2$, the area decreases.

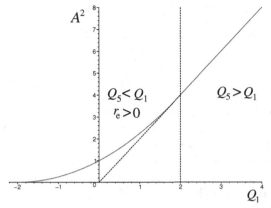

Fig. 17.2. The square of the maximal horizon area as a function of Q_1, for fixed Q_P (=1) and Q_5 (=2). For $N_1 > 2N_5$, the (area)^2 increases linearly. For $N_1 < 2N_5$, to maximise the area, one must use only $N_1/2$ of the available D5-branes (see figure 17.1), and therefore the dependence on N_1 is quadratic.

Note that in this regime, the maximum entropy is greater than one would calculate from equation (17.57). Assuming the excess D5-branes have accumulated in an enhançon shell around the black hole, the maximum entropy configuration corresponds to precisely that where the K3 volume is frozen at V_* throughout the interior region.

Let us return to the curve in figure 17.1. If we were to begin with a black hole with a 'large' number of D1-branes, we would be on a point on the left hand side of the ellipse in the figure. We may now consider increasing the number of D5-branes in the system by bringing them one at a time from infinity. As a result the black hole moves up the ellipse to the extremum at $N_5 = N_1/2$. At this point, however, if we were to add one more D5-brane, we we see that we will in fact *decrease* the horizon area, and hence the entropy of the resulting system. In principle we can bring this D5-brane up to the black hole horizon as slowly as we like, and so we have found a way of reducing the entropy of the hole by an adiabatic process. This is a violation of the second law of thermodynamics.

Actually, there is a very satisfying resolution of this problem[299]. It is precisely for this class of black holes that *the enhançon appears above the horizon*. So an attempt to bring our extra D5-brane into the hole is thwarted by the fact that it will be forced to stop at the enhançon radius $\hat{r}_e$ just above the horizon. We could bind the extra D5-brane with an extra D1-brane to bring it in, but in this case Q_1 remains fixed while

Q_5 increases. Therefore adding the D1/D5 bound state to the black hole increases the entropy.

If we begin with a black hole on the right half of the ellipse ($N_1/2 < N_5 < N_1$), the enhançon again ensures that we cannot move further to the right decreasing the horizon area by dropping D5-branes into the black hole. These were configurations where the black hole is already surrounded by a region where $V(r) < V_*$ and hence the extra D5-branes are restrained from reaching the horizon by the enhançon mechanism.

However, we have seen in section 17.5.3 that D1/D5 bound states can move through such regions where $V(r) < V_*$ and so we must still investigate if we are able to decrease the entropy by sending in a bound D1/D5 probe brane. Adopting the previous notation, let the probe consist of a bound state with n_1 D1-branes and n_5 D5-branes. Assuming that the black hole already contains many more of each type of brane, i.e. $n_1, n_5 << N_1, N_5$, dropping in such a probe would cause an infinitesimal shift in the entropy (squared) given by

$$\delta S^2 = 4\pi^2 Q_P \left(N_5 n_1 + (N_1 - 2N_5)n_5 \right). \qquad (17.59)$$

Note that implicitly we are assuming $N_1, N_5, n_1, n_5 > 0$. Even so the expression in parentheses has the potential to be negative, which would signal a decrease in the black hole entropy. However, we found that this expression also appears in the numerator of equation (17.54) for the radius of vanishing probe tension, but with the opposite sign. Hence the probe-brane finds no obstacle to dropping inside the horizon only in those situations where the entropy increases. Precisely in those cases where second law would be violated, the enhançon locus filters out the wrong type of D1–D5 bound states from reaching the event horizon. Thus the enhançon provides string theory with precisely the mechanism needed to maintain consistency with the second law of black hole thermodynamics[299, 300].

18

D-branes, gravity and gauge theory

As we learned in section 10.2, there are effectively two descriptions of the low energy dynamics of branes. One description uses the collective dynamics of the effetive world-volume field theory. In the case of N coincident D-branes, this is captured in string theory by the open string sectors which give a $U(N)$ gauge theory with sixteen supercharges. The other description treats the brane as a soliton-like source of the various low energy closed string fields in the superstring theory. As such it has a description in terms of a classical solution of the low energy field equations. In both cases, we must remember that there is a whole tower of stringy dynamics which sits on top of this low energy physics, and we must understand in which situations this tower can be made irrelevant, or at least kept under control by a sensible expansion. To control string loops, we must work in a regime where g_s is small, so that we can trust the classical action that we wrote down for the supergravity. Similarly, working in the $\alpha' \to 0$ limit ensures that we can safely ignore the possibility of the massive string states introducing corrections to our supergravity, and in the open string sector, that the truncation to gauge theory of the full Born–Infeld, etc., action, is sensible. In this chapter, we will follow this limit quite some way, and the two complementary descriptions will lead to a sharp statement of a duality between two traditionally disparate fields: large N gauge field theory and gravity. This is a natural and logical outcome of many of the gauge theory and geometry connections we have already been noticing throughout this book.

18.1 The AdS/CFT correspondence

18.1.1 Branes and the decoupling limit

We have already learned from our moduli space probe computations that the specialisation of the results to gauge theory can be achieved by taking $\alpha' \to 0$ while keeping finite some characteristic gauge theory quantity of interest, such as a typical vacuum expectation value of a massless 'Higgs' field. In geometries already considered, this corresponded to, for example, keeping fixed a scaled radial coordinate $u = r/\alpha'$ as we send $\alpha' \to 0$, which also meant that $r \to 0$. In other words, we must approach the core or horizon of the solution closely. In these limits, we found that the remaining supergravity quantities which survived the limit in combinations have physical meaning in the gauge theory on the probe, such as the gauge coupling, etc. Let us see if we can take this further.

Let us consider the case of the extremal D3-brane, initially. At low energy, on the world-volume (ignoring the overall $U(1)$ corresponding to the centre of mass) there is a $U(N)$ gauge theory with $\mathcal{N} = 4$ supersymmetry in four dimensions, i.e. sixteen supercharges. The gauge coupling is $g_{\mathrm{YM}}^2 = 2\pi g_{\mathrm{s}}$. The gauge multiplet contains the vector, A_μ, six scalars ϕ_i, $i = 1, \ldots, 6$ (representing the transverse motions), and four two-component Weyl fermions, λ_a, $a = 1, \ldots, 4$, the fermionic superpartners of the eight physical bosonic degrees of freedom. There is a $SO(6) \simeq SU(4)$ R-symmetry under which the scalars transform as the **6** and the fermions transform as the **4**. The theory is conformally invariant, (i.e. it has vanishing β-function) with the conformal group being $SO(2, 4)$, which contains the Poincaré group, the dilatations, etc., as discussed in sections 3.1 and 10.1.9.

The low energy supergravity solution is:

$$ds^2 = H_3^{-1/2}\eta_{\mu\nu}dx^\mu dx^\nu + H_3^{1/2}dx^i dx^i,$$
$$e^{2\Phi} = g_{\mathrm{s}}^2,$$
$$C_4 = (H_3^{-1} - 1)g_{\mathrm{s}}^{-1}dx^0 \wedge \cdots \wedge dx^3, \tag{18.1}$$

where $\mu = 0, \ldots, 3$, and $i = 4, \ldots, 9$, and the harmonic function H_3 is

$$H_3 = 1 + \frac{4\pi g_{\mathrm{s}} N \alpha'^2}{r^4}. \tag{18.2}$$

We are instructed to send $\alpha' \to 0$, and hold a quantity $u = r/\alpha'$ fixed. This limit focuses on the neighbourhood of the horizon of the brane, and a computation gives the following metric[270]:

$$\frac{ds^2}{\alpha'} = \frac{u^2}{L^2}(-dt^2 + dx_1^2 + dx_2^2 + dx_3^2) + L^2\frac{du^2}{u^2} + L^2 d\Omega_5^2;$$
$$^*dC_{(4)} = 16\pi\alpha'^2 N\epsilon_{(5)}.$$
$$\text{where}\quad L^2 = \sqrt{2g_{\text{YM}}^2 N} = \frac{\ell^2}{\alpha'}. \tag{18.3}$$

We have written it such that we can see the lengths measured by the metric in units of the string length.

From our work in section 10.1.9, we recognise this solution as the metric of $\text{AdS}_5 \times S^5$, where the cosmological constant and the radius of the sphere is set by $\ell^2 = \alpha'\sqrt{2g_{\text{YM}}^2 N}$, a combination of the supergravity/string theory parameters which gives the gauge coupling. Just as in the case of the Reissner–Nordström horizon, the near-horizon geometry of the D3-brane is a smooth 'throat' geometry, with size set by the charge of the solution. Since, as was discussed in chapter 10, $\text{AdS}_5 \times S^5$ is a maximally symmetric solution, just like Minkowski space, we see that the extremal D3-brane is an interpolating soliton solution, like extremal Reissner–Nordström[65] (see insert 1.4). The extremal M-brane solutions of section 12.6.1 also share this property[68].

Let us observe that the limit of small r is also a sensible restriction to low energy from the point of view of supergravity. Recall an effect which is familiar from considerations of ordinary gravity solutions such as black holes. There is a redshift effect, which means that the energy, as measured at asymptotic infinity, of a signal originating at radius r is decreased due to a multiplicative factor $\sqrt{g_{tt}(r)} = H_3^{-1/4}$, arising from having to climb out of the gravitational well produced by the solution. This redshift is infinite as $r \to 0$, and so the throat is decoupled from the asymptotic regime in the low energy limit.

Now we should ask about the regime of validity of this geometry. We have to examine the amount of curvature this solution has, and this is set by the size of a typical squared curvature invariant, R^2. We have sent α' to zero and also are keeping g_{s} small, to remain in the supergravity regime (string tree level). Looking at the essential controlling function (18.2), we see that we have one more parameter we can adjust, and this is N. If we make N large, we can keep the curvatures low. More properly, if we keep the effective coupling $\lambda = g_{\text{YM}}^2 N$ large enough, we can ensure that we stay at closed string tree level and decouple the higher string modes by sending $g_{\text{s}}, \alpha' \to 0$. Notice that this limit is the regime that open string and hence gauge theory perturbation theory breaks down. So we have a useful complementarity. The large N, strong 't Hooft coupling limit of the gauge theory has a description in terms of the supergravity solution above. This is the '*AdS/CFT correspondence*'[270, 271, 272]. The corrections to this in a $1/N$ expansion are the usual stringy loop corrections to the supergravity. In fact, the string coupling is to be identified with $1/N$, just as was anticipated long ago in general terms for gauge theories at large N (see insert 18.1). We have a concrete realisation of this conjectured string theory as type IIB on $AdS_5 \times S^5$. Notice that the $SO(4,2)$ and $SO(6)$ isometries of each space become the conformal group and the R-symmetry of the gauge theory.

Before we go any further, let us therefore compute the five dimensional Newton constant, G_5 in terms of our compactification on an S^5 of radius ℓ. We get $1/G_5 = (\text{Vol}(S^5)\ell^5)/G_{10}$. Looking at our expression for G_{10} in equation (7.44), and the one for ℓ in equation (18.3), it is prudent to substitute for ℓ^8, giving our first precise entry in the AdS/CFT dictionary:

$$G_5 = \frac{\pi \ell^3}{2N^2},\qquad(18.4)$$

since the volume of a unit S^5 is π^3. This will be useful to us many times later, since we will want to convert gravitational quantities to gauge theory ones. Notice that this formula also confirms for us in five dimensional terms that for fixed string length, the large N limit keeps us at tree level in the effective string theory, and hence just gravity. The effective closed string coupling is $g_{\text{eff}} \sim 1/N$.

18.1.2 Sphere reduction and gauged supergravity

We have arrived at a remarkable connection between a particular large N gauge theory (pure $\mathcal{N} = 4$ supersymmetric $D = 4$ $SU(N)$) and a truncation of type IIB string theory on $AdS_5 \times S^5$. For many purposes, it is useful to think of this as simply a five dimensional theory, obtained by the analogue of Kaluza–Klein reduction on S^5. The resulting five dimensional theory is in fact a five dimensional $\mathcal{N} = 8$ 'gauged supergravity', with 15 vector fields acting as gauge bosons of an $SO(6)$ gauge symmetry. There are in fact 42 scalars in the theory, which in general are charged under the $SO(6)$.

One way to think of how to arrive at the gauged supergravity theory and the resulting solution we are studying is as follows[302]. Start with the T^5 reduction of type IIB, which gives an $\mathcal{N} = 8$ theory in five dimensions with a *global* $E_{6(6)}$ symmetry. The 42 scalars ϕ_i in the resulting theory live on the coset $E_{6(6)}/USp(8)$. We discussed this theory in the context of U-duality in section 12.7, where we saw that wrapped branes filled out the various multiplets of the symmetry. Starting with this theory, it is possible

Insert 18.1. The large N limit and string theory

Quite general considerations[301] can lead to a search for a string theory description of gauge theory at large N. In the commonly used scaling, a power of g_{YM} is absorbed into the fields in order to write the Lagrangian as $\mathcal{L} \sim -\mathrm{Tr}F^2/(4g_{YM}^2)$, and this is the only appearance of g_{YM}^2. So there is an overall N/λ (where $\lambda = g_{YM}^2 N$ is the "'t Hooft coupling') in front of the entire Lagrangian. Hence, vertices in Feynman graphs come with a factor N, while propagators come with $1/N$. Feynman graphs are drawn with a double line, one line carrying an index in the fundamental, the other an antifundamental: the full propagator is in the adjoint. (This might remind the reader of open string diagrams of chapter 2.) A closed loop will contribute an N, since a free index can run over all its N values. The reader might like to consider some vacuum graphs (appropriate to any field theory with adjoint fields):

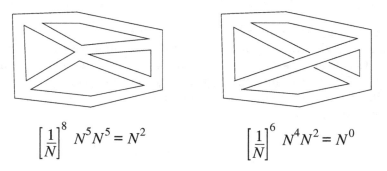

$$\left[\frac{1}{N}\right]^8 N^5 N^5 = N^2 \qquad\qquad \left[\frac{1}{N}\right]^6 N^4 N^2 = N^0$$

A graph with E edges (propagators), V vertices (interactions) and F faces (closed loops) can be drawn on a surface of Euler number $\chi = F - E + V = 2 - 2h$. The second equality is familiar from chapter 2, relating to the genus (number of handles h) of a closed Riemann surface. Overall, a graph comes with a factor N^χ and is some polynomial in λ, and so planar (sphere) diagrams dominate at large N, followed by toroidal, etc. As this is reminiscent of a closed string theory diagram of the same topology, this suggests the identification $g_s \sim 1/N$. With reasoning along these lines, it was suspected that there might be stringy descriptions of large N gauge theories, where the string coupling is related to N as above. The difficulty was trying to find such a string theory. It surely could not be one of the strings used for 'theories of everything', since those would be too simple, it was thought. Now we see that we can use such strings, but propagating on interesting spacetimes, as we shall uncover.

to make some of the global symmetry local, gauging by letting some of the vectors of $E_{6(6)}$ enter into the covariantised versions of the derivatives. In fact, to achieve gauge invariance, such a procedure ultimately leads one to go beyond minimal coupling and generate potentials for the scalars, and the largest subgroup that can be consistently gauged turns out to be $SO(6) \subset E_{6(6)}$. There is a non-trivial potential, $V(\phi_i)$, for the scalars, coming from the non-minimal coupling procedure, and the effective theory is of the form (looking just at the bosonic gravity and scalar sector):

$$ S = \frac{1}{16\pi G_5} \int d^5x \sqrt{-G} \left\{ R - \mathcal{G}_{ij} \partial_\mu \phi^i \partial^\mu \phi^j - V(\phi_i) \right\}. \qquad (18.5) $$

AdS$_5$ with the particular value of the cosmological constant $\Lambda = -6/\ell^2$ and with an $SO(6)$ gauge symmetry is a very special solution to this. It is a *fixed point* for the scalars, and so $\partial \phi_i / \partial x^\mu = 0$ and they all vanish $\phi_i = 0$, and so $SO(6)$ is preserved, since there are no non-zero fields charged under it in this case. The value of the potential is $V(\phi_i = 0) = -12/\ell^2$ and so we have:

$$ S = \frac{1}{16\pi G_5} \int d^5x \sqrt{-G} \left\{ R - 2\Lambda \right\}, \qquad (18.6) $$

for which the maximally symmetric solution is AdS$_5$ with $\Lambda = -6/\ell^2$.

This way of thinking of AdS is quite useful, since it leads immediately to an intuitive understanding of what is going on in the gauge theory in more complicated situations we will encounter in chapter 19.

The other way to think of our $SO(6)$ symmetric solution is in ten dimensional terms, which is how we began. However, it can be thought of as a Kaluza–Klein truncation of the ten dimensional theory by placing it on an S^5. The ansatz that is used is that the five-form $F_{(5)}$ is set by some constant times the volume five-form $\epsilon_{(5)}$ of the S^5:

$$ F_{(5)} = 4\ell^4 \epsilon_{(5)} + 4\ell^{4*} \epsilon_{(5)}. $$

This is the 'Freund–Rubin ansatz'[19], and with this choice, the ten dimensional equations of motion decompose into two sectors:

$$ R_{\mu\nu} = -\frac{4}{\ell^2} g_{\mu\nu}, \qquad R_{mn} = +\frac{4}{\ell^2} g_{mn}, $$

(with μ, ν having Lorentzian signature $(-++++)$ and m, n having Euclidean $(+++++)$) which is the precise generalisation of that which we saw happen for the Reissner–Nordström solution in section 10.1.11. The maximally symmetric solutions are of course AdS$_5$ and S^5.

18.1.3   Extracts from the dictionary

Recall that we identified the scaled radial coordinate u as representing an energy scale in the gauge theory. It is natural to therefore to consider the limit $u \to 0$ as the infra-red (IR) and $u \to \infty$ as the ultra-violet (UV). We must not forget that our theory is defied as strongly coupled even in the UV, since it is conformal, and we must keep $\lambda = g_{\mathrm{YM}}^2 N$ large to remain within gravity.

The limit $u \to \infty$ defines a natural boundary of AdS. In the coordinates used in (18.3) any radial slice is in fact four dimensional Minkowski space, but $u = \infty$ is special for us, since on the one hand it takes a finite time for massless particles to propagate from $u = 0$, reflect at $u = \infty$, and return. On the other hand, the UV is the natural limit in which to discuss intuitive objects in gauge theory, like pointlike operator insertions.

Notice that large u seems like an IR limit for the AdS side of the duality, while it is UV on the side of the CFT. This is a feature of what is known as the 'UV/IR correspondence'. (See also the discussion before equation 6.1.)

When the common phrase 'the dual theory lives on the boundary', or some variation of it, is used, it should be understood that it is a shorthand for this UV identification. There are many properties (or quantities within) the dual which cannot unambiguously be placed at the boundary, and so we should be careful. It is better to think of the dual theory as being *everywhere**, and a slice at some value of u simply focuses on the effective theory obtained by working at a cutoff defined by the energy u, and the background geometry has metric $\gamma_{\mu\nu} = (\ell^2/u^2)h_{\mu\nu}$, where $h_{\mu\nu}$ is the metric induced on the boundary by restricting the five dimensional metric to constant u.

Recall that our coordinates inherited from the brane solution put us on AdS in local coordinates. We know from section 10.1.9 that we can consider this as a local patch of global AdS_5, and so it is natural to consider the same field theory dual to AdS in these coordinates. For example, for global AdS_5 we write:

$$ds^2 = -\left(1 + \frac{r^2}{\ell^2}\right)dt^2 + \left(1 + \frac{r^2}{\ell^2}\right)^{-1} dr^2 + r^2 d\Omega_3^2. \tag{18.7}$$

Going to the radial slice at infinity, we see that the dual theory is on a background $\mathbb{R} \times S^3$ with metric $ds^2 = -dt^2 + \ell^2 d\Omega_3^2$, which is the Einstein static universe. The local coordinates before had us studying a system for which the dual is on $\mathbb{R} \times \mathbb{R}^3$. There is the temptation to be confused at

---

* Frustratingly, perhaps, even better is not to think of the dual gauge theory as anywhere in the five or ten dimensional spacetime at all. It is simply the dual.

this point, since we are supposed to think of the physics as independent of the coordinates, but somehow this seems to matter here. In fact, we must recall that in making the choices of coordinates here, we are also choosing a specific time slicing. This means that we are making choices which will affect our definition of the Hamiltonian of the theory. Further, since the radial coordinate seems to refer to the energy scales within the dual theory, being able to choose alternative radial foliations would seem to make good physical sense, since it refers to a choice of regulator at a given energy scale.

A large part of the rest of the dictionary of the AdS/CFT correspondence comes from equating the partition functions of the two theories:

$$Z_{\text{AdS}}(\phi_{0,i}) = Z_{\text{CFT}}(\phi_{0,i}). \tag{18.8}$$

Here the quantities $\phi_{0,i}$ have two interpretations: On the gravity side, these fields correspond to the boundary values (i.e. at $r = \infty$) for the bulk fields ϕ_i which propagate in AdS. This includes not just the 42 scalars, but *all fields*, including the graviton and the gauge fields. On the field theory side, the $\phi_{0,i}$ correspond to external sources or currents coupled to operators in the CFT. We can then obtain insertions of these operators by differentiation of the partition function with respect to the sources. This fits rather nicely, since all fields on the gravity side have specific $SO(6)$ gauge charges, which matches the corresponding R-charge of the inserted operator.

In fact, there is a specific[271, 272, 274] relation which can be derived between the dimension Δ of an operator which a scalar couples to and the mass m of the scalar in the bulk spacetime. As a solution to the wave equation in AdS$_5$, a scalar $\phi_i(r, \mathbf{x})$'s asymptotic behaviour is in fact:

$$\phi_i(r, \mathbf{x}) = e^{\frac{r}{\ell}(\Delta_i - 4)} \phi_m(\mathbf{x}) + e^{-\frac{r}{\ell}\Delta_i} \phi_v(\mathbf{x}), \tag{18.9}$$

where

$$\Delta_i = 2 + \sqrt{4 + m_i^2 \ell^2}. \tag{18.10}$$

In fact, the first term is a non-normalisabile solution while the other term is normalisable. They both have a meaning in the theory. The first term is interpreted as switching on or 'inserting' the operator, while the latter term has the interpretation as controlling the vacuum expectation value of the operator. We shall see this interpretation in action with specific examples later on. In fact, there is a generalisation of this to the case of a p-form field in AdS$_D$. It couples to a $(D - p - 1)$-form operator and the dimension is:

$$\Delta = \frac{D - 1 - 2p}{2} + \left\{ \frac{(D - 1 - 2p)^2}{4} + m^2 \ell^2 - p(p - D + 1) \right\}^{1/2}. \tag{18.11}$$

We list how the basic 42 scalars are interpreted in the gauge theory in table 18.1. In the table, the trace is over the adjoint of the gauge group,

Table 18.1. *An extract from the dictionary of the AdS/CFT correspondence*

Scalar	$m^2\ell^2$	Operator	$SO(6)$ charge	Δ
Φ	0	$\mathrm{Tr}\,[{}^*F \wedge F]$	**1**	4
$C_{(0)}$	0	$\mathrm{Tr}\,[F \wedge F]$	**1**	4
φ_1	-3	$\mathrm{Tr}[\lambda_{(a}\lambda_{b)}]$	**10**	3
$\bar\varphi_1$	-3	$\mathrm{Tr}[\bar\lambda_{(a}\bar\lambda_{b)}]$	$\overline{\mathbf{10}}$	3
φ_2	-4	$\mathrm{Tr}[\phi_{(i}\phi_{j)}] - \frac{1}{6}\delta_{ij}\mathrm{Tr}\,[\phi_k\phi_k]$	**20**	2

under which every field transforms. In the first two lines we recognise our two friends from ten dimensions, the dilaton and the R–R scalar. As is to be expected, the dilaton couples to the basic Yang–Mills Lagrangian of dimension four, since its asymptotic value sets the string coupling (a fact we know from way back in chapter 2), and $g_{\mathrm{YM}}^2 = 2\pi g_{\mathrm{s}}$. Similarly, we know that the R–R scalar couples to the topological term of dimension four, controlling instanton, from our studies in chapter 9. These fields were not involved in the sphere reduction and so do not have and non-trivial $SO(6)$ charges.

The rest of the 42 scalars couple a different class of operators. The first set, with $m^2 = -2/\ell^2$ couples to a symmetrised product of the scalars ϕ_i, with the trace removed.

> N.B. The reader may be disturbed by the fact that the scalars can have a negative mass squared. It turns out that the presence of negative cosmological constant requires us to reexamine the issue of stable scalar fluctuations about the vacuum. The result is that there is a window of squared masses below zero up to the value $-4/\ell^2$ which is stable. This lower bound is the '*Breitenlohner–Freedman*' bound[303], and its negativity is a crucial feature which helps the dictionary to work.

Recalling that the scalars are in the **6** of $SO(6)$, a little group theory confirms that $\mathbf{6} \otimes \mathbf{6} = \mathbf{20} \oplus \mathbf{15} \oplus \mathbf{1}$, where the latter is the trace, and the previous two are the symmetric and antisymmetric combinations. In fact, there is a whole tower of Kaluza–Klein harmonics which can be made by such symmetrised products of the scalars. The reader might recognise these from group theory as the spherical harmonics of S^5, which are indeed made this way most commonly. These operators, which are the *'chiral primaries'* of the conformal field theory, couple to the basic tower of scalars which arise in the Kaluza–Klein spectrum with these same $SO(6)$ transformation properties.

The other set of scalars with $m^2 = -3/\ell^2$, couple to an antisymmetrised product of the fermions λ_a. These transform in the **4**. Representation multiplication gives $\mathbf{4} \otimes \mathbf{4} = \mathbf{10} \oplus \mathbf{6}$, and since the λ_a are fermions, it is the **10** which is picked out. A similar structure exists for the $\bar{\lambda}_a$, which are in the $\bar{\mathbf{4}}$, and hence give the $\overline{\mathbf{10}}$.

Correlation functions of the various operators in the CFT can be determined through calculation involving the dynamics of the various scalars to which they couple, propagating in the AdS spacetime. This is a very powerful technique which we do not have time to explore here.

Just as we did in the case of black hole studies at the beginning of chapter 17, one can consider evaluating the AdS partition function in a saddle-point approximation:

$$e^{-I_{\mathrm{AdS}}(\phi_i)} = \left\langle e^{\int \phi_{0,i} \mathcal{O}^i} \right\rangle_{\mathrm{CFT}}, \qquad (18.12)$$

where $I_{AdS}(\phi_i)$ is the classical gravitational action as a functional of the (super)gravity fields, and $\mathcal{O}^i$ are the dual CFT operators. Hence, in this approximation the AdS action becomes the generating function of the connected correlation functions in the CFT[271, 272]. This framework is also naturally extended to considering CFT states for which certain operators acquire expectation values by considering solutions of the gravitational equations which are only asymptotically AdS[273]. We shall do that below, but first we will explore a little of the technology of evaluating the action.

18.1.4 The action, counterterms, and the stress tensor

We need to make sense of the path integral of gravity on AdS, given on the left hand side of the correspondence dictionary in equation (18.12). This returns us to the issue of calculating the action of the theory, from which we can compute such quantities as the stress-energy-tensor, and if (as we did for asymptotically flat black holes in chapter 17) we were to Euclideanise, various thermodynamic quantities such as the energy, entropy, etc.

Recall from earlier discussions in chapter 17 that the action in D dimensions is defined as follows:

$$I_{\mathrm{bulk}} + I_{\mathrm{surf}} = -\frac{1}{16\pi G_5} \int_{\mathcal{M}} d^D x \sqrt{g}\,(R - 2\Lambda) - \frac{1}{8\pi G_5} \int_{\partial\mathcal{M}} d^{D-1} x \sqrt{h} K,$$

$$(18.13)$$

where, as we've seen in section 10.1.9, the cosmological constant is related to the length scale ℓ by

$$\Lambda = -\frac{(D-1)(D-2)}{2\ell^2}.$$

Recall that the second integral is the Gibbons–Hawking boundary term, which is required so that upon variation with metric fixed at the boundary, the action yields the Einstein equations. Here, K is the trace of the extrinsic curvature of the boundary $\partial\mathcal{M}$ as embedded in $\mathcal{M}$, which was discussed in insert 10.2.

Remember also that both of these expressions are divergent because the volumes of both $\mathcal{M}$ and $\partial\mathcal{M}$ are infinite (and the integrands are non-zero). The approach we used in section 17.2, (there, the first term vanished and the second term was divergent) to avoid this problem is to perform a 'background subtraction', producing a finite result by subtracting from equation (18.13) the contribution of a background reference spacetime, so that one can compare the properties of the solution of interest relative to those of the reference state. In our computation we ensured that the asymptotic boundary geometries of the two solutions can be matched in order to render the surface contribution finite.

In AdS, we can in fact follow a different approach, which has a natural meaning in the dual gauge theory[304]. We can supplement the action by a finite set of boundary integrals which depend only on the curvature scalar R (and its derivatives) of the induced boundary metric $h_{\mu\nu}$, which itself diverges as $r \to \infty$. These integrals look like a family of counterterms in the dual field theory, and with appropriate coefficients, they cancel the divergences (IR in AdS, UV in the gauge theory) as $r \to \infty$, giving a intrinsic definition of the action for asymptotically AdS spacetimes, with no reference to a background spacetime. Calling the set of boundary integrals[†] I_{ct}, we define the complete action to be $I = I_{\mathrm{bulk}} + I_{\mathrm{surf}} + I_{\mathrm{ct}}$.

One of the useful quantities which we will extract from the action is the stress tensor, which is obtained by the standard expression:

$$T^{\mu\nu} = \frac{2}{\sqrt{-\gamma}}\frac{\delta I}{\delta\gamma_{\mu\nu}} = \lim_{r\to\infty}\left(\frac{r^2}{\ell^2}\frac{2}{\sqrt{-h}}\frac{\delta I}{\delta h_{\mu\nu}}\right), \qquad (18.14)$$

where in the second expression we have used $h_{\mu\nu}$ which is the metric on the boundary induced by restricting the bulk metric by setting r to a

---

[†] That this construction is unique to asymptotically AdS spaces is apparent because the AdS curvature scale ℓ is essential in defining the counterterms. We are excluding non-polynomial terms, which could be introduced in the absence of a cosmological constant[305], giving a definition that is applicable to spacetimes with other asymptotic behaviour.

constant. In the first expression, $\gamma_{\mu\nu}$ is the metric obtained by removing a conformal factor r^2/ℓ^2 to get the dual field theory's natural metric in the UV. From this stress-tensor we can extract quantities like the energy density, etc., in the usual way, for example $\rho = T_{\mu\nu}u^\mu u^\nu$, where u^μ are the components of a timelike unit vector.

It turns out that the counterterm action is[304, 306]:

$$I_{\text{ct}} = \frac{1}{8\pi G_5} \int_{\partial\mathcal{M}} d^{D-1}x\sqrt{h} \left[\frac{D-2}{\ell} + \frac{\ell}{2(D-3)}\mathcal{R} \right.$$
$$\left. + \frac{\ell^3}{2(D-5)(D-3)^2} \left(\mathcal{R}_{ab}\mathcal{R}^{ab} - \frac{D-1}{4(D-2)}\mathcal{R}^2 \right) + \cdots \right].$$
$$(18.15)$$

Here, $\mathcal{R}$ and $\mathcal{R}_{ab}$ are the Ricci scalar and Ricci tensor for the boundary metric, respectively. The three counterterms are sufficient to cancel divergences for $D \leq 7$.

Let us consider an example. Take AdS$_5$ in global coordinates, as given in equation (18.7). As stated beneath that equation, the metric $\gamma_{\mu\nu}$ is that of the Einstein static universe of radius ℓ. Computing with the first two counterterms in equation (18.15), the stress tensor becomes:

$$T_{tt} = \frac{1}{8\pi G_5} \left(\frac{3}{8\ell} \right) + O\left(\frac{1}{r} \right),$$
$$T_{ij} = \frac{1}{8\pi G_5} \left(\frac{1}{8\ell} \right) G_{ij} + O\left(\frac{1}{r} \right), \qquad (18.16)$$

where G_{ij} refer to the metric components in the angular directions for an S^3 of unit radius. In the $r \to \infty$ limit we see that we get a finite result, which can be written in the suggestive form:

$$T_{\mu\nu} = \frac{1}{64\pi G_5\ell} (4u_\mu u_\nu + \gamma_{\mu\nu}) = \frac{N^2}{32\pi^2\ell^4} (4u_\mu u_\nu + \gamma_{\mu\nu}), \qquad (18.17)$$

using the conversion formula (18.4). This is the standard form (see equation (10.23)) for a *conformally invariant* perfect fluid's stress tensor (since it is traceless) of density $\rho = 3p$ in four dimensions with a spacetime of metric $\gamma_{\mu\nu}$. The overall prefactor is $\rho/3$, as written. We have used the conversion formula (18.4) to change gravitational quantities to field theory ones. Note that the stress tensor is traceless, as expected for a conformally invariant theory. The field theory is in an S^3 box of radius ℓ, and so we can integrate the energy density to give the total energy (the dual to the spacetime's gravitational mass) which is:

$$E = \frac{3\pi\ell^2}{32G} = \frac{3N^2}{16\ell}. \qquad (18.18)$$

In fact, this result matches expectations from field theory[304]. Since it is defined in a box, there is a Casimir energy. For free fields, the Casimir energy on $S^3 \times R$, the Einstein static universe of radius ℓ, may be found in the literature[293] to be:

$$E_{\text{Cas}} = \frac{1}{960\ell}(4n_0 + 17n_{1/2} + 88n_1) = \frac{3(N^2 - 1)}{16\ell}, \qquad (18.19)$$

where n_0 denotes the number of real scalars, $n_{1/2}$ is the number of Weyl fermions, and n_1 the number of vectors. We have inserted the correct values for the dual $SU(N)$ gauge theory, $n_0 = 6(N^2 - 1)$, $n_{1/2} = 4(N^2 - 1)$ and $n_1 = N^2 - 1$, giving an expression which agrees with the result (18.18) that we got from the stress tensor in the large N limit. (See also insert 17.2 for comments on the AdS$_3$ case.)

18.2 The correspondence at finite temperature

We arrived at the correspondence between the supersymmetric gauge theory and pure AdS by taking the near horizon limit of the extremal D3-brane solution. It is natural to try to give an interpretation of the non-extremal solution. A key difference between the two is that the latter solution is at finite temperature. As we shall see, these properties relate to those of the field theory.

18.2.1 Limits of the non-extremal D3-brane

Taking the decoupling limit of the solution given in equation (10.34) for $p = 3$, we see that $\alpha_3 \to 1$ and so $H_3 \to \ell^4/r^4$ again, and so we can write the solution as[307, 271]:

$$ds^2 = -\left(\frac{r^2}{\ell^2} - \frac{r_H^4}{\ell^2 r^2}\right) dt^2 + \frac{r^2}{\ell^2} \sum_{i=1}^{3} dx_i^2 + \left(\frac{r^2}{\ell^2} - \frac{r_H^4}{\ell^2 r^2}\right)^{-1} dr^2 + \ell^2 d\Omega_5^2, \qquad (18.20)$$

where $\ell^2 = \alpha_3^{\frac{1}{2}} \hat{r}_3^2 \to \hat{r}_3^2$. This is in fact the AdS$_5$-Schwarzschild black hole, in local coordinates (its horizon is $\mathbb{R}^3$ instead of S^3), times an S^5 of radius ℓ. It is sometimes called a 'flat' black hole. In fact, its mass and temperature are easily computed to be:

$$M = \frac{3\pi r_H^4}{8G_5 \ell^2}, \qquad T = \frac{r_H}{\pi \ell^2}. \qquad (18.21)$$

Interpreting the mass as an energy in the field theory[271], we see that in fact that there is a familiar energy-temperature relation following the

Stephan–Boltzmann law:

$$E = \frac{3\pi^4 \ell^3 N^2}{4} T^4. \tag{18.22}$$

18.2.2 The AdS–Schwarzschild black hole in global coordinates

It is easy to write a global version of the AdS–Schwarzschild black hole solution:

$$ds^2 = -\left(1 + \frac{r^2}{\ell^2} - \frac{r_0^4}{\ell^2 r^2}\right) dt^2 + \left(1 + \frac{r^2}{\ell^2} - \frac{r_0^4}{\ell^2 r^2}\right)^{-1} dr^2 + r^2 d\Omega_3^2, \tag{18.23}$$

and we have relabelled r_H as r_0 since this will in general *not* be an horizon radius. A computation of the stress tensor gives:

$$T_{tt} = \frac{1}{8\pi G_5} \left(\frac{3}{8\ell} + \frac{3r_0^4}{2\ell^5}\right) + O\left(\frac{1}{r}\right),$$

$$T_{ij} = \frac{1}{8\pi G_5} \left(\frac{1}{8\ell} + \frac{r_0^4}{2\ell^5}\right) G_{ij} + O\left(\frac{1}{r}\right),$$

and so: $\quad T_{\mu\nu} = \frac{1}{8\pi G_5} \left(\frac{1}{8\ell} + \frac{r_0^4}{2\ell^5}\right) (4u_\mu u_\nu + \gamma_{\mu\nu}). \tag{18.24}$

In the last line we have taken the $r \to \infty$ limit and put it into the perfect fluid form. The mass-energy can be written as

$$M = \frac{3\pi \ell^2}{32 G_5} + \frac{3\pi r_0^4}{8 G_5 \ell^2}, \qquad \Longrightarrow \qquad E = \frac{3N^2}{16\ell} + \frac{3N^2 r_0^4}{4\ell^5}, \tag{18.25}$$

which after conversion using equation (18.4), gives the Casimir energy we derived before, since we are in the same box, together with an energy over extremality which matches the energy density derived for the flat black hole in the previous subsection.

The horizon of the solution is located at the largest root, r_+, of the equation $G^{rr} = 0$:

$$V \equiv \left(1 + \frac{r^2}{\ell^2} - \frac{r_0^4}{\ell^2 r^2}\right) = 0 \qquad \Longrightarrow \qquad r_+ = \frac{1}{2}\left(\sqrt{\ell^4 + 4r_0^2} - \ell^2\right). \tag{18.26}$$

Notice that $r_+ \neq r_0$ for the global case, in general. The temperature of the solution can be computed to be:

$$T = \frac{V'}{4\pi}\Big|_{r=r_+} = \frac{1}{2\pi\ell^2}\left(r_+ + \frac{r_0^4}{r_+^3}\right). \qquad (18.27)$$

This expression is very interesting, since for a given temperature T, there are in fact *two* values of r_+ which solve the above relation, as is clear from figure 18.1. Notice that there is a minimum temperature below which there are no black hole solutions. We see also that there are two classes of black hole solutions. There is one branch which, for large r_+, the temperature goes linearly with the radius. The other branch goes at small r_+ as the inverse cube of r_+. These 'small' black holes have the familiar behaviour of five dimensional black holes in asymptotically flat spacetime, since their temperature decreases with increasing size. The term 'small' is appropriate, since they are smaller than the characteristic size set by the AdS scale ℓ, and so they have the characteristics of the asymptotically Minkowskian holes. Similarly, the 'large' black holes are obtained when ℓ is small compared to the horizon size. These cases are most apparent when taking the large or small ℓ limit of the equation (18.26). The large black hole limit gives the case $r_+ = r_0$ (which we previously called $r_{\rm H}$) and the linear temperature behaviour seen in the case of the planar black holes obtained in local coordinates in equation (18.21).

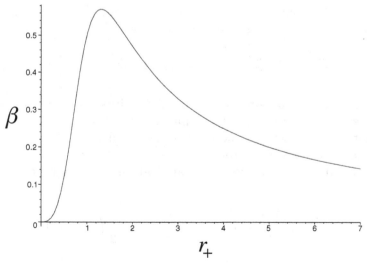

Fig. 18.1. The inverse temperature *vs.* horizon radii, r_+, for AdS black holes. There are two classes of holes, small and large, and a minimum temperature.

18.3 The correspondence with a chemical potential

It is extremely natural, given what we saw in the previous sections, to ask about the role of charged black holes in this AdS scenario. There are $SO(6)$ gauge fields in the supergravity and so a black hole can be a charged source of them. We will focus on the Abelian case, and the $U(1)^3$ Cartan subalgebra is the maximal case. It is easy to see what in the dual gauge theory such a black hole would correspond to. An electric field will be supported by a potential of the form $A_t \sim r^{-2}$. Since this is a rank one massless field in AdS with this asymptotic, it must correspond, by the dictionary of equation (18.11), to a dimension four operator or current in the gauge theory. This is just what we would expect for an R-current, to which the gauge fields correspond. In other words, putting in a charged source is equivalent to switching on an external current source or chemical potential in the theory. It is instructive to construct the precise geometry, as our first example of non-trivial ten dimensional geometries which are dual to gauge theory.

18.3.1 Spinning D3-branes and charged AdS black holes

Given that the $SO(6)$ R-charge comes from Kaluza–Klein reduction from ten dimensions on an S^5, it is natural to guess that the appropriate geometry to seek is one which has some momentum in the compact directions which will be equivalent to the conserved R-charges in the theory. The internal velocities – which couple to momenta in a canonical formalism – will be the chemical potentials in the gauge theory, and hence conjugate to conserved R-charges[308, 309].

So we seek a 'spinning' D3-brane solution[308, 311]. There are six dimensions transverse to a D3-brane, and so we have three independent planes in which a rotation axis can be placed, to define three different angular momenta.

Let us first review some geometrical parameterisations which will be useful[312]. Using the angles θ, ψ on an S^2, we may introduce three direction cosines μ_i, with $\sum_i \mu_i^2 = 1, 0 < \theta \le 2\pi, 0 < \psi \le \pi$:

$$\mu_1 = \sin\theta, \qquad \mu_2 = \cos\theta\sin\psi, \qquad \mu_3 = \cos\theta\cos\psi. \qquad (18.28)$$

In terms of these, and three more angles ϕ_i, $i = 1, 2, 3$, the metric on a round S^5 of unit radius can be written as follows ($0 \le \phi_i < 2\pi$):

$$d\Omega_5^2 = \sum_i^3 (d\mu_i^2 + \mu_i^2 \, d\phi_i^2). \qquad (18.29)$$

Now we can write the metric for the rotating solution, and it is[311]:

$$ds_{10}^2 = H_3^{-1/2} \left(- \left[1 - \frac{r_H^4}{r^4 \Delta} \right] dt^2 + dx_1^2 + dx_2^2 + dx_3^2 \right)$$

$$+ H_3^{1/2} \left[\frac{\Delta dr^2}{\mathcal{H}_1 \mathcal{H}_2 \mathcal{H}_3 - r_H^4/r^4} - \frac{2 r_H^4 \cosh \beta_3}{r^4 H_3 \Delta} dt \left(\sum_{i=1}^3 \ell_i \mu_i^2 d\phi_i \right) \right.$$

$$\left. + r^2 \sum_{i=1}^3 \mathcal{H}_i \left(d\mu_i^2 + \mu_i^2 d\phi_i^2 \right) + \frac{r_H^4}{r^4} \frac{1}{H_3 \Delta} \left(\sum_{i=1}^3 \ell_i \mu_i^2 d\phi_i \right)^2 \right],$$

$$C_{(4)} = \frac{g_s^{-1}(H_3^{-1} - 1)}{\sinh \beta_3} \left(\cosh \beta_3 \, dt - \sum_{i=1}^3 \ell_i \mu_i^2 d\phi_i \right) \wedge dx_1 \wedge dx_2 \wedge dx_3,$$

$$e^\Phi = g_s, \tag{18.30}$$

where the functions Δ, H_3, and $\mathcal{H}_i$ are given by

$$\Delta = \mathcal{H}_1 \mathcal{H}_2 \mathcal{H}_3 \sum_{i=1}^3 \frac{\mu_i^2}{\mathcal{H}_i}, \qquad H_3 = 1 + \frac{r_H^4}{r^4} \frac{\sinh^2 \beta_3}{\Delta} = 1 + \frac{\alpha_3}{\Delta} \frac{r_3^4}{r^4},$$

$$\mathcal{H}_i = 1 + \frac{\ell_i^2}{r^2}, \qquad i = 1, 2, 3. \tag{18.31}$$

It will be useful at this point to refer to the expressions given for the boost form for the non-extremal solution given in section 10.2.2. The structure of the solution can be seen to be closely related to our non-extremal solution presented there, the key difference being that there is a deformation of the round S^5 produced by spoiling the three directions ϕ_i with deformations controlled by the three parameters ℓ_i. The $SO(6)$ isometry of rotation invariance is broken to $U(1)^3$ generated by the obvious Killing vectors $\partial/\partial\phi_i$. There are a number of interesting limits of this solution. One of them is discussed in insert 18.2, where we find an interesting form to which we will return later.

Most pertinent to this section is the decoupling limit of the solution, where we scale the rotation parameters in order to keep them finite in the limit. We write $r = \alpha' u$, $r_H = \alpha' u_H$, and since $r_3 = \alpha'^2 \hat{r}_3$, in the limit $\alpha' \longrightarrow 0$, we see:

$$\sinh \beta_3 \quad \text{and} \quad \cosh \beta_3 \longrightarrow \frac{1}{\alpha'} \frac{\hat{r}_3^2}{u_H^2},$$

$$H_3 \longrightarrow \frac{1}{\alpha'^2} \frac{\hat{r}_3^4}{u^4} \frac{1}{\Delta},$$

$$\ell_i \longrightarrow \alpha' q_i,$$

$$\mathcal{H}_i \longrightarrow 1 + \frac{q_i^2}{u^2},$$

$$\Delta(\ell_i, r) \longrightarrow \Delta(q_i, u). \tag{18.32}$$

Insert 18.2.   D3-brane distributions

Particularly interesting is the straightforward extremal limit $\beta_3 \to \infty$, $r_H \to 0$, holding $r_H^4 e^{2\beta_3}/4 = r_3^4$ fixed, giving

$$ds_{10}^2 = H_3^{-1/2}\left(-dt^2 + dx_1^2 + dx_2^2 + dx_3^2\right)$$

$$+ H_3^{1/2}\left[\frac{\Delta dr^2}{\mathcal{H}_1\mathcal{H}_2\mathcal{H}_3} + r^2\sum_{i=1}^{3}\mathcal{H}_i\left(d\mu_i^2 + \mu_i^2 d\phi_i^2\right)\right],$$

$$H_3 = 1 + \frac{1}{\Delta}\frac{\ell^4}{r^4},$$

$$C_{(4)} = g_s^{-1}(H_3^{-1} - 1)\,dt \wedge dx_1 \wedge dx_2 \wedge dx_3, \qquad e^{\Phi} = g_s.$$

$$(18.33)$$

The terms corresponding to rotation have disappeared, leaving a solution which is supersymmetric[311, 313]. It, of course, still has N D3-branes composing it, but it is not spherically symmetric. The change of variables[314]:

$$y_1 = \sqrt{(r^2 + \ell_1^2)}\,\mu_1\cos\phi_1 = \sqrt{(r^2 + \ell_1^2)}\,\sin\theta\cos\phi_1,$$

$$y_2 = \sqrt{(r^2 + \ell_1^2)}\,\mu_1\sin\phi_1 = \sqrt{(r^2 + \ell_1^2)}\,\sin\theta\sin\phi_1,$$

$$y_3 = \sqrt{(r^2 + \ell_2^2)}\,\mu_2\cos\phi_2 = \sqrt{(r^2 + \ell_2^2)}\,\cos\theta\sin\psi\cos\phi_2,$$

$$y_4 = \sqrt{(r^2 + \ell_2^2)}\,\mu_2\sin\phi_2 = \sqrt{(r^2 + \ell_2^2)}\,\cos\theta\sin\psi\sin\phi_2,$$

$$y_5 = \sqrt{(r^2 + \ell_3^2)}\,\mu_3\cos\phi_3 = \sqrt{(r^2 + \ell_3^2)}\,\cos\theta\cos\psi\cos\phi_3,$$

$$y_6 = \sqrt{(r^2 + \ell_3^2)}\,\mu_3\sin\phi_3 = \sqrt{(r^2 + \ell_3^2)}\,\cos\theta\cos\psi\sin\phi_3,$$

$$(18.34)$$

places the solution back into the familiar form:

$$ds^2 = H_3^{-1/2}\left(-dt^2 + dx_1^2 + dx_2^2 + dx_3^2\right) + H_3^{1/2}\left(d\vec{y}\cdot d\vec{y}\right).$$

Now, H_3 is not of our simple forms discussed in chapter 10. It is still harmonic, as it ought to be, and we may write it in the integral form:

$$H_3(\vec{y}) = L^4\int d^6w\,\frac{\rho_3(\vec{w})}{|\vec{y} - \vec{w}|^4}, \quad \text{with} \quad \int d^6w\,\rho_3(\vec{w}) = 1, \quad (18.35)$$

where the density function ρ_3 – which may be derived implicitly from the change of variables (18.34) – encodes a general *distribution* of branes[313], which we shall study in more detail later in section 19.2.4.

The last term in the metric in equations (18.30) vanishes in this limit, and after some algebra, the metric can be written as:

$$\frac{ds^2}{\alpha'} = \sqrt{\Delta}\left(-(\mathcal{H}_1\mathcal{H}_2\mathcal{H}_3)^{-1} f\, dt^2 + f^{-1} du^2 + \frac{u^2}{\ell^2}[dx_1^2 + dx_2^2 + dx_3^2]\right)$$
$$+ \frac{1}{\sqrt{\Delta}}\sum_{i=1}^{3}\mathcal{H}_i\left(\ell^2 d\mu_i^2 + \mu_i^2\left(\ell d\phi_i + (\mathcal{H}_i^{-1} - 1)dt)^2\right),\right. \tag{18.36}$$

where

$$f = \frac{u^2}{\ell^2}\mathcal{H}_1\mathcal{H}_2\mathcal{H}_3 - \frac{1}{\ell^2}\frac{u_{\mathrm{H}}^4}{u^2}. \tag{18.37}$$

Now we can consider dimensional reduction to five dimensions of this solution. Pulling a factor $(\mathcal{H}_1\mathcal{H}_2\mathcal{H}_2)^{-1/3}$ into $\sqrt{\Delta}$ puts it into the standard Kaluza–Klein form for reduction to five dimensions, and we get:

$$\frac{ds^2}{\alpha'} = -(\mathcal{H}_1\mathcal{H}_2\mathcal{H}_3)^{-2/3} f\, dt^2 + (\mathcal{H}_1\,\mathcal{H}_2\,\mathcal{H}_3)^{1/3}(f^{-1}\,du^2 + \frac{u^2}{\ell^2}\,d\vec{x}\cdot d\vec{x}),$$
$$X_i = \mathcal{H}_i^{-1}\,(\mathcal{H}_1\mathcal{H}_2\mathcal{H}_3)^{1/3}, \qquad A_t^i = 1 - \mathcal{H}_i^{-1}. \tag{18.38}$$

We have two scalar fields from the reduction, since $X_1 X_2 X_3 = 1$. There are three $U(1)$ gauge fields since there are are three independent isometry directions, ϕ_i.

The meaning of this solution might be more apparent if one sets all the q_i, and hence the $\mathcal{H}_i$, to be equal. Then comparing to equation (17.19), we recognise this a family of charged five dimensional black holes, written in isotropic coordinates. One difference is that, just as earlier in section 18.2 these are actually 'flat' black holes, in the sense that the horizons are of $\mathbb{R}^3$ topology. There are spherical and hyperbolic versions which can be readily written down. Similarly, there are such generalisations in the case of the full ten dimensional rotating solution. In the case where all of the charges are different, we see that it is a quite general family, with three charges under the $U(1)^3$, and two scalar fields.

They are solutions of a $U(1)^3$ truncation of the $\mathcal{N} = 8$ $SO(6)$ gauged supergravity, with action:

$$I = -\frac{1}{16\pi G_5}\int d^5x\sqrt{-G}\left(R - \frac{1}{2}(\partial\varphi_1)^2 - \frac{1}{2}(\partial\varphi_2)^2 - \frac{1}{4}\sum_i X_i^{-2}\,(F^i)^2\right.$$
$$\left. + \frac{4}{\ell^2}\sum_i X_i^{-1} + \frac{1}{4}\epsilon^{\mu\nu\rho\sigma\lambda} F_{\mu\nu}^1 F_{\rho\sigma}^2 A_\lambda^3\right). \tag{18.39}$$

In the above, the gauge fields and their field strengths are labelled $1, 2$ or 3 for each of the three $U(1)$ sectors. The final term is a Chern–Simons type term, which only will be non-zero if there are both magnetic and electric charges present, which will not be the case here.

The two scalars φ_1 and φ_2 are contained in the three X_i, via a generalisation of the exponential ansatz that we used in simpler Kaluza–Klein cases. We write them as components of a two-vector, $\vec{\varphi} = (\varphi_1, \varphi_2)$, and:

$$X_i = e^{-\frac{1}{2}\vec{a}_i \cdot \vec{\varphi}}, \tag{18.40}$$

where the $\vec{a}_i$ sum to zero in order to ensure $X_1 X_2 X_3 = 1$, and we make the conventional choice[311]:

$$\vec{a}_1 = \left(\tfrac{2}{\sqrt{6}}, \sqrt{2}\right), \qquad \vec{a}_2 = \left(\tfrac{2}{\sqrt{6}}, -\sqrt{2}\right), \qquad \vec{a}_3 = \left(-\tfrac{4}{\sqrt{6}}, 0\right), \tag{18.41}$$

where $\vec{a}_i$ satisfy the dot products $\vec{a}_i \cdot \vec{a}_j = 4\delta_{ij} - \tfrac{4}{3}$.

18.3.2 The AdS–Reissner–Nordström black hole

A special case of this is to set all of the angular momenta to be equal, $q_i = q$ which makes all the $X_i = 1$, setting all of the scalars to zero. Then with $F^i_{(2)} = F_{(2)}/\sqrt{3}$, the action becomes[310, 308]:

$$I = -\frac{1}{16\pi G_5} \int d^5x \sqrt{-G} \left(R - \frac{1}{4}F^2 + \frac{12}{\ell^2}\right), \tag{18.42}$$

where the cosmological constant is $\Lambda = -6/\ell^2$ (we omit the Chern–Simons term, since we only have electric charges present) and the solution is:

$$\begin{aligned}
ds_5^2 &= -\mathcal{H}^{-2} f \, dt^2 + \mathcal{H}\left(f^{-1} dr^2 + \frac{u^2}{\ell^2} d\vec{x} \cdot d\vec{x}\right), \\
A_t &= 1 - \mathcal{H}^{-1}, \\
\mathcal{H} &= 1 + \frac{q^2}{r^2}, \quad f = \frac{u^2}{\ell^2}\mathcal{H}^3 - \frac{u_{\mathrm{H}}^4}{\ell^2 u^2}.
\end{aligned} \tag{18.43}$$

As stated before, this is the form of our old friend from chapter 17, the Reissner–Nordström black hole in five dimensions (17.19), but now in anti-de Sitter spacetime and with an horizon with topology $\mathbb{R}^3$. We can make the global cousin of this which would have a spherical horizon by replacing $\ell^{-2}d\vec{x} \cdot d\vec{x})$ by $d\Omega_3^2$ and adding a 1 to the function f. We shall study this solution shortly[308].

18.3.3 Thermodynamic phase structure

By changing to a new radial coordinate r, in the same manner in which we did for equation (17.19), we write the black holes we have obtained in static coordinates in the form in which we have previously done our thermodynamic studies. For comparison to the earlier case of AdS–Schwarzschild in section 18.2, we shall, as promised, work with the spherical cousins, obtained as stated below equation (18.43):

$$ds^2 = -V(r)dt^2 + V(r)^{-1}dr^2 + r^2d\Omega_3, \qquad (18.44)$$

where

$$V(r) = 1 - \frac{m}{r^2} + \frac{q^2}{r^4} + \frac{r^2}{\ell^2}. \qquad (18.45)$$

Here, m is related to the mass M of the solution as

$$M = \frac{3\pi}{8G}m. \qquad (18.46)$$

The $U(1)$ charge Q is related to q by

$$Q = \frac{\sqrt{3}\pi}{8\pi G}q. \qquad (18.47)$$

Let r_+ denote the largest real positive root of $V(r)$. This defines the horizon:

$$r_+^6 + \ell^2 r_+^4 - \ell^2 m r_+^2 + q^2\ell^2 = 0.$$

The derivative of V is

$$V' = \frac{1}{r_+^5\ell^2}\left[2r_+^6 + 2mr_+^2\ell^2 - 4q^2\ell^2\right] = \frac{2}{r_+^5\ell^2}\left[2r_+^6 + r_+^4\ell^2 - q^2\ell^2\right],$$

and so for a non-singular horizon we must have $2r_+^6 + r_+^4\ell^2 \geq q^2\ell^2$. Now, as we've seen many times before, V' controls the temperature of the black hole via

$$\beta = \frac{4\pi}{V'} = \frac{2\pi r_+^5\ell^2}{2r_+^6 + r_+^4\ell^2 - q^2\ell^2}. \qquad (18.48)$$

When the inequality above is saturated the horizon is degenerate, β diverges, and we get the zero temperature extremal black hole[‡].

As before, we will choose a gauge in which A is regular at the horizon:

$$A = \sqrt{\frac{3}{4}}\left(-\frac{q}{r^2} + \Phi\right)dt, \quad \text{where} \quad \Phi = \frac{q}{r_+^2}. \qquad (18.50)$$

[‡] Note that this extremal case is *not* supersymmetric, as in asymptotically flat cases. The supersymmetric case is $m = 2q$, and then

$$V(r) = \left(1 - \frac{q}{r^2}\right)^2 + \frac{r^2}{l^2}, \qquad (18.49)$$

which is clearly positive everywhere. This means that the curvature singularity at $r = 0$ is *naked* for this value[310].

It is useful to rewrite the temperature in terms of the potential:

$$\beta = \frac{4\pi l^2 r_+}{2\ell^2(1 - \Phi^2/\Phi_c^2) + 4r_+^2},$$ (18.51)

which will be useful later. Here, $\Phi_c = \sqrt{3/4}$. It is useful to observe the behaviour of the temperature as a function of black hole size r_+, as we did previously for the AdS–Schwarzschild case.

The reader may notice that there are qualitatively two distinct types of behaviour, determined by whether Φ is less than or greater than the critical value Φ_c. In particular, for $\Phi \geq \Phi_c$, β diverges (T vanishes) at $r_+^2 = \ell^2(\Phi^2/\Phi_c^2 - 1)/2$. This regime of large potential has a unique black hole radius associated with each temperature. Meanwhile for $\Phi < \Phi_c$, β goes smoothly towards zero as $r_+ \to 0$. This latter behaviour is just like that we observed in the case of AdS–Schwarzschild in figure 18.1. This small potential regime has two branches of allowed black hole solutions, a branch with larger radii and one with smaller. Correspondingly, the smaller branch of holes is unstable, having negative specific heat. Both cases are plotted in figure 18.2.

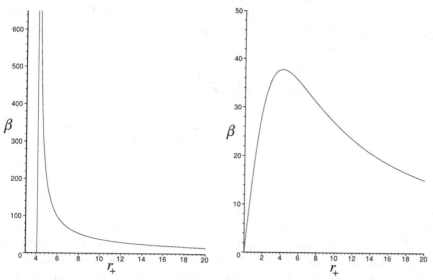

Fig. 18.2. The inverse temperature *vs.* horizon radii, r_+, at fixed potential for $\Phi \geq \Phi_c$, $\Phi < \Phi_c$. The divergence in the first graph (shown with a vertical line) is at zero temperature, where the black hole is extremal. This divergence goes away for $\Phi < \Phi_c$, in general, and the curve is similar to that of the situation with zero potential.

We will study the Euclidean section $(t \to i\tau)$ of the solution, at fixed temperature set by the period, β, of the imaginary time. We will work with fixed temperature and potential, defining thus the grand canonical thermodynamic ensemble using the Euclidean version of the action given in equation (18.42).

In fact, as both spaces we use are asymptotically AdS, it turns out that we need not consider the Gibbons–Hawking boundary term, since its contributions vanishes. The boundary terms from the gauge field will vanish if we keep the potential A_t fixed at infinity. Imposing the equations of motion we can obtain:

$$I^E = \frac{1}{16\pi G} \int_M d^5 x \sqrt{g} \left[\frac{F^2}{6} + \frac{8}{\ell^2} \right],$$ (18.52)

and we get, after substitution and integrating:

$$
\begin{aligned}
I &= \frac{2\pi^2}{16\pi G\ell^2} \beta \left(\ell^2 r_+^2 - r_+^4 - \frac{q^2\ell^2}{r_+^2} \right) \\
&= \frac{2\pi^2}{16\pi G\ell^2} \beta \left(\ell^2 r_+^2 \frac{4}{3}(\Phi_c^2 - \Phi^2) - r_+^4 \right).
\end{aligned}
$$ (18.53)

This is the grand canonical ensemble, at fixed temperature and fixed potential. The grand canonical (Gibbs) potential is $W = I^E/\beta = E - TS - \Phi Q$. Using the expression in equation (18.53), we may compute the state variables of the system as follows:

$$
E = \left(\frac{\partial I^E}{\partial \beta} \right)_\Phi - \frac{\Phi}{\beta} \left(\frac{\partial I^E}{\partial \Phi} \right)_\beta = \frac{3\pi}{8\pi G_5} m = M,
$$

$$
S = \beta \left(\frac{\partial I^E}{\partial \beta} \right)_\Phi - I^E = \frac{2\pi^2 r_+^3}{4G_5} = \frac{A_H}{4G_5}, \quad \text{and}
$$

$$
Q = -\frac{1}{\beta} \left(\frac{\partial I^E}{\partial \Phi} \right)_\beta = \frac{\sqrt{3}\pi}{8G_5} q.
$$ (18.54)

Together, they satisfy: $dE = TdS + \Phi dQ$.

In order to study the phase structure we must study the free energy $W = I^E/\beta$ as a function of the temperature. It is shown in figure 18.3. The interpretation of this is as follows. At any non-zero temperature, for large potential $(\Phi > \Phi_c)$ the charged black hole is thermodynamically preferred, as its free energy (relative to the background of AdS with a fixed potential) is strictly negative for all temperatures.

This behaviour differs sharply from the small potential $(\Phi < \Phi_c)$ situation, which is qualitatively the same as the uncharged case. In that situation, the free energy is positive for some range $0 < T < T_c$, and it is only

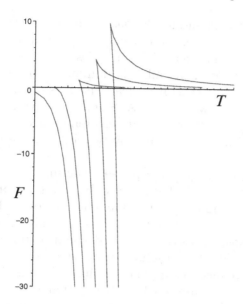

Fig. 18.3. A graph of the free energy vs. temperature for fixed potential ensemble. There is a crossover from the cusp behaviour in the case $\Phi < \Phi_c$ to the single branch ($\Phi > \Phi_c$) behaviour. The two branches consisting of smaller (unstable) and large (stable) black holes are visible. The entire unstable branch has positive free energy while the stable branch's free energy goes negative.

above T_c that the thermodynamics is dominated by AdS–Schwarzschild black holes (the larger, stable branch), as their free energy is negative.

So for high enough temperature in all cases the physics is dominated by non-extremal black holes[§]. This phase represents a sort of 'unconfined' phase of the dual gauge theory, while AdS without a black hole is a 'confined' phase[271]. There is a lot of evidence for this which we cannot uncover here due to lack of space. However, a clear sign of this an examination of the behaviour of the physical quantities we have computed, such as the energy and entropy. One can take the quantities in equations (18.54), converting them to the gauge theory quantities using equation (18.4), and find that there is an overall factor N^2. In an unconfined gauge theory, all of the N^2 adjoint degrees of freedom contribute on the same footing, and we see this here are an overall factor of N^2 in extensive quantities.

[§] The $\Phi = 0$ special case of this transition, from AdS to AdS–Schwarzschild black holes, was studied first by Hawking and Page[291]. The more general phase diagram was worked out later in the AdS/CFT context[308].

At low temperatures, and for $\Phi > \Phi_c$, we have something very new. Notice that as we go to $T = 0$, the free energy curve approaches a maximum value which is less than zero. This implies that even at zero temperature the thermodynamic ensemble is dominated by a black hole. From the temperature curve (18.2) it is clear that it is the extremal black hole. For $\Phi = \Phi_c$, at $T = 0$ we recover AdS space, returning to the 'confined' phase. So this suggests that even at zero temperature the system prefers to be in a state with non-zero entropy (given by the area of the extremal black hole)¶.

The resulting thermodynamic phase structure for the fixed potential ensemble is summarised in figure 18.4. It represents in the dual gauge theory the phase diagram for the introduction of a chemical potential into the gauge theory, and there is a phase boundary across which there is a first order phase transition to the deconfined phase. It is intriguing that this may be a (highly simplified) prototype computation for the phase structure of more realistic gauge theories in analogous situations. One can imagine the chemical potential here being analogous to baryon number in QCD. This would then be an analogue of the finite temperature and density phase diagram, a subject of some current experimental interest, at the time of writing. Perhaps one future use of this gauge/gravity duality might be to model the generic phase structure of more realistic gauge theories using black hole and other objects within the gauge dual. On the one hand, it seems unrealistic to expect a direct connection, but on the other, there may be universality classes of behaviour which are quite robust to modification of the details, and so may be captured by studies of the sort presented here.

18.4 The holographic principle

As we have seen there is a close relationship between the physical properties of five dimensional AdS backgrounds and those of a four dimensional conformally invariant gauge theory. It is a remarkable duality, and is in fact the sharpest known example of what is called *holographic* behaviour[286, 287]: the physics involving gravity in a given number of dimensions is conjectured to be completely captured by a non-gravitational description in fewer dimensions.

¶ Notice that this $T = 0$ situation can be seen to display the 'confined' behaviour characteristic of the ordinary zero-temperature phase, despite the presence of the black hole. This follows from the fact that the horizon at extremality is infinitely far away down a throat. There is the possibility that the extremal black hole might decay away by emission of charged quanta, which is possible since it it not supersymmetric, and so this $T = 0$ part of the story should be studied further.

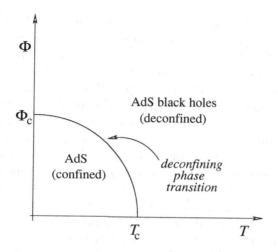

Fig. 18.4. The phase diagram for charged spherical black holes in global AdS. There is a transition from pure AdS to the black hole at finite temperature and potential. In the dual gauge theory, the black holes represent a deconfined phase of the theory. There is a boundary across which there is a first order phase transition between the two phases. The $\Phi = 0$ axis is the Schwarzschild case, with the Hawking–Page transition[291]. The $T = 0$ axis is the extremal charged case, also confining in the gauge theory.

The idea of why such a conjectured phenomenon should be a reality is motivated by the behaviour of black holes. They seem to represent all their degrees of freedom on their horizon, from the point of view of an observer who remains outside, and the universal result that their entropy is one quarter of the area of the horizon is a precise statement that the number of degrees of freedom within the volume that is occupied by the black hole is in fact only of order one per unit area of the horizon, as measured in Planck units.

The idea then is that in any quantum theory of gravity, the number of degrees of freedom in any volume are again just of order one per unit area of the surface surrounding the volume. This is enforced by the expectation that an attempt to examine the structure of the theory right down to the shortest distances, in order to learn about the microscopic degrees of freedom, will eventually probe energy densities which will dynamically favour the formation of a black hole, for which we believe the result is true. The largest obtainable entropy for a given volume is that held by a black hole which fills that volume. This puts an upper limit on the number of degrees of freedom as that given by the total surrounding area.

The AdS/CFT correspondence can be examined in the light of just this type of argument and seen to realise precisely this type of arrangement[315]. In this case, it takes the physics of gravity in five dimensional anti-de Sitter spacetime and makes a hologram of it in terms of a gauge theory. This is also true for anti-de Sitter spacetimes of other dimension too: the hologram is again a non-gravitational conformal field theory in one dimension fewer. Some of the best known examples are as follows. There is AdS_3, which is dual to the 1+1 dimensional gauge theory arising from D1- and D5-branes intersecting. This was responsible for controlling a number of universal properties of five dimensional black holes which we uncovered in chapter 17. The cases of AdS_4 and AdS_7 are also natural in this context. They arise as near-horizon limits (times S^7 and S^4 respectively) of the M2- and M5-brane geometries discussed in chapter 12 (the reader can check this directly). In fact, as hinted at previously (see section 12.6.2), there are important conformal field theories, with sixteen supercharges (in 2+1 and a 5+1 dimensions), on the world-volumes of these branes, whose direct Lagrangian definitions are not known. However, the theories certainly exist as limits of more familiar theories, and the AdS/CFT relation can be taken as a definition of the properties of these theories via the holographic duality.

The holographic expectation has been elevated to the status of a principle, although at present there is a scarcity of well-understood examples outside the AdS/CFT examples and their close cousins. A very active area of research is the endeavour to find further examples, since this is clearly an important clue regarding the nature of fundamental physics about which we should learn more.

19

The holographic renormalisation group

We saw in the previous chapter that the 'holographic'[286, 287] duality[315] between AdS$_5$ physics and the physics of the conformally invariant four dimensional Yang–Mills theory can be extended to the properties of solutions which are only asymptotically AdS$_5$, in keeping with the basic dictionary of the correspondence. We studied the properties of Schwarzschild and Reissner–Nordstrom black holes in AdS, arising naturally as limits of non-extremal and spinning D3-branes, and found that their properties make considerable physical sense in the holographically dual field theory.

It is very clear that this duality between gravitational physics and that of gauge theory is potentially a powerful tool for studying gauge theory. The prototype example is, of course, a highly specialised sort of gauge theory, since it has sixteen supercharges, and is conformally invariant. Of great interest is the study of gauge theories which might be closer to the theories we use to model interactions in particle physics, such as QCD. Perhaps there are gravitational duals of such theories. More generally, of course, we would like to also find and study full string theory duals, if we want to study more than just very large N. At the time of writing, this is subject of considerable research effort.

In this chapter we shall have a brief look at extending the intuition we have developed about the AdS/CFT correspondence a bit further, and address the issue of studying less symmetric gauge theories by deforming the AdS/CFT example.

19.1 Renormalisation group flows from gravity

Recall that, in section 18.1.2, we took a five dimensional perspective, and recognised AdS$_5$ with gauge symmetry $SO(6)$ as a special fixed point solution of the gauged supergravity which preserves the full gauge symmetry. It should be clear from that discussion that other fixed points of the potential will have an intuitive explanation as other conformally invariant theories with fewer supersymmetries. We will again have $\partial\phi_i/\partial x^\mu = 0$, and some set of the scalars approaching some non-zero constants. Since the scalars are charged under the $SO(6)$, such non-zero expectation values will mean that some amount of the $SO(6)$ will be broken, leaving a subgroup G. The scalar potential will take some value $-C/\ell^2$. The solution will be AdS$_5$ with a new value of the cosmological constant and hence the AdS radius for this solution, $\hat{\ell}$, will be given by: $\Lambda = -C/\ell^2 = -6/\hat{\ell}^2$. The expectation is that this defines a dual conformal field theory, with fewer supersymmetries and global symmetry G.

We can imagine a solution that is asymptotically AdS$_5$, with all of the scalars being asymptotically zero, but at smaller radius, approaches this new solution. Since the radial parameter has been identified with an energy scale in the theory, we have the intuitive picture that this solution represents a collection of snapshots (one for each radial slice) of the evolution of the gauge theory as a function of energy scale. It begins in the UV with the symmetric theory and then at lower energies approaches a new theory, which is less symmetric. This picture is just what we would call *renormalisation group flow* (RG flow) [319, 320] in the context of the field theory. Our example is one of flowing from a UV fixed point, using a 'relevant operator' (see insert 3.1, p. 84), to an IR fixed point. The five dimensional gravitational dual picture of this (and its ten dimensional extension) therefore deserves to be called *holographic renormalisation group flow*, and we shall do so.

In fact, we can be even bolder than this. There may be other solutions which are viable vacua which are not AdS$_5$ in the interior. If they are connected at large radius to the familiar $SO(6)$ AdS$_5$, we can also think of them as the result of deforming the UV fixed point by relevant operators and undergoing RG flow to some new non-conformal field theory. Evidently, the utility of such a tool is worth exploring. Ultimately, we can see that this leads us to even consider the existence of well-defined solutions that are not AdS$_5$ in either radial limit, which are nevertheless holographic duals of gauge theories. In fact, gauge theories of considerable phenomenological interest – perhaps the entire Standard Model – may perhaps be represented in this way. It is prudent to develop the tools to find and study these holographic duals.

The flow between fixed points has a precise example which we shall study in section 19.3. It breaks the supersymmetry from $\mathcal{N} = 4$ to $\mathcal{N} = 1$. First, we shall study a simpler RG flow, which is just the switching on of an operator which preserves supersymmetry and merely takes the theory out onto its Coulomb branch. Last, we shall exhibit a flow to a theory which is non-conformal and $\mathcal{N} = 2$ supersymmetric.

First, we will uncover a little of the basic technology that we will need, and emphasise a bit further aspects of the physics of the gravitational side. Before proceeding, we should note that many of the powerful techniques which underlie the construction of the solutions we present here are well beyond the scope of this book, and we must refer the reader to the literature for the details. We shall merely exhibit solutions and hope that our discussion will at least make their properties seem natural and reasonable in the present context. Also, we will not have space to introduce in a self-contained manner some of the more advanced dual field theory properties that we examine. The reader should not regard this as discouragement, but instead as an opportunity to see some of these advanced field theory concepts and properties emerging in an interesting setting, which may, in some cases, serve as a useful introduction.

19.1.1 A BPS domain wall and supersymmetry

Since every radial slice should be dual to the gauge theory at some energy scale set by the radius, we expect that the metric should be of the form:

$$ds_{1,4}^2 = e^{2A(r)} \left(-dt^2 + dx_1^2 + dx_2^2 + dx_3^2 \right) + dr^2, \qquad u = \frac{\ell}{\alpha'} e^{r/\ell}, \quad (19.1)$$

where we have preserved the Poincaré invariant form of the metric. The function $A(r)$ is chosen such that as $r \to \infty$, $A(r) \to r/\ell$, and so we recover the metric of AdS$_5$, where we show how to return to the more familiar local AdS$_5$ coordinates in terms of the variable u.

Let us consider the possibility that one of the 42 scalars, φ, has been switched on, and has a non-trivial profile as we go in to smaller r. The function $A(r)$ will deviate from the AdS behaviour of r/ℓ to some non-trivial behaviour. Generically, it is useful to think of $A(r)$ as parametrising some interpolating region, with AdS$_5$ located at $r \to +\infty$ being one region. On the other side, there are a number of possibilities for what $A(r)$ might do, and we shall see three types by example as we proceed. One possibility is that we get $A(r) \propto r$ again, (with the scalar running to a constant) giving an AdS region in the interior. As discussed before, this is another fixed point, and is expected to be dual to a conformal field theory again. We shall see this later. Away from the asymptotic behaviour, we should still think of A as giving us an interpolating solution, forming a 'domain wall' separating two types of asymptotic behaviour. See figure 19.1, and recall the kink example of insert 1.4 (p. 18).

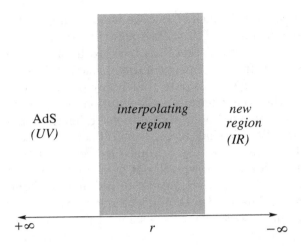

Fig. 19.1. The function $A(r)$ in the metric parametrises a departure from the UV's AdS behaviour, and may be thought of in terms of a 'domain wall' separating it from a new region in the IR.

Let us study some of the physics of this wall[316]. The supergravity action is:

$$S = \frac{1}{16\pi G_5} \int d^5x \sqrt{-G}\,[R - 2\partial_\mu\varphi\,\partial^\mu\varphi - V(\varphi)]. \qquad (19.2)$$

If we insert the form of the metric given in equation (19.1), we get the following equations of motion:

$$12\dot{A}^2 - 2\dot{\varphi}^2 + V = 0$$
$$6\ddot{A} + 12\dot{A}^2 + 2\dot{\varphi}^2 + V = 0$$
$$\ddot{\varphi} + 4\dot{A}\dot{\varphi} - \frac{1}{4}\frac{\partial V}{\partial \varphi} = 0, \qquad (19.3)$$

where a dot denotes a derivative with respect to r. It is interesting to note that differentiating the first equation and then using the third equation gives the second, and in fact

$$\ddot{A} = -\frac{2}{3}\dot{\varphi}^2. \qquad (19.4)$$

It is most interesting to substitute the equation (19.1) into the action itself. Since the only non-trivial behaviour of the metric is as a function of r, the problem reduces to a one dimensional one, since the integral over the directions (t, x_1, x_2, x_3) is trivial. Throwing away the (infinite) constant from performing that integral, the action reduces to an energy functional:

$$\mathcal{E} = \frac{1}{16\pi G_5} \int_{-\infty}^{+\infty} dr\, e^{4A} \left[2\dot{\varphi}^2 - 12\dot{A}^2 + V(\varphi) \right].$$ (19.5)

Let us consider the possibility that V depends upon an auxiliary function W, in the following manner*:

$$V(\varphi) = \frac{4}{\ell^2} \left[\frac{1}{2} \left(\frac{\partial W}{\partial \varphi} \right)^2 - \frac{4}{3} W^2 \right].$$ (19.6)

Let us substitute this into the energy functional, to get

$$\mathcal{E} = \frac{1}{16\pi G_5} \int_{-\infty}^{+\infty} dr\, e^{4A} \left\{ 2\dot{\varphi}^2 - 12\dot{A}^2 + \frac{4}{\ell^2} \left(\frac{\partial W}{\partial \varphi} \right)^2 - \frac{16}{3\ell^2} W^2 \right\}$$

$$= \frac{1}{16\pi G_5} \int_{-\infty}^{+\infty} dr\, e^{4A} \left\{ 2\left(\dot{\varphi} \pm \frac{1}{\ell} \frac{\partial W}{\partial \varphi} \right)^2 - 12 \left(\dot{A}^2 \mp \frac{2}{3\ell} W \right)^2 \right.$$

$$\left. \mp \frac{4}{\ell} \dot{\varphi} \frac{\partial W}{\partial \varphi} \mp \frac{16}{\ell} \dot{A} W \right\}.$$ (19.7)

We have obligingly completed the square, as suggested by four of the terms, and collected the remainder at the end. Since $\dot{\varphi}(\partial W/\partial \varphi) = \dot{W}$, we can write the last two terms under the integral as $\mp d(12 e^{4A} W)/dr$, and therefore it may be integrated and replaced by a boundary term.

So the functional is extremised if the following *first order* equations are satisfied:

$$\frac{\partial A}{\partial r} = -\frac{2}{3\ell} W, \qquad \frac{\partial \varphi}{\partial r} = \frac{1}{\ell} \frac{\partial W}{\partial \varphi}.$$ (19.8)

In fact, (by analogy with many other cases in earlier chapters) the reader should expect that finding a solution to these equations means that we have found a BPS solution of the system, preserving some of the supersymmetries of the original $\mathcal{N} = 8$ supergravity. The precise number of unbroken supersymmetries depends upon the details of W and the dependence on the scalars.

---

* The function W is called the 'superpotential' in the context of supersymmetric domain wall technology. It should not be confused with the W we shall later use for field theory superpotentials.

19.2 Flowing on the Coulomb branch

Recall that the $\mathcal{N} = 4$ supersymmetric Yang–Mills theory's gauge multiplet has bosonic fields (A_μ, ϕ_i), $i = 1, \ldots, 6$, where the scalars ϕ_i transform as a vector of the $SO(6)$ R-symmetry, and fermions λ_i, $i = 1, \ldots, 4$ which transform as the **4** of the $SU(4)$ covering group of $SO(6)$.

As we know from other examples in chapters 13 and 15, it is interesting to give vacuum expectation values of the scalars in the gauge multiplet. Generically, switching on vevs in the Cartan subalgebra of the $SU(N)$ gauge group will break the theory to $U(1)^{N-1}$, while keeping the scalar potential $\sum_{i,j} \mathrm{Tr}[\phi_i, \phi_j]$ vanishing and hence preserving supersymmetry. This is the Coulomb branch of vacua of the theory.

In the AdS/CFT context, the 42 $\mathcal{N} = 8$ gauged supergravity scalars decompose as $\mathbf{1} + \mathbf{1} + \mathbf{10} + \overline{\mathbf{10}} + \mathbf{20}$ of the $SO(6) \simeq SU(4)$ gauge group, coupling to operators which have those R-charges in the gauge theory. Their translation is given in the dictionary extracts in table 18.1. Let us consider a family of vacua which are dual to the case of having switched on some components of this operator. If the AdS/CFT dictionary is to be believed, we should expect to find a non-trivial five dimensional supergravity solution which is asymptotically AdS$_5$ (since in the UV any relevant operator's vev should be negligible), and in the bulk there should be a non-trivial profile for supergravity scalars in the **20**. In ten dimensional type IIB supergravity terms, since we are exciting an $SO(6)$ spherical harmonic, we expect that the supergravity solution is asymptotically AdS$_5 \times S^5$, but in the interior, it deviates from it. In particular, the S^5 should be deformed in such a way which represents the turning on of the **20**.

19.2.1 A five dimensional solution

The scalar which will have a non-trivial profile will be called α. It should be zero as $r \to \infty$, and according to the dictionary entry (18.11), it should go as

$$\alpha \to \frac{a_1}{\sqrt{6}} e^{-2r/\ell} + \cdots, \qquad (19.9)$$

since the **20** is an operator of dimension two.

In fact, there are complete solutions which can be written down[329, 330]. One of them is as follows. The scalar α will correspond to a particular part of the **20**:

$$\alpha: \quad \sum_{i=1}^{4} \mathrm{Tr}(\phi_i \phi_i) - 2 \sum_{i=5}^{6} \mathrm{Tr}(\phi_i \phi_i). \qquad (19.10)$$

This operator, which we can write as $\mathrm{diag}(1, 1, 1, 1, -2, -2)$, in an $SO(6)$

basis, splits the $\mathbb{R}^6$ transverse to the brane into an $\mathbb{R}^4$ and an $\mathbb{R}^2$, and so we expect that the supergravity solution will preserve an $SO(4) \times SO(2)$ of the $SO(6)$. The dependence of α and A can be written as first order differential equations representing a flow from their initial values at $r \to \infty$ to the interior, all the way to $r \to -\infty$. Defining $\rho = e^\alpha$, we have:

$$\frac{\partial \rho}{\partial r} = \frac{1}{6\ell}\rho^2 \frac{\partial W}{\partial \rho} = \frac{1}{3\ell}\left(\frac{1}{\rho} - \rho^5\right), \qquad \frac{\partial A}{\partial r} = -\frac{2}{3\ell}W = \frac{2}{3\ell}\left(\frac{1}{\rho^2} + \frac{\rho^4}{2}\right),$$

(19.11)

where the auxiliary function

$$W = -\left(\frac{1}{\rho^2} + \frac{\rho^4}{2}\right)$$

can be used to write the scalar potential:

$$V = \frac{4}{\ell^2}\left[\frac{1}{2}\left(\frac{\partial W}{\partial \varphi}\right)^2 - \frac{4}{3}W^2\right] = \frac{1}{3\ell^2}\left(\frac{\partial W}{\partial \alpha}\right)^2 - \frac{16}{3\ell^2}W^2.$$

(19.12)

The functions W and V are plotted in figure 19.2.

In fact, the flow equations can be solved explicitly. Since we can write a differential equation for ρ in terms of A:

$$\frac{\partial \rho}{\partial A} = \left(\frac{\rho - \rho^7}{2 + \rho^6}\right),$$

we can write

$$e^{2A} = \frac{l^2}{\ell^2}\frac{\rho^4}{\rho^6 - 1},$$

(19.13)

where l is a conveniently chosen integration constant. This initial value flow problem completely specifies the five dimensional supergravity solution.

Recall that we have two pictures, a five dimensional one in which we just have the gauged supergravity, and a ten dimensional one in which we have some type IIB solution. The first can be obtained from the latter, of course, although as we get more complicated gauged supergravity solutions, it will be harder to find the 'lift' to the full ten dimensional geometry. We shall, in a number of examples, wish to probe the geometry with D3-branes in order to determine more information about the physics. This is appropriate since the solutions are, after all, supposed to be made of D3-branes, in the sense that we discussed as early as in chapter 10. The D3-brane itself is best understood in a ten dimensional setting, and so the full ten dimensional picture is very useful to have, in

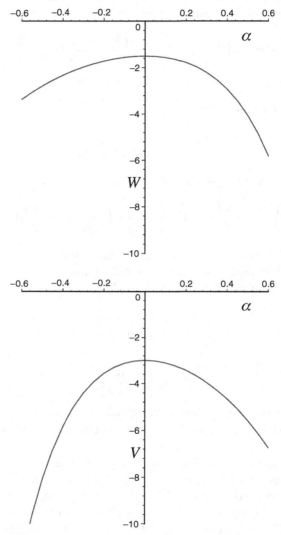

Fig. 19.2. The superpotential and potential, W and V, as a function of the scalar α.

order to do the probe computation. Notice also that ultimately we would like to study the full string theory beyond the tree level gravitational limit, which is again naturally done in a ten dimensional setting. Let us therefore write the ten dimensional lift of that which we have uncovered here.

19.2.2 A ten dimensional solution

We expect that the ten dimensional solution will be of the form,

$$ds_{10}^2 = \Omega^2 ds_{1,4}^2 + ds_5^2, \tag{19.14}$$

where Ω is a 'warp' factor, which can depend upon the angles on the S^5 and r, and ds_5^2 is a deformed metric on the transverse space. Since we expect an $SO(4) \times SO(2)$ invariance, it is sensible to break things up into the metric $d\Omega_3^2$ on a round S^3, and two other angles θ and ϕ which control the rest of the S^5, which is now deformed. The solution is[329, 330]:

$$\Omega^2 = \frac{\bar{X}_1^{1/2}}{\rho}$$
$$\bar{X}_1 = \cos^2\theta + \rho^6 \sin^2\theta, \tag{19.15}$$

with

$$ds_5^2 = \Omega^2 \frac{\ell^2}{\rho^2} \left[d\theta^2 + \frac{\sin^2\theta}{\bar{X}_1} d\phi^2 + \frac{\rho^6 \cos^2\theta}{\bar{X}_1} d\Omega_3^2 \right]. \tag{19.16}$$

The other supergravity fields all vanish except:

$$e^\Phi = g_s, \qquad C_{(4)} = \frac{e^{4A}\bar{X}_1}{g_s\rho^2} dt \wedge dx^1 \wedge dx^2 \wedge dx^3. \tag{19.17}$$

19.2.3 Probing the geometry

The geometry above is very interesting, but there is more physics to be uncovered. It is meant to govern the physics of the Coulomb branch of the moduli space of the $\mathcal{N} = 4$ gauge theory. Going onto the Coulomb branch, recall, is merely the process of pulling the N branes apart, away from the origin at $u = 0$. Recall also our result from chapter 10 that because the branes are all BPS, there is no potential for an individual brane's motion transverse to all the other branes and, furthermore, because we have sixteen supercharges, the actual metric on this moduli space should be flat. This should be true here. It is a simple exercise (see e.g. section 10.3) to probe the supergravity geometry presented in the previous subsection with a D3-brane[318]. In Einstein frame, some of the terms in the D3-brane world-volume action are:

$$S = -\tau_3 \int_{\mathcal{M}_4} d^4\xi \, \det{}^{1/2}[G_{ab} + e^{-\Phi/2}\mathcal{F}_{ab}]$$

$$+ \mu_3 \int_{\mathcal{M}_4} \left(C_{(4)} + C_{(2)} \wedge \mathcal{F} + \frac{1}{2}C_{(0)} \, \mathcal{F} \wedge \mathcal{F} \right), \qquad (19.18)$$

where $\mathcal{F}_{ab} = B_{ab} + 2\pi\alpha' F_{ab}$, and $\mathcal{M}_4$ is the world-volume of the D3-brane, with coordinates $\xi^0, \ldots, \xi^3$. As usual, the parameters μ_3 and τ_3 are the basic R–R charge and tension of the D3-brane:

$$\mu_3 = \tau_3 g_s = (2\pi)^{-3}(\alpha')^{-2}. \qquad (19.19)$$

Also, G_{ab} and B_{ab} are the pulls-back of the ten dimensional metric (in Einstein frame) and the NS–NS two-form potential, respectively.

A quick computation shows that the potential vanishes, and the result for the metric on its moduli space is

$$ds_{\mathcal{M}}^2 = \frac{\tau_3}{2} \frac{\bar{X}_1 e^{2A}}{\rho^2} \left[dr^2 + \frac{\ell^2}{\rho^2} \left(d\theta^2 + \frac{\sin^2\theta}{\bar{X}_1}d\phi^2 + \frac{\rho^6 \cos^2\theta}{\bar{X}_1}d\Omega_3^2 \right) \right],$$
$$(19.20)$$

which looks very far from being flat. The way around this problem must simply be an issue of coordinates. There must be new coordinates more clearly adapted to the dual gauge theory physics in which this geometry is manifestly flat space. We expect to be able to find a new radial coordinate v and a new angle ψ which replace r and θ so that the metric is simply[318]:

$$ds_{\mathcal{M}}^2 = \frac{\tau_3}{2} \left[dv^2 + v^2 \left(d\psi^2 + \sin^2\psi d\phi^2 + \cos^2\psi d\Omega_3^2 \right) \right]$$
$$= \frac{\tau_3}{2} \left[dv^2 + v^2 d\Omega_5^2 \right]. \qquad (19.21)$$

Equating coefficients requires us to show that the following equations can be solved:

$$\frac{\bar{X}_1 e^{2A}}{\rho^2}dr^2 = dv^2,$$

$$\frac{\bar{X}_1 e^{2A}\ell^2}{\rho^4}d\theta^2 = d\psi^2,$$

$$\frac{e^{2A}\ell^2}{\rho^4}\sin^2\theta = v^2 \sin^2\psi,$$

$$e^{2A}\ell^2\rho^2 \cos^2\theta = v^2 \cos^2\psi. \qquad (19.22)$$

In fact, we can now perform this change of variables on the supergravity solution itself. After some algebra, and after using the flow equations themselves, the result is:

$$ds_{10}^2 = \left(\frac{\rho^2}{\bar{X}_1 e^{4A}}\right)^{-1/2} \left(-dt^2 + dx_1^2 + dx_2^2 + dx_3^2\right)$$

$$+ \left(\frac{\rho^2}{\bar{X}_1 e^{4A}}\right)^{1/2} \left(dv^2 + v^2 d\Omega_5^2\right). \tag{19.23}$$

Looking at the form of the other supergravity fields in equation (19.17), we see that we have returned to the standard form for the brane solution, where we now have[318]

$$H_3 = \frac{\rho^2}{\bar{X}_1 e^{4A}} = \frac{\ell^4}{l^2 v^2} \left[\frac{\rho^6 - 1}{(v^2 + l^2)\rho^6 + 2v^2 \cos^2 \psi(\rho^6 - 1)}\right], \tag{19.24}$$

which we have partially translated into the new coordinates using the change of variables (19.22). A useful equation from there is a quadratic in ρ^6 obtained by eliminating θ from the last two lines in equations (19.22), to give:

$$\sin^2 \psi \rho^{12} + \left[\cos^2 \psi - \sin^2 \psi - \frac{l^2}{v^2}\right] \rho^6 - \cos^2 \psi = 0.$$

In the new variables, $H_3(\vec{v})$ is in fact harmonic. One way to see its explicit form is to expand the above equation for ρ^6 in large v. To do this, observe first that to a first approximation, the third term in the square braces vanishes, and so we have the solution $\rho^6 = 1$. Substitute $\rho = 1 + (l^2/v^2)g(l, \psi, v)$, and solve at the next order. The result is $g = 1 + \mathcal{O}(l^2/v^2)$. Recursive substitution like this will give[318]:

$$\rho^6 = 1 + \frac{l^2}{v^2} + (1 - \sin^2 \psi)\frac{l^4}{v^4} + (1 - 3\sin^2 \psi + 2\sin^4 \psi)\frac{l^6}{v^6} + \cdots.$$

Using this, H_3 may be expanded to give:

$$H_3(v) = \frac{\ell^4}{v^4}\left(1 + \frac{l^2}{v^2}(3\sin^2 \psi - 1) + \frac{l^4}{v^4}(1 - 8\sin^2 \psi + 10\sin^4 \psi) + \cdots\right), \tag{19.25}$$

which suggests the form:

$$H_3(v) = \frac{\ell^4}{v^4}\left(1 + \sum_{n=0}^{\infty} \frac{c_{2n}}{|\vec{v}|^{2n}} Y_{2n}(\cos^2 \psi)\right), \qquad c_{2n} = (-1)^n l^{2n}, \tag{19.26}$$

where the $Y_k(\cos^2 \psi)$ (with $Y_k(1) = 1$) are the scalar spherical harmonics on S^5. In the above, we see explicitly the **20** ($n = 1$), and the **50** ($n = 2$).

This is remarkable, since we are seeing explicitly that non-zero l turns on precisely the operator which we want, with subleading mixing with higher order harmonics[318].

19.2.4 Brane distributions

The analysis we carried out above, where we found variables which took us from a complicated solution to one of the simple D3-brane standard form (but with a complicated harmonic function H_3), should remind the reader of the discussion presented in insert 18.2. Let us take the case where we only have one of the ℓ_i, say $\ell_1 = l$ non-zero. This corresponds (before the limit of insert 18.2) to a rotation in only one plane, and hence, after the limit, we expect an $SO(4) \times SO(2)$ invariant configuration. Let us study this.

The metric before the change of variables is:

$$ds_{10}^2 = H_3^{-1/2} \left(-dt^2 + dx_1^2 + dx_2^2 + dx_3^2 \right)$$
$$+ H_3^{1/2} \left[\frac{r^2 + l^2 \cos^2 \theta}{r^2 + l^2} dr^2 + (r^2 + l^2 \cos^2 \theta) d\theta^2 \right.$$
$$\left. + (r^2 + l^2) \sin^2 \theta d\phi_1^2 + r^2 \cos^2 \theta d\Omega_3^2 \right],$$

$$H_3 = 1 + \frac{\ell^4}{r^2(r^2 + l^2 \cos^2 \theta)},$$
$$C_{(4)} = g_{\rm s}^{-1}(H_3^{-1} - 1)\, dt \wedge dx_1 \wedge dx_2 \wedge dx_3, \quad e^\Phi = g_{\rm s}. \quad (19.27)$$

The change of variables[314]:

$$y_1 = \sqrt{(r^2 + l^2)}\, \mu_1 \cos\phi_1 = \sqrt{(r^2 + l^2)} \sin\theta \cos\phi_1$$
$$y_2 = \sqrt{(r^2 + l^2)}\, \mu_1 \sin\phi_1 = \sqrt{(r^2 + l^2)} \sin\theta \sin\phi_1$$
$$y_3 = r\, \mu_2 \cos\phi_2 = r \cos\theta \sin\psi \cos\phi_2$$
$$y_4 = r\, \mu_2 \sin\phi_2 = r \cos\theta \sin\psi \sin\phi_2$$
$$y_5 = r\, \mu_3 \cos\phi_3 = r \cos\theta \cos\psi \cos\phi_3$$
$$y_6 = r\, \mu_3 \sin\phi_3 = r \cos\theta \cos\psi \sin\phi_3, \quad (19.28)$$

places the solution back into the familiar form:

$$ds^2 = H_3^{-1/2} \left(-dt^2 + dx_1^2 + dx_2^2 + dx_3^2 \right) + H_3^{1/2} \left(d\vec{y} \cdot d\vec{y} \right).$$

Let us examine the harmonic (in the y_i) function:

$$H_3 = 1 + \frac{\ell^4}{r^2(r^2 + l^2\cos^2\theta)}. \tag{19.29}$$

Notice that when $r = 0$ there is a quadratic singularity for all θ. From the coordinate transformation (19.28), it is clear that this singularity occurs on the flat plane $\mathbb{R}^4$ given by $y_3 = y_4 = y_5 = y_6 = 0$, and the locus of points $y_1^2 + y_2^2 \le l^2$. This is a disk.

The singularity in the harmonic function should signal the presence of the source – the D3-branes themselves – and it is tempting to conclude that they are distributed on that disk, and we can write[313] the appropriate uniform density function to go into the integral form (18.35):

$$\rho_3(\vec{v}) = \frac{1}{\pi l^2}\Theta\left(l - \sqrt{y_1^2 + y_2^2}\right)\delta^{(4)}(\vec{y}_\perp). \tag{19.30}$$

In fact, since a pointlike source in six dimensions produces a quartic singularity, a smeared two dimensional source should indeed produce a quadratic singularity so we are clearly on the right track. See figure 19.3.

We can check that our density function is correct by working perpendicular to the (y_1, y_2) plane of the disc, $\theta = 0$, to show that we recover

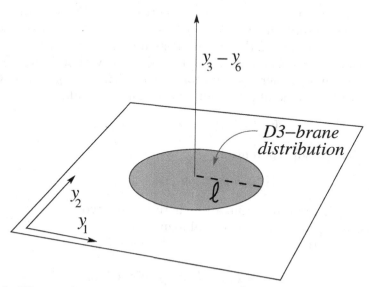

Fig. 19.3. The uniform disc distribution in $\mathbb{R}^6$ of D3-branes produced by switching on an operator in the **20**. This is a part of the Coulomb branch of the dual gauge theory.

expression (19.29) by explicitly integrating the the integral form (18.35). The θ dependence is forced to come out right by standard harmonic analysis: separation of variables, and the uniqueness of the expansion in terms of spherical harmonics.

There remains to establish a direct connection to the geometry of the previous subsection. So far, they have some of the same symmetries, but we have not shown that they are directly related. In fact, despite the differing form of the harmonic functions, they contain precisely the same physics. This can be shown by explicit computation. Notice that from the change of variables (19.28), we can write that

$$r^2 = y^2 - l^2 \sin^2 \theta.$$

We can easily expand the harmonic function in $1/r^2$, and then use

$$\frac{1}{r^2} = \frac{1}{y^2} \left(1 - \frac{l^2 \sin^2 \theta}{y^2} \right)^{-1} = \frac{1}{y^2} \sum_{m=0}^{\infty} \left(\frac{l^2}{y^2} \sin^2 \theta \right)^m.$$

After some algebra, we find precisely the expression (19.26) we wrote earlier in terms of spherical harmonics, with (ψ, v) replaced by (θ, y).

19.3 An $\mathcal{N} = 1$ gauge dual RG flow

To recapitulate, the $\mathcal{N} = 4$ supersymmetric Yang–Mills theory's gauge multiplet has bosonic fields (A_μ, ϕ_i), $i = 1, \dots, 6$, where the scalars ϕ_i transform as a vector of the $SO(6)$ R-symmetry, and fermions λ_i, $i = 1, \dots, 4$ which transform as the **4** of the $SU(4)$ covering group of $SO(6)$. In $\mathcal{N} = 1$ language, there is a vector supermultiplet (A_μ, λ_4), and three chiral multiplets made of a fermion and a complex scalar ($k = 1, 2, 3$):

$$\Phi_k \equiv (\lambda_k, \varphi_k = \phi_{2k-1} + i\phi_{2k}), \tag{19.31}$$

and they have a superpotential

$$W = h\mathrm{Tr}(\Phi_3[\Phi_1, \Phi_2]) + \text{h.c.}, \tag{19.32}$$

('h.c.' means Hermitian conjugate) where h is related to g_{YM} in a specific way consistent with superconformal symmetry.

Let us study the case of giving a mass to Φ_3,

$$L_{\mathrm{ft}} \to L_{\mathrm{ft}} + \int d^2\theta \frac{1}{2} m\Phi_3^2 + \text{h.c}, \tag{19.33}$$

where 'h.c.' is the hermitian conjugate. The resulting spectrum (both massive and massless) can now have at most an $\mathcal{N} = 1$ multiplet structure.

The resulting $SU(N)$ theory has matter multiplets in two flavours, Φ_1 and Φ_2, transforming in the adjoint of $SU(N)$. The $SU(4) \simeq SO(6)$ R-symmetry of the $\mathcal{N} = 4$ gauge theory is broken to $SU(2)_{\rm F} \times U(1)_{\rm R}$, the latter being the R-symmetry of the $\mathcal{N} = 1$ theory, and the former a flavour symmetry under which the matter multiplet forms a doublet.

This mass perturbation is a relevant one and so upon flowing to the IR it becomes more significant. Eventually we fall to scales where the mass is effectively infinite, and we are close to the pure $\mathcal{N} = 1$ theory we discussed in the previous paragraph.

In a supergravity dual, via the dictionary this maps to turning on certain scalar fields in the supergravity, their values being close to zero in the UV ($r \to +\infty$), they develop non-trivial profiles as a function of r, becoming more significantly different from zero as one goes deeper into the IR, $r \to -\infty$. The supergravity equations of motion require that there be a non-trivial back-reaction on the geometry, which deforms the spacetime metric in a way given by $A(r)$, in equation (19.1)

There is a supergravity dual which achieves this[322]. It turns on two scalars, which turn on a combination of the operator which we want, and a vev of the operator we discussed in the previous section:

$$\alpha: \quad \sum_{i=1}^{4} {\rm Tr}(\phi_i \phi_i) - 2 \sum_{i=5}^{6} {\rm Tr}(\phi_i \phi_i)$$

$$\chi: \quad {\rm Tr}(\lambda_3 \lambda_3 + \varphi_1 [\varphi_2, \varphi_3]) + \text{h.c.}, \tag{19.34}$$

At a low enough scale, we can legitimately integrate out the massive scalar Φ_3, and this results in the quartic superpotential[325, 326]

$$W = \frac{h^2}{4m} {\rm Tr}([\Phi_1, \Phi_2]^2), \tag{19.35}$$

which is in fact a marginal operator of the theory[325]. So the theory we get in the IR is also a conformal field theory, as is confirmed by the following considerations. If the operator, represented by the sum of the terms in equations (19.32) and (19.33), is marginal in the IR, then as it is a superpotential, it must have dimension three. This can be achieved if the fields developed anomalous dimensions γ_i (the fields's dimension is $1 + \gamma_i$ in this notation) once they left the UV and went to the IR. Since Φ_3 was treated differently from Φ_1 and Φ_2, it can have a different value for its anomalous dimension. An appropriate assignment is[325, 326]:

$$\gamma_1 = \gamma_2 = -\frac{1}{4}, \gamma_3 = \frac{1}{2}. \tag{19.36}$$

We should also check that the β-function vanishes. In fact, it is proportional [331] to $3 - \sum_i (1 - 2\gamma_i)$, and so we see that it vanishes, showing that our operator is in fact exactly marginal[325].

From what we have already learned about AdS/CFT, it is natural to expect that the gravity dual to this conformal field theory is again AdS_5. It cannot be the same AdS_5 as before, and so it must have a different value for the cosmological constant and for the gauge symmetry associated to the supergravity. In the language of the discussion presented at the beginning of this chapter, it must simply be another fixed point of the $\mathcal{N}{=}8$ gauged supergravity, which has $\mathcal{N} = 2$ supersymmetry and $SU(2) \times U(1)$ gauge symmetry. In the ten dimensional language, it must be that the transverse geometry is no longer S^5, but some deformation of the sphere which preserves $SU(2) \times U(1)$ isometry.

19.3.1   *The five dimensional solution*

Just as in the previous sections, the radial dependences of scalars and the function $A(r)$ are given in terms of first order 'flow' equations (recall that $\rho \equiv e^\alpha$):

$$\frac{d\rho}{dr} = \frac{1}{6\ell}\rho^2 \frac{\partial W}{\partial \rho} = \frac{1}{6\ell}\left(\frac{\rho^6(\cosh(2\chi) - 3) + 2\cosh^2\chi}{\rho}\right)$$

$$\frac{d\chi}{dr} = \frac{1}{\ell}\frac{\partial W}{\partial \chi} = \frac{1}{2\ell}\left(\frac{(\rho^6 - 2)\sinh(2\chi)}{\rho^2}\right)$$

$$\frac{dA}{dr} = -\frac{2}{3\ell}W = -\frac{1}{6\ell\rho^2}\left(\cosh(2\chi)(\rho^6 - 2) - (3\rho^6 + 2)\right), \quad (19.37)$$

where the function

$$W = \frac{1}{4\rho^2}\left(\cosh(2\chi)(\rho^6 - 2) - (3\rho^6 + 2)\right)$$

can be used to construct the potential via:

$$V = \frac{4}{\ell^2}\left[\frac{1}{2}\sum_{i=1}^{2}\left(\frac{\partial W}{\partial \varphi_i}\right)^2 - \frac{4}{3}W^2\right] = \frac{1}{3\ell^2}\left(\frac{\partial W}{\partial \alpha}\right)^2 + \frac{1}{2}\left(\frac{\partial W}{\partial \chi}\right)^2 - \frac{16}{3\ell^2}W^2.$$
$$(19.38)$$

The functions W and V are plotted as contour maps in figure 19.4, and as three dimensional figures in figure 19.5.[†]

It is clear that the values $\chi = 0$, $\alpha = 0$ ($\rho = 1$) define a stationary point for the scalars. After a bit of thought, one can find another fixed point

---

[†] The reader should not take the small scale variations of the contours near the fixed points seriously. They are due to loss of numerical accuracy.

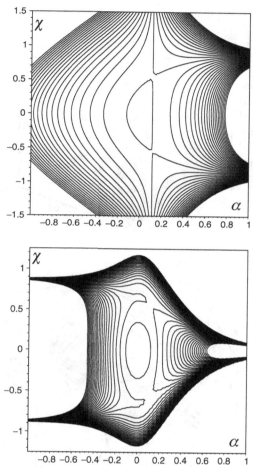

Fig. 19.4. Contour plots of the superpotential and potential, W and V, as functions of the scalars α, χ. This is dual to the RG flow from the $\mathcal{N} = 4$ conformally invariant gauge dual (the fixed point at $\chi = 0, \alpha = 0$) to an $\mathcal{N} = 1$ conformally invariant gauge dual (either of the fixed points at $\alpha = \log 2^{1/6}, \chi = \pm \log 3^{1/2}$).

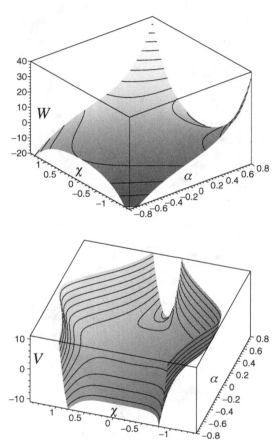

Fig. 19.5. Three dimensional figures of the superpotential and potential, W and V, as functions of the scalar α, χ, for the RG flow from the $\mathcal{N} = 4$ gauge dual to an $\mathcal{N} = 1$ gauge dual.

solution: $\chi = \pm \log 3^{1/2}$, $\alpha = \log 2^{1/6}$. In the first case, those values give

$$\frac{\partial A}{\partial r} = \frac{1}{\ell} \quad \longrightarrow \quad A(r) = e^{r/\ell},$$

after throwing away an integration constant, which gives AdS$_5$ with cosmological constant $\Lambda_{\mathrm{UV}} = -6/\ell^2$. In the second case, the fixed point values give:

$$\frac{\partial A}{\partial r} = \frac{2^{5/3}}{3\ell} \quad \longrightarrow \quad A(r) = e^{r/\tilde{\ell}} \quad \text{where} \quad \tilde{\ell} = \frac{3\ell}{2^{5/3}},$$

after throwing away in integration constant, which gives AdS$_5$ with cosmological constant $\Lambda_{\mathrm{IR}} = -6/\tilde{\ell}^2$. So the ratio between the two cosmological constants is in fact

$$\frac{\Lambda_{\mathrm{UV}}}{\Lambda_{\mathrm{IR}}} = \frac{9}{32^{2/3}}. \tag{19.39}$$

This flow can be recognised as a generalisation of the pure ρ Coulomb branch case from before, by setting $\chi = 0$. Unlike that case, there is no known exact solution for these particular equations, but much can be deduced about the structure of the solution by resorting to numerical methods which we shall not explore much here. It is possible to extract that the asymptotic UV ($r \to +\infty$) behaviour of the fields $\chi(r)$ and $\alpha(r) = \log(\rho(r))$ is given by:

$$\chi(r) \to a_0 e^{-r/\ell} + \cdots; \qquad \alpha(r) \to \frac{2}{3} a_0^2 \frac{r}{\ell} e^{-2r/\ell} + \frac{a_1}{\sqrt{6}} e^{-2r/\ell} + \cdots. \tag{19.40}$$

This behaviour of χ is, according to the dictionary 18.11, characteristic of an operator of dimension three representing a mass term (controlled by a_0), while that of α represents a *mixture* of both a dimension two mass operator (again through a_0) and a vacuum expectation value (vev) of an operator of mass two (through a_1).

Actually, the values of the constant

$$\hat{a} = \frac{a_1}{a_0^2} + \sqrt{\frac{8}{3}} \log a_0 \tag{19.41}$$

characterise a family of different solutions for $(\rho(r), \chi(r), A(r))$ representing different flows to the gauge theory in the IR. Meanwhile, in the IR ($r \to -\infty$) the asymptotic behaviour is:

$$\chi(r) \to \frac{1}{2} \log 3 - b_0 e^{\lambda r/\ell} + \cdots,$$

$$\alpha(r) \to \frac{1}{6} \log 2 - \frac{\sqrt{7} - 1}{6} b_0 e^{\lambda r/\ell} + \cdots,$$

$$\text{where} \quad \lambda = \frac{2^{5/3}}{3}(\sqrt{7} - 1). \tag{19.42}$$

At this end of the flow, there is also a combination which is characteristic of the flow, and this is $b_0 a_0^\lambda$. This may be thought of as characterising the width of the region interpolating between the two AdS asymptotes[322].

The critical value $\hat{a}_c \simeq -1.4694$ represents the particular flow which starts out at the $\mathcal{N} = 4$ critical point and ends precisely on the $\mathcal{N} = 1$ critical point. It has been proposed[317] that the solutions with $\hat{a} > \hat{a}_c$ describe the gauge theory at different points on the Coulomb branch of moduli space. The combination $\hat{a}_c$ then, is pure mass and no vev, while other values are a mixture of both. The vev is that of a combination of massless fields which take us out onto the Coulomb branch.

For the flows with $\hat{a} < \hat{a}_c$, the five dimensional supergravity potential is no longer bounded above by the asymptotic UV value. They are believed to correspond to attempting to give a positive vev to the massive field.

19.3.2   The ten dimensional solution

The ten dimensional solution can be parameterised in the same way as before, in equation (19.14). This time we have[323, 324]:

$$ds_5^2 = \ell^2 \frac{\Omega^2}{\rho^2 \cosh^2 \chi} \left[d\theta^2 + \rho^6 \cos^2 \theta \left(\frac{\cosh \chi}{\bar{X}_2} \sigma_3^2 + \frac{\sigma_1^2 + \sigma_2^2}{\bar{X}_1} \right) \right.$$
$$\left. + \frac{\bar{X}_2 \cosh \chi \sin^2 \theta}{\bar{X}_1^2} \left(d\phi + \frac{\rho^6 \sinh \chi \tanh \chi \cos^2 \theta}{\bar{X}_2} \sigma_3 \right)^2 \right], \quad (19.43)$$

with

$$\Omega^2 = \frac{\bar{X}_1^{1/2} \cosh \chi}{\rho}$$
$$\bar{X}_1 = \cos^2 \theta + \rho^6 \sin^2 \theta$$
$$\bar{X}_2 = \text{sech}\chi \cos^2 \theta + \rho^6 \cosh \chi \sin^2 \theta. \quad (19.44)$$

The σ_i are the standard $SU(2)$ left-invariant forms (see insert 7.4, p. 180), the sum of the squares of which give the standard metric on a round three-sphere. They are normalised such that $d\sigma_i = \epsilon_{ijk}\sigma_j \wedge \sigma_k$. For future use, we shall denote the coordinates on the S^3 as (ψ_1, ψ_2, ψ_3).

It is easily seen that the non-trivial radial dependences of $\rho(r)$ and $\chi(r)$ deform the metric of the supergravity solution from $AdS_5 \times S^5$ at $r = +\infty$ where there is an obvious $SO(6)$ symmetry (the round S^5 is restored), to a spacetime which only has an $SU(2) \times U(1)$ symmetry, which is manifest in the metric of equation (19.43). The $SU(2)$ is the left-invariance of the σ_i and the $U(1)$ rotates σ_1 into σ_2. The obvious extra $U(1)$ symmetry, as

$\partial/\partial\phi$ is also a Killing vector, but this is not a symmetry of the other fields in the full solution.

The fields Φ and $C_{(0)}$, the ten dimensional dilaton and R–R scalar, are gathered into a complex scalar field $\lambda = C_{(0)} + ie^{-\Phi}$, which is constant all along the flow. There are non-zero parts of the two–form potential, $C_{(2)}$, and the NS–NS two-form potential $B_{(2)}$ also, but for our study we won't need them.

Part of $C_{(4)}$ may be written as:

$$C_{(4)} = -\frac{4}{g_s}w(r,\theta)\,dx_0 \wedge dx_1 \wedge dx_2 \wedge dx_3, \tag{19.45}$$

where $w(r,\theta) = \dfrac{e^{4A}}{8\rho^2}[\rho^6 \sin^2\theta(\cosh(2\chi)-3) - \cos^2\theta(1+\cosh(2\chi))]$.

We have only displayed the part of it which will be pertinent to the physics of a D3-brane probe. The part that is missing does not give a non-zero contribution to the probe Lagrangian.

19.3.3 Probing with a D3-brane

In order to understand this geometry a bit better, we shall do what we did in the previous example, and probe the geometry with a D3-brane. Again, this has a natural interpretation[321]. The Coulomb branch moduli space of the $\mathcal{N} = 1$ $SU(N)$ gauge theory is parameterised by the vevs of the complex adjoint scalars $\phi_{1,2}$ which set the potential $\mathrm{Tr}([\phi_1,\phi_2]^2)$ to zero. This generically breaks the theory to a product of $U(1)$s. Probing with a D3-brane will single out a four dimensional subspace of the full moduli space here since our moduli space is the space of allowed zero-cost transverse movements of our single D3-brane probe. These directions are parameterised by the scalars $(\phi_1,\phi_2,\phi_3,\phi_4)$, which make up the complex doublet (φ_1,φ_2). That hyperplane corresponds to the choice $\theta = 0$.

Using the very familiar probe methods from before (see e.g. section 10.3), we get the following result for the effective Lagrangian for the probe moving *slowly* in the transverse directions $y^m = (r,\phi,\theta,\psi_1,\psi_2,\psi_3)$ (we restrict ourselves to considering $F_{ab} = 0$ here):

$$\mathcal{L} \equiv T - V = \frac{\tau_3}{2}\Omega^2 e^{2A}G_{mn}\dot{y}^m\dot{y}^n - \tau_3\sin^2\theta e^{4A}\rho^4(\cosh(2\chi)-1). \tag{19.46}$$

The G_{mn} refer to the Einstein frame metric components.

It is clear that the case $\theta = 0$ indeed makes the potential vanish, picking out the four dimensional moduli space of the probe. The case $\rho = 0$, which is $\alpha = -\infty$, lies outside the physically allowed values of the flow.

19.3.4 The Coulomb branch

It is worthwhile considering the case of large vevs. This should correspond to large r, and we should get a familiar result, flatness in all four (moduli space) transverse directions to the brane. The metric on this moduli space is simply the flat metric on $\mathbb{R}^4$:

$$ds^2_{\mathcal{M}_{\mathrm{UV}}} = \frac{1}{8\pi^2 g^2_{\mathrm{YM}}} [dv^2 + v^2 d\Omega_3^2], \qquad \text{with} \quad v = \frac{\ell}{\alpha'} e^{r/\ell}, \qquad (19.47)$$

where we have defined the energy scale v.

A general point on the flow has $\theta = 0$ as the family of flat directions. This moduli space is the Coulomb branch of the gauge theory anywhere along the flow. We see that we have movement on a (stretched) S^3, with coordinates (ψ_1, ψ_2, ψ_3), and the radial direction r. These give an $\mathbb{R}^4$, topologically, exploring the vevs of the complex scalar fields in the adjoint, ϕ_1 and ϕ_2. The metric on this moduli space for arbitrary $(r, \psi_1, \psi_2, \psi_3)$ is:

$$ds^2 = \frac{\tau_3}{2} \frac{\cosh^2 \chi}{\rho^2} e^{2A} dr^2 + \frac{\tau_3}{2} \ell^2 e^{2A} \rho^2 \left(\cosh^2\chi \, \sigma_3^2 + \sigma_1^2 + \sigma_2^2 \right). \qquad (19.48)$$

We can study this metric in the limit of small vevs: $r \to -\infty$. Inserting the IR values of the functions and defining:

$$u = \frac{\rho_0 \ell}{\alpha'} e^{r/\tilde{\ell}}, \qquad \tilde{\ell} = \frac{3}{2^{5/3}} \ell, \qquad \rho_0 \equiv \rho_{\mathrm{IR}} = 2^{1/6} \qquad (19.49)$$

we get

$$ds^2_{\mathcal{M}_{\mathrm{IR}}} = \frac{1}{8\pi^2 g^2_{\mathrm{YM}}} \left[\frac{3}{4} du^2 + u^2 \left(\frac{4}{3} \sigma_3^2 + \sigma_1^2 + \sigma_2^2 \right) \right]. \qquad (19.50)$$

This is an interesting result[321] which encodes information about the filed theory in a way which it would be nice to understand better. In order to do this, we ought to find better coordinates in which various field theory quantities are more manifest.[321] In a low-energy sigma model, the metric on the moduli space is the quantity which controls the kinetic terms for the scalar fields. In superspace, the kinetic terms are written in terms of a single function, the Kähler potential K:

$$\mathcal{L} = \int d^4\theta \, K(\Phi^i, \Phi^{j\dagger}) - \left\{ \int d^2\theta \, W(\Phi^i) + \mathrm{h.c.} \right\}, \qquad (19.51)$$

where Φ^i are chiral superfields whose lowest components are the scalars whose vevs we are exploring and $W(\Phi)$ is the superpotential. Our next task is to prove the existence of a Kähler potential for the probe metric. It is not at all manifest that this is the case, so we should spend some time on this next.

19.3.5 Kähler structure of the Coulomb branch

Let us start again with some new assignments of coordinates. The moduli space is parametrised by the vevs of the complex massless scalars, which we shall write as z_1 and z_2. The z_i transform as an $SU(2)$ doublet (i.e. in the fundamental), while their complex conjugates transform in the antifundamental. The $SU(2)$ flavour symmetry implies that the Kähler potential is a function of u^2 only where we define,

$$u^2 = z_1 \bar{z}_1 + z_2 \bar{z}_2. \qquad (19.52)$$

This is *not* necessarily the coordinate u we used as the AdS coordinate, or in the small vev presentation of the moduli space in the previous subsection. We shall see how they are related in various limits later.

We can divide the coordinates (and indices) into holomorphic and antiholomorphic (those without and those with a bar). If the Kähler structure exists then the metric is given by

$$ds^2 = g_{\mu\bar{\nu}} dz^\mu dz^{\bar{\nu}} = g_{1\bar{1}} dz_1 d\bar{z}_1 + g_{1\bar{2}} dz_1 d\bar{z}_2 + g_{2\bar{1}} dz_2 d\bar{z}_1 + g_{2\bar{2}} dz_2 d\bar{z}_2,$$

where

$$g_{\mu\bar{\nu}} = \partial_\mu \partial_{\bar{\nu}} K(u^2) = \partial_\mu(\partial_{\bar{\nu}}(u^2) K') = \partial_\mu(\partial_{\bar{\nu}}(u^2)) K' + \partial_\mu(u^2) \partial_{\bar{\nu}}(u^2) K',$$

where the primes denote differentiation with respect to u^2, and we have inserted our assumption about the u dependence of K. Notice that since

$$\partial_i(u^2) = \bar{z}_i \quad \text{and} \quad \bar{\partial}_i(u^2) = z_i, \qquad (19.53)$$

we have

$$g_{1\bar{1}} = \partial_1 \bar{\partial}_1 K = K' + z_1 \bar{z}_1 K'',$$
$$g_{1\bar{2}} = \bar{z}_1 z_2 K', \qquad (19.54)$$

and so on. Some algebra shows that the metric can be written as

$$ds^2 = (dz_1 d\bar{z}_1 + dz_2 d\bar{z}_2) K' + (\bar{z}_1 dz_1 + \bar{z}_2 dz_2)(z_1 d\bar{z}_1 + z_2 d\bar{z}_2) K'.$$

Now notice that[82]

$$du = \frac{1}{2u}(\bar{z}_1 dz_1 + \bar{z}_2 dz_2 + z_1 d\bar{z}_1 + z_2 d\bar{z}_2) \quad \text{and}$$

$$u\sigma_3 = \frac{1}{2u}(-i\bar{z}_1 dz_1 - i\bar{z}_2 dz_2 + iz_1 d\bar{z}_1 + iz_2 d\bar{z}_2). \qquad (19.55)$$

This is convenient, since we can write

$$du + iu\sigma_3 = \frac{1}{u}(\bar{z}_1 dz_1 + \bar{z}_2 dz_2) \quad \text{and} \quad du - iu\sigma_3 = \frac{1}{u}(z_1 d\bar{z}_1 + z_2 d\bar{z}_2).$$

Some more algebra puts the metric in the following form:

$$ds^2 = (K' + u^2 K')du^2 + u^2(K'(\sigma_1^2 + \sigma_2^2) + (K' + u^2 K')\sigma_3^2). \quad (19.56)$$

Looking at the form of the probe result in equation (19.48), we see that in order to put the metric into Kähler form we need a change of radial coordinate relating r and u. Equating coefficients, we obtain three equations:

$$(K' + u^2 K')du^2 = \frac{\tau_3}{2}\frac{\cosh^2 \chi}{\rho^2}e^{2A}dr^2, \quad (19.57)$$

$$u^2(K' + u^2 K') = \frac{\tau_3}{2}\ell^2\rho^2 e^{2A}\cosh^2 \chi, \quad (19.58)$$

$$u^2 K' = \frac{\tau_3}{2}\ell^2\rho^2 e^{2A}. \quad (19.59)$$

Using the first two equations we find

$$dr^2 = \frac{\ell^2\rho^4}{u^2}du^2, \quad (19.60)$$

with solution:

$$u = \frac{\ell}{\alpha'}e^{f(r)/\ell}, \quad \text{with} \quad \frac{df}{dr} = \frac{1}{\rho^2}. \quad (19.61)$$

Since the latter is always positive it defines a sensible radial coordinate u. We can now define K by the differential equation (19.59):

$$K' = \frac{dK}{d(u^2)} = \frac{\tau_3}{2}\frac{\ell^2\rho^2 e^{2A}}{u^2}, \quad (19.62)$$

and we have to check that such a K obeys equation (19.58), which can be written as

$$u^2\frac{d}{d(u^2)}(u^2 K') = \frac{\tau_3}{2}\ell^2\rho^2 e^{2A}\cosh^2 \chi. \quad (19.63)$$

From the definition of u in equation (19.61), we have that

$$\frac{d}{d(u^2)} = \frac{\ell\rho^2}{2u^2}\frac{d}{dr}, \quad (19.64)$$

and so we need to show that

$$\frac{\ell\rho^2}{2}\frac{d}{dr}(u^2 K') = \frac{\tau_3}{2}\ell^2\rho^2 e^{2A}\cosh^2 \chi. \quad (19.65)$$

From our definition of K in equation (19.62) this amounts to requiring us to show that:

$$\frac{d}{dr}(\rho^2 e^{2A}) = \frac{2}{\ell}e^{2A}\cosh^2 \chi. \quad (19.66)$$

We can achieve this by performing the derivative on the left hand side and substituting the flow equations for $\rho(r)$ and $A(r)$ listed in equations (1.13) gives precisely the result on the right[321].

We have demonstrated the existence of the Kähler potential. In fact, using the equation (19.64) we can write an alternative form for the definition of K, to accompany (19.62), which is:

$$\frac{dK}{dr} = \tau_3 \ell e^{2A(r)}. \tag{19.67}$$

N.B. This remarkably simple equation has been shown[321] to be satisfied by the Kähler potentials of all of the holographic RG flow examples that are (currently) known in ten dimensions. It would be interesting to learn what lies beneath this apparent universality, and the direct meaning of this equation in field theory.

In fact, one can readily write down an exact solution to this equation everywhere along the flow. Up to additive constants, it is:

$$K = \frac{\tau_3 \ell^2 e^{2A}}{4} \left(\rho^2 + \frac{1}{\rho^4} \right). \tag{19.68}$$

Let us unpack some of the content of this solution[321]. For large u (i.e. in the limit of large vevs), $\rho \sim 1$ so that, from equation (19.61), we have $u \sim \frac{\ell}{\alpha'} \exp(r/\ell)$, and to leading order:

$$K \sim \frac{\tau_3}{2} \ell^2 e^{2r/\ell} = \frac{1}{8\pi^2 g_{YM}^2} u^2, \tag{19.69}$$

which implies the expected flat four dimensional metric (19.47) that we obtained before. We can also look at next-to-leading order corrections to the Kähler potential. Recalling the asymptotic solutions for α and χ in equations (19.40) and also the flow equations (19.37) gives:

$$A(r) \simeq \frac{r}{\ell} - \frac{a_0^2}{6} e^{-2r/\ell} + O(e^{-4r/\ell}), \tag{19.70}$$

so that

$$K \simeq \tau_3 \ell^2 \left(\frac{1}{2} e^{2r/\ell} - \frac{a_0^2}{3} \frac{r}{\ell} \right). \tag{19.71}$$

We have discarded terms of order $\exp(-2r/\ell)$ as well as constant terms. Similarly, the corresponding expression for u^2 is from (19.61):

$$u^2 \simeq \frac{\ell^2}{\alpha'^2}\left(e^{2r/\ell} + \frac{4a_0^2}{3}\frac{r}{\ell}\right). \tag{19.72}$$

Returning to the Kähler potential, we find that:

$$K \simeq \frac{1}{8\pi^2 g_{YM}^2}\left[u^2 - \frac{a_0^2\ell^2}{\alpha'^2}\ln\left(\frac{\alpha'^2 u^2}{\ell^2}\right)\right], \tag{19.73}$$

an expression which looks like a one-loop field theory result. Further comparison requires some knowledge of how a_0^2 corresponds to the mass for Φ_3. To deduce this we can look at the probe result at large u more closely. The result of the probe calculation was given in equation (19.46). To leading order, we have

$$|z_1|^2 + |z_2|^2 = \frac{\ell^2}{\alpha'^2}e^{2r/\ell}\cos^2\theta, \quad \text{and}$$

$$|z_3|^2 = \frac{\ell^2}{\alpha'^2}e^{2r/\ell}\sin^2\theta, \tag{19.74}$$

and so

$$\mathcal{L} = \frac{1}{8\pi^2 g_{YM}^2}\left((|\dot{z}_1|^2 + |\dot{z}_2|^2 + |\dot{z}_3|^2) - \frac{4a_0^2}{\ell^2}|z_3|^2\right), \tag{19.75}$$

where we used the asymptotic solution (19.40) for α and for χ. The mass of Φ_3 is therefore

$$m_3 = \frac{2a_0}{\ell}. \tag{19.76}$$

Inserting this into the Kähler potential, we obtain

$$K \simeq \frac{1}{8\pi^2 g_{YM}^2}u^2 - \frac{Nm_3^2}{16\pi^2}\ln\left(\frac{\alpha'^2 u^2}{\ell^2}\right), \tag{19.77}$$

which is of the form expected for the tree level plus one loop correction, since (it turns out that) the $\mathcal{N} = 4$ field content ensures that $u^2\ln u^2$ terms cancel exactly. For small u, $\rho \to 2^{1/6}$ and we have

$$u \sim \frac{\ell}{\alpha'}\exp\left(\frac{r}{2^{1/3}\ell}\right). \tag{19.78}$$

This gives us:

$$K \sim \frac{\tau_3}{2}\ell^2\frac{3}{2^{5/3}}\left(\frac{u^2\alpha'^2}{\ell^2}\right)^{4/3} = \frac{1}{8\pi^2 g_{YM}^2}\frac{3}{2^{5/3}}\left(\frac{\alpha'^2}{\ell^2}\right)^{1/3}(u^2)^{4/3}, \tag{19.79}$$

and so the metric in this limit is:

$$ds^2 \sim \frac{1}{8\pi^2 g_{YM}^2} 2^{1/3} \left(\frac{u^2 \alpha'^2}{\ell^2}\right)^{1/3} \left(\frac{4}{3} du^2 + u^2 \left(\sigma_1^2 + \sigma_2^2 + \frac{4}{3}\sigma_3^2\right)\right),$$

(19.80)

which can be converted to the original form (19.50) after the redefinition $u \to u^{3/4}$ and an overall rescaling.

So now we understand that the curious form of this metric is simply a consequence of the power, $4/3$, of u^2 which appears in the Kähler potential. This power in turn follows from simple supergravity scaling, which translates nicely into the field theory data we already discussed.

At the UV end of the flow we have the standard $AdS_5 \times S^5$ geometry. The AdS_5 part of the metric given in equation (19.1) with $A = \frac{r}{\ell}$ which has a scaling symmetry under

$$x \to \frac{1}{\alpha} x \qquad e^A \to \alpha e^A \qquad u \to \alpha u,$$

(19.81)

where we have used that $u \sim e^A$ for large r. In other words the fields on moduli space have scaling dimension one, and so match with the dual field theory values for the scalar components of these chiral superfields in the $\mathcal{N} = 4$ theory. Next we consider the IR end of the flow. Here the solution again has the scaling symmetry (19.81) except that $A = 2^{5/3}r/3\ell$ in this case. The coordinate u goes like $u \sim \exp\left(\frac{r}{2^{1/3}\ell}\right) \sim (e^A)^{3/4}$ and thus the scaling symmetry becomes

$$x \to \frac{1}{\alpha} x \qquad u \to \alpha^{3/4} u.$$

(19.82)

Therefore, we see that the massless fields have scaling dimension $3/4$ here. Again this agrees with the field theory, as it includes the anomalous dimensions discussed earlier in equation 19.36.

Let's put it another way. Consider the Kähler potential at either end (UV or IR) of the flow. From the $SU(2)$ flavour symmetry we know that K is a function of u^2 only. We also know the scaling dimension of u^2 at each end of the flow. The action's kinetic term is:

$$S = \int d^4x \, \partial_\varphi \partial_{\bar\varphi} K \, \partial_\mu \varphi \, \partial^\mu \bar\varphi,$$

(19.83)

where φ are the massless scalars with some scaling dimension. For the action to be invariant under scaling, $K(u^2)$ must have scaling dimension 2. At the UV end of the flow u has scaling dimension 1, so $K \sim u^2$, as expected. At the IR end of the flow, u has scaling dimension $3/4$ and so $K \sim (u^2)^{4/3}$, matching our earlier results.

19.4 An $\mathcal{N} = 2$ gauge dual RG flow and the enhançon

It is worthwhile studying just one more flow example. This time it will not flow to a fixed point, and will preserve twice the supersymmetry as the previous example. This is achieved by turning on operators which correspond to giving equal masses to the $\mathcal{N} = 1$ multiplets Φ_1, Φ_2. Together, these form an $\mathcal{N} = 2$ hypermultiplet. This leaves one adjoint chiral multiplet Φ_3, together with the vector $\mathcal{N} = 1$ supermultiplet (A_μ, λ_4), forming the $\mathcal{N} = 2$ vector supermultiplet. So the deformation preserves an $\mathcal{N} = 2$ structure.

As before, this should correspond to an appropriate combination of scalars being switched on in supergravity, and the solution is known[322]. Again there are two scalars, and they correspond to the following operators:

$$\alpha : \qquad \sum_{i=1}^{4} \mathrm{Tr}(\phi_i \phi_i) - 2 \sum_{i=5}^{6} \mathrm{Tr}(\phi_i \phi_i)$$

$$\chi : \qquad \mathrm{Tr}(\lambda_1 \lambda_2 + \lambda_2 \lambda_2) + \mathrm{h.c.} \tag{19.84}$$

Moving around on the accessible part of the Coulomb branch of the $\mathcal{N}=2$ theory corresponds to giving a vacuum expectation value (vev) to $\varphi_3 = \phi_5 + i\phi_6$, which is the plane $\theta = \pi/2$.

The Coulomb branch of the moduli space of the $\mathcal{N} = 2$ $SU(N)$ gauge theory is parametrised by the vevs of the complex adjoint scalar φ_3 which set the potential $\mathrm{Tr}[\phi_3, \phi_3^\dagger]^2$ to zero. This generically breaks the theory to $U(1)^{N-1}$. This moduli space is of course an $N - 1$ complex dimensional space, but we are just focusing on the one-complex dimensional subspace corresponding to $SU(N - 1) \times U(1)$. The low energy effective action of the theory is described in terms of a low energy field u with an effective complex coupling $\tau(u)$:

$$\tau(u) = \tau_c + \frac{\theta}{2\pi} + i\frac{4\pi}{g_{\mathrm{YM}}^2}, \tag{19.85}$$

where the classical value is $\tau_c = \theta_s/2\pi + i/g_s$ in our case. The quantities θ_s and g_s are of course set by the R–R scalar $C_{(0)}$ and the dilaton Φ. Recall that $C_{(0)}$ couples to $F \wedge F$ on the D3-brane world volume, contributing to the θ-angle in the $\mathcal{N} = 2$ effective low energy theory.

19.4.1 *The five dimensional solution*

As before, at $r \to \infty$, the various functions in the solution have the following asymptotic behaviour[322]:

$$\rho(r) \to 1, \ \chi(r) \to 0, \ A(r) \to r/\ell. \tag{19.86}$$

For arbitrary r, the values of the functions are determined by the following flow equations:

$$\frac{d\alpha}{dr} = \frac{1}{\ell}\frac{\partial W}{\partial \alpha} = \frac{1}{3\ell}\left(\frac{1}{\rho^2} - \rho^4 \cosh(2\chi)\right)$$

$$\frac{d\chi}{dr} = \frac{1}{\ell}\frac{\partial W}{\partial \chi} = -\frac{1}{2\ell}\rho^4 \sinh(2\chi)$$

$$\frac{dA}{dr} = -\frac{2}{3\ell}W = \frac{2}{3\ell}\left(\frac{1}{\rho^2} + \frac{1}{2}\rho^4 \cosh(2\chi)\right), \qquad (19.87)$$

where the function

$$W = -\left(\frac{1}{\rho^2} + \frac{1}{2}\rho^4 \cosh(2\chi)\right),$$

can be used to construct the potential via:

$$V = \frac{4}{\ell^2}\left[\frac{1}{2}\sum_{i=1}^{2}\left(\frac{\partial W}{\partial \varphi_i}\right)^2 - \frac{4}{3}W^2\right] = \frac{1}{3\ell^2}\left(\frac{\partial W}{\partial \alpha}\right)^2 + \frac{2}{\ell^2}\left(\frac{\partial W}{\partial \chi}\right)^2 - \frac{16}{3\ell^2}W^2.$$

$$(19.88)$$

The functions W and V are plotted as contour maps in figure 19.6, and as three dimensional figures in figure 19.7.[‡]

By using the middle equation of (19.87), we can write expressions for $d\alpha/d\chi$ and $dA/d\chi$, which we can integrate (with some manipulation) to give:

$$e^A = k\frac{\rho^2}{\sinh(2\chi)}$$

$$\rho^6 = \cosh(2\chi) + \sinh^2(2\chi)\left(\gamma + \log\left[\frac{\sinh \chi}{\cosh \chi}\right]\right). \qquad (19.89)$$

Here, k is a constant we shall fix later, while γ is a constant whose values characterise a family of different solutions for $(\rho(r), \chi(r))$ representing different flows to the $\mathcal{N} = 2$ gauge theory in the IR. See figure 19.8.

- For $\gamma < 0$, equation (19.89) yields a *finite* value, $\chi_0 = \frac{1}{2}\cosh^{-1} c_0$, of χ in the IR, while ρ goes to *zero*. The supergravity solution has a naked singularity as a result.

- For $\gamma = 0$, χ *diverges* in the IR and again ρ goes to zero. Supergravity again has singular behaviour, coming from both the divergence and the zero.

- For $\gamma > 0$ both χ and ρ diverge, and the supergravity is singular.

[‡] As mentioned before, the reader should not take the small scale variations of the contours near the fixed points seriously. They are due to loss of numerical accuracy.

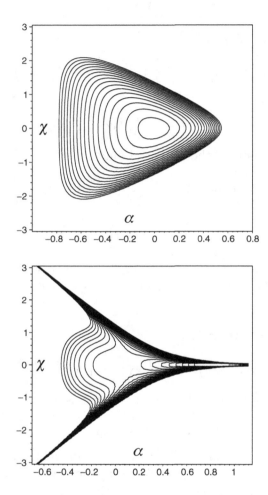

Fig. 19.6. Contour plots of the superpotential and potential, W and V, as functions of the scalars α, χ, for the RG flow to an $\mathcal{N} = 2$ gauge dual. The flows depicted in figure 19.8 are centred on the ridges to the left, the case $\gamma = 0$ being precisely along the ridge.

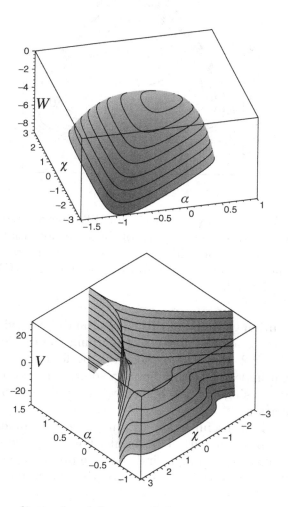

Fig. 19.7. Three dimensional figures of the superpotential and potential, W and V, as functions of the scalar α, χ, for the RG flow to an $\mathcal{N} = 2$ gauge dual.

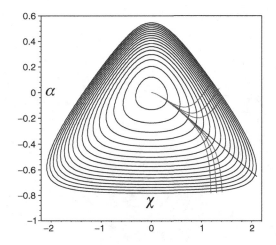

Fig. 19.8. The families of (χ, α) curves for differing γ, given by equation (19.89), superimposed on the contours of the superpotential W. There are three classes of curves. The middle curve is $\gamma = 0$, the $\gamma < 0$ curves are below it, and the $\gamma > 0$ curves are above. The flow from UV to IR along each curve is to the right.

All of our intuition gathered in this chapter and the previous one points towards there being sensible physics concerning the Coulomb branch of the expected $\mathcal{N} = 2$ dual gauge theory to be found at the end of the flow. We see that instead, the supergravity solution flows to regions which produce unphysical singularities. Somehow, this must be obscuring actual physical information.

This is where it is useful again to study the ten dimensional lift of the solution and probe it with a D3-brane. Following the wisdom of the previous two examples, we might find that the probe has a better handle on what are the right variables to use for the extraction of meaningful physics.

19.4.2   The ten dimensional solution

The ten dimensional solution written in the form (19.14), with[327, 328]:

$$ds_5^2 = \ell^2 \frac{\Omega^2}{\rho^2} \left[\frac{d\theta^2}{c} + \rho^6 \cos^2 \theta \left(\frac{\sigma_1^2}{cX_2} + \frac{\sigma_2^2 + \sigma_3^2}{X_1} \right) + \sin^2 \theta \frac{d\phi^2}{X_2} \right], \quad (19.90)$$

where $c = \cosh(2\chi)$, and

$$\Omega^2 = \frac{(cX_1X_2)^{1/4}}{\rho}$$
$$X_1 = \cos^2\theta + \rho^6\cosh(2\chi)\sin^2\theta$$
$$X_2 = \cosh(2\chi)\cos^2\theta + \rho^6\sin^2\theta. \tag{19.91}$$

It is easily seen that the non-trivial radial dependences of $\rho(r)$ and $\chi(r)$ deform the supergravity solution from $AdS_5 \times S^5$ at $r = \infty$ where there is an obvious $SO(6)$ symmetry (the round S^5 is restored), to a spacetime which only has an $SU(2) \times U(1)^2$ symmetry, which is manifest in the metric (19.90).

There are also explicit solutions for the R–R two-form potential, $C_{(2)}$, and the NS–NS two-form potential $B_{(2)}$. We will not need them here. The fields $(\Phi, C_{(0)})$ are gathered into a complex scalar field which we shall denote as $\lambda = C_{(0)} + ie^{-\Phi}$, and the solution for them is as follows:

$$\lambda = i\left(\frac{1-B}{1+B}\right), \tag{19.92}$$

with

$$B = \left[\frac{b^{1/4} - b^{-1/4}}{b^{1/} + b^{-1/4}}\right]e^{2i\phi}, \quad \text{where } b \equiv c\frac{X_1}{X_2}. \tag{19.93}$$

We shall extract the specific form for the dilaton, which we shall need, a bit later.

We will need the explicit form for the R–R four-form potential $C_{(4)}$, to which the D3-brane naturally couples. It is

$$C_{(4)} = e^{4A}\frac{X_1}{g_s\rho^2}dx_0 \wedge dx_1 \wedge dx_2 \wedge dx_3. \tag{19.94}$$

As is clear from the behaviour displayed in figure 19.8, it is evident that in the IR the supergravity becomes singular. This makes it hard to interpret the physics which is supposed be telling us about a dual gauge theory. Again, it is prudent to probe the geometry with a D3-brane to see if we can determine more about the physics.

19.4.3 Probing with a D3-brane

Following on what we did before, it is again straightforward to probe the geometry[332, 333], and the reader is urged to carry out the computation.

The result is an effective Lagrangian $\mathcal{L} = T - V$, where:

$$T = \frac{\mu_3}{2g_s} \Omega^4 e^{2A}$$

$$\times \left\{ \dot{r}^2 + \frac{\ell^2}{\rho^2} \left(\frac{\dot{\theta}^2}{c} + \rho^6 \cos^2\theta \left(\frac{v_1^2}{c\bar{X}_2} + \frac{v_2^2 + v_3^2}{\bar{X}_1} \right) + \sin^2\theta \frac{\dot{\phi}^2}{\bar{X}_2} \right) \right\},$$

$$= \frac{\mu_3}{2g_s} \frac{\ell^2 \left(c\bar{X}_1\bar{X}_2\right)^{1/2}}{(c^2 - 1)}$$

$$\times \left\{ \frac{\dot{c}^2}{\rho^6(c^2 - 1)^2} + \left(\frac{\dot{\theta}^2}{c} + \rho^6 \cos^2\theta \left(\frac{v_1^2}{c\bar{X}_2} + \frac{v_2^2 + v_3^2}{\bar{X}_1} \right) + \sin^2\theta \frac{\dot{\phi}^2}{\bar{X}_2} \right) \right\},$$

$$V = \frac{\mu_3}{g_s} e^{4A} \left(\Omega^4 - \frac{\bar{X}_1}{\rho^2} \right) = \frac{\mu_3}{g_s} \frac{k^4\rho^6}{(c^2 - 1)^2} \left(\sqrt{c\bar{X}_1\bar{X}_2} - \bar{X}_1 \right), \qquad (19.95)$$

where the v_i are the natural velocities associated to the one-forms σ_i given in insert 7.4 (p. 180), and in the last line we have used the first of the results in equations (19.89). The penultimate line was arrived at by using the fact that the second flow equation in (19.87) allows us to replace $\dot{r}^2$ by $\dot{c}^2\ell^2/[\rho^8(c^2 - 1)^2]$.

19.4.4   The moduli space

In order to make the potential vanish, there are two independent conditions: $c\bar{X}_2 = \bar{X}_1$, which means $\theta = \pi/2$, or $\rho = 0$. Notice that for the cases of $\gamma > 0$, the second situation does not exist, since (as is clear from figure 19.8) $\rho \to \infty$ and $\chi \to \infty$, while for $\gamma < 0$, the flow assigns a specific value, $\chi_0 = \frac{1}{2}\cosh^{-1} c_0$, for χ while $\rho \to 0$. The moduli space is parameterised by the coordinates (θ, ϕ), with metric:

$$ds^2_{\mathcal{M}2}(\gamma < 0) = \frac{\mu_3}{2g_s} \frac{\ell^2}{(c_0^2 - 1)} \left(\cos^2\theta d\theta^2 + \sin^2\theta d\phi^2 \right).$$

Notice that at $\gamma = 1$, the value of c_0 diverges, and so the metric vanishes. In the first situation there is a sensible metric for all classes of γ. It is parameterised by the (c, ϕ) space and the metric is:

$$ds^2_{\mathcal{M}1}(\gamma) = \frac{\mu_3}{2g_s} \frac{\ell^2 c k^2}{(c^2 - 1)} \left(\frac{dc^2}{(c^2 - 1)^2} + d\phi^2 \right). \qquad (19.96)$$

As discussed earlier, the $\gamma < 0$ flows lead to $\rho = 0$ and some finite value of c, which we call c_0. The supergravity is singular there, but the probe metric is perfectly smooth there. This situation is similar to ones we have encountered before, and is suggestive of a the edge of a disklike D3-brane source.

In the case $\gamma \geq 0$, c diverges, and the probe metric vanishes again. This is a signal of an enhançon-like locus, which we encountered in chapter 15. It appears here to be a circle, as would appear in the case of wrapped D7-branes, but we must be careful before we interpret this in gauge theory. As in the previous two examples that we have studied in this chapter, we must be careful to ensure that we are using the right coordinates.

We have two scalars, c and ϕ, but we must recall that these should be the components of a complex scalar, the adjoint scalar in the low energy effective low energy $U(1)$ action on the brane. So they should have the same coefficient[332, 333]. So we must find a complex coordinate z in which the metric is conformal to $dzd\bar{z}$. This is achieved by finding a new radial coordinate v such that

$$\frac{dc^2}{(c^2 - 1)^2} = \frac{dv^2}{v^2},$$

which has solution

$$v = \sqrt{\frac{c+1}{c-1}},$$

and so our putative enhançon circle at $c \to \infty$ on the $\gamma \geq 0$ branches is at $z = 1$. We can write the metric as:

$$ds^2_{\mathcal{M}1}(\gamma) = \frac{\mu_3}{2g_s} \frac{\ell^2 c k^2}{(c+1)^2} dz d\bar{z}. \tag{19.97}$$

In the low energy theory, the scalar field Y, being part of the $\mathcal{N} = 2$ gauge multiplet on the brane's world-volume, should have the same functional dependence for the kinetic term that the $U(1)$ gauge field on the probe has[333]. This translates into a kinetic term for Y:

$$ds^2 = \frac{\mu_3}{2} e^{-\Phi} dY d\bar{Y}, \tag{19.98}$$

where the dilaton may be extracted from the equation (19.93) as:

$$e^{-\Phi} = \frac{c}{g_s |\cos\phi + ic\sin\phi|}.$$

We must therefore change variables to the complex coordinates Y. Writing

$$dz d\bar{z} = dY d\bar{Y} \frac{\partial z}{\partial Y} \frac{\partial \bar{z}}{\partial \bar{Y}},$$

we get an equation

$$\left| \frac{\partial Y}{\partial z} \right|^2 = k^2 \ell^2 \left| \frac{\cos\phi + ic\sin\phi}{c+1} \right|^2 = \frac{k^2 \ell^2}{4} \left| 1 + \frac{1}{z^2} \right|,$$

and therefore

$$Y = \frac{k\ell}{2}\left(z + \frac{1}{z}\right). \tag{19.99}$$

We should write our complex coupling τ in terms of this coordinate. Some substitution gives the holomorphic result:

$$\tau = \frac{i}{g_s}\left(\frac{Y^2}{Y^2 - k^2\ell^2}\right)^{1/2} + \frac{\theta_s}{2\pi}. \tag{19.100}$$

We have a branch cut forming a segment[333] of the real line: $-k\ell \leq Y \leq k\ell$. We see from the change of variables in equation (19.99) that this branch cut is the circle $z = 1$, which is the enhançon, appearing in the $\gamma \leq 0$ flows. The $\gamma > 0$ flows are currently believed to be unphysical.

So we see that in fact the singular behaviour of the supergravity was hiding valuable physics which we uncovered by probing with a D3-brane. Just as in chapter 15, we find a region of the moduli space of large N gauge theory with eight supercharges where the constituent D-branes have spread out into a locus which we call the enhançon. Just as there, were this not to have happened (as the naive supergravity would allow), the constituent branes would have attained negative values for their tension. In the dual gauge theory, this negativity is a negative value for the kinetic term in the low energy action one moduli space, or alternatively, a negative value for the effective squared gauge coupling g_{eff}^2. In any of those pictures, this would be unphysical, and the brane physics protects itself against this case by moving the constituent branes to quantum corrected positions. In the language of the gauge theory, this is of course a large N manifestation of the Seiberg–Witten locus[240], which owes its origin to the same positivity requirements[§].

19.5 Beyond gravity duals

The last example is a situation where a supergravity solution, in attempting to reveal the physics about highly non-trivial behaviour of the dual gauge theory, needs to be supplemented with information about the string theory. We found this by probing with D3-branes by hand.

This is the expected sign that our ability to extract useful information about dual gauge theories will rely on our success in understanding more about the full string theory in the background. The D3-brane probe method, while powerful in its own right, is only a hybrid method, and

---

[§] See also section 16.1.12, for examples where quantum corrections to brane geometry in F-theory correlate with underlying Seiberg–Witten theory.

much more progress will be made when some way of computing in the full string theory for these backgrounds is found. The difficulty here is that one of the most crucial features of the solution is that it is supported by N units of R–R flux. This cannot be described as a small perturbation of an NS–NS background, and so the string theory must be phrased directly in terms of the R–R data. It should have been apparent, however, from the many studies that we have carried out in this book that it is in fact difficult to describe fully the strings propagating in such backgrounds. Looking back, it should be clear that we have only ever described string propagation in these backgrounds in the supergravity limits. The full conformal field theories that we described or alluded to were only ever for propagation in non-trivial NS–NS fields (like K3 geometry, or the NS5-brane's core). For the R–R p-branes, or the non-trivial F-theory or other such fascinating backgrounds, we were never in a position to present a world-sheet model (like a σ-model of chapter 2) which corresponded to the full string theory in the background, even perturbatively. The problem is that in the formalism described in chapter 7, the vertex operators corresponding to R–R states introduce world-sheet branch cuts in the presence of the superconformal generators, making them non–local with respect to each other[1], and hence outside the realm of the local conformal field theories that we have been studying.

Tools for the description of string theory propagating in R–R backgrounds need to be developed further, with some urgency. Results in this area will be especially interesting in view of the variety of physical phenomena that we have learned about from D-branes throughout this book. We learned all of this by indirect arguments combined with powerful technology in various limits (such as open strings, conformal field theory, and supergravity). Imagine what we might learn, and what useful tools we could develop if we could formulate things more directly.

A tantalising glimpse in this direction has been obtained recently. In addition to Minkowski space and $AdS_5 \times S^5$, it has been realised[344] that a certain type of pp-wave (see pp. 422–423) with R–R flux is also maximally supersymmetric. Furthermore, the pp-wave can be obtained[345,347] as a certain limit of $AdS_5 \times S^5$ that focuses on trajectories with large angular momentum in the S^5. String propagation in this pp-wave is exactly solvable, despite the R–R flux, in the light-cone gauge [346].

Remarkable, a class of gauge theory operators from the original dual CFT with R-charge going as $\sqrt{N}$ as $N \to \infty$ can be directly identified with the full tower of string states[347]. Properties of the light-cone string can be reconstructed from the gauge theory, and vice versa. This is an exciting development that will undoubtedly be explored further.

20

Taking stock

It is hoped that we have learned rather a lot about string theory in this book, and that the role of D-branes and other extended objects has been fascinating, entertaining, and instructive. It was with great pain that we had to sacrifice a tremendous amount of material in order to keep this book close to a sensible length, while retaining enough to succeed in telling a coherent story.

It is tempting to sit and reflect upon what great lessons we have learned, although it is not clear that this is a useful exercise at this stage, so we will be brief in our remarks. The main and most unambiguous lesson is that extended objects are vitally important to our understanding of string theory, and possibly (probably) whatever the final form of the fundamental quantum theory of space and time turns out to be. While extended objects are universally accepted as important, it is still (at a stretch) a matter of taste whether someone wants to go further and accept that it is unambiguously true that 'string theory is not a theory of strings'. The author believes it to be so, but will not insist that the reader take a position, since it seems that nobody can yet say what string or M-theory actually are theories of.

Whatever the final theory turns out to be, and whether or not once it is found it turns out be directly relevant to nature at all, it is clear that we have many new tools to work with which should keep us busy for some time to come in various areas. There are still many very specific questions that might be partly answered with the present technology which would have considerable benefits. For example, various gravity duals of increasingly complex gauge theory phenomena are being found from time to time, and a useful body of knowledge is being assembled about how such tools work in some detail. As a by-product, valuable lessons about strongly coupled gauge theory are being learned. This is despite the lack of a viable technology for studying string theory in R–R backgrounds, (an issue discussed at the end of chapter 19) with which considerable leaps in our understanding will be likely.

A topic that we have not touched upon at all is the whole *'Brane World'* discussion. This topic fruitfully borrows many ideas from the string theory constructions which we have discussed here, applying them to phenomenology. One class of models is that we simply live on a three-brane embedded in higher dimensions, to which are confined the gauge interactions which give rise to the standard model physics (so this is rather like a D-brane). Meanwhile, gravity lives in the whole spacetime and its relative weakness as compared to the other forces is apparently then attributable to the fact that it lives in more dimensions[335]. The other sort of scenario is again the idea that we live on a brane in higher dimensional spacetime, but that gravity is localised in the neighbourhood of the brane, due to the properties of the larger spacetime[336].

These both lead to interesting toy models of our world, and may find a home one day within a fundamental theory. The efforts to move these ideas forward are often mistakenly identified with research in string theory, but it should be clear that although there is some overlap, these are entirely different endeavours. It is safe to say that at the time of writing the many models which are being studied in these genres are not anywhere near constrained enough by being embedded in the (relatively) tight framework of string or M-theory (as far as we understand the latter two). On the other hand, this might not turn out to be entirely a bad thing, but it is too early to say.

Both of the topics above (and much of the content of the body of research described in this book) rely on the fact that although we do not know the details of the theory, we can learn a lot about things by working with low energy truncations. Of course, any small child educated in the modern field theory era will rightly immediately speak up at this point and mention that this is not special to string theory, but is a foundation of quantum field theory in general. The remarkable thing that seems to be available to us in the stringy arena is the wide variety of different ways of embedding various low energy phenomena into string and M-theory. This inevitably leads to new effective and often geometrical tools for studying these low energy phenomena, and sometimes powerful dual descriptions of the same physics, as we have seen many times.

This urges us to begin looking around for more examples, and possibly applications to other fields of physics where strongly coupled phenomena and interesting effective field theories of various sorts abound, like condensed matter physics. An example of this is the recent activity in embedding the physics of the quantum hall effect into string theory[334], following on from recasting it in terms of tools (such as non-commutative geometry[338]) sharpened in the string context. This may not be the only class of examples, and in fact there may be a lot to gain by deliberately exploring connections. There is also of course great likelihood that some of these connections will enrich understanding of string and M-theory.

While it is all well and good to discuss the elegant tools that we have uncovered, perhaps for useful application to difficult (nearly) phenomenological questions, we must not sidestep the issue of the direct search for a definition of M-theory. It becomes apparent when preparing a book of this sort, which surveys a large portion of the field, that it is perhaps not surprising at all that we have not yet stumbled on the definition. At nearly every turn of a page there seems to be a host of interesting unexplored connections and directions which might lead to interesting new physics. To save embarrassment, no attempt will be made to list them, since the large number may simply be a result of the author's ignorance, rather than his profound insight. In any case, the reader has probably their own list to be getting on with.

There are a number of seemingly very interesting features of D-branes which apparently have very deep roots, however. Whether or not they are a signal of the right variables for a dynamical formulation of the underlying theory is quite possibly an entirely different matter, but they are intriguing. For example, it is very striking that D-branes seem to supply the right variables for many very elegant descriptions of various geometries such as that of instantons, monopoles, the moduli spaces of these objects, ALE spaces, etc. Essentially, these 'right variables' are all of the charged hypermultiplets of various sorts which constitute the D-branes' collective coordinates, together with an appropriate set of constraints and/or projections.

Further to this (and closely related of course) is the fact that the world-volume couplings of D-branes seem to be naturally written in terms of very powerful geometrical objects: characteristic classes of various sorts, which enable them to enumerate topology so naturally*. This is an awfully generous circumstance and makes one wonder whether we should look deliberately for other powerful tools by starting with some of our other favourite geometrical or topological devices (other characteristic classes, etc.) and attempting to make them dynamical, perhaps by designing a world-volume interaction around them. Of course, it is not clear what the best candidates are, and what guiding principle one should use. Furthermore, this has in many senses been tried before, but perhaps

---

* Part of the outcome of this was the K-theory[113] description of D-branes alluded to but (sadly) not described in any detail in this book.

some of this approach could be revisited in the light of recent developments.

An interesting endeavour which is firmly underway at the time of writing is the revisitation of string field theory. As was clear to some people long ago (back at the time when string duality was generating a heady excitement, and hasty and unkind (but perhaps forgivable) things were said about string field theory), it may yet have its uses. One such use has turned out to be the detailed study of local portions of the potential in which the string theories which we know are special vacua. Various ideas are afloat concerning the suggestion that the decay of unstable D-branes (via 'tachyon condensation') takes the theory to a new but familiar place.

So for example the space-filling D25-brane of bosonic open string theory decays away leaving a closed string theory vacuum. A relatively simple string field theory computation shows that the energy difference between the starting vacuum and the ending vacuum well approximates the tension of the D25-brane, which is intriguing[337]. Similar suggestions for the unstable branes of other theories have be tested also, with encouraging results. This has led to renewed vigour in the matter of understanding formulations of string field theory, and the interpretation of mysterious aspects of current formulations. In the latter regard for example, still puzzling to some extent is the fact that the computation done above is within the purely open string field theory. However, since the endpoint of the process is not an open string theory at all (the D25-brane has disappeared), the appearance in that framework of the closed strings which remain (if that is what remains) is not well understood. Overall, this is certainly a very interesting area which may well sharpen our understanding of aspects of the string vacua we know and how they are connected.

There has also been recent progress in understanding the fate of tachyonic purely closed string backgrounds and the vacua to which they decay[348]. For example, supersymmetry-breaking orbifold projections of $\mathbb{C}_2$ can be constructed, generalising the ALE spaces of chapters 13 and 14. Tachyons arise in the twisted sectors, giving interesting models in which the supersymmetry breaking is localised at the conical singularity. The decay channels involve the spaces radiating some of their curvature to infinity, becoming less singular and settling to stable vacua with curvature, or to flat space. Even in these tachyonic models, techniques closely related to some discussed in chapter 13 show that there are fascinating connections to a rich mathematical framework of singularity theory[349].

Included in the above are topics which incorporate the fact that we are very much at home with the idea of non-BPS D-branes, and in fact some of the outcome of that research may be to find more useful techniques for finding them and working with them. Non-BPS D-branes were not discussed at all in this book, unfortunately, but the tachyon condensation techniques can be used to point to their existence (and sometimes stability)[18, 21]. Furthermore, the K-theory organisation[113] of D-branes in fact does not care whether they are supersymmetric or not. This is very encouraging but we need more than a classification, we need a technology. With a good handle on the properties of non-BPS D-branes we can address two large areas of research. One is the area of probing various dualities beyond the supersymmetric sector. This applies to field theory dualities, where we might learn about more realistic strongly coupled gauge theory phenomena with such tools, and also string theory dualities, where connections to non-supersymmetric string vacua can be explored. This is a major motivation (in part) for some of the endeavours mentioned above.

The whole area of 'holography'[286, 287] has certainly only just begun to be uncovered. It is perhaps clear that the collection of ideas surrounding that topic is an intriguing part of a profound story about spacetime, quantum mechanics, gravity and field theory, but the problem so far is the lack of a constructive way of phrasing the Holographic Principle: it tells one what the count of degrees of freedom ought to be, but gives (at time of writing) no insights as to how to implement the hologram. It may be again that this is a result of working with the wrong basic objects: as a result, it seems that the few working examples of holography that we have, like the AdS/CFT correspondence and perhaps matrix theory, are too different from each other in order to teach us anything general but yet specific enough.

On the whole, it remains a very exciting area in which to be working. In fact, one has a feeling of anticipation that there is something just around the corner which will put us back into the remarkable situation we were in a few years ago. Up to late 1994 or early 1995 we used to dream about various scenarios in string theory (perturbative and non-perturbative) which we had scarcely any tools to help us realise. Branes, and particularly D-branes, came along as the sharp tools that were needed and some of those dreams were made concrete, others discarded. For a while, D-branes were so intrinsically rich with new physics that they supplied us with new scenarios that we had not dreamed of at all, and gave us further ideas about how to concretely study those new situations.

Now we are in the situation where things have become hard again. Since the Second Revolution, we now dream in technicolour: branes are still telling us that there are remarkable places to which the theory can go (eg. by their changing shape and expanding into other branes, mingling with spacetime geometry or describing it as non-commutative, etc.) but it is harder for them to take us there. In other words, it is perhaps time for them to hand over to another tool that can take us to these new places. One has a feeling that this new tool (or tools) may well be within our grasp, perhaps just at the edge of our collective peripheral vision. It may be that, as happened with D-branes initially, the new tool has already been discovered, but has not yet been recognised. Perhaps it is time to squint more quizzically at some of the objects which lurk in our notebooks.

There are many more topics of considerable importance which we have only touched upon, or not mentioned at all. An obvious one is the emergence of non-commutative geometry in both field theory and string theory which has been a topic of much research[338]. We touched upon it in one of its guises earlier in section 13.6, but have left most of it unexplored. There are other topics too, such as compactifications of string theory and M-theory to four dimensions using the some of the new ideas and techniques that D-branes have supplied, perhaps giving new insights into phenomenology and/or other important four dimensional physics. Any of those topics could well be the area in which the next major breakthrough occurs, which is exciting. The hope is that, regardless of where the next breakthrough might be, this book will serve as a useful guide to some of the ideas and tools which have brought us to this point, and which may well help in moving things further.

References

[1] J. Polchinski, *String Theory*, Vols. 1 and 2; Cambridge University Press (1998) (Cambridge Monographs on Mathematical Physics).

[2] J. Polchinski, S. Chaudhuri and C. V. Johnson, *Notes on D-Branes*, hep-th/9602052.

[3] C. V. Johnson, *D-Brane primer*, in TASI 1999, *Strings, Branes and Gravity*, World Scientific (2001), hep-th/0007170.

[4] C. V. Johnson, *Etudes on D-branes*, hep-th/9812196.

[5] M. B. Green, J. H. Schwarz and E. Witten, *Superstring Theory*, Vols. 1 and 2; Cambridge University Press (1987) (Cambridge Monographs on Mathematical Physics).

[6] This is an example: E. Kiritsis, *Introduction to Superstring Theory*, hep-th/9709062. Leuven University Press (1998) 315 pp. (Belgium) (Leuven Notes in Mathematical and Theoretical Physics, B9).

[7] A. Strominger and C. Vafa, *Phys. Lett.* **B379**, 99 (1996), hep-th/9601029.

[8] J. Dai, R. G. Leigh and J. Polchinski, *Mod. Phys. Lett.* **A4**, 2073 (1989).

[9] A. Chodos and C. B. Thorn, *Nucl. Phys.* **B72**, 509 (1974); W. Siegel, *Nucl. Phys.* **B109**, 244 (1976); S. M. Roy and V. Singh, *Pramana* **26**, L85 (1986); *Phys. Rev.* **D35**, 1939 (1987); J. A. Harvey and J. A. Minahan, *Phys. Lett.* **B188**, 44 (1987).

[10] N. Ishibashi and T. Onogi, *Nucl. Phys.* **B318**, 239 (1989); G. Pradisi and A. Sagnotti, *Phys. Lett.* **B216**, 59 (1989); A. Sagnotti, *Phys. Rept.* **184**, 167 (1989); P. Horava, *Nucl. Phys.* **B327**, 461 (1989).

[11] J. H. Schwarz, *Nucl. Phys.* **B65**, 131 (1973); E. F. Corrigan and D. B. Fairlie, *Nucl. Phys.* **B91**, 527 (1975); M. B. Green, *Nucl. Phys.* **B103**, 333 (1976); M. B. Green and J. A. Shapiro, *Phys. Lett.* **64B**, 454 (1976); A. Cohen, G. Moore, P. Nelson, and J. Polchinski, *Nucl. Phys.* **B267**, 143 (1986); **B281**, 127 (1987).

[12] M. Dine, P. Huet, and N. Seiberg, *Nucl. Phys.* **B322**, 301 (1989).

[13] P. Horava, *Phys. Lett.* **B231**, 251 (1989); M. B. Green, *Phys. Lett.* **B266**, 325 (1991).

[14] K. Kikkawa and M. Yamanaka, *Phys. Lett.* **B149**, 357 (1984); N. Sakai and I. Senda, *Prog. Theor. Phys.* **75**, 692 (1986).

[15] V.P. Nair, A. Shapere, A. Strominger, and F. Wilczek, *Nucl. Phys.* **B287**, 402 (1987).

[16] P. Ginsparg and C. Vafa, *Nucl. Phys.* **B289**, 414 (1987).

[17] T. H. Buscher, *Phys. Lett.* **B194B**, 59 (1987); **B201**, 466 (1988).

[18] A. Sen, *JHEP* **9806**, 007 (1998), hep-th/9803194; *JHEP* **9808**, 010 (1998), hep-th/9805019; *JHEP* **9808**, 012 (1998), hep-th/9805170; *JHEP* **9809**, 023 (1998), hep-th/9808141; *JHEP* **9810**, 021 (1998), hep-th/9809111; Reviews can be found in: A. Sen, *Non-BPS states and branes in string theory*, hep-th/9904207; A. Lerda and R. Russo, *Int. J. Mod. Phys.* **A15**, 771 (2000), hep-th/9905006.

[19] P. G. Freund and M. A. Rubin, *Phys. Lett.* B **97**, 233 (1980).

[20] D. J. Gross, J. A. Harvey, E. Martinec and R. Rohm, *Phys. Rev. Lett.* **54**, 502 (1985); *Nucl. Phys.* **B256**, 253 (1985); *Nucl. Phys.* **B267**, 75 (1986).

[21] Two useful reviews are: J. H. Schwarz, *TASI lectures on non-BPS D-brane systems*, in TASI 1999: *Strings, Branes and Gravity*, World Scientific (2001), hep-th/9908144; K. Olsen and R. Szabo, *Constructing D-branes From K-theory*, hep-th/9907140.

[22] J. Paton and Chan Hong–Mo, *Nucl. Phys.* **B10**, 519 (1969).

[23] L. Dixon, J. A. Harvey, C. Vafa and E. Witten, *Nucl. Phys.* **B261**, 678 (1985).

[24] J. H. Schwarz, in *Lattice Gauge Theory, Supersymmetry and Grand Unification*, 233, Florence 1982, *Phys. Rept.* **89**, 223 (1982); N. Marcus and A. Sagnotti, *Phys. Lett.* **119B**, 97 (1982).

[25] J. Polchinski, *Phys. Rev.* **D50**, 6041 (1994), hep-th/9407031.

[26] E. Witten, *Nucl. Phys.* **B460**, 335 (1996), hep-th/9510135.

[27] A. Sagnotti, in *Non-Perturbative Quantum Field Theory*, eds. G. Mack *et al.* (Pergamon Press, 1988), 521; V. Periwal, unpublished; J. Govaerts, *Phys. Lett.* **B220**, 77 (1989).

[28] A. Dabholkar, *Lectures on orientifolds and duality*, hep-th/9804208.

[29] S. P. de Alwis, *A Note on Brane Tension and M Theory*, hep-th/9607011.

[30] C. Lovelace, *Phys. Lett.* **B34**, 500 (1971); L. Clavelli and J. Shapiro, *Nucl. Phys.* **B57**, 490 (1973); M. Ademollo, R. D' Auria, F. Gliozzi, E. Napolitano, S. Sciuto, and P. di Vecchia, *Nucl. Phys.* **B94**, 221 (1975); C. G. Callan, C. Lovelace, C. R. Nappi, and S. A. Yost, *Nucl. Phys.* **B293**, 83 (1987).

[31] J. Polchinski and Y. Cai, *Nucl. Phys.* **B296**, 91 (1988); C. G. Callan, C. Lovelace, C. R. Nappi and S. A. Yost, *Nucl. Phys.* **B308**, 221 (1988).

[32] A. Abouelsaood, C. G. Callan, C. R. Nappi and S. A. Yost, *Nucl. Phys.* **B280**, 599 (1987).

[33] This is a vast subject by now. Ref. [31], and the last of ref. [30] are some of the originals, but there are many more. Some good reviews are: P. Di Vecchia and A. Liccardo, *D branes in string theory. I*, hep-th/ 9912161; P. Di Vecchia and A. Liccardo, *D-branes in string theory. II*, hep-th/9912275. I. V. Vancea, *Introductory lectures to D-branes*, hep-th/ 0109029.

[34] R. G. Leigh, Mod. *Phys. Lett.* **A4**, 2767 (1989).

[35] S. Coleman and E. Weinberg, *Phys. Rev.* **D7**, 1888 (1973).

[36] J. Polchinski, *Comm. Math. Phys.* **104**, 37 (1986).

[37] M. Douglas and B. Grinstein, *Phys. Lett.* **B183**, 552 (1987); (E) **187**, 442 (1987); S. Weinberg, *Phys. Lett.* **B187**, 278 (1987); N. Marcus and A. Sagnotti, *Phys. Lett.* **B188**, 58 (1987).

[38] See refs. [39, 40, 41].

[39] C. Bachas, *Phys. Lett.* **B374**, 37 (1996), hep-th/9511043.

[40] C. Bachas, *Lectures on D-branes*, hep-th/9806199.

[41] E. Bergshoeff, M. de Roo, M. B. Green, G. Papadopoulos, and P. K. Townsend, *Nucl. Phys.* **B470**, 113 (1996), hep-th/9601150; E. Alvarez, J. L. F. Barbon, and J. Borlaf, *Nucl. Phys.* **B479**, 218 (1996), hep-th/ 9603089; E. Bergshoeff and M. De Roo, *Phys. Lett.* **B380**, 265 (1996), hep-th/9603123.

[42] E. S. Fradkin and A. A. Tseytlin, *Phys. Lett.* **B163**, 123 (1985).

[43] A. A. Tseytlin, *Nucl. Phys.* **B501**, 41 (1997), hep-th/9701125.

[44] See refs. [46, 47, 48, 49, 50].

[45] See, for example, refs. [43, 46, 50, 51].

[46] D. Brecher and M. J. Perry, *Nucl. Phys.* **B527**, 121 (1998), hep-th/9801127.

[47] D. Brecher, *Phys. Lett.* **B442**, 117 (1998), hep-th/9804180.

[48] M. R. Garousi and R. C. Myers, *Nucl. Phys.* **B542**, 73 (1999), hep-th/ 9809100.

[49] A. Hashimoto and W. I. Taylor, *Nucl. Phys.* **B503**, 193 (1997), hep-th/ 9703217; P. Bain, hep-th/9909154.

[50] A. A. Tseytlin, *Born–Infeld action, supersymmetry and string theory*, hep-th/9908105.

[51] R. C. Myers, *JHEP* **9912**, 022 (1999), hep-th/9910053.

[52] W. I. Taylor and M. Van Raamsdonk, *Nucl. Phys.* **B573**, 703 (2000), hep-th/9910052; *Nucl. Phys.* **B558**, 63 (1999), hep-th/9904095.

[53] There are many good treatments of anomalies. The field theory treatment should begin with a good modern text. See vol. 2 of ref. [54], vol. 2 of ref. [5], and: S. B. Treiman, E. Witten, R. Jackiw and B. Zumino, *Current Algebra and Anomalies*, World Scientific, Singapore (1985).

[54] S. Weinberg, *The Quantum Theory Of Fields*, Vols. 1, 2, and 3, Cambridge University Press (2000) (Cambridge Monographs on Mathematical Physics).

[55] See refs. [56, 57, 58, 59, 60].

[56] G. W. Gibbons, *Nucl. Phys.* **B514**, 603 (1998), hep-th/9709027.

[57] C. G. Callan and J. M. Maldacena, *Nucl. Phys.* **B513**, 198 (1998), hep-th/9708147.

[58] P. S. Howe, N. D. Lambert and P. C. West, *Nucl. Phys.* **B515**, 203 (1998), hep-th/9709014; S. Lee, A. Peet and L. Thorlacius, *Nucl. Phys.* **B514**, 161 (1998), hep-th/9710097.

[59] R. Emparan, *Phys. Lett.* **B423**, 71 (1998), hep-th/9711106.

[60] J. P. Gauntlett, J. Gomis and P. K. Townsend, *JHEP* **9801**, 003 (1998), hep-th/9711205.

[61] E. B. Bogomolny, *Sov. J. Nucl. Phys.* **24**, 449 (1976).

[62] M. K. Prasad and C. M. Sommerfield, *Phys. Rev. Lett.* **35**, 760 (1975).

[63] Two good books from which to learn this sort of construction are: J. Bagger and J. Wess, *Supersymmetry and Supergravity*, Princeton University Press, 1991; P. West, *Introduction to Supersymmetry and Supergravity*, World Scientific, Singapore, 1990.

[64] E. Witten and D. Olive, *Phys. Lett.* **78B**, 97 (1978).

[65] G. W. Gibbons, in *Proc. Heisenberg Memorial Symp. 1981*, eds. P. Breitenlohner and H. P. Dürr (*Lecture notes in Physics* **160**, Springer–Verlag 1982); G. W. Gibbons and C. M. Hull, *Phys. Lett.* B **109**, 190 (1982).

[66] A. Papapetrou, *Proc. R. Irish Acad.* **A51** (1947) 191; S.D. Majumdar, *Phys. Rev.* **72**, 930 (1947).

[67] C. Teitelboim, *Phys. Lett.* B **69**, 240 (1977).

[68] G. W. Gibbons and P. K. Townsend, *Phys. Rev. Lett.* **71**, 3754 (1993), hep-th/9307049.

[69] A useful review, with references, is: R. Kallosh, *From Gravity to Supergravity*, Lectures given at Theoretical Advanced Study Institute in Elementary Particle Physics (TASI 97): *Supersymmetry, Supergravity and Supercolliders*, Boulder, CO, 1–7 June 1997.

[70] J. A. Shapiro and C. B. Thorn, *Phys. Rev.* **D36**, 432 (1987); J. Dai and J. Polchinski, *Phys. Lett.* **B220**, 387 (1989).

[71] F. Gliozzi, J. Scherk and D. Olive, *Nucl. Phys.* **B122**, 253 (1977); *Phys. Lett.* **B65**, 282 (1976).

[72] A. Strominger, *Nucl. Phys.* **B343**, 167 (1990); *Erratum: ibid.*, **353**, 565 (1991); S.-J. Rey, in *Superstrings and Particle Theory: Proceedings*, eds. L. Clavelli and B. Harms, (World Scientific, 1990); S.-J. Rey, *Phys. Rev.* **D43**, 526 (1991); I. Antoniades, C. Bachas, J. Ellis and D. Nanopoulos, *Phys. Lett.* **B211**, 393 (1988); *ibid.*, *Nucl. Phys.* **328**, 117 (1989); C. G. Callan, J. A. Harvey and A. Strominger, *Nucl. Phys.* **B359**, 611 (1991).

[73] C. G. Callan, J. A. Harvey and A. Strominger, in Trieste 1991, *String Theory and Quantum Gravity*, hep-th/9112030.

[74] D. Friedan, E. Martinec, and S. Shenker, *Nucl. Phys.* **B271**, 93 (1986).

[75] For example, see the conventions in: E. Bergshoeff, C. Hull and T. Ortin, *Nucl. Phys.* **B451**, 547 (1995), hep-th/9504081.

[76] M. B. Green, C. M. Hull and P. K. Townsend, *Phys. Lett.* **B382**, 65 (1996), hep-th/9604119.

[77] P. Meessen and T. Ortin, *Nucl. Phys.* **B541**, 195 (1999), hep-th/9806120.

[78] See refs. [79, 80, 81, 88, 89].

[79] D. N. Page, *Phys. Lett.* **B80**, 55 (1978).

[80] M. A. Walton, *Phys. Rev.* **D37**, 377 (1988).

[81] A very useful reference for the properties of string theory on ALE spaces is: D. Anselmi, M. Billó, P. Fré, L. Girardello and A. Zaffaroni, *Int. J. Mod. Phys.* **A9**, 3007 (1994), hep-th/9304135.

[82] An excellent reference for various relevant geometrical facts is: T. Eguchi, P. B. Gilkey and A. J. Hanson, *Gravitation, Gauge Theories And Differential Geometry*, *Phys. Rept.* **66**, 213 (1980).

[83] G. W. Gibbons and S. W. Hawking, *Commun. Math. Phys.* **66**, 291 (1979).

[84] T. Eguchi and A. J. Hanson, *Ann. Phys.* **120**, 82 (1979).

[85] N. J. Hitchin, *Polygons and gravitons*, in Gibbons, G. W. (ed.), Hawking, S. W. (ed.): *Euclidean quantum gravity*, World Scientific (1993), pp. 527–538.

[86] F. Klein, *Vorlesungen Über das Ikosaeder und die Auflösung der Gleichungen vom fünften Grade*, Teubner, Leipzig 1884; F. Klein, *Lectures on the Icosahedron and the Solution of an Equation of Fifth Degree*, Dover, New York (1913).

[87] J. Mckay, *Proc. Symp. Pure Math.* **37**, 183 (1980), American Mathematical Society.

[88] N. Seiberg, *Nucl. Phys.* **B303**, 286 (1988).

[89] P. S. Aspinwall and D. R. Morrison, *String theory on K3 surfaces*, in Greene, B. (ed.), Yau, S. T. (ed.): Mirror symmetry II* 703–716, hep-th/9404151.

[90] P. Aspinwall, *K3 Surfaces and String Duality*, in TASI 1996, World Scientific 1997, hep-th/9611137; *Compactification, Geometry and Duality: N = 2*,

in TASI 1999: *Strings, Branes and Gravity*, World Scientific (2001), hep-th/0001001.

[91] G. W. Gibbons and S. W. Hawking, *Comm. Math. Phys.* **66**, 291 (1979).

[92] M. K. Prasad, *Phys. Lett.* **B83**, 310 (1979).

[93] M. B. Green, *Phys. Lett.* **B329**, 435 (1994), hep-th/9403040.

[94] G. T. Horowitz and A. Strominger, *Nucl. Phys.* **B360**, 197 (1991).

[95] For a review of string solitons, see: M. J. Duff, Ramzi R. Khuri and J. X. Lu, *String Solitons*, *Phys. Rept.* **259**, 213 (1995), hep-th/9412184.

[96] A. Strominger, *Nucl. Phys.* **B451**, 96 (1995), hep-th/9504090.

[97] L. J. Romans, *Phys. Lett.* **B169**, 374 (1986).

[98] J. Polchinski and A. Strominger, *Phys. Lett.* **B388**, 736 (1996), hep-th/9510227.

[99] M. B. Green, *Phys. Lett.* **B69**, 89 (1977); **B201**, 42 (1988); **B282**, 380 (1992).

[100] S. H. Shenker, *The Strength of Non-Perturbative Effects in String Theory*, in Cargese 1990, Proceedings: *Random Surfaces and Quantum Gravity* (1990), p. 191.

[101] T. Banks and L. Susskind, *Brane–Anti-Brane Forces*, hep-th/9511194.

[102] S. H. Shenker, *Another Length Scale in String Theory?*, hep-th/9509132.

[103] D. Kabat and P. Pouliot, *Phys. Rev. Lett.* **77**, 1004 (1996), hep-th/9603127; U. H. Danielsson, G. Ferretti and B. Sundborg, *Int. J. Mod. Phys.* **A11**, 5463 (1996), hep-th/9603081.

[104] M. R. Douglas, D. Kabat, P. Pouliot and S. H. Shenker, *Nucl. Phys.* **B485**, 85 (1997), hep-th/9608024.

[105] W. Fischler and L. Susskind, *Phys. Lett.* **B171**, 383 (1986); **173**, 262 (1986).

[106] See refs. [102, 103, 104, 233, 234, 237].

[107] M. B. Green and J. H. Schwarz, *Phys. Lett.* **B149**, 117 (1984); **B151**, 21 (1985); *Nucl. Phys.* **B255**, 93 (1985).

[108] M. B. Green and J. H. Schwarz, *Phys. Lett.* B **136**, 367 (1984); *Nucl. Phys.* B **243**, 285 (1984).

[109] M. B. Green, J. H. Schwarz and P. C. West, *Nucl. Phys.* B **254**, 327 (1985).

[110] C. G. Callan and J. A. Harvey, *Nucl. Phys.* **B250**, 427 (1985); S. G. Naculich, *Nucl. Phys.* **B296**, 837 (1988); J. M. Izquierdo and P. K. Townsend, *Nucl. Phys.* **B414**, 93 (1994), hep-th/9307050; J. D. Blum and J. A. Harvey, *Nucl. Phys.* **B416**, 119 (1994), hep-th/9310035.

[111] M. B. Green, J. A. Harvey and G. Moore, *Class. Quant. Grav.* **14**, 47 (1997), hep-th/9605033.

[112] Y. E. Cheung and Z. Yin, *Nucl. Phys.* **B517**, 69 (1998), hep-th/9710206.

[113] R. Minasian and G. Moore, *JHEP* **9711**, 002 (1997), hep-th/9710230; E. Witten, *JHEP* **9812**, 019 (1998), hep-th/9810188; P. Horava, *Adv. Theor. Math. Phys.* **2**, 1373 (1999), hep-th/9812135; D. Diaconescu, G. Moore and E. Witten, hep-th/0005091 and hep-th/0005090.

[114] R. I. Nepomechie, *Phys. Rev.* **D31**, 1921 (1985); C. Teitelboim, *Phys. Lett.* **B167**, 63, 69 (1986).

[115] M. Bershadsky, C. Vafa, and V. Sadov, *Nucl. Phys.* **B463**, 420 (1996), hep-th/9511222.

[116] M. Bershadsky, C. Vafa and V. Sadov, *Nucl. Phys.* **B463**, 398 (1996), hep-th/9510225.

[117] S. Katz and C. Vafa, *Nucl. Phys.* **B497**, 196 (1997), hep-th/9611090; S. Katz, A. Klemm and C. Vafa, *Nucl. Phys.* **B497**, 173 (1997), hep-th/9609239.

[118] M. Li, *Nucl. Phys.* **B460**, 351 (1996), hep-th/9510161.

[119] M. R. Douglas, *Branes within Branes*, hep-th/9512077.

[120] A. A. Belavin, A. M. Polyakov, A. S. Shvarts and Y. S. Tyupkin, *Phys. Lett.* B **59**, 85 (1975).

[121] See also the very useful refs. [125, 124, 126, 127, 123].

[122] K. Dasgupta, D. P. Jatkar and S. Mukhi, *Nucl. Phys.* **B523**, 465 (1998), hep-th/9707224.

[123] K. Dasgupta and S. Mukhi, *JHEP* **9803**, 004 (1998), hep-th/9709219. C. A. Scrucca and M. Serone, *Nucl. Phys.* **B556**, 197 (1199), hep-th/9903145.

[124] B. Craps and F. Roose, *Phys. Lett.* **B445**, 150 (1998), hep-th/9808074; B. Craps and F. Roose, *Phys. Lett.* **B450**, 358 (1999), hep-th/9812149.

[125] J. F. Morales, C. A. Scrucca and M. Serone, *Nucl. Phys.* **B552**, 291 (1999), hep-th/9812071; B. Stephanski, *Nucl. Phys.* **B548**, 275 (1999), hep-th/9812088.

[126] S. Mukhi and N. V. Suryanarayana, *JHEP* **9909**, 017 (1999), hep-th/9907215.

[127] J. F. Ospina Giraldo, *Gravitational couplings for generalized orientifold planes*, hep-th/0006076; *Gravitational couplings for yGOp-planes*, hep-th/0006149.

[128] C. P. Bachas, P. Bain and M. B. Green, *JHEP* **9905**, 011 (1999), hep-th/9903210.

[129] M. Berkooz, M. R. Douglas and R. G. Leigh, *Nucl. Phys.* **B480**, 265 (1996), hep-th/9606139.

[130] E. Witten, *Nucl. Phys.* **B460**, 541 (1996), hep-th/9511030.

[131] See refs. [132, 193, 194, 192, 28, 195, 196].

[132] E. G. Gimon and J. Polchinski, *Phys. Rev.* **D54**, 1667 (1996), hep-th/9601038.

[133] J. H. Schwarz, *Phys. Lett.* **B360** 13 (1995); (E) **B364**, 252 (1995), hep-th/9508143.

[134] P. S. Aspinwall, *Nucl. Phys. Proc. Suppl.* **46**, 30 (1996), hep-th/9508154.

[135] J. H. Schwarz, *Nucl. Phys. Proc. Suppl.* **55B**, 1 (1997), hep-th/9607201.

[136] P. K. Townsend, *M-theory from its superalgebra*, hep-th/9712004.

[137] O. Aharony, J. Sonnenschein and S. Yankielowicz, *Nucl. Phys.* **B474**, 309 (1996), hep-th/9603009. M. R. Gaberdiel and B. Zwiebach, *Nucl. Phys.* B **518**, 151 (1998), hep-th/9709013.

[138] A. Sen, *JHEP* **9803**, 005 (1998), hep-th/9711130.

[139] E. Witten, *Nucl. Phys.* **B500**, 3 (1997), hep-th/9703166.

[140] K. Dasgupta and S. Mukhi, *Phys. Lett.* **B423**, 261 (1998), hep-th/9711094.

[141] A. Sen, *Phys. Rev.* **D54**, 2964 (1996), hep-th/9510229.

[142] Here is a selection of papers in this topic: J. Froehlich and J. Hoppe, *Commun. Math. Phys.* **191**, 613 (1998), hep-th/9701119; P. Yi, *Nucl. Phys.* **B505**, 307 (1997), hep-th/9704098; S. Sethi and M. Stern, *Commun. Math. Phys.* **194**, 675 (1998), hep-th/9705046; M. Porrati and A. Rozenberg, *Nucl. Phys.* **B515**, 184 (1998), hep-th/9708119; M. B. Green and M. Gutperle, *JHEP* **9801**, 005 (1998), hep-th/9711107; M. B. Halpern and C. Schwartz, *Int. J. Mod. Phys.* **A13**, 4367 (1998), hep-th/9712133; G. Moore, N. Nekrasov and S. Shatashvili, *Commun. Math. Phys.* **209**, 77 (2000), hep-th/9803265; N. A. Nekrasov, *On the size of a graviton*, hep-th/9909213; S. Sethi and M. Stern, *Adv. Theor. Math. Phys.* **4**, 487 (2000), hep-th/0001189.

[143] P. K. Townsend, *Phys. Lett.* **B373**, 68 (1996), hep-th/9512062.

[144] A. Sen, *Phys. Rev.* **D53**, 2874 (1996), hep-th/9511026.

[145] C. Vafa, *Nucl. Phys.* **B463**, 415 (1996), hep-th/9511088.

[146] S. Sethi and M. Stern, *Phys. Lett.* B398 47 (1997), hep-th/9607145; *Nucl. Phys.* **B578**, 163 (2000), hep-th/0002131.

[147] G. Papadopoulos and P. K. Townsend, *Phys. Lett.* **B393**, 59 (1997), hep-th/9609095.

[148] U. H. Danielsson and G. Ferretti, *Int. J. Mod. Phys.* **A12**, 4581 (1997), hep-th/9610082; S. Kachru and E. Silverstein, *Phys. Lett.* **B396**, 70 (1997), hep-th/9612162; D. Lowe, *Nucl. Phys.* **B501**, 134 (1997), hep-th/9702006; T. Banks, N. Seiberg and E. Silverstein, *Phys. Lett.* **B401**, 30 (1997), hep-th/ 9703052; T. Banks and L. Motl, *JHEP* **12**, 004 (1997), hep-th/9703218; D. Lowe, *Phys. Lett.* **B403**, 243 (1997), hep-th/9704041; S.-J. Rey, *Nucl. Phys.* **B502**, 170 (1997), hep-th/9704158.

[149] See refs. [151, 152, 153]. There are also excellent reviews available, some of which are listed in refs. [156, 135, 136].

[150] C. M. Hull, *Nucl. Phys.* **B468**, 113 (1996), hep-th/9512181.

[151] C. M. Hull and P. K. Townsend, *Nucl. Phys.* **B438**, 109 (1995), hep-th/9410167.

[152] P. K. Townsend, *Phys. Lett.* **B350**, 184 (1995), hep-th/9501068.

[153] E. Witten, *Nucl. Phys.* **B443**, 85 (1995), hep-th/9503124.

[154] See for example refs. [143, 135, 156].

[155] See refs. [170, 171, 143].

[156] For other reviews, see: M. J. Duff, *M-Theory (the Theory Formerly Known as Strings)*, *Int. J. Mod. Phys.* **A11**, 5623 (1996), hep-th/9608117; A. Sen, *An Introduction to Non-perturbative String Theory*, hep-th/9802051.

[157] T. Banks, W. Fischler, S. H. Shenker and L. Susskind, *Phys. Rev.* **D55**, 5112 (1997), hep-th/9610043.

[158] For reviews, see: T. Banks, *TASI lectures on matrix theory*, in TASI 1999, *Strings, Branes and Gravity*, World Scientific (2001), hep-th/9911068. T. Banks, *Matrix Theory, Nucl. Phys. Proc. Suppl.* **B67**, 180 (1998), hep-th/9710231; D. Bigatti and L. Susskind, *Review of Matrix Theory*, hep-th/9712072; H. Nicolai and R. Helling, *Supermembranes and M(atrix) Theory*, hep-th/9809103; W. I. Taylor, *The M(atrix) model of M-theory*, hep-th/0002016; A. Bilal, *M(atrix) theory: A pedagogical introduction*, Fortsch. *Phys.* **47**, 5 (1999), hep-th/9710136.

[159] C. G. Callan, J. A. Harvey, and A. Strominger, *Nucl. Phys.* **B367**, 60 (1991).

[160] E. Witten, in the proceedings of *Strings 95*, USC, 1995, hep-th/9507121.

[161] P. S. Aspinwall, *Phys. Lett.* B **357**, 329 (1995), hep-th/9507012.

[162] J. Polchinski and E. Witten, *Nucl. Phys.* **B460**, 525 (1996), hep-th/9510169.

[163] A. Dabholkar and J. A. Harvey, *Phys. Rev. Lett.* **63**, 478 (1989); A. Dabholkar, G. Gibbons, J. A. Harvey and F. Ruiz Ruiz, *Nucl. Phys.* **B340**, 33 (1990).

[164] A. Dabholkar, *Phys. Lett.* **B357**, 307 (1995); C. M. Hull, *Phys. Lett.* **B357**, 545 (1995).

[165] C. V. Johnson, N. Kaloper, R. R. Khuri and R. C. Myers, *Phys. Lett.* **B368**, 71 (1996), hep-th/9509070.

[166] E. Bergshoeff, E. Sezgin and P. K. Townsend, *Phys. Lett.* **B189**, 75 (1987); M. J. Duff and K. S. Stelle, *Phys. Lett.* **B253**, 113 (1991).

[167] R. Güven, *Phys. Lett.* **B276**, 49 (1992).

[168] R. Sorkin, *Phys. Rev. Lett.* **51**, 87 (1983); D. J. Gross and M. J. Perry, *Nucl. Phys.* **B226**, 29 (1983).

[169] P. Horava and E. Witten, *Nucl. Phys.* **B460**, 506 (1996), hep-th/9510209.

[170] M. J. Duff and J. X. Lu, *Nucl. Phys.* **B390**, 276 (1993), hep-th/9207060; S. P. de Alwis and K. Sato, *Phys. Rev.* **D53**, 7187 (1996), hep-th/9601167; A. A. Tseytlin, *Nucl. Phys.* **B469**, 51 (1996), hep-th/9602064.

[171] C. Schmidhuber, *Nucl. Phys.* **B467**, 146 (1996), hep-th/9601003.

[172] K. Hori, *Nucl. Phys.* **B539**, 35 (1999), hep-th/9805141; K. Landsteiner and E. Lopez, *Nucl. Phys.* **B516**, 273 (1998), hep-th/9708118; E. Witten, *JHEP* **9802**, 006 (1998), hep-th/9712028; E. G. Gimon, *On the M-theory interpretation of orientifold planes*, hep-th/9806226; C. Ahn, H. Kim and H. S. Yang, *Phys. Rev.* **D59**, 106002 (1999), hep-th/9808182; S. Sethi, *JHEP* **9811**, 003 (1998), hep-th/9809162; C. Ahn, H. Kim, B. Lee and H. S. Yang, *Phys. Rev.* **D61**, 066002 (2000), hep-th/9811010; A. Hanany, B. Kol and A. Rajaraman, *JHEP* **9910**, 027 (1999), hep-th/9909028; A. M. Uranga, *JHEP* **0002**, 041 (2000), hep-th/9912145; A. Hanany and B. Kol, *JHEP* **0006**, 013 (2000), hep-th/0003025.

[173] K. S. Narain, *Phys. Lett.* **169B**, 41 (1986).

[174] P. Ginsparg, *Phys. Rev.* **D35**, 648 (1987).

[175] K. S. Narain, M. H. Sarmadi and E. Witten, *Nucl. Phys.* **B279**, 369 (1987).

[176] B. Julia, in *Supergravity and Superspace*, ed. S. W. Hawking and M. Rocek (Cambridge University Press, Cambridge, 1981).

[177] C. Vafa and E. Witten, *Nucl. Phys.* **B431**, 3 (1994), hep-th/9408074.

[178] C. Vafa, *Nucl. Phys.* **B463**, 435 (1996), hep-th/9512078.

[179] A. Strominger, *Phys. Lett.* **B383**, 44 (1996), hep-th/9512059.

[180] S. Sethi and L. Susskind, *Phys. Lett.* B **400**, 265 (1997), hep-th/9702101.

[181] An excellent review can be found in D. R. Morrison, *TASI Lectures on Compactification and Duality*, in TASI 1999, *Strings, Branes and Gravity*, World Scientific (2001).

[182] See refs. [132, 191, 194, 188].

[183] K. Kodaira, *Ann. of Math.* (2) **77**, 563 (1963); *ibid.*, **78** (1963) 1.

[184] M. R. Douglas, *J. Geom. Phys.* **28**, 255 (1998), hep-th/9604198.

[185] L. Alvarez-Gaume and D. Z. Freedman, *Commun. Math. Phys.* **80**, 443 (1981).

[186] S. W. Hawking, *Phys. Lett.* **60A** 81, (1977).

[187] J. Polchinski, *Phys. Rev.* **D55**, 6423 (1997), hep-th/9606165.

[188] M. R. Douglas and G. Moore, *D-Branes, Quivers, and ALE Instantons*, hep-th/9603167.

[189] A. Hitchin, A. Karlhede, U. Lindstrom, and M. Roček, *Comm. Math. Phys.* **108**, 535 (1987).

[190] P. Kronheimer, *J. Diff. Geom.* **28**, 665 (1989); **29**, 685 (1989).

[191] C. V. Johnson and R. C. Myers, *Phys. Rev.* **D55**, 6382 (1997), hep-th/9610140.

[192] A. Dabholkar and J. Park, *Nucl. Phys.* **B477**, 701 (1996), hep-th/9604178; *Nucl. Phys.* **B472**, 207 (1996), hep-th/9602030.

[193] See also: G. Pradisi and A. Sagnotti, *Phys. Lett.* **B216**, 59 (1989); M. Bianchi and A. Sagnotti, *Phys. Lett.* **B247**, 517 (1990).

[194] E. G. Gimon and C. V. Johnson, *Nucl. Phys.* **B477**, 715 (1996), hep-th/9604129.

[195] J. D. Blum, *Nucl. Phys.* **B486**, 34 (1997), hep-th/9608053; J. D. Blum and K. Intriligator, *Nucl. Phys.* **B506**, 223 (1997), hep-th/9705030; P. Berglund and E. Gimon, *Nucl. Phys.* **B525**, 73 (1998), hep-th/9803168; R. Blumenhagen, L. Gorlich and B. Kors, *Nucl. Phys.* **B569**, 209 (2000), hep-th/9908130.

[196] M. Berkooz, R. G. Leigh, J. Polchinski, J. H. Schwarz, N. Seiberg and E. Witten, *Nucl. Phys.* **B475**, 115 (1996), hep-th/9605184.

[197] A. Sen, *Nucl. Phys.* **B475**, 562 (1996), hep-th/9605150.

[198] K. Dasgupta and S. Mukhi, *Phys. Lett.* **B385**, 125 (1996), hep-th/9606044.

[199] C. Vafa, *Nucl. Phys.* **B469**, 403 (1996), hep-th/9602022.

[200] D. R. Morrison and C. Vafa, *Nucl. Phys.* **B473**, 74 (1996), hep-th/9602114; *Nucl. Phys.* **B476**, 437 (1996), hep-th/9603161.

[201] D. Diaconescu, M. R. Douglas and J. Gomis, *JHEP* **9802**, 013 (1998), hep-th/9712230.

[202] K. Dasgupta and S. Mukhi, *JHEP* **9907**, 008 (1999), hep-th/9904131.

[203] H. Ooguri and C. Vafa, *Nucl. Phys.* **B463**, 55 (1996), hep-th/9511164.

[204] I. Brunner and A. Karch, *JHEP* **9803**, 003 (1998), hep-th/9712143; A. Karch, D. Lust and D. Smith, *Nucl. Phys.* **B533**, 348 (1998), hep-th/9803232; B. Andreas, G. Curio and D. Lust, *JHEP* **9810**, 022 (1998), hep-th/9807008.

[205] R. Gregory, J. A. Harvey and G. Moore, *Adv. Theor. Math. Phys.* **1**, 283 (1997), hep-th/9708086.

[206] A. Hanany and E. Witten, *Nucl. Phys.* **B492**, 152 (1997), hep-th/9611230.

[207] E. Witten, *J. Geom. Phys.* **15**, 215 (1995), hep-th/9410052.

[208] M. F. Atiyah, V. G. Drinfeld, N. J. Hitchin, and Y. I. Manin, *Phys. Lett.* **A65**, 185 (1978).

[209] D. Diaconescu, *Nucl. Phys.* **B503**, 220 (1997), hep-th/9608163.

[210] D. Tsimpis, *Phys. Lett.* **B433**, 287 (1998), hep-th/9804081.

[211] S. K. Donaldson, *Commun. Math. Phys.* **96**, 387 (1984).

[212] S. Elitzur, A. Giveon and D. Kutasov, *Phys. Lett.* **B400**, 269 (1997), hep-th/9702014; S. Elitzur, A. Giveon, D. Kutasov, E. Rabinovici and A. Schwimmer, *Nucl. Phys.* **B505**, 202 (1997), hep-th/9704104.

[213] This is a useful review: A. Giveon and D. Kutasov, *Brane dynamics and gauge theory*, *Rev. Mod. Phys.* **71**, 983 (1999), hep-th/9802067.

[214] G. t'Hooft, *Nucl. Phys.* **B79**, 276 (1974); A. M. Polyakov, *JETP Lett.* **20**, 194 (1974).

[215] B. Julia and A. Zee, *Phys. Rev.* **D11**, 2227 (1975).

[216] W. Nahm, *The Construction Of All Selfdual Multi-Monopoles By The ADHM Method (Talk)*, in N. S. Craigie, P. Goddard and W. Nahm, *Monopoles In Quantum Field Theory. Proceedings, Monopole Meeting, Trieste, Italy, December 11–15, 1981*, World Scientific (1982).

[217] For reviews, the appendix of ref. [232] is useful, and also: P. M. Sutcliffe, *BPS monopoles*, *Int. J. Mod. Phys.* **A12**, 4663 (1997), hep-th/9707009.

[218] E. J. Weinberg, *Phys. Rev.* **D20**, 936 (1979); W. Nahm, *Phys. Lett.* **B85**, 373 (1979).

[219] H. Nakajima, *Monopoles and Nahm's Equations*, in Sanda 1990, Proceedings, *Einstein metrics and Yang-Mills connections*, pp. 193–211.

[220] A. Hashimoto, *Phys. Rev.* **D57**, 6441 (1998), hep-th/9711097.

[221] J. Schwinger, *Phys. Rev.* **144**, 1087 (1966); **173**, 1536 (1968); D. Zwanziger, *Phys. Rev.* **176**, 1480, 1489 (1968); B. Julia and A. Zee, *Phys. Rev.* **D11**, 2227 (1974); F. A. Bais and J. R. Primak, *Phys. Rev.* **D13**, 819 (1975).

[222] P. Fayet and J. Iliopoulos, *Phys. Lett.* **51B**, 461 (1974).

[223] There are many good references. Some examples are: vol. 3 of ref. [54]; M. F. Sohnius, *Phys. Rept.* **128**, 39 (1985).

[224] M. Melvin, *Phys. Lett.* **8**, 65 (1963).

[225] F. Dowker, J. P. Gauntlett, D. A. Kastor and J. Traschen, *Phys. Rev.* D **49**, 2909 (1994), hep-th/9309075; F. Dowker, J. P. Gauntlett, G. W. Gibbons and G. T. Horowitz, *Phys. Rev.* D**52**, 6929 (1995), hep-th/9507143; D**53**, 7115 (1996), hep-th/9512154.

[226] R. Emparan, *Nucl. Phys.* B **610**, 169 (2001), hep-th/0105062; M. S. Costa, C. A. Herdeiro and L. Cornalba, *Nucl. Phys.* B **619**, 155 (2001), hep-th/0105023; D. Brecher and P. M. Saffin, *Nucl. Phys.* B **613**, 218 (2001), hep-th/0106206.

[227] H. E. Rauch and A. Lebowitz, *Elliptic Functions, Theta Functions, and Riemann Surfaces*, Williams and Wilkins (1973).

[228] N. Seiberg, *Nucl. Phys.* **B303**, 286 (1988).

[229] M. J. Duff, R. Minasian and E. Witten, *Nucl. Phys.* B **465**, 413 (1996), hep-th/9601036.

[230] M. J. Duff, *TASI lectures on branes, black holes and anti-de Sitter space*, in TASI 1999: *Strings, Branes and Gravity*, World Scientific (2001), hep-th/9912164.

[231] See refs. [232, 237, 248, 249].

[232] M. F. Atiyah and N. J. Hitchin, *Phys. Lett.* **A107**, 21 (1985); *Phil. Trans. R. Soc. Lond.* **A315**, 459 (1985); *The Geometry And Dynamics Of Magnetic Monopoles.*, M. B. Porter Lectures, Princeton University Press (1988).

[233] G. Lifschytz, *Phys. Lett.* **B388**, 720 (1996), hep-th/9604156.

[234] M. Douglas, J. Polchinski and A. Strominger, *JHEP* **9712**, 003 (1997), hep-th/9703031.

[235] See refs. [237, 238, 248, 172].

[236] See refs. [132, 193, 194, 192].

[237] N. Seiberg, *Phys. Lett.* **B384**, 81 (1996), hep-th/9606017.

[238] A. Sen, *JHEP* **9709**, 001 (1997), hep-th/9707123; *JHEP* **9710**, 002 (1997) hep-th/9708002.

[239] C. V. Johnson, A. W. Peet and J. Polchinski, *Phys. Rev.* **D61**, 086001 (2000), hep-th/9911161.

[240] N. Seiberg and E. Witten, *Nucl. Phys.* **B431**, 484 (1994), hep-th/9408099; *ibid.*, **B426** (1994) 19; *Erratum: ibid.*, **B430**, 485 (1994), hep-th/9407087.

[241] M. R. Douglas and S. H. Shenker, *Nucl. Phys.* **B447**, 271 (1995), hep-th/9503163.

[242] R. R. Khuri, *Phys. Lett.* **B294**, 325 (1992), hep-th/9205051; *Nucl. Phys.* **B387**, 315 (1992), hep-th/9205081; J. P. Gauntlett, J. A. Harvey and J. T. Liu, *Nucl. Phys.* **B409**, 363 (1993), hep-th/9211056.

[243] M. Krogh, *JHEP* **9912**, 018 (1999), hep-th/9911084.

[244] The following has a nice discussion of the appearances of monopoles in string and gauge theory: A. Hanany and A. Zaffaroni, *JHEP* **9912**, 014 (1999), hep-th/9911113.

[245] L. Järv and C. V. Johnson, *Phys. Rev.* D62, 126010 (2000), hep-th/0002244.

[246] G. Chalmers and A. Hanany, *Nucl. Phys.* **B489**, 223 (1997), hep-th/9608105.

[247] N. S. Manton, *Phys. Lett.* **B110**, 54 (1982).

[248] N. Seiberg and E. Witten, in *Saclay 1996, The mathematical beauty of physics*, hep-th/9607163.

[249] N. Dorey, V. V. Khoze, M. P. Mattis, D. Tong and S. Vandoren, *Nucl. Phys.* **B502**, 59 (1997), hep-th/9703228.

[250] A. S. Dancer, *Commun. Math. Phys.* **158**, 545 (1993).

[251] G. W. Gibbons and N. S. Manton, *Nucl. Phys.* **B274**, 183 (1986).

[252] J. Madore, *Class. Quant. Grav.* **9**, 69 (1992); *Annals Phys.* **219**, 187 (1992); *Phys. Lett.* **B263**, 245 (1991).

[253] D. Kabat and W. I. Taylor, *Adv. Theor. Math. Phys.* **2**, 181 (1998, hep-th/9711078.

[254] S. Rey, *Gravitating M(atrix) Q-balls*, hep-th/9711081.

[255] B. de Wit, J. Hoppe and H. Nicolai, *Nucl. Phys.* **B305**, 545 (1988).

[256] J. McGreevy, L. Susskind and N. Toumbas, *JHEP* **0006**, 008 (2000), hep-th/0003075; J. Polchinski and M. J. Strassler, hep-th/0003136.

[257] K. Behrndt, *Nucl. Phys.* **B455**, 188 (1995), hep-th/9506106; R. Kallosh and A. Linde, *Phys. Rev.* **D52**, 7137 (1995), hep-th/9507022; See also: M. Cvetič and D. Youm, *Phys. Lett.* **B359**, 87 (1995), hep-th/9507160.

[258] C. V. Johnson, *Phys. Rev.* **D63**, 065004 (2001), hep-th/0004068; *Int. J. Mod. Phys.* A **16**, 990 (2001), hep-th/0011008 (talk at Strings 2000).

[259] W. Israel, *Nuovo Cim.* **44B**, 1 (1966).

[260] C. W. Misner, K. S. Thorne and J. A. Wheeler, *Gravitation*, San Francisco, Freeman (1973).

[261] S. W. Hawking, *Commun. Math. Phys.* **43**, 199 (1975).

[262] J. D. Bekenstein, *Phys. Rev.* D **7**, 2333 (1973), *Phys. Rev.* D **9**, 3292 (1974).

[263] G. T. Horowitz and J. Polchinski, *Phys. Rev.* D **55**, 6189 (1997), hep-th/9612146.

[264] C. V. Johnson, R. C. Myers, A. W. Peet and S. F. Ross, *Phys. Rev.* D **64**, 106001 (2001), hep-th/0105077.

[265] J. Polchinski, *Phys. Rev. Lett.* **75**, 4724 (1995), hep-th/9510017.

[266] M. Douglas, J. Polchinski and A. Strominger, *JHEP* **9712**, 003 (1997), hep-th/9703031.

[267] S. Ferrara, R. Kallosh and A. Strominger, *Phys. Rev.* D **52**, 5412 (1995), hep-th/9508072; S. Ferrara and R. Kallosh, *Phys. Rev.* D **54**, 1514 (1996), hep-th/9602136; *Phys. Rev.* D **54**, 1525 (1996), hep-th/9603090; S. Ferrara, G. W. Gibbons and R. Kallosh, *Nucl. Phys.* B **500**, 75 (1997), hep-th/9702103.

[268] C. V. Johnson, R. R. Khuri and R. C. Myers, *Phys. Lett.* B **378**, 78 (1996), hep-th/9603061; J. M. Maldacena and A. Strominger, *Phys. Rev. Lett.* **77**, 428 (1996), hep-th/9603060; N. R. Constable, C. V. Johnson and R. C. Myers, *JHEP* **0009**, 039 (2000), hep-th/0008226; N. R. Constable, *Phys. Rev.* D **64**, 104004 (2001), hep-th/0106038.

[269] C. G. Callan and J. M. Maldacena, *Nucl. Phys.* B **472**, 591 (1996), hep-th/9602043; G. T. Horowitz and A. Strominger, *Phys. Rev. Lett.* **77**, 2368 (1996), hep-th/9602051.

[270] J. Maldacena, *Adv. Theor. Math. Phys.* **2**, 231 (1988), hep-th/9711200.

[271] E. Witten, *Adv. Theor. Math. Phys.* **2**, 253 (1998), hep-th/9802150.

[272] S. S. Gubser, I. R. Klebanov and A. M. Polyakov, *Phys. Lett.* B **428**, 105 (1998), hep-th/9802109.

[273] E. Witten, *Adv. Theor. Math. Phys.* **2**, 505 (1998), hep-th/9803131.

[274] This is a review: O. Aharony, S. S. Gubser, J. Maldacena, H. Ooguri and Y. Oz, *Phys. Rept.* **323**, 183 (2000), hep-th/9905111.

[275] Two very useful treatments are: P. Ginsparg, *Applied Conformal Field Theory*, Les Houches, France, June 28–Aug 5, 1988, eds. E. Brezin and J. Zinn-Justin, North-Holland, (1990); P. Di Francesco, P. Mathieu and D. Senechal, *Conformal Field Theory*, New York, Springer (1997).

[276] G. W. Gibbons and D. A. Rasheed, *Phys. Lett.* B **365**, 46 (1996), hep-th/9509141. M. B. Green and M. Gutperle, *Phys. Lett.* B **377**, 28 (1996), hep-th/9602077. A. A. Tseytlin, *Nucl. Phys.* **B469**, 51 (1996), hep-th/9602064.

[277] C. Montonen and D. I. Olive, *Phys. Lett.* B **72**, 117 (1977).

[278] A. W. Peet, *TASI lectures on black holes in string theory*, in TASI 1999: *Strings, Branes and Gravity*, World Scientific (2001), hep-th/0008241.

[279] B. R. Greene, A. D. Shapere, C. Vafa and S. T. Yau, *Nucl. Phys.* B **337**, 1 (1990).

[280] L. Susskind, *Another conjecture about M(atrix) theory*, hep-th/9704080.

[281] H. C. Pauli and S. J. Brodsky, *Phys. Rev.* D **32**, 1993 (1985).

[282] W. I. Taylor, Phys. Lett. B **394**, 283 (1997), hep-th/9611042; O. J. Ganor, S. Ramgoolam and W. I. Taylor, *Nucl. Phys.* B **492**, 191 (1997), hep-th/9611202.

[283] L. Motl, hep-th/9701025. T. Banks and N. Seiberg, *Nucl. Phys.* B **497**, 41 (1997), hep-th/9702187.

[284] R. Dijkgraaf, E. Verlinde and H. Verlinde, *Nucl. Phys.* B **500**, 43 (1997), hep-th/9703030.

[285] For a review, see: R. Dijkgraaf, E. Verlinde and H. Verlinde, *Nucl. Phys. Proc. Suppl.* **62**, 348 (1998), hep-th/9709107.

[286] G. 't Hooft, *Dimensional Reduction In Quantum Gravity*, gr-qc/9310026.

[287] L. Susskind, *J. Math. Phys.* **36**, 6377 (1995), hep-th/9409089.

[288] G. W. Gibbons and S. W. Hawking, *Phys. Rev.* D **15**, 2752 (1977); G. W. Gibbons and M. J. Perry, *Proc. R. Soc. Lond.* A **358**, 467 (1978); G. W. Gibbons, S. W. Hawking and M. J. Perry, *Nucl. Phys.* B **138**, 141 (1978).

[289] J. M. Bardeen, B. Carter and S. W. Hawking, *Commun. Math. Phys.* **31**, 161 (1973).

[290] S. W. Hawking and G. T. Horowitz, *Class. Quant. Grav.* **13**, 1487 (1996) gr-qc/9501014.

[291] S. W. Hawking and D. N. Page, *Commun. Math. Phys.* **87**, 577 (1983).

[292] An excellent reference for many matters of this nature is: R. M. Wald, *General Relativity*, Chicago University Press (1984), 491p.

[293] N. D. Birrell and P. C. Davies, *Quantum Fields in Curved Space*, Cambridge University Press (1982), 340p.

[294] S. Hyun, *U-duality between three and higher dimensional black holes*, hep-th/9704005.

[295] K. Sfetsos and K. Skenderis, *Nucl. Phys.* B **517**, 179 (1998), hep-th/9711138.

[296] M. Bañados, C. Teitelboim and J. Zanelli, *Phys. Rev. Lett.* **69**, 1849 (1992), hep-th/9204099.

[297] J. D. Brown and M. Henneaux, *Commun. Math. Phys.* **104**, 207 (1986).

[298] O. Coussaert and M. Henneaux, *Phys. Rev. Lett.* **72**, 183 (1994), hep-th/9310194.

[299] C. V. Johnson and R. C. Myers, *Phys. Rev.* D **64**, 106002 (2001), hep-th/0105159.

[300] See the last reference in ref. [268], for the four dimensional version of this.

[301] G. 't Hooft, *Nucl. Phys.* B **72**, 461 (1974).

[302] M. Gunaydin, L. J. Romans and N. P. Warner, *Nucl. Phys.* B **272**, 598 (1986); *Phys. Lett.* B **154**, 268 (1985); M. Pernici, K. Pilch and P. van Nieuwenhuizen, *Nucl. Phys.* B **259**, 460 (1985).

[303] P. Breitenlohner and D. Z. Freedman, *Phys. Lett.* B **115**, 197 (1982).

[304] V. Balasubramanian and P. Kraus, *Commun. Math. Phys.* **208**, 413 (1999), hep-th/9902121.

[305] S. R. Lau, *Phys. Rev.* D **60**, 104034 (1999), gr-qc/9903038. R. B. Mann, *Phys. Rev.* D **60**, 104047 (1999), hep-th/9903229.

[306] R. Emparan, C. V. Johnson and R. C. Myers, *Phys. Rev.* D **60**, 104001 (1999), hep-th/9903238.

[307] G. T. Horowitz and S. F. Ross, *JHEP* **9804**, 015 (1998), hep-th/9803085.

[308] A. Chamblin, R. Emparan, C. V. Johnson and R. C. Myers, *Phys. Rev.* D **60**, 064018 (1999), hep-th/9902170.

[309] M. Cvetic and S. S. Gubser, *JHEP* **9904**, 024 (1999), hep-th/9902195.

[310] L. J. Romans, *Nucl. Phys.* B **383**, 395 (1992), hep-th/9203018.

[311] M. Cvetic, M. J. Duff, P. Hoxha, J. T. Liu, H. Lu, J. X. Lu, R. Martinez-Acosta, C. N. Pope, H. Sati, T. A. Tran, *Nucl. Phys.* B **558**, 96 (1999), hep-th/9903214.

[312] R. C. Myers and M. J. Perry, *Annals Phys.* **172**, 304 (1986).

[313] P. Kraus, F. Larsen and S. P. Trivedi, *JHEP* **9903**, 003 (1999), hep-th/9811120.

[314] J. G. Russo, *Nucl. Phys.* B **543**, 183 (1999), hep-th/9808117.

[315] L. Susskind and E. Witten, *The holographic bound in anti-de Sitter space*, hep-th/9805114.

[316] A very clear presentation of this sort of technology can be found in: P. K. Townsend, *Phys. Lett.* B **148**, 55 (1984).

[317] S. S. Gubser, *Curvature singularities: The good, the bad, and the naked*, hep-th/0002160.

[318] J. Babington, N. Evans and J. Hockings, *JHEP* **0107**, 034 (2001), hep-th/0105235.

[319] L. Girardello, M. Petrini, M. Porrati and A. Zaffaroni, *JHEP* **9812**, 022 (1998), hep-th/9810126.

[320] J. Distler and F. Zamora, *Adv. Theor. Math. Phys.* **2**, 1405 (1998), hep-th/9810206.

[321] C. V. Johnson, K. J. Lovis and D. C. Page, *JHEP* **0105**, 036 (2001), hep-th/0011166; *JHEP* **0110**, 014 (2001), hep-th/0107261.

[322] D. Z. Freedman, S. S. Gubser, K. Pilch and N. P. Warner, *Adv. Theor. Math. Phys.* **3**, 363 (1999), hep-th/9904017.

[323] K. Pilch and N. P. Warner, *N= 1 Supersymmetric Renormalization Group Flows from IIB Supergravity*, hep-th/0006066.

[324] A. Khavaev, K. Pilch and N. P. Warner, *Phys. Lett.* **B487**, 14 (2000), hep-th/9812035

[325] R. G. Leigh and M. J. Strassler, *Nucl. Phys.* **B447**, 95 (1995), hep-th/9503121.

[326] A. Karch, D. Lust and A. Miemiec, *Phys. Lett.* B **454**, 265 (1999) hep-th/9901041.

[327] K. Pilch and N. P. Warner, *Nucl. Phys.* B **594**, 209 (2001), hep-th/0004063.

[328] A. Brandhuber and K. Sfetsos, *Phys. Lett.* B **488**, 373 (2000), hep-th/0004148.

[329] D. Z. Freedman, S. S. Gubser, K. Pilch and N. P. Warner, *JHEP* **0007**, 038 (2000), hep-th/9906194.

[330] I. Bakas and K. Sfetsos, *Nucl. Phys.* B **573**, 768 (2000), hep-th/9909041.

[331] M. A. Shifman and A. I. Vainshtein, *Nucl. Phys.* B **277**, 456 (1986) [*Sov. Phys. JETP* **64**, 428 (1986)]; *Nucl. Phys.* B **359**, 571 (1991).

[332] N. Evans, C. V. Johnson and M. Petrini, *JHEP* **0010**, 022 (2000), hep-th/0008081.

[333] A. Buchel, A. W. Peet and J. Polchinski, *Phys. Rev.* D **63**, 044009 (2001), hep-th/0008076.

[334] L. Susskind, *The quantum Hall fluid and non-commutative Chern Simons theory*, hep-th/0101029; S. Hellerman and M. Van Raamsdonk, *JHEP* **0110**, 039 (2001), hep-th/0103179; A. P. Polychronakos, *JHEP* **0104**, 011 (2001), hep-th/0103013; J. H. Brodie, L. Susskind and N. Toumbas, *JHEP* **0102**, 003 (2001), hep-th/0010105; O. Bergman, Y. Okawa and J. H. Brodie, *JHEP* **0111**, 019 (2001), hep-th/0107178; S. Hellerman and L. Susskind, *Realizing the quantum Hall system in string theory*, hep-th/0107200.

[335] N. Arkani-Hamed, S. Dimopoulos and G. Dvali, *Phys. Lett.* **B429**, 263 (1998), hep-th/9803315; *Phys. Rev.* **D59**, 086004 (1999), hep-th/9807344. I. Antoniadis, N. Arkani-Hamed, S. Dimopoulos and G. Dvali, *Phys. Lett.* **B436**, 257 (1998), hep-th/9804398.

[336] L. Randall and R. Sundrum, *Phys. Rev. Lett.* **83**, 3370 (1999), hep-th/9905221; *Phys. Rev. Lett.* **83**, 4690 (1999), hep-th/9906064.

[337] A. Sen and B. Zwiebach, *JHEP* **0003:002** (2000), hep-th/9912249.

[338] A. Connes, M. R. Douglas and A. Schwarz, *JHEP* **9802**, 003 (1998), hep-th/9711162; M. R. Douglas and C. M. Hull, *JHEP* **9802**, 008 (1998), hep-th/9711165; C. S. Chu and P. M. Ho, *Nucl. Phys.* B **550**, 151 (1999), hep-th/9812219; *Nucl. Phys.* **B568**, 447 (2000), hep-th/9906192; V. Schomerus, *JHEP* **9906**, 030 (1999), hep-th/9903205; A. Y. Alekseev, A. Recknagel and V. Schomerus, *JHEP* **9909**, 023 (1999), hep-th/9908040; *JHEP* **0005**, 010 (2000), hep-th/0003187. N. Seiberg and E. Witten, *JHEP* **9909**, 032 (1999), hep-th/9908142.

[339] L. Alvarez-Gaumé and E. Witlen, *Nucl. Phys.* **B234**, 269 (1983).

[340] D. Tong, *JHEP* **0207**, 013 (2002), hep-th/0204186.

[341] O. J. Ganor and A. Hanany, *Nucl. Phys.* **B474**, 122 (1996), hep-th/9602120.

[342] N. Ishibashi, H. Kawai, Y. Kitzawa and A. Tsuchiya, *Nucl. Phys.* **B498**, 467 (1997), hep-th/9612115.

[343] See also: P. Bain, 'On the non-Abelian Born–Infeld action', hep-th/9909154; F. Denef, A. Sevrin and J. Troost, *Nucl. Phys.* **B581**, 135 (2000), hep-th/0002180; A. Sevrin, J. Troost and W. Troost, *Nucl. Phys.* **B603**, 389 (2001), hep-th/0101192.

[344] M. Blau, J. Figueroa–O'Farrill, C. Hull and G. Papadopoulos, *JHEP* **0201**, 047 (2002), hep-th/0110242.

[345] M. Blau, J. Figueroa–O'Farrill, C. Hull and G. Papadopoulos, *Class. Quant. Grav.* **19**, L87 (2002) hep-th/0201081.

[346] R. R. Metsaev, *Nucl. Phys.* **B625**, 70 (2002), hep-th/0112044.

[347] D. Berenstein, J. M. Maldacena and H. Nastase, *JHEP* **0204**, 013 (2002), hep-th/0202021.

[348] A. Adams, J. Polchinski and E. Silverstein, *JHEP* **0110**, 029 (2001), hep-th/ 0108075.

[349] J. A. Harvey, D. Kutasov, E. J. Martinec and G. Moore, *Localized Tachyons and RG Flows*, hep-th/0111154.

Index

Page numbers in sloping type denote items treated in an insert.

$\hat{A}$-genus, 215–218, 220
A–D–E classification
  of ALE spaces, 187, 289
  of discrete $SU(2)$ subgroups, 289
  of simply laced Lie algebras, *112*
  of singularities of K3, 386–390
action
  Born–Infeld, 134
  Dirac–Born–Infeld, 135, 205
  Einstein frame, bosonic, 61
  Einstein–Hilbert, 5, 69, 234, 272, 410
  Einstein–Hilbert–Maxwell, 224
  Euclidean, for gravity, 410
  for particle motion, 24
  from boundary counterterms in AdS, 449–452
  low energy effective, bosonic, 59
  low energy effective, heterotic, 175–176
  low energy effective, superstrings, 174–175
  Maxwell, 7
  Nambu–Goto, 27
  of D-brane world-volume, 131–140, *198*, 220
  of Euclidean Reissner–Nordström, 416–417
  of Euclidean Schwarzschild, 414–416
  Polyakov, 29
  string frame, bosonic, 59
  σ-model, 58
adjoint representation, 109
affine connection
  computed for Reissner–Nordström, 226
  from derivatives of metric, 3
affine coordinates, 379–390
  for torus T^2, 379–383
  S^2 or $\mathbb{CP}^1$ example, *380*
amplitude
  D-brane exchange of graviton and dilaton, 146–147
  vacuum, cylinder, 142–144, 197–200, 324–333
  vacuum, Klein bottle, 324–333
  vacuum, Möbius strip, 148–150, 324–333
annihilation operators
  coherent states and, *152*
  in description of fermionic states, 159, 231
  string modes as, 40
anomalous dimensions, 481
anomaly
  conformal, 45, 170
  conformal, and Virasoro central term, 45, 79
  for K3 orientifolds, 341–344

anomaly (*cont.*)
    gauge and gravitational, 160, *161*, 341–344
    Green–Schwarz mechanism, 165–169, 206, 220, 344
    inflow mechanism, 219
    miraculous cancellation, 164, 183–184, 342
    polynomials, *161, 162*
    world-volume curvature couplings and, 206–220
anti-de Sitter (AdS)
    AdS/CFT correspondence, 22, 243, 441–466
    AdS/CFT correspondence and holography, 464–466
    AdS/CFT correspondence dictionary, *see* dictionary
    as a hyperbolic slice, 236
    as solution of gauged supergravity, 445, 467–471
    black holes in, 452–466
    charged black hole and, 455–459
    domain wall in, 469–471
    extremal black branes and, 442
    extremal black hole and, 233
    Freund–Rubin ansatz and, 237–238, 445
    holographic renormalisation group flow and, 467–471
    in various coordinate systems, 235
    local vs global coordinates, 235–237
antisymmetric tensor fields
    as generalisations of photon, 12
    coupling to branes, *see under* p-form
    in world-volume dynamics of M5-branes, 277–278
    in world-volume dynamics of NS5-branes, 268–270
    R–R sector, 163
asymptotic freedom, *84*
asymptotically locally Euclidean (ALE) space, 187–191
    A–D–E classification of, 187, 289

    as Higgs branch, 285–289
    D-brane probes of, 282–291
    Eguchi–Hanson as, 187, 287
    Euler characteristic, 189
    Gibbons–Hawking metric and, 187, 287
    hyper-Kähler property, 289
    hyper-Kähler quotient and, 289–291
    T-duality of, 295–296
    topology of, 292–294
Atiyah–Hitchin manifold
    as two-monopole moduli space, 364–366
    hyper-Kähler property, 351, 365
    O6-plane and, 351–352
    relation to Taub–NUT, 364–366
Atiyah–Hitchin–Drinfeld–Manin (ADHM) construction, 304, 307
attractor mechanism, 429

β-function
    of $\mathcal{N} = 1$, $D = 4$ Yang–Mills, 481
    of $\mathcal{N} = 2$, $D = 4$ Yang–Mills, 393
    of $\mathcal{N} = 4$, $D = 4$ Yang–Mills, 373, 441
    of string world-sheet σ-model, 59
    RG flow and, 84
    Yang–Mills and asymptotic freedom, *84*
background fields
    D-branes in R–R, 315–318
    dielectric effect and R–R, 315–318
    in D-brane tension computation, 145–147
    string propagation in, 12, 56–61
basis
    coordinate vs orthonormal, 65–67
Bañados–Teitelboim–Zanelli (BTZ)
    black hole, *see under* black hole
Bekenstein–Hawking entropy, *see under* entropy
Bernoulli numbers, 216
Bertotti–Robinson solution, 233
bifundamental matter, 284, 288

BIons
as BPS states, 138–140
stretched D-branes and, 308
stretched fundamental strings and, 139–140
black brane
D-branes vs, 241–245
extremal, *see* extremal black brane
non-extremal, 238–240
black hole
Bañados–Teitelboim–Zanelli (BTZ), *431*
BTZ and CFT dual, *431*
BTZ as near horizon geometry, 429–430
Bekenstein–Hawking entropy, 21
charged, *see also* Reissner–Nordström
charged in AdS, from spinning branes, 455–459
constructing extremal Reissner–Nordström, 418–425
constructing non-extremal Reissner–Nordström, 427–428
construction from branes and momentum, 420–425
D1/D5 bound states and, 425–427, 433–439
entropy, microscopic description, 425–428, 433–434
flat, in AdS, 452, 458, 459
holographic principle and, 465
large vs small in AdS, 454, 461
multicentre solutions, 232
non-extremal Reissner–Nordström, *see* non-extremal Reissner–Nordström
Reissner–Nordström, 225, *226*, 228–233, 412–414
Reissner–Nordström and super-symmetry, 231
Reissner–Nordström–AdS, 459
Schwarzschild, 225–228, 412–414
Schwarzschild–AdS, 452
temperature of, 412, 413, 452, 454, 460
thermodynamics, 409–414

thermodynamics in AdS, 459
Bogomol'nyi condition, *312*
Bogomol'nyi–Prasad–Summerfield (BPS) states
AdS domain wall as, 469–471
BIons as, 138–140
BPS monopole limit, *312*
D-branes as, 195–197
extremal branes as, 240, 245
extremal Reissner–Nordström as, 228–232, 413
monopoles as, 311–314
no-force condition and, 232, 240, 245
O-planes as, 200
bolt singularity
of Atiyah–Hitchin manifold, 364–365
of Eguchi–Hanson space, *188*
of Euclidean black hole, 412
bosonisation, 113–116, 163
bound states
of D-branes; mass formulae, 253
of D-branes; various, 258–260
of D1- and D5-branes, and black holes, 425–427, 433–439
of F-strings and D-strings, 254–255, 369–371
dyons and, 314, 374–375
of F-strings and D-strings, tension, 253, 254, 370, 371, 397
of NS5-branes and D5-branes, 375–376, 398
$SL(2, \mathbb{Z})$ and five-branes, 375–376
$SL(2, \mathbb{Z})$ and strings, 257, 263, 369–371
tension formula and Pythagoras, 397
boundary conditions
closed string, 30
fermions, R vs NS, 115, 156
open string mixed Dirichlet and Neumann, 250
open string Neumann, 30
boundary counterterm prescription, 450–452

boundary state, 150–154
    as coherent state, 151, *152*
brane distributions
    from spinning branes, *457*
branes ending on branes, *see*
    stretched branes
Breitenlohner–Freedman bound, 448
bundle
    instanton, *211, 212*
    monopole, *209, 210*
    normal, 220
    tangent, 215, 216, 220
    toroidal or elliptic fibration of K3,
    383–392

Cartan subalgebra, 110
Casimir energy
    from boundary counterterm
    prescription, 452–453
    world-sheet zero point energy
    and, 45–46
central charge or extension
    as charge of gauge field, 232
    of supersymmetry algebra, 230–
    231
    of Virasoro algebra, 41, 79
Chan–Paton factors, 51–54, 159
    for oriented strings, 52
    for unoriented strings, 53–54
    orbifold action on, 283–284, 323
characteristic classes, 210–216
    Chern, *see* Chern, class
    Euler, 215, *216*
    Pontryagin, 214–215, *216*
chemical potential
    AdS/CFT correspondence and,
    455–464
Chern
    character, 208, 219
    class, first, 208, 211
    class, first: of Dirac monopole,
    209, 210
    class, second, 211
    class, second: of instanton, *211,
    212*

Chern–Simons three-form, *167*, 168,
    211
circle
    closed strings on, 96–116
    fields on, *see* Kaluza–Klein
    self-dual radius of, 100–103
Clifford algebra, 68, 116, 158
coherent state
    boundary state and, 151, *152*
cohomology, 64, 210–216
Coleman–Weinberg formula, 143, *144*
complex
    affine coordinates, 379
    coordinate transverse to seven-
    brane, 376
    coordinates on world-sheet, 48
    coupling of $D = 4$ Yang–Mills
    theory, 393, 502
    coupling of type IIB, 368, 393, 395
confining/deconfining phase transition
    large N Yang–Mills and, 462–464
conformal anomaly, 83
conformal dimension, *see* conformal
    weight
conformal factor
    world-sheet geometry and, 47–48
conformal gauge
    fixing by reparams and Weyl, 36
    fixing with Faddeev–Popov
    ghosts, 85–87
conformal ghosts, 85, *87*, 157
    and critical dimension, 86–87
conformal group, 72–73
    anti-de Sitter isometries and, 236,
    443
    four dimensions, 373, 441
    two dimensions, 73
conformal invariance, 70–80
    as residual symmetry, 37
    at fixed point, *84*
    of $\mathcal{N} = 1$, $D = 4$ Yang–Mills,
    480–482
    of $\mathcal{N} = 2$, $D = 4$ Yang–Mills, 393
    of $\mathcal{N} = 4$, $D = 4$ Yang–Mills, 373,
    441

conformal weight, 79, 93
constraints
    Virasoro, 39, 41
coset
    as target space of scalars, 279,
    369, 443
cosmic string
    seven-brane as, 377
cosmological constant
    Einstein's equations and, 233
    negative, *see* anti-de Sitter
    positive, *see* de Sitter
Coulomb branch, 285, 346, 349,
    426
    Atiyah–Hitchin as, 351–352,
    364–366
    enhançon and, 362–366, 502
    fractional D-branes and, 291–292
    lifting by tilting D-branes, 298
    lifting by unwrapping D-branes,
    292–294
    of large N Yang–Mills, from
    holographic RG flow, 472–480
    Seiberg–Witten theory and,
    393–394, 502
    singularities on, 351–352
    stretched D-branes and, 297
    Taub–NUT as, 349–352
    touching Higgs branch, 351–352
covariantly constant spinor, *see* Killing
    spinor
creation operators
    coherent states and, *152*
    in description of fermionic states,
    159, 231
    string modes as, 40
critical dimension
    bosonic string, 43–44
    heterotic string, 170
    supersymmetric string, 157
critical string theory, *see* citical
    dimension
current algebra, *see* Lie algebras, affine
curvature
    couplings in DBI action, 222–223

    extrinsic, *see* extrinsic curvature
    two-form, *66*, 67–69
    world-volume couplings, *see also*
    world-volume
cylinder
    amplitude, *see* amplitude,
    vacuum
    fundamental region, 148

D-flatness condition, 285, 303–304,
    307
D-manifolds, 191
D-term, 285, 303–304, 307
de Sitter
    and the sphere, 234
decoupling limit, 349, 363, 441
    AdS/CFT and, *see under* anti-de
    Sitter
    of non-extremal D3-brane
    solution, 452
    of spinning branes, 456
    on D3-brane supergravity
    solution, 441–442
Dedekind's η-function, 91, 378
deficit angle, *90*
    of $\mathbb{Z}_N$ orbifold, 388
    of bolt, *188*
    of Euclidean black hole's bolt,
    412, 460
    of nut, *350*, 351
    of seven-branes, 378–379, 388–390
dictionary, AdS/CFT
    action and energy-momentum
    tensor, 449–452
    AdS$_3$ and $D = 2$ CFT, *431*
    boundary metric, 446–447
    Breitenlohner–Freedman bound,
    448
    BTZ black holes and $D = 2$ CFT
    sectors, *431*
    chemical potential, 455–464
    fields and operators, 447–449
    finite temperature and black
    holes, 452–453
    Kaluza–Klein reduction, 449

dictionary, AdS/CFT (*cont.*)
  partition function, 447, 449
  semi-classical action, 449
  spherical harmonics, 448
  symmetries and isometries, 443
  ultraviolet and infrared, 446
dielectric effect, 314–321
  background R–R fields and,
    315–318
  non-Abelian Dirac–Born–Infeld
    and, 314–316
  non-Abelian R–R couplings and,
    314–316
differential forms, *see* p-forms
dilatations, 72, 441
dilaton
  as massless closed string mode,
    12
  propagator, 146
  string coupling and, 12, 60
dilute gas limit
  non-extremal black holes and,
    427–428
dimension
  critical, *see* critical dimension
  eleventh, and Kaluza–Klein, 271–
    273, *274*, 346–348, 400–404
  eleventh, from brane world-volume,
    269, 277–278, 346–348
  eleventh, from strong coupling,
    271–273, 275, 400–404
  of spacetime, and conformal ghosts,
    86–87, 170
  of spacetime, for bosonic string,
    43–44, 86–87
  of spacetime, for heterotic strings,
    170
  of spacetime, for strings, 14
  of spacetime, for supersymmetric
    strings, 157, 170
Dirac genus, *see* $\hat{A}$-genus
Dirac monopole, *see* bundle, monopole
Dirac string, 201–202, *210*
Dirac–Born–Infeld, *see* action, Dirac–
  Born–Infeld

Dirac–Nepomechie–Teitelbiom charge
  quantisation, 201–202
  for D-branes, 202
  for M-branes, 277
discrete light cone quantisation
  (DLCQ)
  matrix string theory and, 407
  matrix theory and, 403
discriminant of cubic, 380–383,
  387–390
dissolving $D(p-2)$ into Dp, 206, 294
domain wall
  in AdS, *see* holographic
  renormalisation group
duality
  Hodge, *see* Hodge duality
  S-, or strong/weak coupling; for
    field theory, *see* field theory duality
  S-, or strong/weak coupling; for
    string theory, *see* string duality
  strong/weak coupling, *see* strong/
    weak coupling duality
  T-, or target space, *see* T-duality
  U-, *see* U-duality
Dynkin diagram, 111
  extended, for simply laced Lie
    algebras, 288
  for simply laced Lie algebras, *112*
  U-duality groups and, *280*
dyons
  D/F-string bound states and, 314,
    374–375

Eguchi–Hanson space, 187, *188*
  Euler charcteristic of, *188*
  Gibbons–Hawking metric and,
    287
eleven dimensional supergravity, *see
  also* dimension, eleventh, 261
  bosonic form of, 272
  string duality and, 271–273
eleventh dimension, *see* dimension,
  eleventh
elliptic fibration, 383–392
energy-momentum tensor

and Einstein's equations, 3
as generator of conformal
transformations, 76–78
fermionic part, 156
for electromagnetism, 7, 225
from boundary counterterms in
AdS, 449–452
in two dimensions, 35–36, 38–42,
76–82
of excision, 360–362
perfect fluid form, 234, 451
string world-sheet, 35–36, 38–42
Virasoro algebra and, *see*
Virasoro
enhançon
as a filter, 439
black hole and, 433–439
black hole horizon and, 439
BPS monopole and, 356–359
Coulomb branch moduli space and,
362–366, 502
mechanism, 358–362
positivity of coupling and, 502
positivity of tension and, 362,
502
second law of theormodynamics
and, 437–439
enhanced gauge symmetry
affine Lie algebras and, 113
F-theory and, 386–390
from D4-branes wrapped on K3,
358–359
from D5-branes wrapped on K3,
359
from M-theory on K3, 352
from type IIA on K3, 298, 400
heterotic string and, 171, 386–390
of coincident NS5-branes, 268–
270, 299
of string on circle, 100–103, 358
small instantons and, 305–306
entropy
from Euclidean path integral, 411
of black hole, Higgs branch and,
426, 434

of Reissner–Nordström, 417,
424
of Schwarzschild, 416
Euclidean, *see also* signature
path integral, 410, 411
path integral and temperature,
410
quantum gravity, 410
Euler angles, *180*
Euler characteristic
of ALE space, 189
of graphs, *444*
of K3, 187–189
of sphere S^2, *216*
of torus T^2, 380
Euler class, 215
excision techniques, 360–362
energy-momentum tensor,
360–362
exponential corrections
Atiyah–Hitchin manifold and,
365
of Taub–NUT, as instanton
corrections, 363
exponential map, 47
exterior derivative, 64
extremal black brane, 240–246
distributions of, 248
extremal black hole vs, 240
multicentre solutions and, 240
near horizon geometry, 241
no-force condition and, 240, 245
extremal black hole, *see also* extremal
Reissner–Nordström
extremal Reissner–Nordström
as BPS state, 228–232, 413
as interpolating solution, 233
construction from branes and
momentum, 418–425
in AdS, 460
extrinsic curvature, *229*
from wrapping branes, 360
of Reissner–Nordström, 416
of Schwarzschild, 414

F-theory, 263, 367
  enhanced gauge symmetry and,
    386–390
  from M-theory, 394–396
  orientifold limit of, 390–392
  torus and, 394–396
  twelve dimensions and, 394
Faddeev–Popov ghosts, 85–87,
  see conformal ghosts
Fayet–Iliopoulos term, 286
fermionisation, 113–116
fermions
  in curved spacetime, 68
  R vs NS boundary conditions, 115,
    156
Feynman graphs
  of large N gauge theory, *444*
fibre bundle, *see* bundle
field equations
  Einstein's gravitational, 3
  from vanishing of world-sheet
    β-function, 59
field theory duality
  Montonen–Olive, 374–375
  $SL(2, \mathbb{Z})$ and, 373–375
fixed points
  $SL(2, \mathbb{Z})$; fundamental domain
    and, 371
  of $SL(2, \mathbb{Z})$; fundamental domain
    and, *90*
  of orbifold, 117–119, 126, 388–390
  of RG flows, *84*
  orientifolds as, 126
flat direction, 83
fractional D-branes, 291–292
  from wrapped D-branes,
    292–294
  T-duality to stretched D-branes,
    296–300
frame
  dual tangent, 67, 220
  inertial, *6*
  string vs Einstein, 60
  tangent, 61–69, 220
Freund–Rubin ansatz, 237–238

fundamental domain, 89, 379
  j-function and, 382, 387–390
  $SL(2, \mathbb{Z})$ fixed points, *90*
  $SL(2, \mathbb{Z})$ fixed points, 371
  special orbifold points, *90*, 371
fuzzy sphere, 317–318

gauge fields
  as massless open string modes, 12
gauge theory/geometry
  correspondence, 243
  AdS/CFT, *see* anti-de Sitter
geodesic
  analysis of repulson, 354–356
  equation, 2
gerbes, 270
ghosts
  Faddeev–Popov, *see* conformal
    ghosts
  negative norm states, 43
Gibbons–Hawking boundary term,
  410, 450, 462
Gibbons–Hawking metric, 187
  as moduli space metric, 287
  Eguchi–Hanson space and, 187,
    287
  from D-brane probe, 287
  hyper-Kähler property, 287
Gibbs
  thermodynamic potential,
    416, 462
Gliozzi–Scherk–Olive (GSO)
  projection
    D-string and, 262, 302
    heterotic string and, 172, 265
    superstring and, 158–159
    tachyon and, 158
graviton
  as massless closed string mode, 12
  linearised gravity and, 7–11, 145–
    147
  propagator, 146
Green–Schwarz form, 262, 405
Green–Schwarz mechanism, *see under*
  anomaly

Hamiltonian
  Euclidean path integral and, 410
  of matrix string theory, 405
  of matrix theory, 402
  on world-sheet, 39
Hanany–Witten, 300
Hawking temperature, *see* black hole,
  temperature of
heterotic strings, *see* supersymmetric
  strings
Higgs branch, 346, 349
  ALE space as, 285
  black hole entropy and, 426, 434
  touching Coulomb branch,
  351–352
Hirzebruch $\hat{\mathcal{L}}$-polynomial, 215–220
Hodge duality, 65
  D-branes and, 196, *198*, 201,
  348
  fundamental string and
  NS5-brane, *198*
  M-branes and, 276
  on D2-brane world-volume, 278,
  347
  on K3 wrapped D6-brane world-
  volume, 362–363
holographic principle, 467
  a rough statement of, 22, 464–466
  AdS/CFT correspondence and,
  464–466
  black holes and, 464–466
  remarks, 508
holographic renormalisation group,
  467–471
  brane distributions and flow
  geometries, 478–480
  domain wall and, 469–471
  fixed points and, 467–471
  from gravity, 467–471
  Kähler potential of
  supersymmetric Yang–Mills,
  489–493
Hopf fibration
  $S^3 : S^1 \hookrightarrow S^2$; and Taub–NUT,
  349

$S^3 : S^1 \hookrightarrow S^2$, *209, 210*
$S^7 : S^3 \hookrightarrow S^4$, *211, 212*
horizon
  as null surface, 228
  enhançon and, 439
  flat vs round, 452, 458
  geometry near, 233, 241, 429–430,
  441–442
  of black branes, 239, 240
  of BTZ black hole, *431*
  of Reissner–Nordström, 228, 412,
  419–420, 424, 439
  of Reissner–Nordström–AdS, 460
  of Schwarzschild, 227–228, 412
  of Schwarzschild–AdS, 452
hyper-Kähler property
  of $\mathbb{R}^4$, 289
  of ALE spaces, 289
  of Atiyah–Hitchin manifold, 351,
  365
  of Gibbons–Hawking metric, 287
  of K3 manifold, 289
  of Taub–NUT metric, 349
hyper-Kähler quotient, 289–291
  ADHM construction as, 304
  ALE spaces and, 289–291
hypersurface technology, *229*

induced metric
  on World-sheet, 27
  on World-volume, 131–132
inertial frame, *6*
infinite momentum frame (IMF)
  matrix theory and, 402–404,
  407
inner product, 65
instanton
  contributions to moduli space,
  363, 365
  core size, *212*, 267, 305
  corrections of Taub–NUT, 363,
  365
  D(p+4)-Dp system and, 208, 300–
  301
  D-brane as, 208, 300–306

instanton (*cont.*)
    from D-brane probe, 301–305
    in large N gauge theory, *see under* large N gauge theory
    moduli space, and black hole entropy, 426, 434
    NS5-brane as, 267
    number, *212*
    positivity of Coulomb branch metric and, 363, 365, 393, 394, 502
    positivity of gauge coupling and, 363, 365, 393, 394, 502
    world-volume coupling, 208
    zero core size, *see* small instantons
intercept
    critical dimensions and, 42–44
    numerical value, 43–44, 157, 170
interpolating solution
    extremal black branes as, 442
    extremal black hole as, 233
    kink as, *18*
invariant polynomials, 210–216
isotropic coordinates, 226–227

j-function
    fundamental domain and, 382, 387–390
    modular invariance and, 382, 387–390
Jacobi
    identity for Lie algebras, 108
    ϑ-function identities, *328*
    ϑ-functions, *327*
    ϑ-function identities and supersymmetry, 198

K-theory
    cohomology vs, 221
    other remarks, 507
    tachyon condensation and, 220–221
    world-volume couplings and, 220–221

K3 manifold, 322–344
    compactification and black hole, 432–439
    elliptic, moduli space of, 383
    enhanced gauge symmetry from wrapped branes, 298, 352, 399–400
    Euler characteristic of, 187–191
    heterotic string on, *343*
    hyper-Kähler property, 289
    Kodaira classification of singularities, 390
    string duality and, *186*, *343*
    topology of, 184–185, 187–191
    torus or elliptic fibration, 383–392
    type IIA superstring on, 184–185
    various orbifold limits of, 189–191, 388–390
K3 orientifolds, 191, 322–344
    anomalies of, 341–344
Kähler potential
    of $\mathcal{N} = 1$ $D = 4$ Yang–Mills, 489–493
Kac–Moody algebra, *see* Lie algebras, affine
Kaluza–Klein, 94–96
    charged black holes and, 455–459
    D0-branes and string duality, 271, 400–404
    in construction of $D = 5$ black hole, 421–425
    monopole, 357
    monopole; six-brane as, 348–352
    reduction formulae, 95–96, *274*
    sphere reduction and gauged supergravity, 443–445
    type IIA from eleven dimensions, 271–273, *274*, 400–404
Killing
    spinor, 232, 233
    vectors, 225–228
kink solution, *18*
Klein bottle
    amplitude, *see* amplitude, vacuum
    fundamental region, 148

Kodaira
  classification of singularities of
  elliptic K3 manifold, 390

$\hat{\mathcal{L}}$-polynomial, *see* Hirzebruch
large N gauge theory
  as a string theory, *444*
  confining/deconfining phase
  transition, 462–464
  Feynman graphs, *444*
  instantons in, 365–366
  RG flow from gravity, 467–471
  't Hooft coupling, 363, 365, 442,
  444
lattice
  Euclidean signature, 171–172
  even, self-dual, 104, 106, 171–172,
  176–177
  Lorenzian signature, 103, 106,
  176–177
  modular invariance and even,
  self-dual, 103, 106
Laurent expansion
  of open and closed strings,
  48
left-invariant one-forms, 180
level matching
  from translational invariance,
  42, 91
  modular invariance and, 91
  of closed string modes, 42
Lie algebras
  adjoint representation of,
  109
  affine, *102*
  Cartan subalgebra, 110
  classical, 111
  Lie groups and, 108–111
  simple, 109
  simply laced, *112*
  Yang–Mills and, *66*
linearised gravity
  graviton from, 7–11, 145–147
Lorentz group, 68
  as a gauge group, 67–68

inertial frames and, *6*
  representation using gamma
  matrices, 68, 159
  tangent frame and, 62, 67–68

M-branes
  descending to D- and NS5-branes,
  277–278
  descending to odd D-branes and
  NS5-branes, 396–399
  Dirac–Nepomechie–Teitelbiom
  charge quantisation, 277
  supergravity solutions, 276–277
  tensions, 277
M-theory, 367, 400–408
  and strongly coupled strings, *see*
  string duality
  eleven dimensional supergravity
  and, 271–273
  F-theory and, 394–396
  matrix theory formulation of, 20,
  400–408
Möbius strip
  amplitude, *see* amplitude, vacuum
  fundamental region, 148
magnetic monopole, *see* monopole
Majumdar–Papapetrou solutions,
  232
matrix string theory, 404–408
  Hamiltonian of, 405
  interactions from irrelevant
  operator, 407
  long strings from twisted sector,
  406
  orbifold of $(\mathbb{R}^8)^N$ by $\mathcal{S}_N$,
  405–407
matrix theory, 367, 400–408
  and non-commutativity of
  spacetime, 20, 403
  Hamiltonian of, 402
  toroidal compactification of, 404–
  408
Maurer–Cartan one-forms, *180*
McKay correspondence, 187, 289
Melvin solution, 320

mode expansion
 of open and closed strings, 37, 48
 various open string boundary
 conditions, 250
modular invariance, 91, 103–104, 116,
 158
 even self-dual lattices and, 103, 106
 j-function and, 382, 387–390
 level matching and, 91
 seven-brane metric and, 378–379
modular transformations, 88–89, *145*,
 379
 as generators of $SL(2, \mathbb{Z})$, 88
 fundamental domain and, 89, 379
moduli space
 Coulomb branch, *see* Coulomb
 branch
 Higgs branch, *see* Higgs branch
 of BPS monopoles, 363–366
 of elliptic K3 manifolds, 383
 of instantons, and black hole
 entropy, 426, 434
moduli space metric, *see also* moduli
 space
 from Dp-brane probe of p-brane,
 245–246
 from D-brane probe of RG flow
 geometry, 476, 487, 488, 493, 500,
 502
 from D1-brane probe of ALE
 space, 286–289
 instanton contributions to, 363,
 365
 of $D = 2 + 1$ susy gauge theory,
 348–352
monodromy
 $SL(2, \mathbb{Z})$ and, 376–382, 384–396
monopole
 BPS saturated, 311, *312*, 313,
 314, 363–366
 BPS, enhançon and, 356–359
 D-brane as, 306–314
 Dirac, *see* bundle, monopole
 H-, as wrapped N5-brane, 357
 Kaluza–Klein, 357

 Kaluza–Klein; six-brane as,
 348–352
 moduli space of, 363–366
 't Hooft–Polyakov, 310
Montonen–Olive duality, 374–375
multiple D-branes
 boundary conditions, 249–252
 unboken supersymmetry, 252–254
Myers effect, 314–321

Nahm
 monopole data, 311
 monopole equations, 307
naked singularity
 as limit of charged black hole in
 AdS, 460
 enhançon and, *see* enhançon
 mechanism
 from holographic RG flows, 495
 repulson as, 354
near-horizon, *see* horizon, geometry
 near
Neveu–Schwarz fermions, 115, 156
Newton's constant
 from low energy effective action,
 61, 175
 relation to central charge of 2D
 CFT, 431
 relation to large N gauge theory
 quantities, 443
no-force condition, *see* Bogomol'nyi–
 Prasad–Summerfield states
non-Abelian
 Dirac–Born–Infeld, 136–137
 Dirac–Born–Infeld, and dielectric
 effect, 314–316
 gauge theory, *see also* Yang–Mills
 R–R couplings, 221–222
 R–R couplings, and dielectric
 effect, 314–316
 Taylor expansion, 316
 tensor multiplet on coincident
 NS5-branes, 270
 vector multiplet on coincident
 NS5-branes, 268

non-commutative geometry
  fuzzy sphere and, 317–318
  matrix theory and, 20, 403
  remarks, 508
non-compact Lie groups
  by continuation, 111
  T-duality and, 108
  U-duality and, 279, *280*
non-extremal Reissner–Nordström
  construction from branes and
  momentum, 427–428
  dilute gas and, 427–428
non-perturbative strings, *see under*
  duality
  and extended objects, 17
normal ordering, 40
  time ordering vs, 76
NS–NS sector, 160
NS5-brane, 241
  as dual of fundmental string,
  198
  as instanton, 267
  branes ending on, 268–270, 297
  coincident, 268–270
  from D5-brane, 266
  heterotic, 267–268
  T-duality of, 267–268
  type II, 268–270
  world-volume dynamics, 268–270
nut singularity, *350*
  of Taub–NUT metric, 350–352

operator
  irrelevant, and matrix string
  theory, 407
  marginal, 83, *84*
  relevant and irrelevant, *84*
  string state correspondence,
  48–51
  vertex, *see* vertex operator
operator product expansion (OPE),
  75–76
operators
  spherical harmonics and, 448, 478
  states and, 74–75

orbifold, 117
  action on Chan–Paton factors,
  283–284, 323
  blow-up to smooth manifold, 185
  fixed point, 117–119, 126,
  388–390
  from T-duality of Type I, 193
  K3 limit for superstring, 179–191
  of $(\mathbb{R}^8)^N$ by $\mathcal{S}_N$, and matrix string
  theory, 405–407
  of circle (bosonic), 117
  of superstring on T^4, 179–191
  points of fundamental domain,
  90, 371
  superstring on $\mathbb{R}^4/\mathbb{Z}_2$, 282
  $T^4/\mathbb{Z}_2$ spectrum, 180–184, 336–
  339
orientifold
  as orbifold fixed point, 126, 193–
  194
  at strong coupling; O6, 352
  at strong coupling; O7, 392, 394
  at strong coupling; O8, 275, 277
  from T-duality, 125–126, 193–194
  group, 126, 166, 194, 324
  limit of F-theory, 390–392
  making type I from type IIB, 165–
  166, 201
  parity and, 125
  tension of O-plane, *see* tension
  world-volume curvature
  couplings, 217, 220
oxidation
  of RG flow geometries to ten
  dimensions, 475, 486–487, 499
  reduction as, in compactified
  matrix theory, 405, 408
  reduction vs, 405, 408

(p, q)-five-branes, *see* bound states of
  NS5-branes and D5-branes
(p, q)-seven-branes, 384–386
(p, q)-strings, *see* bound states of
  F-strings and D-strings
p-branes, *see* black branes

p-forms, 63–65
    Dp-branes charge of, 197–200
    NS-NS coupling to F-string and
    NS5-brane, *198*
    Op-planes charge of, 200
    R-R coupling to p-branes, 196,
    198
    R-R coupling to p-branes,
    non-Abelian, 221–222
    Yang–Mills theory and, *66*
parity
    right-handed, and T-duality, 100
    world-sheet combined with
    spacetime, 125
    world-sheet, closed strings, 54,
    165
    world-sheet, open strings, 52–54
partition function, 87–93
    at bosonisation radius, 113–115
    black hole entropy and, 427
    of closed string on circle, 103–104
    of open string, 142–144, 326–330
    of orbifolded circle, 118
    simple computation, *92*
Pauli matrices, 283
perfect fluid, 234, 451
periodic time
    temperature and, 410
phase transitions
    AdS/CFT correspondence and,
    462–464
    confining/deconfining, 462–464
physical state conditions, *see* Virasoro
    constraints
Poincaré
    form of AdS, 235, 237
    group, 72, 236, 441
Poisson brackets
    of classical strings, 38
    replaced by commutators, 40
Poisson resummation formula, 106,
    107
Poisson's equation
    seven-branes and, 378
Pontryagin class, 214–215

positivity of coupling
    enhançon and, 502
    instantons and, 363, 365, 393, 394,
    502
positivity of metric
    instantons and, 363, 365, 393, 394,
    502
positivity of tension
    enhançon and, 362, 502
    instantons and, 502
pp-wave, *423*
    and gauge/string duals, 503
    in construction of $D = 5$ black
    hole, 422
primary field, 74, 79
probing
    black hole with D-branes, 434–436
    extremal p-branes with D($p − 4$)-
    branes, 345–348
    extremal six-branes with
    D2-branes, 346–348
    extremal black p-branes with
    Dp-branes, 243–246
    holographic RG flows with
    D-brane, 475–478, 487–493,
    500–502
    of ALE space by D-branes,
    282–291
    of Dp–D(p+4) system by D-brane,
    301–305
    wrapped six-branes with wrapped
    D6-brane, 356–359
propagator
    closed string, 153
    dilaton, 146
    graviton, 146
Pythagoras
    D-brane bound state tension
    formula and, 397–399

quasi-primary field, 74
quiver diagram, 284, 288

Ramond fermions, 115, 156
Reissner–Nordström, *see also under*
    black hole

embedded in string theory, 424

extremal, *see* extremal Reissner–Nordström

renormalisation group flow, *84*

    AdS/CFT and, 467–471

reparametrisation invariance

    of classical string action, 29, 31, 36

repulson geometry

    excision of, 360–362

    geodesic analysis, 354–356

    naked singularity and, 354

    wrapped D-branes and, 354–356

Revolution

    First Superstring, 15, 169–170

    Second Superstring, 15, 17, 170, 261, 508

right-invariant one-forms, *180*

Romans' massive supergravity, 275

roots

    of cubic, torus and, 380–383, 387–390

    simple, 110

R–R sector, 163

R–R charge

    D-branes and, 197–199, 201

    p-branes and, 238–240

    orientifolds and, 200

$SL(2,\mathbb{C})$ invariance, 73

    fixing with 3 points, 74, 383

$SL(2,\mathbb{R})$ invariance, 73

    fixing with 3 points, 74

$SL(2,\mathbb{R})$

    vs $SL(2,\mathbb{Z})$, 368

    supergravity action and, 368–369

$SL(2,\mathbb{Z})$

    and bound states of F-strings and D-strings, 257, 263, 369–371

    bound states of NS5-branes and D5-branes and, 375–376

    field theory duality and, 373–375

    fundamental domain and, 89, 379

generators, *see* modular transformations

    monodromy or jump, seven-branes and, 376–382, 384–396

    string duality and, 263, 367–400

    torus and, 88–89

    U-duality and, 278

scaling dimension, 79

Schrödinger picture

    Euclidean path integral and, 410

Schwarzian derivative, 80

Schwarzschild, *see under* black hole

    mass from Euclidean action, 415

Seiberg–Witten theory

    $\mathcal{N} = 2\ D = 4$ Yang–Mills, 353, 393, 394, 502

    Coulomb branch and, 393–394, 502

    enhançon and, 502

    from F-theory, 393–394

    torus and, 393–394

semiclassical quantum gravity, 411–412

signature

    Euclidean, 410, 412

    Euclidean de Sitter as sphere, 234

    Euclidean world-sheet, 47

    mostly plus convention, 2

    of lattice, Euclidean, 171–172

    of lattice, Lorenzian, 103, 106, 176–177

singularity

    conical, *see* deficit angle

    of black branes, 239, 240

    of Reissner–Nordström, 228

    of Schwarzschild, 228

small instantons

    $E_8 \times E_8$, 305–306

    enhanced tensor gauge symmetry, 305

    enhanced vector gauge symmetry, 305

    $SO(32)$, 305

smeared brane, 206, 248, 295

special conformal transformations, 72, 441

spherical harmonics
    operators and, 448, 478
spin connection, 67–69
    fermions and, 68
    Killing spinor and, 231
spin field, 163
spinning branes
    brane distributions from, *457*
    charged black holes from, 455–459
states
    operators and, 74–75
    spurious; spacetime gauge
    invariance and, 43–44
    vertex operator correspondence,
    48–51
static gauge, 134, 136
static solutions, 227
Stefan–Boltzmann law
    AdS/CFT correspondence and, 453
stress tensor, *see* energy-momentum
    tensor
stretched D-branes, 294–300, 375
    as monopoles, 306–314
    BIons and, 308
    T-duality to fractional D-branes,
    296–300
    wrapped D-branes and, 294–300
stretched fundamental strings, 314, 375
string coupling
    dilaton and, 12, 60
    from eleven dimensions, 273, 395
    large N gauge theory and, 443,
    444
    world-sheet topology and, 34
string duality
    $D = 10$ supergravity and,
    175–176
    $D = 6$ supergravity and, *186*
    $E_8 \times E_8$ heterotic $\leftrightarrow$ M, 273–276
    $SO(32)$ heterotic $\leftrightarrow$ type I, 264–
    265
    dual branes and, 265–270,
    277–278
    eleven dimensional supergravity
    and, 271–273

F-string $\leftrightarrow$ D-string, 262–263
    heterotic $\leftrightarrow$ heterotic in $D = 6$,
    343
    heterotic $\leftrightarrow$ type IIA in $D = 6$,
    186, 298, 357, 358
    heterotic $\leftrightarrow$ F in $D = 8$, 383–384,
    390–392, 399
    heterotic $\leftrightarrow$ M in $D = 7$, 399–400
    heterotic $\leftrightarrow$ type IIA in $D = 6$,
    400
    K3 manifold and, *186*, *343*, 383–
    384, 399, 400
    M-theory and, 17
    $SL(2, \mathbb{Z})$ and, 263, 367–400
    type IIA $\leftrightarrow$ M, 271–273
    type IIB $\leftrightarrow$ type IIB, 261–263,
    367–400
string field theory
    and background independence, 15
    and non-perturbative issues, 15
    matrix string theory as, 407
    tachyon condensation and, 507
string network, 371–372
    three-string junction and, 371
string spectrum
    infinite tower of excitations, 11
    massless sector, 12
    NS–NS sector, 160
    NS–R and R–NS sectors, 163
    of heterotic string, 172
    of type II strings, 160–164
    on circle, 98
    perturbative, 11
    R–R sector, 163
stringy cosmic string, 377
strong coupling
    and gauge theory, 21
    fate of strings, 17, *see also* string
    duality
strong/weak coupling duality
    for field theory, *see* field theory
    duality
    for string theory, *see* string
    duality
structure constants, 108

supergravity
    gauged, and anti-de Sitter, 443–445, 467–471
    gauged, and sphere reductions, 443–445
    $SL(2, \mathbb{R})$ and, 278
    $SL(2, \mathbb{R})$ invariant form, 368–369
    ten dimensional, 174–176
superpotential
    of supergravity, for AdS domain walls, 471
supersymmetric strings
    heterotic, construction, 169–174
    type I and type II, construction, 155–169, 201
supersymmetry multiplets
    long, 231
    $\mathcal{N} = 1$ in $D = 10$, 159
    $\mathcal{N} = 1$ in $D = 6$, 341
    $\mathcal{N} = 2$ in $D = 10$, 163
    $\mathcal{N} = 2$ in $D = 4$, 230
    $\mathcal{N} = 2$ in $D = 6$, 183
    short, 231, 255
surface gravity
    black hole temperature and, 413
symmetric polynomials, 213

θ-angle
    in $\mathcal{N} = 2$, $D = 4$ Yang–Mills, 373, 393, 494
't Hooft coupling, *see under* large N gauge theory
T-duality, 19, 94–128
    action on dilaton, 129
    action on Dirac–Born–Infeld, 136
    action on R–R fields, 193
    as $O(d, d + 16, \mathbb{Z})$, 177
    as $O(d, d, \mathbb{Z})$, 108
    as right-handed parity, 100, 105, 125, 192
    boundary conditions and, 120, 125
    discovering D-branes, 119–121, 193
    discovering orientifolds, 125–126, 193

    in background fields, 129–131, 193
    minimum distance and, 19
    of ALE spaces, 295–296
    of black brane solutions, 246–248
    of closed strings, 99–108
    of fractional D-branes, 296–300
    of heterotic strings, 177, 194–195, 273
    of NS5-brane, 267–268
    of NS5-branes, 295–296
    of open strings, 119–125
    of stretched D-branes, 296–300
    of tilted brane, 133–135, 205–206, 294
    of type I superstrings, 193–194
    of type II superstrings, 192–193
    type IA vs type IB, 193, 273
tachyon
    condensation, 220–221
    of bosonic string, 42
    removal by GSO projection, 158
tadpole cancellation, 201, 324, 330–336
tangent space, *see* frame, tangent
Taub–NUT metric
    as moduli space metric, 348–352
    hyper-Kähler property, 349
    instanton corrections, 363, 365
    K3 wrapped six-brane and, 363–366
    negative mass parameter, 363
    relation to Atiyah–Hitchin manifold, 364–366
    six-brane and, 348–352
temperature
    from Euclidean path integral, 410
    of black holes in AdS, 454
    of BTZ black hole, *431*
    of Reissner–Nordström, 412
    of Reissner–Nordström–AdS, 460
    of Schwarzschild, 413
    of Schwarzschild–AdS, 452, 454
    periodic time and, 410
    surface gravity and, 413

tension
  of bound states of NS5-branes and D5-branes, 376
  of D-brane; recursion relation, 132
  of D-branes in bosonic string theory, 142–147
  of D-branes in superstring theory, 197–200
  of F/D-string bound state, 253, 254, 370, 371
  of fundamental strings, 11, *32*
  of O-planes in superstring theory, 200
  of orientifold or O-plane, 148–150
tensionless
  branes, 358–359, 362
  strings, 270, 359
thermodynamics
  entropy from Euclidean path integral, 411
  first law, for black holes, 417
  free energy, 462
  Gibbs potenial, 416, 462
  Hawking–Page phase transition, 462, 465
  of black holes, 409–414
  phase transitions in AdS/CFT, 462–464
  potential, 411
  second law, and enhançon, 437–439
  second law, for black holes, 417
  third law, for black holes, 417
three-string junction, 255–258
  string network and, 371
throat, *see* horizon, geometry near
tilted brane
  and Born–Infeld action, 133–135
  dissolved brane and, 205–206, 294
  $\mathcal{F}$-flux coupling and, 205–206, 294
toroidal compactification
  heterotic string and, 171, 383–390, 399–400
  moduli space of, for bosonic string, 108

  moduli space of, for heterotic string, 177
  of bosonic strings, 104–108
  of heterotic string, 176–177
  of matrix theory, 404–408
  of superstrings, 178–179
  U-duality and, 279, *280*
torsion, 67
torus
  compactification, *see* toroidal compactification
  F-theory and, 394–396
  fibration of K3 manifold, 383–392
  in affine coordinates, 379
  maximal, of Lie algebra, 110
  moduli space of, 88–89
  roots of cubic and, 380–383, 387–390
  Seiberg–Witten theory and, 393–394
  special shapes, *90*
  Weierstrass form, 379–383
transition functions, 208, *209*, *211*
twisted sector
  Fayet–Iliopoulos terms and, 286
  long matrix strings and, 406
  of $(\mathbb{R}^8)^N/\mathcal{S}_N$ orbifold, 406
  of $S^1/\mathbb{Z}_2$ orbifold, 118–119
  of $T^4/\mathbb{Z}_2$ orbifold, 181–183

U-duality, 278–281
  bound states and, 279–281
  near horizon geometry and, 429
  non-compact Lie groups, 279, *280*
  $SL(2,\mathbb{Z})$ and, 278
  toroidal compactification and, 279, *280*
  type II on T^5, 278–279
unorientable
  closed strings, 54
  open strings, 52–54
  orientifolds, *see* orientifolds
  world-sheet diagrams, 55–56, *57*
UV/IR connection, 143, 446, 450

vacuum energy, *see* zero point energy

vacuum interpolation, *see* interpolating solution

vertex operator, 48–51
  correspondence to string states, 48
  enhanced gauge symmetry and, 101, *102*
  for R–R states, 163
  of $G_{\mu\nu}$, $B_{\mu\nu}$, Φ, 50, 58
  of gravitino, 158, 163
  of tachyon, 50
  of vector gauge field, 51, 52

vielbein, *see* frame, tangent

Virasoro
  algebra, and stress tensor, 39, 79
  algebra, supersymmetric, 157
  central charge or extension, 41, 44, 45, 79
  constraints, 39, 41, 157, 163
  zero mode and mass spectrum, 40–42, 157

volume element, 69

volume form, 64

wedge product, 63

Weierstrass
  form of torus T^2, 379–383

weight vector, 110

Wess–Zumino coupling, *see* p-forms, R–R coupling

Weyl invariance
  of classical string action, 30, 36
  of world-sheet σ-model action, 59

Wick
  contraction, 76, 81–82
  rotation, 410

Wilson lines
  Chan–Paton factors and, 121–123
  D-brane positions and, 121
  fractional momentum and, 121, *122*
  on a circle, *122*
  $SO(16) \times SO(16)$, for heterotic T-duality, 194–195, 273

world-line
  of particle, 2, 24

world-sheet
  of string, 13, 27
  various possible topologies, *57*

world-volume
  clues to eleventh dimension, 269, 277–278, 348
  curvature couplings, 205–223
  curvature couplings for D-branes, 217, 220
  curvature couplings for O-planes, 217, 220
  dynamics of D-branes, *see* Yang–Mills and action, Dirac–Born–Infeld
  dynamics of M2-brane, 277–278, 466
  dynamics of M5-brane, 277–278, 466
  dynamics of type II NS5-branes, 268–270, 297
  Hodge duality in $D = 2 + 1$, 278, 347, 362–363
  instantons on, 208
  of D-brane, 131

wrapped D-branes, 292–294
  D6-brane on K3, 353–366
  dual strings from, 390–392
  enhanced gauge symmetries from D4-branes on K3, 358–359
  enhanced gauge symmetries from two-branes on K3, 298, 352
  extrinsic curvature of, 360
  fractional D-brane as, 292–294
  K3-induced charge, 222–223, 352
  K3-induced tension shift, 222–223, 352
  on K3 manifold, 352–353
  repulson geometry and, 354–356
  six-brane as BPS monopole, 357–359
  spacetime metric, 352–353

Yang–Mills
$\mathcal{N} = 1$ $D = 4$, fixed point theory, 480–493
$\mathcal{N} = 2$ $D = 4$, 392–394, 494–502
$\mathcal{N} = 4$ $D = 3$, 348–352, 363–366
$\mathcal{N} = 4$ $D = 4$, 373, 441, 472
at finite temperature via AdS/CFT, 459–464
at large N, *see* large N gauge theory
β-function and asymptotic freedom, *84*
β-function of $\mathcal{N} = 2$, $D = 4$, 393
β-function of $\mathcal{N} = 4$, $D = 4$, 373, 441
complex coupling in $D = 4$, 393, 502
coupled to Higgs, 309
coupling, and D-brane tension, 138
D-brane collective coordinates as, 123–125
from compactified matrix theory, 408
from D-brane world-volume action, 138
Montonen–Olive duality, 374–375
open strings and, 60, 123–125
θ-angle in $\mathcal{N} = 2$, $D = 4$, 373, 393

zero point energy, 45–46, 116, 157, 181, 250, 262, 264
as Casimir energy, 45
BTZ black holes and, *431*
exponential map and, *46*
from AdS/CFT, *431*, 452–453
general formula in $D = 2$, 46
of twisted sector, 118
ζ-function regularisation and, 46
ζ-function regularisation, *see* zero point energy

Printed in the United States
by Baker & Taylor Publisher Services